MW00760636

Lecture Notes in Physics

Springer
Berlin
Heidelberg
New York
Barcelona
Budapest
Hong Kong
London
Milan
Paris
Santa Clara
Singapore
Tokyo

The Editorial Policy for Proceedings

The series Lecture Notes in Physics reports new developments in physical research and teaching – quickly, informally, and at a high level. The proceedings to be considered for publication in this series should be limited to only a few areas of research, and these should be closely related to each other. The contributions should be of a high standard and should avoid lengthy redraftings of papers already published or about to be published elsewhere. As a whole, the proceedings should aim for a balanced presentation of the theme of the conference including a description of the techniques used and enough motivation for a broad readership. It should not be assumed that the published proceedings must reflect the conference in its entirety. (A listing or abstracts of papers presented at the meeting but not included in the proceedings could be added as an appendix.)
When applying for publication in the series Lecture Notes in Physics the volume's editor(s) should submit sufficient material to enable the series editors and their referees to make a fairly accurate evaluation (e.g. a complete list of speakers and titles of papers to be presented and abstracts). If, based on this information, the proceedings are (tentatively) accepted, the volume's editor(s), whose name(s) will appear on the title pages, should select the papers suitable for publication and have them refereed (as for a journal) when appropriate. As a rule discussions will not be accepted. The series editors and Springer-Verlag will normally not interfere with the detailed editing except in fairly obvious cases or on technical matters.
Final acceptance is expressed by the series editor in charge, in consultation with Springer-Verlag only after receiving the complete manuscript. It might help to send a copy of the authors' manuscripts in advance to the editor in charge to discuss possible revisions with him. As a general rule, the series editor will confirm his tentative acceptance if the final manuscript corresponds to the original concept discussed, if the quality of the contribution meets the requirements of the series, and if the final size of the manuscript does not greatly exceed the number of pages originally agreed upon. The manuscript should be forwarded to Springer-Verlag shortly after the meeting. In cases of extreme delay (more than six months after the conference) the series editors will check once more the timeliness of the papers. Therefore, the volume's editor(s) should establish strict deadlines, or collect the articles during the conference and have them revised on the spot. If a delay is unavoidable, one should encourage the authors to update their contributions if appropriate. The editors of proceedings are strongly advised to inform contributors about these points at an early stage.
The final manuscript should contain a table of contents and an informative introduction accessible also to readers not particularly familiar with the topic of the conference. The contributions should be in English. The volume's editor(s) should check the contributions for the correct use of language. At Springer-Verlag only the prefaces will be checked by a copy-editor for language and style. Grave linguistic or technical shortcomings may lead to the rejection of contributions by the series editors. A conference report should not exceed a total of 500 pages. Keeping the size within this bound should be achieved by a stricter selection of articles and not by imposing an upper limit to the length of the individual papers. Editors receive jointly 30 complimentary copies of their book. They are entitled to purchase further copies of their book at a reduced rate. As a rule no reprints of individual contributions can be supplied. No royalty is paid on Lecture Notes in Physics volumes. Commitment to publish is made by letter of interest rather than by signing a formal contract. Springer-Verlag secures the copyright for each volume.

The Production Process

The books are hardbound, and the publisher will select quality paper appropriate to the needs of the author(s). Publication time is about ten weeks. More than twenty years of experience guarantee authors the best possible service. To reach the goal of rapid publication at a low price the technique of photographic reproduction from a camera-ready manuscript was chosen. This process shifts the main responsibility for the technical quality considerably from the publisher to the authors. We therefore urge all authors and editors of proceedings to observe very carefully the essentials for the preparation of camera-ready manuscripts, which we will supply on request. This applies especially to the quality of figures and halftones submitted for publication. In addition, it might be useful to look at some of the volumes already published. As a special service, we offer free of charge LaTeX and TeX macro packages to format the text according to Springer-Verlag's quality requirements. We strongly recommend that you make use of this offer, since the result will be a book of considerably improved technical quality. To avoid mistakes and time-consuming correspondence during the production period the conference editors should request special instructions from the publisher well before the beginning of the conference. Manuscripts not meeting the technical standard of the series will have to be returned for improvement.

For further information please contact Springer-Verlag, Physics Editorial Department II, Tiergartenstrasse 17, D-69121 Heidelberg, Germany

Frieder Lenz   Harald Grießhammer
Dieter Stoll (Eds.)

Lectures on QCD

Applications

Editors

Frieder Lenz
Harald Grießhammer
Dieter Stoll
Institut für Theoretische Physik III
Universität Erlangen-Nürnberg
Staudtstrasse 7
D-91058 Erlangen, Germany

Cataloging-in-Publication Data applied for.

Die Deutsche Bibliothek - CIP-Einheitsaufnahme

Lectures on QCD / Frieder Lenz ... (ed.). - Berlin ; Heidelberg ; New York ; Barcelona ; Budapest ; Hong Kong ; London ; Milan ; Paris ; Santa Clara ; Singapore ; Tokyo : Springer
Applications. - 1997
(Lecture notes in physics ; 496)
ISBN 3-540-63442-8

ISSN 0075-8450
ISBN 3-540-63442-8 Springer-Verlag Berlin Heidelberg New York

Typesetting: Camera-ready by the authors/editors
Cover design: *design & production* GmbH, Heidelberg
SPIN: 10556558 55/3144-543210 - Printed on acid-free paper

Preface

The two volume set "Lectures on QCD" provides an introductory overview of Quantum Chromodynamics, the theory of strong interactions. In a series of articles, the fundamentals of QCD are discussed and significant areas of application are described. Emphasis is put on recent developments. The field-theoretic basis of QCD is the focus of the first volume. The topics discussed include lattice gauge theories, anomalies, finite temperature field theories, sum-rules, the Skyrme model, and supersymmetric QCD. Applications of QCD to the phenomenology of strong interactions form the subject of the second volume. There, investigations of deep inelastic lepton–nucleon scattering, of high energy hadronic reactions and studies of the quark–gluon plasma in relativistic heavy ion collisions are presented.

These articles are based on lectures delivered by internationally well known experts on the occasion of a series of workshops organised by the "Graduiertenkolleg on Strong Interaction Physics" of the Universities of Erlangen-Nürnberg and Regensburg in the years 1992–1995. The workshops were held at "Kloster Banz". Kloster Banz is a former monastery overlooking the valley of the river Main and still serves, for some days of the year, as the stage where certain canons and orthodoxies are vigorously formulated.

Inspired by the atmosphere of the site, the workshops were set up with the aim of introducing novices in the field to the basics of QCD. Accordingly, the character of the lectures was pedagogical rather than technical. With the organisation of these workshops we have attempted to establish a new form in graduate education. Graduate students of the "Graduiertenkolleg" constituted a large fraction of the audience. They have worked out these articles on QCD in collaboration with the lecturers.

Thanks are due to Jutta Geithner and Achim Oppelt for help in the preparation of these proceedings. The support of the "Graduiertenkolleg" by the Deutsche Forschungsgemeinschaft was instrumental in this endeavor and is gratefully acknowledged.

Erlangen, August 1997

F. Lenz
H. W. Grießhammer
D. Stoll

Contents

High Energy Collisions and Nonperturbative QCD*

O. Nachtmann

Institut für Theoretische Physik, Universität Heidelberg, Philosophenweg 16, D-69120 Heidelberg, Germany

Abstract. We discuss various ideas on the nonperturbative vacuum structure in QCD. The stochastic vacuum model of Dosch and Simonov is presented in some detail. We show how this model produces confinement. The model incorporates the idea of the QCD vacuum acting like a dual superconductor due to an effective chromomagnetic monopole condensate. We turn then to high energy, small momentum transfer hadron-hadron scattering. A field-theoretic formalism to treat these reactions is developed, where the basic quantities governing the scattering amplitudes are correlation functions of light-like Wegner-Wilson lines and loops. The evaluation of these correlation functions with the help of the Minkowskian version of the stochastic vacuum model is discussed. A further surprising manifestation of the nontrivial vacuum structure in QCD may be the production of anomalous soft photons in hadron-hadron collisions. We interpret these photons as being due to "synchrotron radiation from the vacuum". A duality argument leads us from there to the expectation of anomalous pieces proportional to $(Q^2)^{1/6}$ in the electric form factors of the nucleons for small Q^2. Finally we sketch the idea that in the Drell-Yan reaction, where a quark-antiquark pair annihilates with the production of a lepton pair, a "chromodynamic Sokolov-Ternov effect" may be at work. This leads to a spin correlation of the $q\bar{q}$ pair, observable through the angular distribution of the lepton pair.

1 Introduction

In these lectures I would like to review some ideas on the way nonperturbative QCD may manifest itself in high energy collisions. Thus we will be concerned with strong interactions where we claim to know the fundamental Lagrangian for a long time now [1]:

$$\mathcal{L}_{\mathrm{QCD}}(x) = -\frac{1}{4}G^a_{\lambda\rho}(x)G^{\lambda\rho a}(x) + \sum_q \bar{q}(x)(i\gamma^\lambda D_\lambda - m_q)q(x). \tag{1.1}$$

Here $q(x)$ are the quark fields for the various quark flavours $(q = u, d, s, c, b, t)$ with masses m_q. We denote the gluon potentials by $G^a_\lambda(x)$ $(a = 1, ..., 8)$ and the gluon field strengths by

$$G^a_{\lambda\rho}(x) = \partial_\lambda G^a_\rho(x) - \partial_\rho G^a_\lambda(x) - g f_{abc} G^b_\lambda(x) G^c_\rho(x), \tag{1.2}$$

---

* Grown out of lectures presented at the workshop "Topics in Field Theory" organised by the Graduiertenkolleg Erlangen–Regensburg, held on October 12th–14th, 1993 in Kloster Banz, Germany

where g is the strong coupling constant and f_{abc} are the structure constants of SU(3). The covariant derivative of the quark fields is

$$D_\lambda q(x) = (\partial_\lambda + igG_\lambda^a \frac{\lambda_a}{2}) q(x), \tag{1.3}$$

with λ_a the Gell-Mann matrizes of the SU(3) group. The gluon potential and field strength matrizes are defined as

$$\begin{aligned} G_\lambda(x) &:= G_\lambda^a(x)\frac{\lambda_a}{2}, \\ G_{\lambda\rho}(x) &:= G_{\lambda\rho}^a(x)\frac{\lambda_a}{2}. \end{aligned} \tag{1.4}$$

The Lagrangian (1.1) is invariant under SU(3) gauge transformations. Let $x \to U(x)$ be an arbitrary matrix function, where for fixed x the $U(x)$ are SU(3) matrices:

$$\begin{aligned} U(x)U^\dagger(x) &= 1, \\ \det U(x) &= 1. \end{aligned} \tag{1.5}$$

With the transformation laws:

$$q(x) \to U(x)q(x), \tag{1.6}$$

$$G_\lambda(x) \to U(x)G_\lambda(x)U^\dagger(x) - \frac{i}{g}U(x)\partial_\lambda U^\dagger(x), \tag{1.7}$$

we find

$$G_{\lambda\rho}(x) \to U(x)G_{\lambda\rho}(x)U^\dagger(x),$$

and invariance of $\mathcal{L}_{\text{QCD}}$:

$$\mathcal{L}_{\text{QCD}}(x) \to \mathcal{L}_{\text{QCD}}(x). \tag{1.8}$$

If we want to derive results from the Lagrangian (1.1), we face problems, the most notable being that $\mathcal{L}_{\text{QCD}}$ is expressed in terms of quark and gluon fields whose quanta have not been observed as free particles. In the real world we observe only hadrons, namely colourless objects; quark and gluons are permanently confined. Nevertheless it has been possible in some cases to derive first principle results which can be compared with experiment, starting from $\mathcal{L}_{\text{QCD}}$ (1.1). These are in essence the following:

(1) Pure short-distance phenomena: Due to asymptotic freedom [2] the QCD coupling constant becomes small in this regime and one can make reliable perturbative calculations. Examples of pure short distance processes are for instance the total cross section for electron-positron annihilation into hadrons and the total hadronic decay rate of the Z-boson.

(2) Pure long-distance phenomena: Here one is in the nonperturbative regime of QCD and one has to use numerical methods to obtain first principle results from $\mathcal{L}_{\text{QCD}}$, or rather from the lattice version of $\mathcal{L}_{\text{QCD}}$ introduced by Wilson

[3]. Today, Monte Carlo simulations of lattice QCD are a big industry among theorists. Typical quantities one can calculate in this way are hadron masses and other low energy hadron properties. (For an up-to-date account of these methods cf. [4]).

There is a third regime of hadronic phenomena, hadron-hadron collisions, which are – apart from very low energy collisions – neither pure long-distance nor pure short-distance phenomena. Thus, none of the above-mentioned theoretical methods apply directly. Traditionally one classifies high-energy hadron-hadron collisions as "hard" and "soft" ones:

(3) High energy hadron-hadron collisions:
  (a) hard collisions,
  (b) soft collisions.

A typical hard reaction is the Drell-Yan process, e.g.

$$\begin{aligned} \pi^- + N \to \ & \gamma^* + X \\ & \hookrightarrow \ell^+ \ell^- \end{aligned} \tag{1.9}$$

where $\ell = e, \mu$. All energies and momentum transfers are assumed to be large. However, the masses of the π^- and N in the initial state stay fixed and thus we are not dealing with a pure short distance phenomenon.

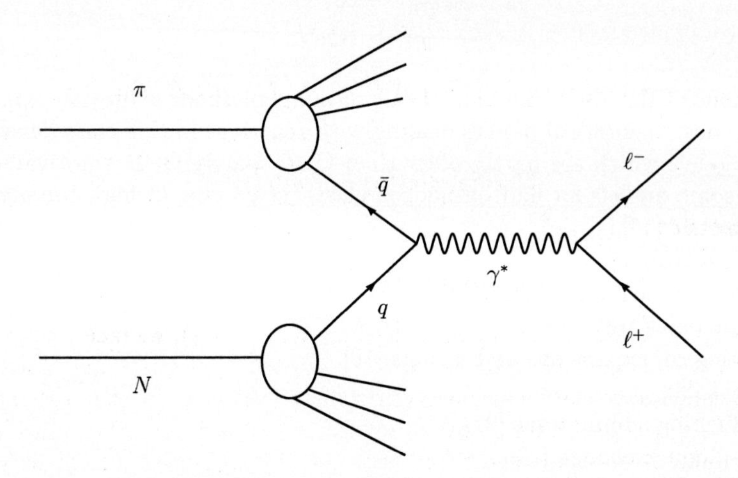

Fig. 1. *The lowest order diagram for the Drell-Yan reaction (1.9) in the QCD improved parton model.*

In the reaction (1.9) we claim to see directly the fundamental quanta of the theory, the partons, i.e. the quarks and gluons, in action (cf. Fig. 1). In the usual theoretical framework for hard reactions, the QCD improved parton model (cf. e.g. [5]), one describes the reaction of the partons, in the Drell-Yan case the $q\bar{q}$ annihilation into a virtual photon, by perturbation theory. This should be reliable, since the parton process involves only high energies and high momentum transfers. All the long distance physics due to the bound state nature of the hadrons is then lumped into parton distribution functions of the participating hadrons. This is called the factorization hypothesis, which after early investigations of soft initial and final state interactions [6] was formulated and studied in low orders of QCD perturbation theory in [7]. Subsequently, great theoretical effort has gone into proving factorization in the framework of QCD perturbation theory [8]-[10]. The result seems to be that factorization is most probably correct there (cf. the discussion in [11]). However, it is legitimate to ask if factorization is respected also by nonperturbative effects. To my knowledge this question was first asked in [12] -[14]. In Sect. 4 of these lectures I will come back to this question and will argue that there may be evidence for a breakdown of factorization in the Drell-Yan reaction due to QCD vacuum effects.

Let us consider now soft high energy collisions. A typical reaction in this class is proton-proton elastic scattering:

$$p + p \to p + p \tag{1.10}$$

at c.m. energies $E_{\rm cm} = \sqrt{s} \gtrsim 5{\rm GeV}$ say and small momentum transfers $\sqrt{|t|} = |q| \lesssim 1{\rm GeV}$. Here we have two scales, one staying finite, one going to infinity:

$$\begin{aligned} E_{\rm cm} &\to \infty, \\ |q| &\lesssim 1{\rm GeV}. \end{aligned} \tag{1.11}$$

Thus, none of the above mentioned calculational methods is directly applicable. Indeed, most theoretical papers dealing with reactions in this class develop and apply models which are partly older than QCD, partly QCD "motivated". Let me list some models for hadron-hadron elastic scattering at high energies:

geometric [15],
eikonal [16],
additive quark model [17],
Regge poles [18],
topological expansions and strings [19],
valons [20],
leading log summations [21],
two-gluon exchange [22],
the Donnachie-Landshoff model for the "soft Pomeron" [23].

It would be a forbidding task to collect all references in this field. The references given above should thus only be considered as representative ones. In addition I would like to mention the inspiring general field theoretic considerations for high energy scattering and particle production by Heisenberg [24] and the impressive work by Cheng and Wu on high energy behaviour in field theories in the framework of perturbative calculations [25].

I will now argue that the theoretical description of measurable quantities of soft high energy reactions like the total cross sections should involve in an essential way nonperturbative QCD. To see this, consider massless pure gluon theory where all "hadrons" are massive glueballs. Then we know from the renormalization group analysis that the glueball masses must behave as

$$m_{\text{glueball}} \propto M e^{-c/g^2(M)} \tag{1.12}$$

for $M \to \infty$, i.e. for $g(M) \to 0$, due to asymptotic freedom. Here M is the renormalization scale, $g(M)$ is the QCD coupling strength at this scale and c is a constant:

$$\begin{aligned} &\frac{g^2(M)}{4\pi^2} \longrightarrow \frac{12}{33\ln(M^2/\Lambda^2)} \quad \text{for} \quad M \to \infty, \\ &c = \frac{8\pi^2}{11}, \\ &\Lambda: \quad \text{QCD scale parameter.} \end{aligned} \tag{1.13}$$

Masses in massless Yang-Mills theory are a purely nonperturbative phenomenon, due to "dimensional transmutation". Scattering of glueball-hadrons in massless pure gluon theory should look very similar to scattering of hadrons in the real world, with finite total cross sections, amplitudes with analytic t dependence etc. At least, this would be my expectation. If the total cross section σ_{tot} has a finite limit as $s \to \infty$ we must have from the same renormalization group arguments:

$$\lim_{s\to\infty} \sigma_{\text{tot}}(s) \propto M^{-2} e^{2c/g^2(M)} \tag{1.14}$$

for $g(M) \to 0$. In this case, the total cross sections in pure gluon theory are also nonperturbative objects! It is easy to see that this conclusion is not changed if $\sigma_{\text{tot}}(s)$ has a logarithmic behaviour with s for $s \to \infty$, e.g.

$$\sigma_{\text{tot}}(s) \to \text{const} \times (\log s)^2. \tag{1.15}$$

I would then expect that also in full QCD total cross sections are nonperturbative objects, at least as far as hadrons made out of light quarks are concerned.

Some time ago P.V. Landshoff and myself started to think about a possible connection between the nontrivial vacuum structure of QCD – a typical nonperturbative phenomenon – and soft high energy reactions [26]. In the following I will first review some common folklore on the QCD vacuum and discuss in more detail the so-called "stochastic vacuum model". I will then sketch possible consequences of these ideas for high energy collisions.

2 The QCD Vacuum

According to current theoretical prejudice the vacuum state in QCD has a very complicated structure [27]-[37]. It was first noted by Savvidy [27] that by introducing a constant chromomagnetic field

$$\mathbf{B}^a = \mathbf{n}\eta^a B, \ (a = 1, \ldots, 8), \tag{2.1}$$

into the <u>perturbative</u> vacuum one can lower the vacuum-energy density $\varepsilon(B)$. Here $\mathbf{n}$ and η^a are constant unit vectors in ordinary and colour space. The result of his one-loop calculation was

$$\varepsilon(B) = \frac{1}{2}B^2 + \frac{\beta_0 g^2}{32\pi^2}B^2 \left[\ln\frac{B}{M^2} - \frac{1}{2}\right] \tag{2.2}$$

where g is the strong coupling constant, M is again the renormalization scale, and β_0 is given by the lowest order term in the Callan-Symanzik β-function:

$$M\frac{\mathrm{d}g(M)}{\mathrm{d}M} =: \beta(g) = -\frac{\beta_0}{16\pi^2}g^3 + \dots \tag{2.3}$$

For 3 colours and f flavours:

$$\beta_0 = 11 - \frac{2}{3}f. \tag{2.4}$$

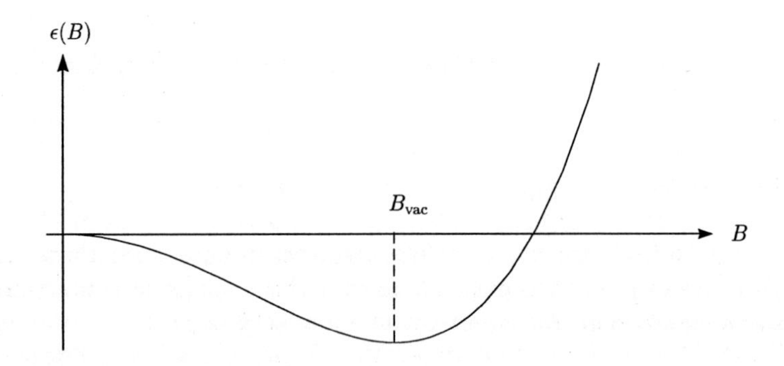

Fig. 2. *The schematic behaviour of the vacuum energy density $\varepsilon(B)$ as function of a constant chromomagnetic field B according to Savvidy's calculation (2.2).*

Thus, as long as we have asymptotic freedom, i.e. for $f \le 16$, the energy density $\varepsilon(B)$ looks as indicated schematically in Fig. 2 and has its minimum for $B = B_{\text{vac}} \neq 0$. Therefore, we should expect the QCD-vacuum to develop spontaneously a chromomagnetic field, the situation being similar to that in a ferromagnet below the Curie temperature where we have spontaneous magnetization.

Of course, the vacuum state in QCD has to be relativistically invariant and cannot have a preferred direction in ordinary space and colour space. What has been considered [33] are states composed of domains with random orientation of the gluon-field strength (Fig. 3). This is analogous to Weiss domains in a ferromagnet. The vacuum state should then be a suitable linear superposition of states with various domains and orientation of the fields inside the domains. This implies that the orientation of the fields in the domains as well as the boundaries of the domains will fluctuate.

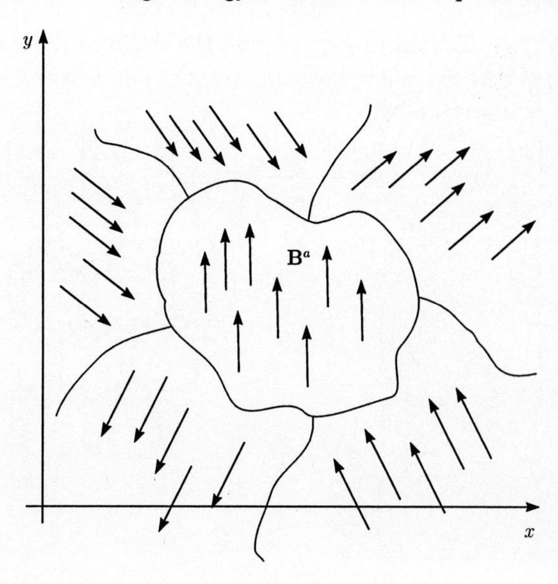

Fig. 3. *A "snapshot" of the QCD vacuum showing a domain structure of spontaneously created chromomagnetic fields.*

A very detailed picture for the QCD vacuum along these lines has been developed in ref. [33]. I cannot refrain from comparing this modern picture of the QCD vacuum (Fig. 4a) with the "modern picture" of the ether developed by Maxwell more than 100 years ago [38] (Fig. 4b). The analogy is quite striking and suggests to me that with time passing on we may also be able to find simpler views on the QCD vacuum. Remember that Einstein made great progress by eliminating the ether from electrodynamics. In the following we will adopt the picture of the QCD vacuum as developed in refs. [27]-[34],[36] and outlined above as a working hypothesis.

Let me now come to the values for the field strengths $\mathbf{E}^a$ and $\mathbf{B}^a$ in the vacuum. These must also be determined by Λ, the QCD scale parameter, the only dimensional parameter in QCD if we disregard the quark masses. Therefore, we must have on dimensional grounds for the renormalization group invariant quantity $(gB)^2$

$$(gB)^2 \propto \Lambda^4. \tag{2.5}$$

But we have much more detailed information on the values of these field strengths due to the work of Shifman, Vainshtein, and Zakharov (SVZ), who introduced the gluon condensate and first estimated its value using sum rules for charmonium states [30]:

$$\begin{aligned} < 0|\frac{g^2}{4\pi^2}G^a_{\mu\nu}(x)G^{\mu\nu a}(x)|0> &\equiv < 0|\frac{g^2}{2\pi^2}\left(\mathbf{B}^a(x)\mathbf{B}^a(x) - \mathbf{E}^a(x)\mathbf{E}^a(x)\right)|0> \\ &\equiv G_2 = (2.4 \pm 1.1)\cdot 10^{-2}\mathrm{GeV}^4 \\ &= (335 - 430\mathrm{MeV})^4. \end{aligned} \tag{2.6}$$

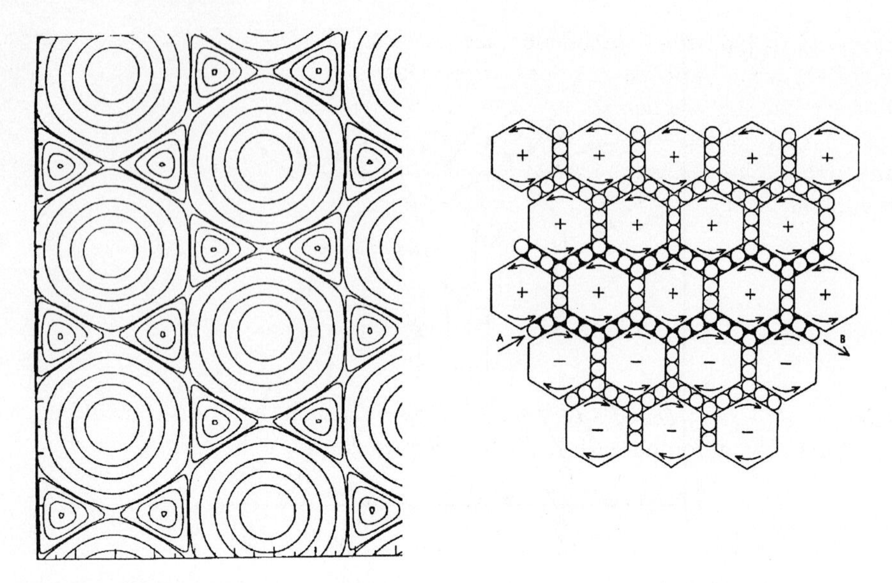

Fig. 4. *The QCD vacuum according to Ambjørn and Olesen [33] (a). The ether according to Maxwell [38] (b).*

Here we quote numerical values as given in the review [39]. A simple analysis shows that this implies

$$< 0|g^2\mathbf{B}^a(x)\mathbf{B}^a(x)|0 >= - < 0|g^2\mathbf{E}^a(x)\mathbf{E}^a(x)|0 >= \pi^2 G_2 \simeq (700\text{MeV})^4. \quad (2.7)$$

To prove (2.7) we note that Lorentz and parity invariance require the vacuum expectation value of the uncontracted product of two gluon field strengths to be of the form

$$< 0|\frac{g^2}{4\pi^2} G^a_{\mu\nu}(x) G^b_{\rho\sigma}(x)|0 >= (g_{\mu\rho}g_{\nu\sigma} - g_{\mu\sigma}g_{\nu\rho})\delta^{ab}\frac{G_2}{96} \quad (2.8)$$

where G_2 is the same constant as in (2.6). Taking appropriate contractions leads to (2.6) and (2.7).

We find that $< 0|\mathbf{B}^a(x)\ \mathbf{B}^a(x)|0 >$ is positive, $< 0|\mathbf{E}^a(x)\ \mathbf{E}^a(x)|0 >$ negative! This can happen because we are really considering products of field operators, normal-ordered with respect to the perturbative vacuum. The interpretation of (2.7) is, therefore, that the B-field fluctuates with bigger amplitude, the E-field with smaller amplitude than in the perturbative vacuum state.

What about the size a of the colour domains and the fluctuation times τ of the colour fields? On dimensional grounds we must have

$$a \sim \tau \sim \Lambda^{-1}. \quad (2.9)$$

A detailed model for the QCD vacuum incorporating the gluon condensate idea and a fall-off of the correlation of two field strengths with distance was proposed in [40]: the "stochastic vacuum model" (SVM). In the following we will discuss the basic assumptions of the model and then apply it to derive the area law for the Wegner-Wilson loop, i.e. confinement of static quarks. For the rest of this section we will work in Euclidean space-time. To accomplish the analytic continuation from Minkowski to Euclidean space-time we have to make the following replacements for x^λ and G^λ (cf. (1.4)):

$$\begin{aligned} x^0 &\to -iX_4, \\ \mathbf{x} &\to \mathbf{X}, \\ G^0 &\to i\mathcal{G}_4, \\ \mathbf{G} &\to -\boldsymbol{\mathcal{G}}. \end{aligned} \tag{2.10}$$

Here $X = (\mathbf{X}, X_4)$ denotes an Euclidean space-time point and $\mathcal{G}_\alpha$ $(\alpha = 1, ..., 4)$ the Euclidean gluon potential. With (2.10) we get

$$\begin{aligned} &(x \cdot y) \to -(X \cdot Y) = -X_\alpha Y_\alpha, \\ &-ig \int \mathrm{d}x^\mu G_\mu(x) \to -ig \int \mathrm{d}X_\alpha \mathcal{G}_\alpha(X), \\ &G^{0j} \to -i\mathcal{G}_{4j}, \\ &G^{jk} \to \mathcal{G}_{jk}, \end{aligned} \tag{2.11}$$

where $1 \le j,k \le 3$, $1 \le \alpha, \beta \le 4$ and

$$\mathcal{G}_{\alpha\beta} = \partial_\alpha \mathcal{G}_\beta - \partial_\beta \mathcal{G}_\alpha + ig[\mathcal{G}_\alpha, \mathcal{G}_\beta] \tag{2.12}$$

is the Euclidean gluon field strength tensor.

2.1 Connectors

Consider classical gluon fields in Euclidean space-time. Let X, Y be two points there and C_X a curve from X to Y (Fig. 5).

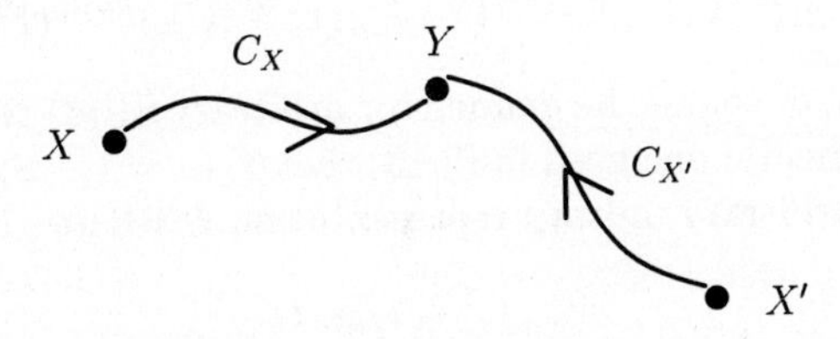

Fig. 5. *Points X, X', Y in Euclidean space time and curves $C_X, C_{X'}$ running from X to Y and X' to Y, respectively.*

We define the connector, the non-abelian generalization of the "Schwinger string" [41] of QED as

$$V(Y, X; C_X) = \mathrm{P}\{\exp[-ig \int_{C_X} \mathrm{d}Z_\alpha \mathcal{G}_\alpha(Z)]\}. \tag{2.13}$$

where P means path ordering. The connector can be obtained as solution of a differential equation. Let

$$\begin{aligned} &\tau \to Z(\tau), \\ &\tau_1 \le \tau \le \tau_2, \\ &Z(\tau_1) = X, \quad Z(\tau_2) = Y, \end{aligned} \tag{2.14}$$

be a parametrization of C_X. Consider the differential equation for a 3×3 matrix function $V(\tau)$:

$$\frac{\mathrm{d}}{\mathrm{d}\tau} V(\tau) = -ig \frac{\mathrm{d}Z_\alpha(\tau)}{\mathrm{d}\tau} \mathcal{G}_\alpha(Z(\tau)) V(\tau), \tag{2.15}$$

with the boundary condition

$$V(\tau_1) = \mathbb{1}. \tag{2.16}$$

The solution of (2.15), (2.16) gives for $\tau = \tau_2$ just the connector (2.13).

Under a gauge transformation

$$\mathcal{G}_\alpha(X) \to U(X)\mathcal{G}_\alpha(X)U^\dagger(X) - \frac{i}{g} U(X)\partial_\alpha U^\dagger(X) \tag{2.17}$$

where $U(X) \in SU(3)$, we have

$$V(Y, X; C_X) \to U(Y)V(Y, X; C_X)U^{-1}(X). \tag{2.18}$$

The connector can be used to "shift" various objects from one space-time point to another in a gauge-covariant way. We define for instance the field strength tensor shifted from X to Y along C_X as

$$\hat{\mathcal{G}}_{\alpha\beta}(Y, X; C_X) := V(Y, X; C_X)\mathcal{G}_{\alpha\beta}(X)V^{-1}(Y, X; C_X). \tag{2.19}$$

Under a gauge transformation $\hat{\mathcal{G}}_{\alpha\beta}(Y, X; C_X)$ transforms like a field strength tensor at Y:

$$\hat{\mathcal{G}}_{\alpha\beta}(Y, X; C_X) \to U(Y)\hat{\mathcal{G}}_{\alpha\beta}(Y, X; C_X)U^{-1}(Y). \tag{2.20}$$

Connectors can, of course, be defined for arbitrary SU(3) representations, not only for the fundamental one used in (2.13). Let $T_a (a = 1, ...8)$ be the generators of SU(3) in some arbitrary unitary representation R where

$$[T_a, T_b] = i f_{abc} T_c. \tag{2.21}$$

We define the connector for this representation by:

$$V_R(Y, X; C_X) := \mathrm{P}\ \exp[-ig \int_{C_X} \mathrm{d}Z_\alpha \mathcal{G}^a_\alpha(Z) T_a]. \tag{2.22}$$

We list some basic properties of connectors:

(i) For 2 adjoining curves C_1, C_2 (Fig. 6), the connectors are multiplied:

$$V_R(X_3, X_1; C_2 + C_1) = V_R(X_3, X_2; C_2).V_R(X_2, X_1; C_1). \tag{2.23}$$

(ii) Let C be a curve from X to Y and $\bar{C}$ the same curve but oriented in inverse direction, running from Y to X (Fig. 6). Then

$$V_R(X, Y; \bar{C}) V_R(Y, X; C) = 1\!\!1, \tag{2.24}$$

$$V_R^\dagger(Y, X; C) = V_R(X, Y; \bar{C}). \tag{2.25}$$

The product of the connectors for the path and the reverse path is equal to the unit matrix. The reversal of the path is equivalent to hermitian conjugation.

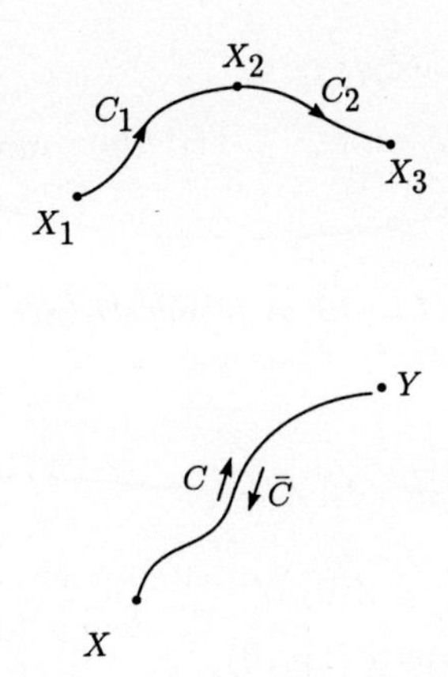

Fig. 6. *Curves in Euclidean space time: C_1 going from X_1 to X_2, C_2 from X_2 to X_3, C from X to Y, and $\bar{C}$ from Y to X.*

We leave the proof of (2.23)-(2.25) as an exercise for the reader.

2.2 The Non-Abelian Stokes Theorem

In this subsection we will derive the non-abelian generalization of the Stokes theorem [42]. Let us consider a surface S in Euclidean space time with boundary $C = \partial S$. Let X be some point on C as indicated in Fig. 7 and consider the connector (2.22) from X back to X along C:

$$V_R(X, X; C) = \mathrm{P}\ \exp[-ig \int_C \mathrm{d}Z_\alpha \mathcal{G}_\alpha^a(Z) T_a]. \tag{2.26}$$

The problem is to transform this line integral into a surface integral.

We start by considering a point Z_1 in S and a small plaquette formed by curves $C_1, .., C_4$ where one corner point is Z_1 (Fig. 7). We choose a coordinate system on S in the neighbourhood of Z_1:

$$(u, v) \rightarrow Z(u, v) \tag{2.27}$$

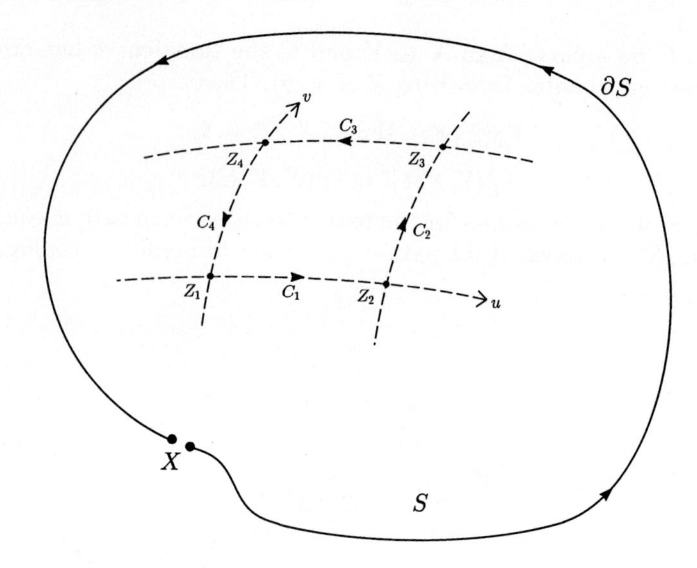

Fig. 7. *A surface S with boundary $C = \partial S$. A plaquette with corner points $Z_1 ..., Z_4$ and boundary formed by lines $u = const$, $v = const$.*

such that

$$
\begin{aligned}
Z_1 &= Z(0,0), \\
Z_2 &= Z(\Delta u, 0), \\
Z_3 &= Z(\Delta u, \Delta v), \\
Z_4 &= Z(0, \Delta v)
\end{aligned}
\tag{2.28}
$$

and the curves $C_1, C_3(C_2, C_4)$ correspond to lines v=const. (u=const.). The matrix representing the line integral around the small plaquette is

$$U_R(\Delta u, \Delta v) := V_R(C_4)V_R(C_3)V_R(C_2)V_R(C_1). \tag{2.29}$$

We want to make a Taylor expansion of $U_R(\Delta u, \Delta v)$ in Δu and Δv. From (2.24) we find immediately that

$$U_R(0, \Delta v) = U_R(\Delta u, 0) = 1\!\!1. \tag{2.30}$$

Thus the lowest order term in the expansion after the zeroth order is proportional to $\Delta u \cdot \Delta v$ and we get easily:

$$U_R(\Delta u, \Delta v) = 1\!\!1 - ig\frac{1}{2}\Delta\sigma_{\alpha\beta}G^a_{\alpha\beta}(Z_1)T_a + O(\Delta u^2 \Delta v, \Delta u \Delta v^2), \tag{2.31}$$

where

$$\Delta\sigma_{\alpha\beta} = \Delta u \Delta v \frac{\partial(Z_\alpha, Z_\beta)}{\partial(u, v)}. \tag{2.32}$$

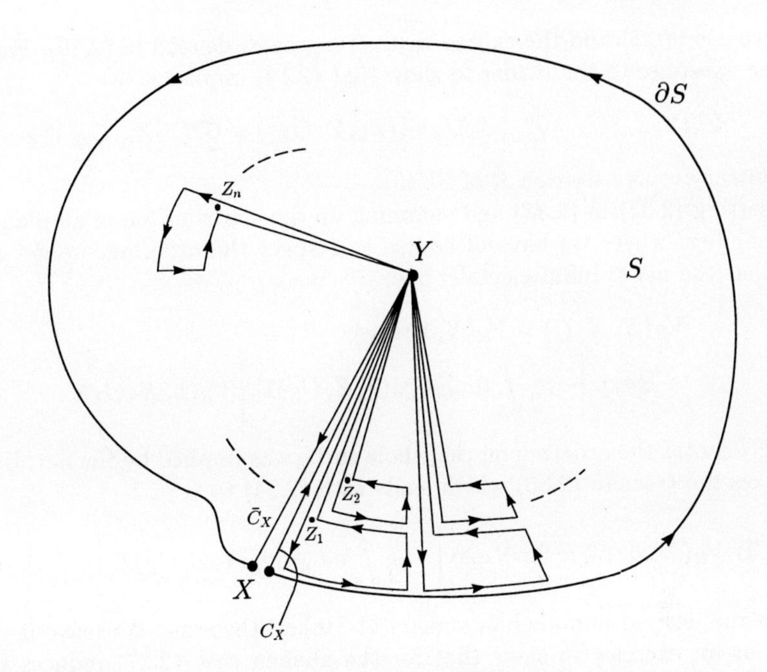

Fig. 8. *A reference point Y on S and the fan-type net with center Y spanned over S.*

In the limit $\Delta u, \Delta v \to 0$ $\Delta\sigma_{\alpha\beta}$ becomes the surface element $d\sigma_{\alpha\beta}$ of the plaquette.

The next step is to choose an arbitrary point Y, the reference point, on the surface S and to draw a fan-type net on S as a spider would do (Fig. 8). The system of curves of the net consists of the curve C_X running from X to Y, then $\bar{C}_{Z_1}$ from Y to Z_1, then around a small plaquette at Z_1, back to Y along C_{Z_1} and so on. The final curve is $\bar{C}_X$ from Y to X. Apart from the initial and final curves C_X and $\bar{C}_X$ we have a system of plaquettes with "handles" connecting them to Y. With the help of (2.23) -(2.25) we see that the connector along the whole net is equivalent to the original connector (2.26).

$$\begin{aligned} V_R(X,X;C) = V_R(X,Y;\bar{C}_X) \cdot \text{ product of connectors} \\ \text{for the plaquettes with handles } \cdot V_R(Y,X;C_X). \end{aligned} \tag{2.33}$$

Let us consider one plaquette with handle, for instance the one at Z_n in Fig. 8. For this contribution to (2.33) we get

$$\begin{aligned} & V_R(Y,Z_n;C_{Z_n})V_R(\text{plaquette at } Z_n)V_R(Z_n,Y;\bar{C}_{Z_n}) \\ & = V_R(Y,Z_n;C_{Z_n})\left[\mathbb{1} - ig\frac{1}{2}\Delta\sigma_{\alpha\beta}\mathcal{G}^a_{\alpha\beta}(Z_n)T_a + \cdots\right] V_R(Z_n,Y;\bar{C}_{Z_n}) \\ & = \mathbb{1} - ig\frac{1}{2}\Delta\sigma_{\alpha\beta}\hat{\mathcal{G}}^a_{\alpha\beta}(Y,Z_n;C_{Z_n})T_a + \cdots . \end{aligned} \tag{2.34}$$

Here we use (2.31) and the shifted field strengths as defined in (2.19). We leave it as an excercise to the reader to show that (2.19) implies

$$V_R(Y, Z_n; C_{Z_n})\mathcal{G}^a_{\alpha\beta}(Z_n)T_a V_R(Z_n, Y; \bar{C}_{Z_n}) = \hat{\mathcal{G}}^a(Y, Z_n; C_{Z_n})T_a \quad (2.35)$$

for arbitrary representation R of $SU(3)$.

Inserting (2.34) in (2.33) and summing up the contribution of all plaquettes with handles, where we have of course to respect the ordering, we get in the limit that the net is infinitesimally fine:

$$\begin{aligned} V_R(X, X; C) &= V_R(X, Y; \bar{C}_X) \cdot \\ &\mathrm{P}\exp\left[-i\frac{g}{2}\int_S \mathrm{d}\sigma_{\alpha\beta}\hat{\mathcal{G}}^a_{\alpha\beta}(Y, Z; C_Z)T_a\right] V_R(Y, X; C_X). \end{aligned} \quad (2.36)$$

Here P denotes the ordering on the whole surface as implied by the net. Usually one takes the trace in (2.36) which leads with (2.24) to

$$\mathrm{Tr}\ V_R(X, X; C) = \mathrm{Tr}\ \mathrm{P}\exp\left[-i\frac{g}{2}\int_S \mathrm{d}\sigma_{\alpha\beta}\hat{\mathcal{G}}^a_{\alpha\beta}(Y, Z; C_Z)T_a\right]. \quad (2.37)$$

This is the desired non-abelian version of Stokes' theorem. We leave it to the reader as an exercise to show that for the abelian case (2.37) reduces to the conventional Stokes theorem.

2.3 The Cumulant Expansion

As a last mathematical tool for making calculations with the SVM we discuss the cumulant expansion [43]. Consider functions

$$\tau \to B(\tau) \quad (2.38)$$

on the interval $O \le \tau \le 1$ where $B(\tau)$ are quadratic matrices. We assume that an averaging procedure over products of the functions $B(\cdot)$ is defined:

$$E(B(\tau_1)), E(B(\tau_1)B(\tau_2)), \ldots .$$

We consider first the case that all averages $E(\cdot)$ are c numbers and that

$$E(1) = 1. \quad (2.39)$$

Let us consider the expectation value of the τ-ordered exponential:

$$f(t) := E(\mathrm{P}\exp\left[t\int_0^1 \mathrm{d}\tau B(\tau)\right]), \quad (2.40)$$

where $t \in C$. The cumulant expansion asserts that $\ln f(t)$ can be expanded as

$$\ln f(t) = \sum_{n=1}^{\infty}\frac{t^n}{n!}\int_0^1 \mathrm{d}\tau_1 \ldots \int_0^1 \mathrm{d}\tau_n K_n(\tau_1, \ldots, \tau_n), \quad (2.41)$$

where the n-th cumulant $K_n(\tau_1, \ldots, \tau_n)$ is a symmetric function of its arguments. A frequently used notation for the K_n is:

$$K_n(\tau_1, \ldots, \tau_n) \equiv \langle\langle B(\tau_1) \ldots B(\tau_n) \rangle\rangle. \tag{2.42}$$

To prove (2.41) we proceed as follows. Expanding in powers of t on the r.h.s. of (2.40) we get

$$f(t) = 1 + \sum_{n=1}^{\infty} \frac{t^n}{n!} \mathcal{B}_n \tag{2.43}$$

where

$$\mathcal{B}_n = \int_0^1 \mathrm{d}\tau_1 \ldots \int_0^1 \mathrm{d}\tau_n E(\mathrm{P}(B(\tau_1) \ldots B(\tau_n))). \tag{2.44}$$

Now we expand $\ln f(t)$:

$$\ln f(t) = \sum_{n=1}^{\infty} \frac{t^n}{n!} \mathcal{K}_n, \tag{2.45}$$

where the expansion coefficients $\mathcal{K}_n$ are obtained from:

$$\begin{aligned} f(t) &= \exp(\ln f(t)), \\ 1 + \sum_{n=1}^{\infty} \frac{t^n}{n!} \mathcal{B}_n &= \exp\left[\sum_{n=1}^{\infty} \frac{t^n}{n!} \mathcal{K}_n\right]. \end{aligned} \tag{2.46}$$

From this we obtain the $\mathcal{K}_n$ as solution of the following system of equations:

$$\begin{aligned} &\mathcal{B}_1 = \mathcal{K}_1, \\ &\mathcal{B}_2 = \mathcal{K}_2 + \mathcal{K}_1^2, \\ &\mathcal{B}_3 = \mathcal{K}_3 + \frac{3}{2}(\mathcal{K}_2\mathcal{K}_1 + \mathcal{K}_1\mathcal{K}_2) + \mathcal{K}_1^3, \\ &\mathcal{B}_4 = \mathcal{K}_4 + 2(\mathcal{K}_3\mathcal{K}_1 + \mathcal{K}_1\mathcal{K}_3) + 3\mathcal{K}_2^2 + 2(\mathcal{K}_2\mathcal{K}_1^2 + \mathcal{K}_1\mathcal{K}_2\mathcal{K}_1 + \mathcal{K}_1^2\mathcal{K}_2) + \mathcal{K}_1^4, \\ &\ldots . \end{aligned} \tag{2.47}$$

Clearly, this system can be inverted and we get $\mathcal{K}_n$ as sum of monomials of the form

$$\mathcal{B}_{i_1} \cdot \mathcal{B}_{i_2} \ldots \mathcal{B}_{i_k}, \tag{2.48}$$

where

$$\sum_{j=1}^{k} i_j = n. \tag{2.49}$$

Using (2.44), every monomial (2.48) can be written as n-fold integral over $\tau_1, \ldots, \tau_n$ with $0 \le \tau_j \le 1$ $(j = 1, \ldots, n)$. Since the integration domain is symmetric under arbitrary permutations of $\tau_1, \ldots, \tau_n$ we can symmetrize the integrand completely. In this way we get

$$\mathcal{K}_n = \int_0^1 \mathrm{d}\tau_1 \ldots \int_0^1 \mathrm{d}\tau_n K_n(\tau_1, \ldots, \tau_n), \tag{2.50}$$

where K_n is a totally symmetric function of its arguments. Inserting (2.50) in (2.45) we get the cumulant expansion (2.41), q.e.d. Explicitly we find for the first few cumulants:

$$\begin{aligned}
K_1(1) &= E(B(1)), \\
K_2(1,2) &= E(\mathrm{P}(B(1)B(2))) \\
&\quad -\frac{1}{2}E(B(1))E(B(2)) - \frac{1}{2}E(B(2))E(B(1)), \\
K_3(1,2,3) &= E(\mathrm{P}(B(1)B(2)B(3))) \\
&\quad -\frac{1}{2}[E(\mathrm{P}(B(1)B(2)))E(B(3)) \\
&\quad +E(B(1))E(\mathrm{P}(B(2)B(3))) \\
&\quad +\text{cycl. perm.}\ ] \\
&\quad +\frac{1}{3}[E(B(1))E(B(2))E(B(3)) + \text{perm.}\ ], \\
&\quad \ldots .
\end{aligned} \tag{2.51}$$

Here we write as a shorthand notation $K_1(1) \equiv K_1(\tau_1)$, $B(1) \equiv B(\tau_1)$ etc.

The cumulant expansion (2.41) has the so-called "cluster" property: Let us assume that the expectation values of the P-ordered products factorize

$$E(\mathrm{P}(B(1)\ldots B(n))) = \frac{1}{n!}\{E(B(1))\ldots E(B(n)) + \text{perm.}\} \tag{2.52}$$

for all $n \geq 2$ and all

$$|\tau_i - \tau_j| \geq \tau_{\min} \quad (i \neq j). \tag{2.53}$$

We can then show that the cumulants for $n \geq 2$ vanish:

$$K_n(1,\ldots,n) = 0, \quad (n \geq 2) \tag{2.54}$$

if the τ_i satisfy (2.53).

To prove (2.54) we show first that it is true for $n = 2$. Indeed, from (2.51) and (2.52) we get for $|\tau_1 - \tau_2| \geq \tau_{\min}$

$$\begin{aligned}
K_2(1,2) &= \frac{1}{2!}\{E(B(1))E(B(2)) + \text{perm.}\} \\
&\quad -\frac{1}{2}E(B(1))E(B(2)) - \frac{1}{2}E(B(2))E(B(1)) \\
&= 0.
\end{aligned} \tag{2.55}$$

Now we proceed by mathematical induction. Assume that (2.54) has been shown for all k with $2 \leq k \leq n-1$. We have from (2.47):

$$\begin{aligned}
K_n(1,\ldots,n) &= E(\mathrm{P}(B(1)\ldots B(n))) \\
&\quad -\frac{1}{n!}[K_1(1)\ldots K_1(n) + \text{perm.}] \\
&\quad + [\text{symmetrized products of cumulants } K_k(1,\ldots,k) \\
&\quad\ \ \text{with } 1 \leq k \leq n-1]
\end{aligned} \tag{2.56}$$

but at least one factor with $k \geq 2$. With (2.52) we get now immediately $K_n(1,\ldots,n) = 0$ in the region defined by (2.53), q.e.d.

Up to now we have assumed the expectation values $E(\cdot)$ to be c-numbers. In this case many of the formulae (2.47) - (2.56) can be simplified by using

$$E(B(1))E(B(2)) = E(B(2))E(B(1)), \quad \text{etc.} \tag{2.57}$$

We have, on purpose, not used such commutativity relations, since we are now going to generalize the cumulant expansion to the case where

$$E(B(1)), E(B(1)B(2)), \ldots \tag{2.58}$$

are themselves quadratic matrix valued expectation values with

$$E(\mathbb{1}) = \mathbb{1}. \tag{2.59}$$

Then we have, of course, in general, no more the commutativity relations (2.57). But all formulae (2.40) - (2.56) are written in such a way that they remain true also for the case of matrix valued expectation values $E(\cdot)$.

2.4 The Basic Assumptions of the Stochastic Vacuum Model

The basic object of the SVM is the correlator of two field strengths shifted to a common reference point. Let X, X' be two points in Euclidean space-time, Y a reference point and $C_X, C_{X'}$ curves from X to Y and X' to Y, respectively (Fig. 5). We consider the shifted field strengths as defined in (2.19) and the vacuum expectation value of their product in the sense of Euclidean QFT:

$$\langle \frac{g^2}{4\pi^2}\left[\hat{\mathcal{G}}^a_{\mu\nu}(Y,X;C_X)\hat{\mathcal{G}}^b_{\rho\sigma}(Y,X';C_{X'})\right]\rangle =: \frac{1}{4}\delta^{ab}F_{\mu\nu\rho\sigma}(X,X',Y;C_X,C_{X'}). \tag{2.60}$$

Here (2.20) and colour conservation allow us to write the r.h.s. of (2.60) proportional to δ^{ab}. It is easy to see that $F_{\mu\nu\rho\sigma}$ depends only on X, X' and the curve $C_X + \bar{C}_{X'}$ connecting them; i.e. the reference point Y can be freely shifted on the connecting curve. In the SVM one makes now the strong assumption that the correlator (2.60) even does not depend on the connecting curve at all:

- **Assumption 1:** $F_{\mu\nu\rho\sigma}$ is independent of the reference point Y and of the curves C_X and $C_{X'}$.

Translational, $O(4)$- and parity invariance require then the correlator (2.60) to be of the following form:

$$F_{\mu\nu\rho\sigma} = F_{\mu\nu\rho\sigma}(Z) = \frac{1}{24}G_2\Big\{(\delta_{\mu\rho}\delta_{\nu\sigma} - \delta_{\mu\sigma}\delta_{\nu\rho})\,\kappa D(-Z^2) \tag{2.61}$$
$$+\frac{1}{2}\Big[\frac{\partial}{\partial Z_\nu}(Z_\sigma\delta_{\mu\rho} - Z_\rho\delta_{\mu\sigma}) + \frac{\partial}{\partial Z_\mu}(Z_\rho\delta_{\nu\sigma} - Z_\sigma\delta_{\nu\rho})\Big](1-\kappa)D_1(-Z^2)\Big\}.$$

Here $Z = X - X'$, G_2 is the gluon condensate, D, D_1 are invariant functions normalized to

$$D(0) = D_1(0) = 1 \tag{2.62}$$

and κ is a parameter measuring the non-abelian character of the correlator.

Indeed, if we consider an abelian theory we have to replace the gluon field strengths $\mathcal{G}_{\mu\nu}$ by abelian field strengths $\mathcal{F}_{\mu\nu}$, which satisfy the homogenous Maxwell equation

$$\epsilon_{\mu\nu\rho\sigma}\partial_\nu \mathcal{F}_{\rho\sigma}(X) = 0, \tag{2.63}$$

if there are no magnetic monopoles present. It is easy to see that this implies $\kappa = 0$ in (2.61). Thus, in an abelian theory, the D-term in (2.61) is absent without magnetic monopoles but would be non-zero if the vacuum contained a magnetic monopole condensate. In the non-abelian theory the D-term has no reason to vanish. In fact, we will see that it dominates over the D_1-term. The abelian analogy suggests an interpretation of the D-term as being due to an effective chromomagnetic monopole condensate in the QCD vacuum.

Two further assumptions are made in the SVM:

- **Assumption 2:** The correlation functions $D(-Z^2)$ and $D_1(-Z^2)$ fall off rapidly for $Z^2 \to \infty$. There exists a characteristic finite correlation length a, which we define as

$$a := \int_0^\infty \mathrm{d}Z\ D(-Z^2). \tag{2.64}$$

A typical ansatz for the function D, incorporating Assumption 2, is as follows:

$$D(-Z^2) = \frac{27}{64}a^{-2}\int \mathrm{d}^4K e^{iKZ} K^2 \left[K^2 + \left(\frac{3\pi}{8a}\right)^2\right]^{-4}, \tag{2.65}$$

which leads to

$$D(-Z^2) \propto \exp\left(-\frac{3\pi|Z|}{8a}\right) \qquad \text{for } Z^2 \to \infty. \tag{2.66}$$

The function D_1 is chosen such that

$$\left(4 + Z_\mu \frac{\partial}{\partial Z_\mu}\right) D_1(-Z^2) = 4D(-Z^2) \tag{2.67}$$

which leads to

$$D_1(-Z^2) = (Z^2)^{-2}\int_0^{Z^2} \mathrm{d}v 2v D(-v). \tag{2.68}$$

With (2.67) the contracted field strength correlator has the form (cf. (2.61):

$$F_{\mu\nu\mu\nu} = \frac{1}{2}G_2 D(-Z^2). \tag{2.69}$$

The ansatz (2.65), (2.67) can be compared to a lattice gauge theory calculation of the gluon field strength correlator [44] in order to fix the parameters. One finds (cf. [44], [45] and Fig. 9):

$$\begin{aligned} a &= 0.35 \text{ fm}, \\ \kappa &= 0.74, \\ G_2 &= (496 \text{ MeV})^4, \end{aligned} \tag{2.70}$$

with an educated guess for the error of $\approx 10\%$. Of course, from Fig. 9 we get only the product $\kappa \cdot G_2$. These quantities are obtained separately by measuring on the

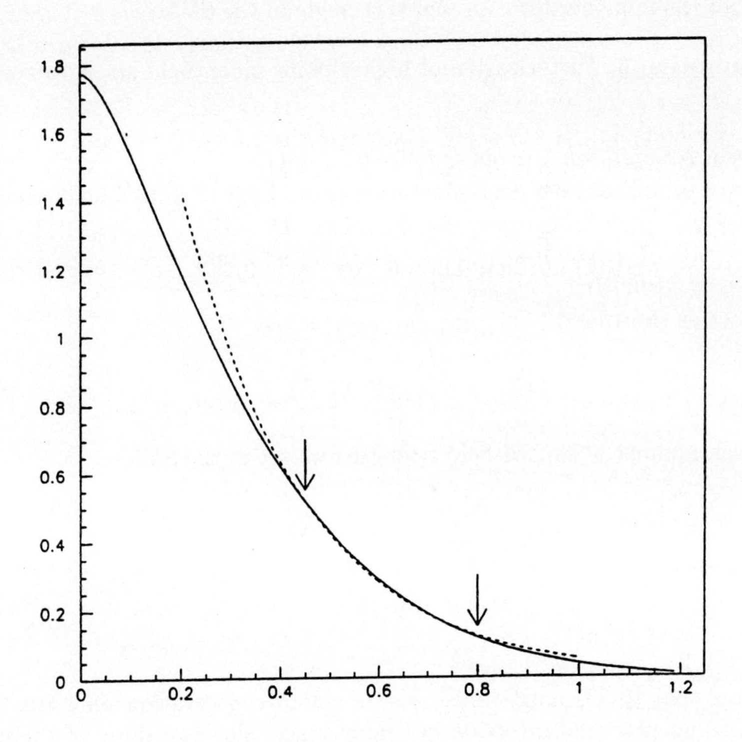

Fig. 9. *The correlator function* $4\pi^2\kappa G_2 D(-Z^2)$ *(cf. (2.61)) as function of* $|Z|$*. The dashed line is the lattice result of [44] with the arrows indicating the range where these results are considered reliable. The solid line corresponds to the ansatz (2.65) with* $a = 0.35$ fm *(cf. Fig. 5.1 of [45]).*

lattice several components of the correlator (2.60). The value for G_2 in (2.70) from the lattice calculations is somewhat larger than from phenomenology. This can be understood as follows. The lattice calculations of [44] are for the pure gluon theory. In the real world light quarks are present. Their effect is estimated to reduce the value of G_2 substantially [46].

We note that the correlation length is smaller, albeit not much smaller than a typical radius R of a light hadron (cf. e.g. [47], [48], [49],[50]):

$$R \sim 0.7 - 1 \text{ fm}. \tag{2.71}$$

Still

$$a^2/R^2 \approx 0.2 - 0.3 \tag{2.72}$$

is a reasonably small number and this will be important for us in the following.

We come now to the third Assumption made in the SVM:

- Assumption 3: Factorization of higher point gluon field strength correlators.

In detail Assumption 3 reads as follows (cf. [45]):

All expectation values of an odd number of products of shifted field strengths vanish:

$$\langle \hat{\mathcal{G}}(1) ... \hat{\mathcal{G}}(2n+1) \rangle = 0 \quad \text{for} \quad n = 0, 1, 2, ... \tag{2.73}$$

Here we set as shorthand

$$\hat{\mathcal{G}}(i) \equiv \hat{\mathcal{G}}^{a_i}_{\alpha_i \beta_i}(Y, X_i; C_{X_i}). \tag{2.74}$$

For an even number of shifted field strengths we set in the SVM:

$$\langle \hat{\mathcal{G}}(1) ... \hat{\mathcal{G}}(2n) \rangle = \sum_{\substack{\text{all pairings} \\ (i_1, j_1) ... (i_n, j_n)}} \langle \hat{\mathcal{G}}(i_1) \hat{G}(j_1) \rangle ... \langle \hat{\mathcal{G}}(i_n) \hat{\mathcal{G}}(j_n) \rangle, \tag{2.75}$$

where $n = 1, 2, ...$.

We note that $\langle \hat{\mathcal{G}}(1) \rangle$ must vanish due to colour conservation since the QCD vacuum has no preferred direction in colour space. The vanishing of the other correlators of odd numbers of field strengths, postulated in (2.73), as well as the factorization property (2.75) are strong dynamical assumptions. They mean that the vacuum fluctuations are assumed to be of the simplest type: a Gaussian random process.

For some applications of the SVM, for instance the calculation of the Wegner-Wilson loop described below, Assumption 3 is not crucial and can be relaxed. But for the applications of the SVM to high energy scattering (cf. Sect. 3) Assumption 3 is crucial. In any case we prefer to specify the model completely, thus giving it maximal predictive power. On the other hand, of course, the model can then more easily run into difficulties in comparison with experiments.

2.5 The Wegner-Wilson Loop in the Stochastic Vacuum Model

We have now specified the SVM completely and can proceed to show how this model produces confinement. We consider a static quark-antiquark pair at distance R from each other and ask for the potential $V(R)$. To calculate $V(R)$ we start with a rectangular Wegner-Wilson loop in the $X_1 - X_4$ plane (Fig. 10): Let C be the loop and S the rectangle, $C = \partial S$. Then

$$W(C) = \frac{1}{3}\langle \text{Tr P} \exp[-ig \int_C \mathrm{d}Z_\mu \mathcal{G}_\mu(Z)] \rangle \tag{2.76}$$

and

$$V(R) = -\lim_{T \to \infty} \frac{1}{T} \ln W(C). \tag{2.77}$$

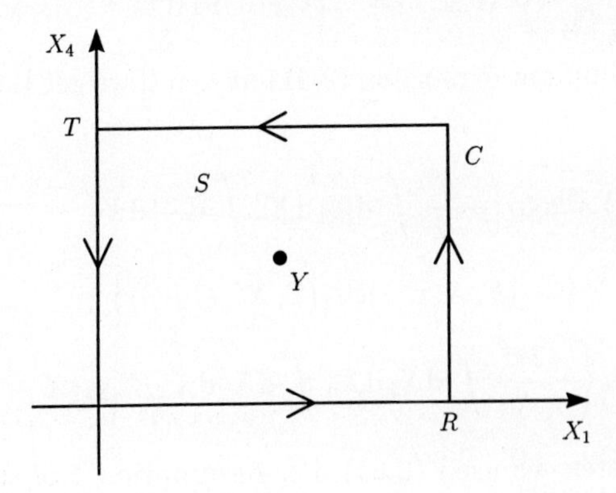

Fig. 10. *Rectangular Wegner-Wilson loop in Euclidean space-time in the $X_1 - X_4$ plane. The linear extensions are R in X_1 direction and T in X_4 direction. Y is the reference point used in the application of the non-abelian Stokes theorem.*

To evaluate $W(C)$ in the SVM we first transform the line integral of the potentials in (2.76) into a surface integral of field strengths, using the non-abelian version of Stokes theorem (cf. Sect. 2.2). For this we choose a reference point Y in S. We get then

$$W(C) = \frac{1}{3}\langle \text{Tr P} \exp[-ig \int_S \mathrm{d}X_1 \mathrm{d}X_4 \hat{\mathcal{G}}_{14}(Y, X, C_X)] \rangle \tag{2.78}$$

where $\hat{\mathcal{G}}_{\mu\nu}$ are the field strengths parallel-transported from X to Y along a straight line C_X. The path-ordered way to integrate over S in a fan-type net (cf. Fig. 8) is indicated by P. To evaluate the expectation value of the path-ordered exponential in (2.78), we will use the technique of the cumulant expansion and the Assumptions 1-3 of the SVM. We make the following replacements in the formulae (2.38) ff. of Sect. 2.3:

$$\begin{aligned}
&\tau \to (X_1, X_4),\\
&B(\tau_i) \to \hat{\mathcal{G}}_{14}(Y, X^{(i)}, C_{X^{(i)}}),\\
&E(B(\tau_1)..B(\tau_n)) \to \frac{1}{3}\mathrm{Tr}\langle \hat{\mathcal{G}}_{14}(Y, X^{(1)}; C_{X^{(1)}})...\hat{\mathcal{G}}_{14}(Y, X^{(n)}; C_{X^{(n)}})\rangle,\\
&t \to -ig,\\
&f(t) \to W(C).
\end{aligned} \tag{2.79}$$

With this we can express $W(C)$ as an exponential of cumulants as given in (2.41). From Assumption 3 of the SVM we get that all cumulants for odd numbers of gluon field strengths vanish. The lowest nontrivial cumulant is the second one, K_2, and from (2.51) we find

$$K_2(1,2) \to \frac{1}{3}\mathrm{Tr}\langle \mathrm{P}(\hat{\mathcal{G}}_{14}(Y, X^{(1)}; C_{X^{(1)}})\hat{\mathcal{G}}_{14}(Y, X^{(2)}; C_{X^{(2)}})\rangle. \tag{2.80}$$

If we cut off the cumulant expansion (2.41) at $n = 2$ we get then for the Wegner-Wilson loop

$$\begin{aligned}
W(C) &= \exp\Big\{-\frac{g^2}{2}\int_S \mathrm{d}X_1\mathrm{d}X_4 \int_S \mathrm{d}X_1'\mathrm{d}X_4'\\
&\langle\frac{1}{3}\mathrm{Tr}\ \mathrm{P}\ [\hat{\mathcal{G}}_{14}(Y, X, C_X)\hat{\mathcal{G}}_{14}(Y, X', C_{X'})]\rangle\Big\}\\
&= \exp\left\{-\frac{2\pi^2}{3}\int_S \mathrm{d}X_1\mathrm{d}X_4 \int_S \mathrm{d}X_1'\mathrm{d}X_4' F_{1414}(X - X')\right\},
\end{aligned} \tag{2.81}$$

where in the last step we used (2.61), i.e. Assumption 1 of the SVM. Next the Assumption 2 of short-range correlation for the field strengths enters in a crucial way. This implies that for large Wegner-Wilson loops the integration over X_1', X_4' keeping X_1, X_4 fixed in (2.81) gives essentially a factor a^2. The remaining X_1, X_4 integration gives then the area of $S = RT$. Thus we arrive at an area law for the Wegner-Wilson loop for $R, T \gg a$:

$$W(C) = e^{-\sigma\cdot R\cdot T}, \tag{2.82}$$

where the constant σ is obtained as

$$\sigma = \frac{\pi^3\kappa G_2}{36}\int_0^\infty \mathrm{d}Z^2 D(-Z^2) = \frac{32\pi\kappa G_2 a^2}{81}. \tag{2.83}$$

We leave the proof of (2.83) as an exercise for the reader. Comparing (2.77) and (2.82) we find that the SVM produces a linearly rising potential

$$V(R) = \sigma\cdot R \quad \text{for} \quad R \gg a, \tag{2.84}$$

where σ is the string tension.

The results (2.82)-(2.84) were derived in the framework of the SVM in [40], where the model was introduced, and they are interesting in quite a number of respects:

Only for $\kappa \neq 0$ does one get an area law and thus confinement. The D_1 term which is the only one present in the abelian theory produces no confinement. In the SVM confinement is related to an effective chromomagnetic monopole condensate in the vacuum.

The short range correlation for the field strengths produces a long range correlation for the potentials if the D term is present in (2.61), i.e. if $\kappa \neq 0$.

The string tension σ is obtained numerically with the input (2.70) as

$$\sigma = (420 \text{ MeV})^2 \tag{2.85}$$

This is very consistent with the phenomenological determination of $\sigma \simeq (430 \text{ MeV})^2$ from the charmonium spectrum [51].

The SVM has been applied in many other studies of low energy hadronic phenomena (cf. [39] for a review). It was, for instance, possible to calculate flux distributions around a static quark-antiquark pair [52]. The results compare well with lattice gauge theory calculations wherever the latter are available.

Let us come back to the calculation of the Wegner-Wilson loop above. It is legitimate to ask about the contribution of higher cumulants to $W(C)$. How do they modify (2.81)? It turns out that higher cumulants may cause some problems, which we discuss in Appendix A together with a proposal for their remedy.

3 Soft Hadronic Reactions

3.1 General Considerations

In this section we will present a microscopic approach towards hadron-hadron diffractive scattering (cf. [26], [53]). Consider as an example elastic scattering of two hadrons h_1, h_2

$$h_1 + h_2 \rightarrow h_1 + h_2 \tag{3.1}$$

at high energies and small momentum transfer. We will look at reaction (3.1) from the point of view of an observer living in the "femto-universe", i.e. we imagine having a microscope with resolution much better than 1 fm for observing what happens during the collision. Of course, we should choose an appropriate resolution for our microscope. If we choose the resolution much too good, we will see too many details of the internal structure of the hadrons which are irrelevant for the reaction considered and we will miss the essential features. The same is true if the resolution is too poor. In [53] we used a series of simple arguments based on the uncertainty relation to estimate this appropriate resolution.

Let $t = 0$ be the nominal collision time of the hadrons in (3.1) in the c.m. system. This is the time when the hadrons $h_{1,2}$ have maximal spatial overlap.

Let furthermore be $t_0/2$ the time when, in an inelastic collision, the first produced hadrons appear. We estimate $t_0 \approx 2$ fm from the LUND model of particle production [54]. Then the appropriate resolution, i.e. the cutoff in transverse parton momenta k_T of the hadronic wave functions to be chosen for describing reaction (3.1) in an economical way is

$$k_T^2 \leq \sqrt{s}/(2t_0) \tag{3.2}$$

where $\sqrt{s}$ is the c.m. energy. Modes with higher k_T can be assumed to be integrated out. With this we could argue that over the time interval

$$-\frac{1}{2}t_0 \leq t \leq \frac{1}{2}t_0 \tag{3.3}$$

the following should hold or better: could be assumed:

(a) The parton state of the hadrons does not change qualitatively, i.e. parton annihilation and production processes can be neglected for this time.

(b) Partons travel in essence on straight lightlike world lines.

(c) The partons undergo "soft" elastic scattering.

The strategy is now to study first soft parton-parton scattering in the femtouniverse. There, the relevant interaction will turn out to be mediated by the gluonic vacuum fluctuations. We have argued at length in Sect. 2 that these have a highly nonperturbative character. In this way the nonperturbative QCD vacuum structure will enter the picture for high energy soft hadronic reactions. Once we have solved the problem of parton-parton scattering we have to fold the partonic S-matrix with the hadronic wave functions of the appropriate resolution (3.2) to get the hadronic S-matrix elements.

We will now give an outline of the various steps in this program.

3.2 The Functional Integral Approach to Parton-Parton Scattering

Consider first quark-quark scattering:

$$q(p_1) + q(p_2) \to q(p_3) + q(p_4), \tag{3.4}$$

where we set

$$\begin{aligned} s &= (p_1 + p_2)^2 \\ t &= (p_1 - p_3)^2 \\ u &= (p_1 - p_4)^2 . \end{aligned} \tag{3.5}$$

Of course, free quarks do not exist in QCD, but let us close our eyes to this at the moment. Now we should calculate the scattering of the quarks over the finite time interval (3.3) of length $t_0 \approx 2$ fm. Let us assume that 2 fm is nearly infinitely long on the scale of the femto universe and use the standard reduction formula, due to Lehmann, Symanzik, and Zimmermann, to relate the S-matrix element for (3.4) to an integral over the 4-point function of the quark fields. We use the following normalization for our quark states

$$\begin{aligned}\langle q(p_j, s_j, A_j) \mid q(p_k, s_k, A_k)\rangle \\ &= \delta_{s_j s_k} \delta_{A_j A_k} (2\pi)^3 \sqrt{2p_j^0 2p_k^0} \delta^3(\mathbf{p}_j - \mathbf{p}_k) \\ &\equiv \delta(j,k),\end{aligned} \tag{3.6}$$

where s_j, s_k are the spin and A_j, A_k the colour indices. With this we get

$$\begin{aligned}&\langle q(p_3, s_3, A_3) q(p_4, s_4, A_4)|S|q(p_1, s_1, A_1) q(p_2, s_2, A_2)\rangle \\ &\equiv \langle 3,4|S|1,2\rangle \\ &= \langle 3,4|1,2\rangle + Z_\psi^{-2} \Big\{ (4|(i\overrightarrow{\partial\!\!\!/} - m_q') \otimes (3|(i\overrightarrow{\partial\!\!\!/} - m_q') \\ &\qquad \langle 0|\mathrm{T}(q(4)q(3)\bar{q}(1)\bar{q}(2))|0\rangle \\ &\qquad (i\overleftarrow{\partial\!\!\!/} + m_q')|1) \otimes (i\overleftarrow{\partial\!\!\!/} + m_q')|2) \Big\}.\end{aligned} \tag{3.7}$$

Here Z_ψ is the quark wave function renormalization constant and m_q' the renormalized quark mass. We use a shorthand notation

$$\begin{aligned}|j) &\equiv u_{s_j,A_j}(p_j) e^{-ip_j x_j}, \\ (j| &= e^{ip_j x_j} \bar{u}_{s_j,A_j}(p_j), \\ q(j) &\equiv q(x_j), \\ (j &= 1,..,4),\end{aligned} \tag{3.8}$$

where u is the spinor in Dirac and colour space. Two repeated arguments $j, k, ...$ imply a space-time integration, for instance

$$\bar{q}(1)(i\overleftarrow{\partial\!\!\!/} + m_q')|1) \equiv \int \mathrm{d}x_1 \bar{q}(x_1)(i\overleftarrow{\partial\!\!\!/} + m_q') e^{-ip_1 x_1} u_{s_1,A_1}(p_1). \tag{3.9}$$

Thus in (3.7) we have four integrations over $x_1, ..., x_4$.

We can represent the 4-point function of the quark fields as a functional integral:

$$\begin{aligned}&\langle 0|\mathrm{T}(q(4)q(3)\bar{q}(1)\bar{q}(2))|0\rangle \\ &= Z^{-1} \int \mathcal{D}(G, q, \bar{q}) \exp\{i \int \mathrm{d}x \mathcal{L}_{\mathrm{QCD}}(x)\} q(4)q(3)\bar{q}(1)\bar{q}(2),\end{aligned} \tag{3.10}$$

where Z is the partition function:

$$Z = \int \mathcal{D}(G, q, \bar{q}) \exp\{i \int \mathrm{d}x \mathcal{L}_{\mathrm{QCD}}(x)\}. \tag{3.11}$$

The QCD Lagrangian (1.1) is bilinear in the quark and antiquark fields. Thus – as is well known – the functional integration over q and $\bar{q}$ can be carried out immediately. After some standard manipulations we arrive at the following expression:

$$\begin{aligned}
&\langle 0|\mathrm{T}(q(4)q(3)\bar{q}(1)\bar{q}(2))|0\rangle \\
&= \frac{1}{Z}\int \mathcal{D}(G)\exp\left\{-i\int \mathrm{d}x \frac{1}{2}\mathrm{Tr}(G_{\lambda\rho}(x)G^{\lambda\rho}(x))\right\} \\
&\det[-i(i\gamma^{\lambda}D_{\lambda} - m_q + i\epsilon)] \\
&\left\{\frac{1}{i}S_F(4,2;G)\frac{1}{i}S_F(3,1;G) - (3 \leftrightarrow 4)\right\}.
\end{aligned} \tag{3.12}$$

Here $S_F(j,k;G) \equiv S_F(x_j, x_k; G)$ is the unrenormalized quark propagator in the given gluon potential $G_\lambda(x)$. We have

$$(i\gamma^{\mu}D_{\mu} - m_q)S_F(x,y;G) = -\delta(x-y). \tag{3.13}$$

Functional integrals as in (3.12) will occur frequently further on. Let $F(G)$ be some functional of the gluon potentials. We will denote the functional integral over $F(G)$ by brackets $\langle F(G)\rangle_G$:

$$\begin{aligned}
\langle F(G)\rangle_G &:= \frac{1}{Z}\int \mathcal{D}(G)\exp\left\{-i\int \mathrm{d}x \frac{1}{2}\mathrm{Tr}(G_{\lambda\rho}G^{\lambda\rho})\right\}\cdot \\
&\det[-i(i\gamma^{\lambda}D_{\lambda} - m_q + i\epsilon)]F(G).
\end{aligned} \tag{3.14}$$

Now we insert (3.12) in (3.7) and get

$$\langle 3,4|S|1,2\rangle = \langle 3,4|1,2\rangle - Z_{\psi}^{-2}\langle \mathcal{M}^F_{31}(G)\mathcal{M}^F_{42}(G) - (3 \leftrightarrow 4)\rangle_G, \tag{3.15}$$

where

$$\begin{aligned}
&\mathcal{M}^F_{kj}(G) = (k|(i\overrightarrow{\partial\!\!\!/} - m'_q)S_F(i\overleftarrow{\partial\!\!\!/} + m'_q)|j), \\
&(k = 3,4;\ j = 1,2).
\end{aligned} \tag{3.16}$$

The term $\mathcal{M}^F_{31} \cdot \mathcal{M}^F_{42}$ on the r.h.s. of (3.15) corresponds to the t-channel exchange diagrams, the second term, where the role of the quarks 3 and 4 is interchanged, to the u-channel exchange diagrams (Fig. 11). The latter term should be unimportant for high energy, small $|t|$ scattering. Thus we neglect it in the following. For the scattering of different quark flavours it is absent anyway. We set, therefore:

$$\langle 3,4|S|1,2\rangle \cong \langle 3,4|1,2\rangle - Z_{\psi}^{-2}\langle \mathcal{M}^F_{31}(G)\mathcal{M}^F_{42}(G)\rangle_G. \tag{3.17}$$

We can interpret $\mathcal{M}^F_{kj}(G)$ as scattering amplitude for quark j going to k in the fixed gluon potential $G_\lambda(x)$. To see this, let us define the wave function

$$|\psi^F_{p_j}) = S_F(i\overleftarrow{\partial\!\!\!/} + m'_q)|j) \tag{3.18}$$

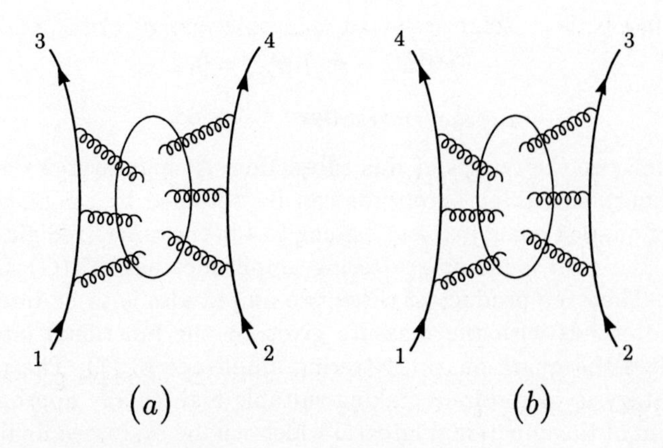

Fig. 11. *The t-channel (a) and u-channel (b) exchange topologies for the diagrams describing quark-quark scattering.*

which satisfies the Dirac equation with the gluon potential $G_\lambda(x)$:

$$(i\gamma^\lambda D_\lambda - m_q)|\psi^F_{p_j}) = 0. \tag{3.19}$$

Furthermore we use the Lippmann-Schwinger equation for S_F:

$$S_F = S_F^{(0)} - S_F^{(0)}(g\not{G} - \delta m_q)S_F \tag{3.20}$$

where $S_F^{(0)}$ is the free quark propagator for mass m'_q and $\delta m_q = m'_q - m_q$ is the quark mass shift. Inserting (3.20) and (3.18) in (3.16) gives after some simple algebra

$$\mathcal{M}^F_{kj}(G) = (p_k|(g\not{G} - \delta m)|\psi^F_{p_j}). \tag{3.21}$$

This represents $\mathcal{M}^F_{kj}$ in the form a scattering amplitude should have: a complete incoming wave is folded with the potential and the free outgoing wave. However, there is a small problem. The wave function $|\psi^F_{p_j})$ defined in (3.18) does not satisfy the boundary condition which we should have for using it in the scattering amplitude, i.e. it does not go to a free incoming wave for time $t \to -\infty$. The wave function with this boundary condition is obtained by replacing the Feynman propagator S_F in (3.18) by the retarded one, S_r. We have shown in [53] that in the high energy limit this replacement can indeed be justified for the calculation of $\mathcal{M}^F_{kj}(G)$ if the gluon potential $G_\lambda(x)$ contains only a limited range of frequencies:

$$\mathcal{M}^F_{kj}(G) \simeq \mathcal{M}^r_{kj}(G) = (p_k|(g\not{G} - \delta m)|\psi^r_{p_j}), \tag{3.22}$$

where

$$|\psi^r_{p_j}) = S_r(i\overleftarrow{\not{\partial}} + m'_q)|j), \tag{3.23}$$

which satisfies

$$(i\gamma^\mu D_\mu - m_q)|\psi^r_{p_j}) = 0, \tag{3.24}$$

$$|\psi^r_{p_j}) \to |j) \quad \text{for} \quad t \to -\infty. \tag{3.25}$$

We summarize the results of this subsection: At high energies and small $|t|$ the quark-quark scattering amplitude can be obtained by calculating first the scattering of quark 1 going to 3 and 2 going to 4 in the same fixed gluon potential $G_\lambda(x)$. Let the corresponding scattering amplitudes be $\mathcal{M}^r_{31}(G)$ and $\mathcal{M}^r_{42}(G)$ (cf. (3.22)). Then the product of these two amplitudes is to be integrated over all gluon potentials with the measure given by the functional integral (3.14) and this gives the quark-quark scattering amplitude (3.17). The point of our further strategy is to continue making suitable high energy approximations in the integrand of this functional integral which will be evaluated finally using the methods of the stochastic vacuum model.

In the following it will be convenient to choose a coordinate system for the description of reaction (3.4) where the quarks 1,3 move with high velocity in essence in positive x^3 direction, the quarks 2,4 in negative x^3 direction. We define the light cone coordinates

$$x_\pm = x^0 \pm x^3 \tag{3.26}$$

and in a similar way the $\pm$ components of any 4-vector. With this we have for the 4-momenta of our quarks:

$$p_j = \begin{pmatrix} \frac{1}{2}p_{j+} & + & \frac{\mathbf{p}^2_{jT}+m'^2_q}{2p_{j+}} \\ & \mathbf{p}_{jT} & \\ \frac{1}{2}p_{j+} & - & \frac{\mathbf{p}^2_{jT}+m^2_{q'}}{2p_{j+}} \end{pmatrix} \tag{3.27}$$

for $j = 1, 3$ with $p_{j+} \to \infty$ and

$$p_k = \begin{pmatrix} \frac{1}{2}p_{k-} & + & \frac{\mathbf{p}^2_{kT}+m'^2_{q'}}{2p_{k-}} \\ & \mathbf{p}_{kT} & \\ -\frac{1}{2}p_{k-} & + & \frac{\mathbf{p}^2_{kT}+m^2_{q'}}{2p_{k-}} \end{pmatrix} \tag{3.28}$$

for $k = 2, 4$ with $p_{k-} \to \infty$.

3.3 The Eikonal Expansion

The problem is now to solve the Dirac equation (3.24) for arbitrary external gluon potential $G_\lambda(x)$. Of course, we cannot do this exactly. But we are only interested in the high energy, small $|t|$ limit. This suggests to use an eikonal type approach. This works indeed, but it is not as straightforward as one would think at first, since the Dirac equation is of first order in the derivatives, whereas the eikonal expansion is easy to make for for a second-order differential equation.

What we did in [53] was to make an ansatz for the Dirac field $\psi^r_{p_j}(x)$ in terms of a "potential" $\phi_j(x)$ as follows:

$$\psi^r_{p_j}(x) = (i\gamma^\lambda D_\lambda + m_q)\phi_j(x). \tag{3.29}$$

A suitable boundary condition for $\phi_j(x)$ which is compatible with (3.25) is

$$\phi_j(x) \to \frac{1}{p_j^0 + m_q} \frac{1+\gamma^0}{2} e^{-ip_j x} u(p_j) \tag{3.30}$$

for $t \to -\infty$. Inserting (3.29) into the Dirac equation (3.24) gives:

$$\{i\gamma^\lambda D_\lambda - m_q\}\{i\gamma^\rho D_\rho + m_q\}\phi_j(x) = 0. \tag{3.31}$$

For the case of no gluon field, $G_\lambda(x) = 0$, the covariant derivatives D_λ degenerate to the ordinary ones, ∂_λ, and (3.31) to the Klein-Gordon equation:

$$(\Box + m_q^2)\phi_j(x) = 0. \tag{3.32}$$

Thus the problem for the Dirac field is in essence reduced to a scalar field-type problem which is handled more easily.

Now it is more or less straightforward to turn the theoretical crank and to obtain the eikonal approximations for $\phi_j(x)$ and $\psi^r_{p_j}(x)$, respectively. Take $j = 1$ as an example. We make the ansatz

$$\phi_1(x) = e^{-ip_1 x}\tilde{\phi}_1(x). \tag{3.33}$$

Inserting p_1 from (3.27) we see that on the r.h.s. of (3.33) the fast varying phase factor is

$$\exp(-i\frac{1}{2}p_{1+}x_-),$$

since $p_{1+} \to \infty$. Assuming all the remaining factors to vary slowly with x we insert (3.33) in (3.31) and order the resulting terms in powers of $1/p_{1+}$. The solution of (3.31) to leading order in $1/p_{1+}$ is then easily obtained. The final formula for $\psi^r_{p_1}(x)$ reads:

$$\psi^r_{p_1}(x) = V_-(x_+, x_-, \mathbf{x}_T). \left\{1 + O\left(\frac{1}{p_{1+}}\right)\right\} e^{-ip_1 x} u(p_1), \tag{3.34}$$

where

$$V_-(x_+, x_-, \mathbf{x}_T) = \mathrm{P}\left\{\exp\left[-\frac{i}{2}g\int_{-\infty}^{x_+} dx'_+ G_-(x'_+, x_-, \mathbf{x}_T)\right]\right\} \tag{3.35}$$

and P means path ordering. When coming in, the quark picks up a non-abelian phase factor, just the ordered integral of G along the path. Of course, V_- is a connector, as studied in Sect. 2.1, but now in Minkowski space and for a straight light-like line running from $-\infty$ to x (Fig. 12).

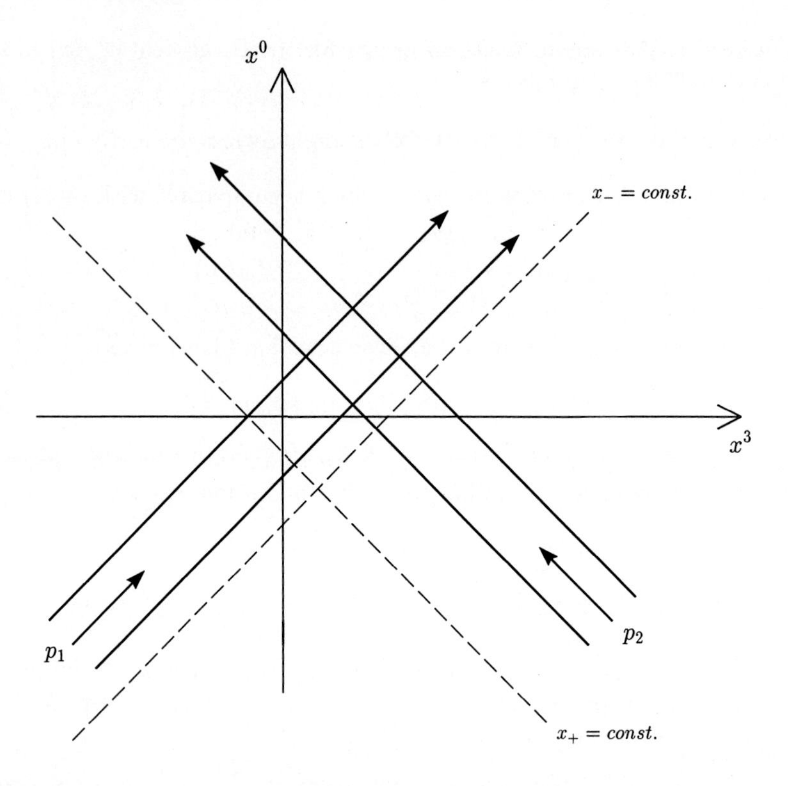

Fig. 12. *Projection of the world lines of the quarks 1(2) moving at high velocity in positive (negative) x^3 direction onto the $x^0 - x^3$ plane in Minkowski space. The non-abelian phase factors V_- in (3.35) (V_+ in (3.37)) are the connectors taken along lines $x_-, \mathbf{x}_T = const.$ ($x_+, \mathbf{x}_T = const.$).*

In a similar way we obtain for the quark with initial momentum p_2, i.e. the one coming in from the right, for $p_{2-} \to \infty$:

$$\psi^r_{p_2}(x) = V_+(x_+, x_-, \mathbf{x}_T) \left\{ 1 + O\left(\frac{1}{p_{2-}}\right) \right\} e^{-ip_2 x} u(p_2), \tag{3.36}$$

where

$$V_+(x_+, x_-, \mathbf{x}_T) = \mathrm{P}\left\{ \exp\left[-\frac{i}{2} g \int_{-\infty}^{x_-} \mathrm{d}x'_- G_+(x_+, x'_-, \mathbf{x}_T) \right] \right\}. \tag{3.37}$$

Recall that

$$G_\pm(x) = \left(G^{0a}(x) \pm G^{3a}(x) \right) \frac{\lambda_a}{2} \tag{3.38}$$

are matrices in colour space. Thus path ordering in (3.35) and (3.37) is essential.

A solution for $\psi^r_{p_1}$ ($\psi^r_{p_2}$) as a series expansion in powers of $1/p_{1+}$ ($1/p_{2-}$) was obtained to all orders in [55].

3.4 The Quark-Quark Scattering Amplitude

We can now insert our high energy approximations (3.34), (3.36) for $\psi^r_{p_{1,2}}(x)$ in the expression for $\mathcal{M}^{(r)}_{kj}$ in (3.22). The resulting integrals are easily done and we get for $p_{1+}, p_{3+} \to \infty$:

$$\begin{aligned}
\mathcal{M}^r_{31}(G) &\to \int \mathrm{d}x\, e^{i(p_3-p_1)x}\,\bar{u}(p_3)(g\not{G}(x) - \delta m)V_-(x)u(p_1) \\
&\to \frac{i}{2}\int \mathrm{d}x_+ \mathrm{d}x_- \mathrm{d}^2x_T\ \exp\left[\frac{i}{2}(p_3-p_1)_+x_- - i(\mathbf{p}_3-\mathbf{p}_1)_T\cdot\mathbf{x}_T\right] \\
&\quad \bar{u}(p_3)\gamma_+\frac{\partial}{\partial x_+}V_-(x_+,x_-,\mathbf{x}_T)u(p_1) \\
&\to i\sqrt{p_{3+}p_{1+}}\cdot\delta_{s_3,s_1}\int \mathrm{d}x_-\mathrm{d}^2x_T \exp[\frac{i}{2}(p_3-p_1)_+x_- - i(\mathbf{p}_3-\mathbf{p}_1)_T\cdot\mathbf{x}_T] \\
&\quad [V_-(\infty,x_-,\mathbf{x}_T) - 1\!\!1]_{A_3,A_1}.
\end{aligned} \tag{3.39}$$

In a similar way we obtain for $p_{2-}, p_{4-} \to \infty$:

$$\begin{aligned}
\mathcal{M}^r_{42}(G) &\to i\sqrt{p_{4-}\cdot p_{2-}}\cdot\delta_{s_4,s_2} \\
&\quad \int \mathrm{d}y_+\mathrm{d}^2y_T \exp[\frac{i}{2}(p_4-p_2)_-y_+ - i(\mathbf{p}_4-\mathbf{p}_2)_T\cdot\mathbf{y}_T] \\
&\quad [V_+(y_+,\infty,\mathbf{y}_T) - 1\!\!1]_{A_4,A_2}.
\end{aligned} \tag{3.40}$$

Here we have written out the spin and colour indices of the in- and outgoing quarks (cf. (3.6), (3.7)). We have used furthermore:

$$\begin{aligned}
\bar{u}_{s_3}(p_3)\gamma^\mu u_{s_1}(p_1) &\to \sqrt{p_{3+}p_{1+}}\cdot\delta_{s_3,s_1}n^\mu_+ \quad \text{for} \quad p_{1,3+}\to\infty, \\
\bar{u}_{s_4}(p_4)\gamma^\mu u_{s_2}(p_2) &\to \sqrt{p_{4-}p_{2-}}\cdot\delta_{s_4,s_2}n^\mu_- \quad \text{for} \quad p_{2,4-}\to\infty,
\end{aligned} \tag{3.41}$$

where

$$n^\mu_\pm = \begin{pmatrix} 1 \\ 0 \\ 0 \\ \pm 1 \end{pmatrix}. \tag{3.42}$$

Finally, the $x_+(x_-)$ integration for $\mathcal{M}^r_{31}(\mathcal{M}^r_{42})$ could be done with the help of (2.15) or rather the analogous equation for connectors in Minkowski space time.

Now we can insert everything in our expression (3.17) for the S-matrix element. This gives:

$$\begin{aligned}
&\langle 3,4|S|1,2\rangle = \langle 3,4|1,2\rangle + \\
&\sqrt{p_{3+}p_{1+}p_{4-}p_{2-}}\cdot\delta_{s_3,s_1}\delta_{s_4,s_2}Z_\psi^{-2}\int \mathrm{d}x_-\mathrm{d}^2x_T\int \mathrm{d}y_+\mathrm{d}^2y_T \\
&\exp\left[\frac{i}{2}(p_3-p_1)_+x_- - i(\mathbf{p}_3-\mathbf{p}_1)_T\cdot\mathbf{x}_T\right] \\
&\exp\left[\frac{i}{2}(p_4-p_2)_-y_+ - i(\mathbf{p}_4-\mathbf{p}_2)_T\cdot\mathbf{y}_T\right] \\
&\langle [V_-(\infty,x_-,\mathbf{x}_T) - 1\!\!1]_{A_3,A_1}[V_+(y_+,\infty,\mathbf{y}_T) - 1\!\!1]_{A_4,A_2}\rangle_G.
\end{aligned} \tag{3.43}$$

From translational invariance of the functional integral we have:

$$\langle [V_-(\infty, x_-, x_T) - \mathbb{1}]_{A_3,A_1} [V_+(y_+, \infty, \mathbf{y}_T) - \mathbb{1}]_{A_4,A_2} \rangle_G$$
$$= \langle [V_-(\infty, 0, \mathbf{x}_T - \mathbf{y}_T) - \mathbb{1}]_{A_3,A_1} [V_+(0, \infty, 0) - \mathbb{1}]_{A_4,A_2} \rangle_G. \quad (3.44)$$

Inserting (3.44) in (3.43), we can pull out the δ-function for the overall energy-momentum conservation and we get finally:

$$\langle 3,4|S|1,2\rangle = \langle 3,4|1,2\rangle + i(2\pi)^4 \delta(p_3 + p_4 - p_1 - p_2)\langle 3,4|T|1,2\rangle,$$

$$\langle 3,4|T|1,2\rangle = i2\sqrt{p_{3+}p_{1+}p_{4-}p_{2-}} \cdot \delta_{s_3,s_1}\delta_{s_4,s_2}(-Z_\psi^{-2}) \int d^2 z_T e^{i\mathbf{q}_T \cdot \mathbf{z}_T}$$
$$\langle [V_-(\infty, 0, \mathbf{z}_T) - \mathbb{1}]_{A_3,A_1} [V_+(0, \infty, 0) - \mathbb{1}]_{A_4,A_2} \rangle_G. \quad (3.45)$$

Here q is the momentum transfer:

$$q = p_1 - p_3 = p_4 - p_2,$$
$$q^2 = t. \quad (3.46)$$

In the high energy limit q is purely transverse

$$q \to \begin{pmatrix} 0 \\ \mathbf{q}_T \\ 0 \end{pmatrix}, \quad q^2 \to -\mathbf{q}_T^2. \quad (3.47)$$

Using different techniques the type of formula (3.45) was also obtained in [56].

In (3.45) we still have to calculate the wave function renormalization constant Z_ψ. This can be done by considering a suitable matrix element of the baryon number current

$$\frac{1}{3}\bar{q}(x)\gamma^\mu q(x)$$

which is conserved and, therefore, needs no renormalization. The result is (cf. [53]):

$$Z_\psi = \frac{1}{3}\langle \mathrm{Tr} V_-(\infty, 0, 0)\rangle_G. \quad (3.48)$$

Let us summarize the results obtained so far:

The quark-quark scattering amplitude (3.45) is diagonal in the spin indices. Thus we get helicity conservation in high energy quark-quark scattering. Using (3.41) we can write the spin factor in (3.45) as

$$2\sqrt{p_{3+}p_{1+}p_{4-}p_{2-}}\delta_{s_3,s_1}\delta_{s_4,s_2} \cong \bar{u}_{s_3}(p_3)\gamma^\mu u_{s_1}(p_1)\bar{u}_{s_4}(p_4)\gamma_\mu u_{s_2}(p_2) \quad (3.49)$$

for $p_{1,3+} \to \infty$ and $p_{2,4-} \to \infty$. This $\gamma^\mu \otimes \gamma_\mu$ structure was postulated for high energy quark-quark scattering in the Donnachie-Landshoff model for the "Pomeron" coupling [23]. However, a study of quark-antiquark scattering (cf. [53] and below) reveals that (3.45) does not allow an interpretation in terms of an effective Lorentz vector exchange between the quarks. The amplitude (3.45) has both, charge conjugation C even and odd contributions. The C even part corresponds to the "Pomeron" and this is not a Lorentz-vector exchange, but the coherent sum of spin 2,4,6,... exchanges (cf. [57], [53]). The C-odd part corresponds to the "Odderon" introduced in [58] and this is indeed a Lorentz vector exchange.

The quark-quark scattering amplitude (3.45) is governed by the correlation function of two connectors or string operators $V_{\pm}$ associated with two light-like Wegner-Wilson lines (Fig. 13).

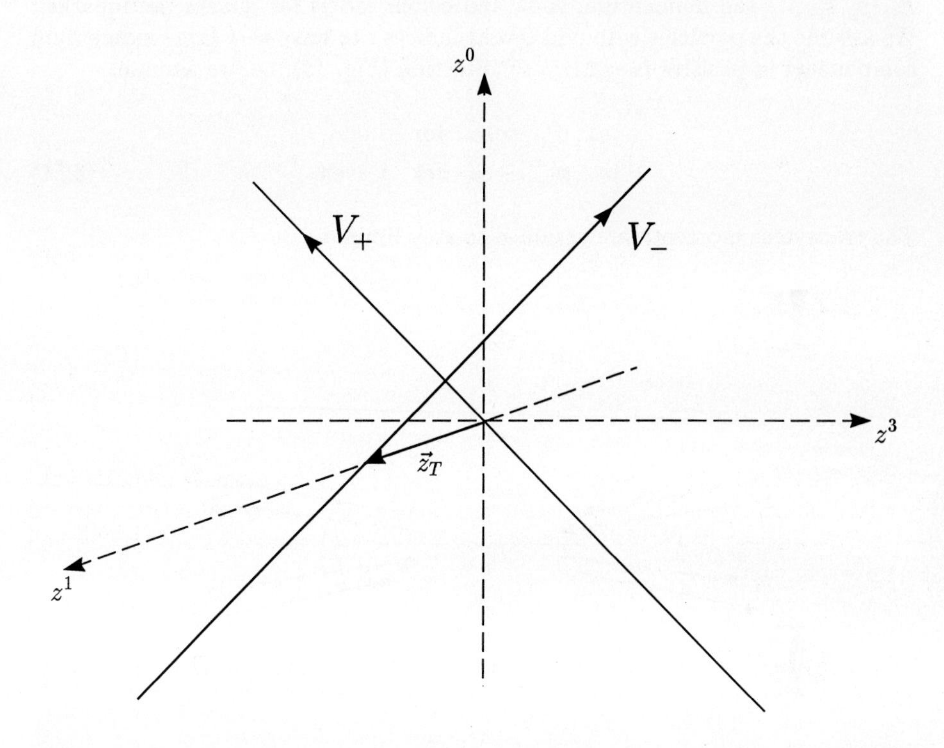

Fig. 13. *Two light-like lines on which the associated string operators $V_{\pm}$ in (3.45) are evaluated. Their correlation function governs quark-quark scattering at high energies.*

The first numerical evaluations of (3.45) using the methods of the SVM were done in [59]. However, it turned out that quark-quark scattering was calculable from (3.45) for abelian gluons only. Indeed, we are embarked on a program to reproduce in this way the results obtained in [25] in the framework of perturbation theory for high energy scattering [60]. For the non-abelian gluons difficulties arose having to do with our neglect of quark confinement. This was really a blessing in disguise and the solution proposed in [59] was to consider directly hadron-hadron scattering, representing the hadrons as $q\bar{q}$ and qqq wave packets for mesons and baryons, respectively. We will see below how this is done.

3.5 The Scattering of Systems of Quarks, Antiquarks, and Gluons

Let us consider now the scattering of systems of partons. As an example we study the scattering of two $q\bar{q}$ pairs on each other:

$$q(1) + \bar{q}(1') + q(2) + \bar{q}(2') \to q(3) + \bar{q}(3') + q(4) + \bar{q}(4'), \tag{3.50}$$

where we set $q(i) \equiv q(p_i, s_i, A_i)$, $\bar{q}(i') \equiv \bar{q}(p'_i, s'_i, A'_i)$ $(i = 1, ..., 4)$ with p_i, s_i, A_i (p'_i, s'_i, A'_i) the momentum, spin, and colour labels for quarks (antiquarks). We assume the particles with odd (even) indices i to have very large momentum components in positive (negative) x^3 direction (Fig. 14), i.e. we assume:

$$\begin{aligned} p_{i+}, p'_{i+} &\to \infty \quad \text{for} \quad i \text{ odd}, \\ p_{i-}, p'_{i-} &\to \infty \quad \text{for} \quad i \text{ even}. \end{aligned} \tag{3.51}$$

The transverse momenta are assumed to stay limited.

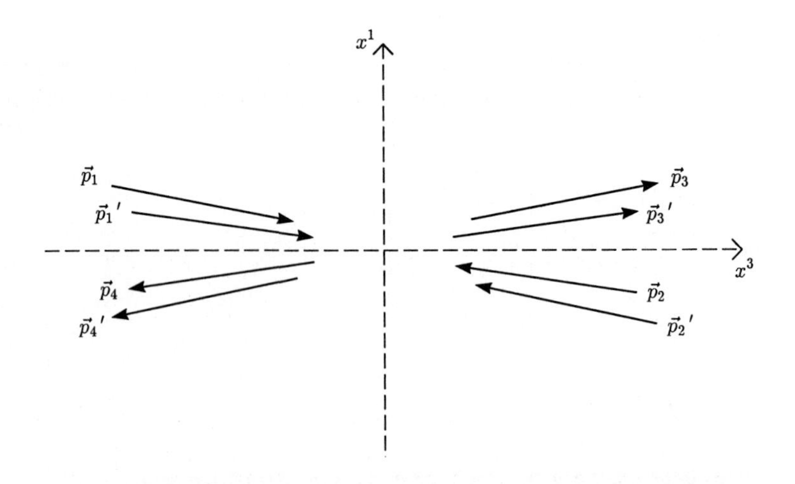

Fig. 14. *Sketch of reaction (3.50) in the overall c.m. system*

Of course, the reduction formula can also be applied for the reaction (3.50). We have to be careful in keeping disconnected pieces. The further strategy is completely analogous to the one employed in Sect. 3.2 for deriving (3.7)-(3.17). As we dropped the u-channel exchange diagrams in Sect. 3.2, we drop now all terms which are estimated to give a vanishing contribution to high energy small momentum transfer scattering. These terms are characterized by large momenta of order of the c.m. energy $\sqrt{s}$ flowing through gluon lines, leading to suppression factors of order $1/s$. Keeping only the t-channel exchange terms and performing all the steps as done in Sects. 3.2 and 3.3 for quark-quark scattering leads finally to a very simple answer for the S-matrix element correponding to reaction (3.50) in the limit (3.51):

$$\begin{aligned}&\langle 3,3',4,4'|S|1,1',2,2'\rangle \rightarrow\\ &\langle[\delta(3,1)-iZ_\psi^{-1}\mathcal{M}^r_{31}(G)][\delta(3',1')-iZ_\psi^{-1}\mathcal{M}'^r_{3'1'}(G)]\\ &[\delta(4,2)-iZ_\psi^{-1}\mathcal{M}^r_{42}(G)][\delta(4',2')-iZ_\psi^{-1}\mathcal{M}'^r_{4'2'}(G)]\rangle_G. \end{aligned}\tag{3.52}$$

Here $\delta(i,j)$ is as defined in (3.6), $\mathcal{M}^r_{31}(G)$ and $\mathcal{M}^r_{42}(G)$ are as in (3.22) and $\mathcal{M}'^r_{3'1'}(G)$ and $\mathcal{M}'^r_{4'2'}(G)$ are the corresponding amplitudes for the scattering of antiquarks on the gluon potential $G_\lambda(x)$. We define

$$(j'| = \bar{v}_{s'_j,A'_j}(p'_j)e^{-ip'_jx},\tag{3.53}$$

where $v_{s'_j,A'_j}(p'_j)$ is the Dirac and colour spinor for the antiquark $\bar{q}(j')$. We have then with S_r the retarded Green's function for quarks in the gluon potential $G_\lambda(x)$:

$$\mathcal{M}'^r_{k'j'}(G) = -(j'|(i\overrightarrow{\not{\partial}}-m'_q)S_r(i\overleftarrow{\not{\partial}}+m'_q)|k'),\tag{3.54}$$

where $(k',j')=(3',1'),(4',2')$.

In the high energy limit (3.51) the scattering amplitudes (3.54) can again be obtained in the eikonal approximation. Indeed, we can just use C-invariance to get:

$$\begin{aligned}&\mathcal{M}'^r_{3'1'}(G)\rightarrow i\sqrt{p'_{3+}p'_{1+}}\cdot\delta_{s'_3,s'_1}\\ &\int dx_-d^2x_T\exp\left[\frac{i}{2}(p'_3-p'_1)_+x_- - i(\mathbf{p}'_3-\mathbf{p}'_1)_T\cdot\mathbf{x}_T\right]\\ &[V^*_-(\infty,x_-,\mathbf{x}_T)-1\!\!1]_{A'_3,A'_1},\end{aligned}\tag{3.55}$$

$$\begin{aligned}&\mathcal{M}'^r_{4'2'}(G)\rightarrow i\sqrt{p'_{4-}p'_{2-}}\cdot\delta_{s'_4,s'_2}\\ &\int dy_+d^2y_T\exp\left[\frac{i}{2}(p'_4-p'_2)_-y_+ - (\mathbf{p}'_4-\mathbf{p}'_2)_T\cdot\mathbf{y}_T\right]\\ &[V^*_+(y_+,\infty,\mathbf{y}_T)-1\!\!1]_{A'_4,A'_2}.\end{aligned}\tag{3.56}$$

In Sect. 3.1 we have argued that over the time interval (3.3) we can neglect parton production and annihilation processes. For the scattering over such a time interval we should have an effective wave function renormalization constant

$$Z_\psi = 1,\tag{3.57}$$

since the deviation of Z_ψ from 1 is just a measure of the strength of quark splitting processes: $q\rightarrow q+G$ etc. In the calculation of Z_ψ in the framework of the SVM to be described below, one finds indeed $Z_\psi=1$, showing the consistency of this approach with the simple physical picture of Sect. 3.1. Anticipating this result we see from (3.6), (3.39), (3.40) and (3.55), (3.56) that in the S-matrix element (3.52) the $\delta(k,j)$ $(\delta(k',j'))$ terms cancel with the $\mathbb{1}$ terms in $\mathcal{M}^r_{kj}(G)$ $(\mathcal{M}'^r_{k'j'}(G))$ in the limit (3.51). This leads us to the following simple rules for obtaining the S-matrix element in the high energy limit: For the right-moving quark $(1 \to 3)$ we have to insert the factor:

$$\mathcal{S}_{q+}(3,1) = \sqrt{p_{3+}p_{1+}} \cdot \delta_{s_3,s_1} \int \mathrm{d}x_- \mathrm{d}^2 x_T$$
$$\exp\left[\frac{i}{2}(p_3 - p_1)_+ x_- - i(\mathbf{p}_3 - \mathbf{p}_1)_T \cdot \mathbf{x}_T\right] V_-(\infty, x_-, \mathbf{x}_T)_{A_3,A_1}. \quad (3.58)$$

For the right-moving antiquark $(1' \to 3')$ we have to insert the factor:

$$\mathcal{S}_{\bar{q}+}(3',1') = \sqrt{p'_{3+}p'_{1+}} \cdot \delta_{s'_3,s'_1} \int \mathrm{d}x_- \mathrm{d}^2 x_T$$
$$\exp\left[\frac{i}{2}(p'_3 - p'_1)_+ x_- - i(\mathbf{p}'_3 - \mathbf{p}'_1)_T \cdot \mathbf{x}_T\right] V^*_-(\infty, x_-, \mathbf{x}_T)_{A'_3,A'_1}. \quad (3.59)$$

For the left-moving quark $(2 \to 4)$ and antiquark $(2' \to 4')$ we have to exchange the $+$ and $-$ labels everywhere in (3.58) and (3.59). This gives:

$$\mathcal{S}_{q-}(4,2) = \sqrt{p_{4-}p_{2-}} \cdot \delta_{s_4,s_2} \int \mathrm{d}y_+ \mathrm{d}^2 y_T$$
$$\exp\left[\frac{i}{2}(p_4 - p_2)_- y_+ - i(\mathbf{p}_4 - \mathbf{p}_2)_T \cdot \mathbf{y}_T\right] V_+(y_+, \infty, \mathbf{y}_T)_{A_4,A_2}, \quad (3.60)$$

$$\mathcal{S}_{\bar{q}-}(4',2') = \sqrt{p'_{4-}p'_{2-}} \cdot \delta_{s'_4,s'_2} \int \mathrm{d}y_+ \mathrm{d}^2 y_T$$
$$\exp\left[\frac{i}{2}(p'_4 - p'_2)_- y_+ - i(\mathbf{p}'_4 - \mathbf{p}'_2)_T \cdot \mathbf{y}_T\right] V^*_+(y_+, \infty, \mathbf{y}_T)_{A'_4,A'_2}. \quad (3.61)$$

Finally we have to multiply together the factors $\mathcal{S}_{q\pm}, \mathcal{S}_{\bar{q}\pm}$ and integrate over all gluon potentials with the functional integral measure (3.14) to get

$$\langle 3,3',4,4'|S|1,1',2,2'\rangle = \langle \mathcal{S}_{q+}(3,1)\mathcal{S}_{\bar{q}+}(3',1')\mathcal{S}_{q-}(4,2)\mathcal{S}_{\bar{q}-}(4',2')\rangle_G. \quad (3.62)$$

Going from quarks to antiquarks corresponds, of course, just to the change from the fundamental representation (3) of $SU(3)_c$ to the complex conjugate representation $(\mathbf{\bar{3}})$ as we see by comparing (3.58) with (3.59) and (3.60) with (3.61).

It is an easy exercise to show that these rules can be generalized in an obvious way for the scattering of arbitrary systems of quarks and antiquarks on each other. Here we always assume that we have one distinguished collision axis and that one group of partons moves with momenta approaching infinity to the right, the other group to the left. The transverse momenta are assumed to stay limited.

In Appendix B we show that these rules can also be extended to gluons participating in the scattering. We simply have to change the colour representation in (3.58), (3.60) from the fundamental to the adjoint one. In detail we find that for a right-moving gluon

$$G(p_1, j_1, a_1) \to G(p_3, j_3, a_3) \tag{3.63}$$

the following factor has to be inserted in the S-matrix element (cf. Appendix B):

$$\begin{aligned}
\mathcal{S}_{G+}(3,1) &= \sqrt{p_{3+}p_{1+}} \cdot \delta_{j_3,j_1} \\
&\int \mathrm{d}x_- \mathrm{d}^2x_T \exp\left[\frac{i}{2}(p_3 - p_1)_+ x_- - i(\mathbf{p}_3 - \mathbf{p}_1)_T \cdot \mathbf{x}_T\right] \\
&\mathcal{V}_-(\infty, x_-, \mathbf{x}_T)_{a_3,a_1}.
\end{aligned} \tag{3.64}$$

Here $j_{1,3}$ are the spin indices which are purely transverse, $1 \leq j_{1,3} \leq 2$. The colour indices are a_1, a_3 with $1 \leq a_{1,3} \leq 8$ and $\mathcal{V}$ is the connector for the adjoint representation of $SU(3)_c$ (cf. (B.23)).

For a left-moving gluon we have again to exchange $+$ and $-$ labels in (3.64).

3.6 The Scattering of Wave Packets of Partons Representing Mesons

In this section we will go from the parton-parton to hadron-hadron scattering. Our strategy will be to represent hadrons by wave packets of partons, where we make simple "Ansätze" for the wave functions. Then the partonic S-matrix element obtained by the rules derived in Sect. 3.5 will be folded with these wave functions to give the hadronic S-matrix elements. Of course, we always work in the limit of high energies and small momentum transfers.

Let us start by considering meson-meson scattering:

$$M_1(P_1) + M_2(P_2) \to M_3(P_3) + M_4(P_4), \tag{3.65}$$

where $M_{1,3}$ are again the right movers, $M_{2,4}$ the left movers. We make simple "Ansätze" for the mesons as $q\bar{q}$ wave packets as follows:

$$\begin{aligned}
|M_j(P_j)\rangle &= \int \mathrm{d}^2 p_T \int_0^1 \mathrm{d}\zeta \frac{1}{(2\pi)^{3/2}} h^j_{s_j,s'_j}(\zeta, \mathbf{p}_T) \\
&\frac{1}{\sqrt{3}} \delta_{A_j,A'_j} |q(p_j, s_j, A_j), \bar{q}(P_j - p_j, s'_j, A'_j)\rangle \quad (j = 1, ..., 4),
\end{aligned} \tag{3.66}$$

where for $j = 1, 3$:

$$\begin{aligned}
P_{j+} &\to \infty \\
p_{j+} &= \zeta P_{j+}, \\
\mathbf{p}_{jT} &= \frac{1}{2}\mathbf{P}_{jT} + \mathbf{p}_T
\end{aligned} \tag{3.67}$$

and for $j = 2, 4$:

$$\begin{aligned} P_{j-} &\to \infty \\ p_{j-} &= \zeta P_{j-}, \\ \mathbf{p}_{jT} &= \frac{1}{2}\mathbf{P}_{jT} + \mathbf{p}_T. \end{aligned} \tag{3.68}$$

Here ζ is the longitudinal momentum fraction of the quark in the meson, $\mathbf{p}_T$ is the relative transverse momentum of q and $\bar{q}$.

We stress that we are not restricting ourselves to spin 0 mesons only. With appropriate functions $h(\zeta, \mathbf{p}_T)$ in (3.66) we can represent states of $q\bar{q}$-mesons of arbitrary spin.

We choose for our states (3.66) the usual continuum normalization:

$$\langle M_j(P')|M_j(P)\rangle = (2\pi)^3 2P^0 \delta^3(\mathbf{P}' - \mathbf{P}). \tag{3.69}$$

With (3.6) this requires:

$$\int \mathrm{d}^2 p_T \int_0^1 \mathrm{d}\zeta \; 2\zeta(1-\zeta) h^{*j}_{s,s'}(\zeta, \mathbf{p}_T) h^j_{s,s'}(\zeta, \mathbf{p}_T) = 1 \tag{3.70}$$

In (3.69) and (3.70) no summation over j is to be taken.

For later use we define the wave functions in transverse position space at fixed longitudinal momentum fraction ζ:

$$\varphi^j_{s,s'}(\zeta, \mathbf{x}_T) := \sqrt{2\zeta(1-\zeta)} \frac{1}{2\pi} \int \mathrm{d}^2 p_T \exp(i\mathbf{p}_T \cdot \mathbf{x}_T) h^j_{s,s'}(\zeta, \mathbf{p}_T)$$
$$(j = 1, ..., 4). \tag{3.71}$$

With this we define profile functions for the transitions $M_j \to M_k$ for right and left movers as:

$$w_{k,j}(\mathbf{x}_T) := \int_0^1 \mathrm{d}\zeta (\varphi^k_{s,s'}(\zeta, \mathbf{x}_T))^* \varphi^j_{s,s'}(\zeta, \mathbf{x}_T), \tag{3.72}$$

where k, j are both odd or even. Clearly we have (cf. (3.70))

$$\begin{aligned} &w_{j,j}(\mathbf{x}_T) \geq 0, \\ &\int \mathrm{d}^2 x_T w_{j,j}(\mathbf{x}_T) = 1, \\ &(\text{no summation over } j). \end{aligned} \tag{3.73}$$

Let us first study the transition of the right movers alone, i.e. the "reaction":

$$M_1(P_1) \to M_3(P_3). \tag{3.74}$$

For stable mesons $M_{1,3}$ we should find that the corresponding S-matrix elements are identical to the matrix elements of the unit operator. Is this borne out in our approach?

From the rules given in Sect. 3.5 we find easily the S-matrix element for the transition

$$q(1)+\bar{q}(1')\rightarrow q(3)+\bar{q}(3') \tag{3.75}$$

in the form

$$\langle 3,3'|S|1,1'\rangle=\langle\mathcal{S}_{q+}(3,1)\mathcal{S}_{\bar{q}+}(3',1')\rangle_G. \tag{3.76}$$

After folding (3.76) with the mesonic wave functions (3.66) we get:

$$\begin{aligned}&\langle M_3(P_3)|S|M_1(P_1)\rangle=(2\pi)^3 2P_1^0\delta^3(\mathbf{P}_3-\mathbf{P}_1)\\&\int d^2z_T\int_0^1 d\zeta_3\int_0^1 d\zeta_1\ \varphi^{3^*}_{s,s'}(\zeta_3,\mathbf{z}_T)\varphi^1_{s,s'}(\zeta_1,\mathbf{z}_T)\\&\frac{1}{2}P_{1+}\int\frac{dz_-}{2\pi}\exp\left[\frac{i}{2}P_{1+}(\zeta_3-\zeta_1)z_-\right]\\&\langle\frac{1}{3}\mathrm{Tr}[V_-(\infty,z_-,\mathbf{z}_T)V_-^\dagger(\infty,0,0)]\rangle_G.\end{aligned} \tag{3.77}$$

Now we remember that we consider the limit $P_{1+}\rightarrow\infty$. Therefore we perform a change of variables in the z_- integral by setting:

$$z'_-:=\frac{1}{2}P_{1+}z_-. \tag{3.78}$$

This gives:

$$\begin{aligned}&\langle M_3(P_3)|S|M_1(P_1)\rangle=(2\pi)^3 2P_1^0\delta^3(\mathbf{P}_3-\mathbf{P}_1)\\&\int d^2z_T\int_0^1 d\zeta_3\int_0^1 d\zeta_1\ \varphi^{3^*}_{s,s'}(\zeta_3,\mathbf{z}_T)\varphi^1_{s,s'}(\zeta_1,\mathbf{z}_T)\int\frac{dz'_-}{2\pi}\exp[i(\zeta_3-\zeta_1)z'_-]\\&\langle\frac{1}{3}\mathrm{Tr}[V_-(\infty,\frac{2}{P_{1+}}z'_-,\mathbf{z}_T)V_-^\dagger(\infty,0,0)]\rangle_G\\&\longrightarrow(2\pi)^3 2P_1^0\delta^3(\mathbf{P}_3-\mathbf{P}_1)\int d^2z_T\ w_{3,1}(\mathbf{z}_T)\\&\langle\frac{1}{3}\mathrm{Tr}[V_-(\infty,0,\mathbf{z}_T)V_-^\dagger(\infty,0,0)]\rangle_G\quad\text{for}\quad P_{1+}\rightarrow\infty.\end{aligned} \tag{3.79}$$

In (3.79) $V_-(\infty,0,\mathbf{z}_T)(V_-^\dagger(\infty,0,0))$ is the quark (antiquark) connector taken along the line $C_q(C_{\bar{q}})$, where:

$$C_q:\tau\rightarrow z_q(\tau)=\begin{pmatrix}\tau\\\mathbf{z}_T\\\tau\end{pmatrix},\quad(-\infty<\tau<\infty), \tag{3.80}$$

$$C_{\bar{q}}:\tau\rightarrow z_{\bar{q}}(\tau)=\begin{pmatrix}\tau\\0\\\tau\end{pmatrix},\quad(-\infty<\tau<\infty). \tag{3.81}$$

But from Sect. 2.1 we know that the connector $V_-^\dagger$ along $C_{\bar{q}}$ is equal to the connector V_- taken along the oppositely oriented line $\bar{C}_{\bar{q}}$ (cf. (2.25)). Now we will allow ourselves to join the lines C_q and $\bar{C}_{\bar{q}}$ at large positive and negative times. We can imagine the gluon potentials to be turned off adiabatically there. We obtain then from the product of the connectors in (3.79) a connector taken along a closed lightlike Wegner-Wilson loop

$$\frac{1}{3}\mathrm{Tr}[V_-(\infty,0,\mathbf{z}_T)V_-^\dagger(\infty,0,0)] \longrightarrow \mathcal{W}_+(\frac{1}{2}\mathbf{z}_T,\mathbf{z}_T). \tag{3.82}$$

Here we define

$$\mathcal{W}_\pm(\mathbf{y}_T,\mathbf{z}_T) = \frac{1}{3}\mathrm{TrP}\exp[-ig\int_{C_\pm} \mathrm{d}x_\mu G^\mu(x)] \tag{3.83}$$

with $C_+(C_-)$ a lightlike Wegner-Wilson loop in the plane $x_- = 0(x_+ = 0)$, where in the transverse space the centre of the loop is at $\mathbf{y}_T$ and the vector from the antiquark to the quark line is $\mathbf{z}_T$ (Fig. 15).

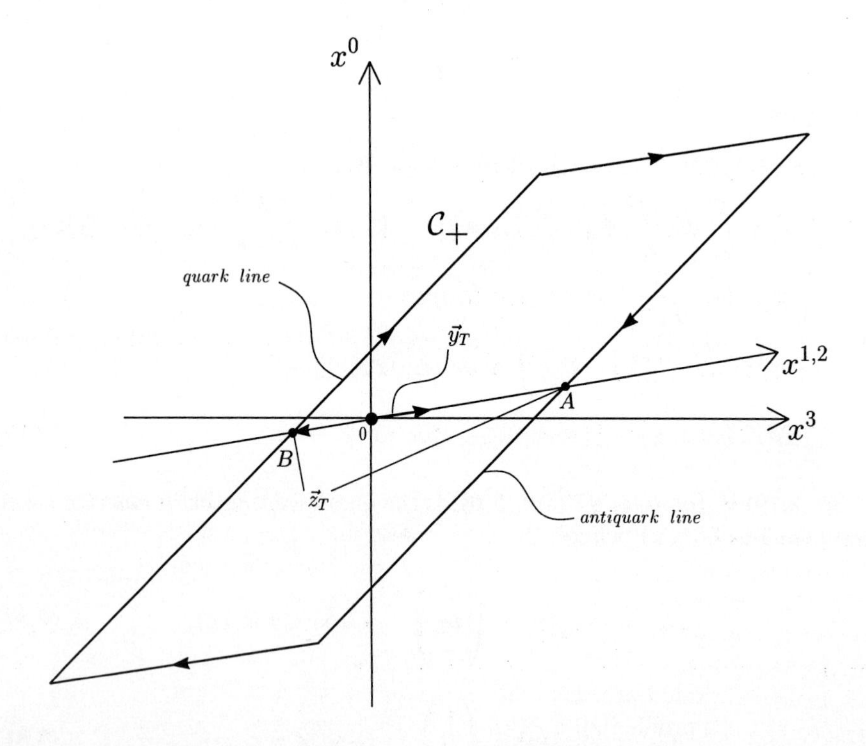

Fig. 15. *The light-like Wegner-Wilson loop in Minkowski space-time, C_+ consisting of two light-like lines in the hyperplane $x_- = 0$ and connecting pieces at infinity. In transverse space the centre of the loop is at $\mathbf{y}_T$, the vector from the antiquark to the quark line is $\mathbf{z}_T$ (from A to B).*

Inserting now everything in (3.79) we get the simple answer:

$$\langle M_3(P_3)|S|M_1(P_1)\rangle = (2\pi)^3 2P_1^0 \delta^3(\mathbf{P}_3 - \mathbf{P}_1)$$
$$\int d^2 z_T \; w_{3,1}(z_T)\langle \mathcal{W}_+(\frac{1}{2}\mathbf{z}_T, \mathbf{z}_T)\rangle_G. \tag{3.84}$$

In the next section we will evaluate the functional integral in (3.84) in the SVM. We will find

$$\langle \mathcal{W}_+(\mathbf{y}_T, \mathbf{z}_T)\rangle_G|_{SVM} = 1. \tag{3.85}$$

Inserting this in (3.84) leads to the expected result (cf. (3.66)-(3.72)):

$$\langle M_3(P_3)|S|M_1(P_1)\rangle = \langle M_3(P_3)|M_1(P_1)\rangle. \tag{3.86}$$

In our approach the $q\bar{q}$ pair in the right-moving meson $M_1 \to M_3$ does not interact. Of course this is only valid over our finite time interval (3.3)!

The techniques developed thus far are now easily employed for the reaction (3.65). After performing similar steps as above we arrive at the following S-matrix element:

$$\begin{aligned} S_{fi} &= \delta_{fi} + i(2\pi)^4 \delta(P_3 + P_4 - P_1 - P_2) T_{fi}, \\ T_{fi} &\equiv \langle M_3(P_3), M_4(P_4)|T|M_1(P_1), M_2(P_2)\rangle \\ &= -2is \int d^2 b_T d^2 x_T d^2 y_T e^{i\mathbf{q}_T\cdot\mathbf{b}_T} w_{3,1}(\mathbf{x}_T) w_{4,2}(\mathbf{y}_T) \\ &\quad \langle \mathcal{W}_+(\frac{1}{2}\mathbf{b}_T, \mathbf{x}_T)\mathcal{W}_-(-\frac{1}{2}\mathbf{b}_T, \mathbf{y}_T) - 1\rangle_G. \end{aligned} \tag{3.87}$$

Here $s = (P_1 + P_2)^2$ is the c.m. energy squared and $\mathbf{q}_T$ is the momentum transfer, which is purely transverse in the high energy limit:

$$\mathbf{q}_T = (\mathbf{P}_1 - \mathbf{P}_3)_T. \tag{3.88}$$

From (3.87) we see that the amplitude for soft meson-meson scattering at high energies is governed by the correlation function of two lightlike Wegner-Wilson loops, where one is in the hyperplane $x_- = 0$, the other in $x_+ = 0$. The transverse separation between the centres of the two loops is given by $\mathbf{b}_T$, the impact parameter. The vectors $\mathbf{x}_T$ and $\mathbf{y}_T$ give the extensions and orientations of the loops in transverse space (Fig. 16). The loop-loop correlation function has to be integrated over all orientations of the loops in transverse space with the (transition) profile functions of the mesons, $w_{3,1}$ and $w_{4,2}$. Finally a Fourier transform in the impact parameter has to be done.

The methods presented here for meson-meson scattering can of course also be employed for scattering reactions involving baryons and antibaryons. This is sketched in Appendix C.

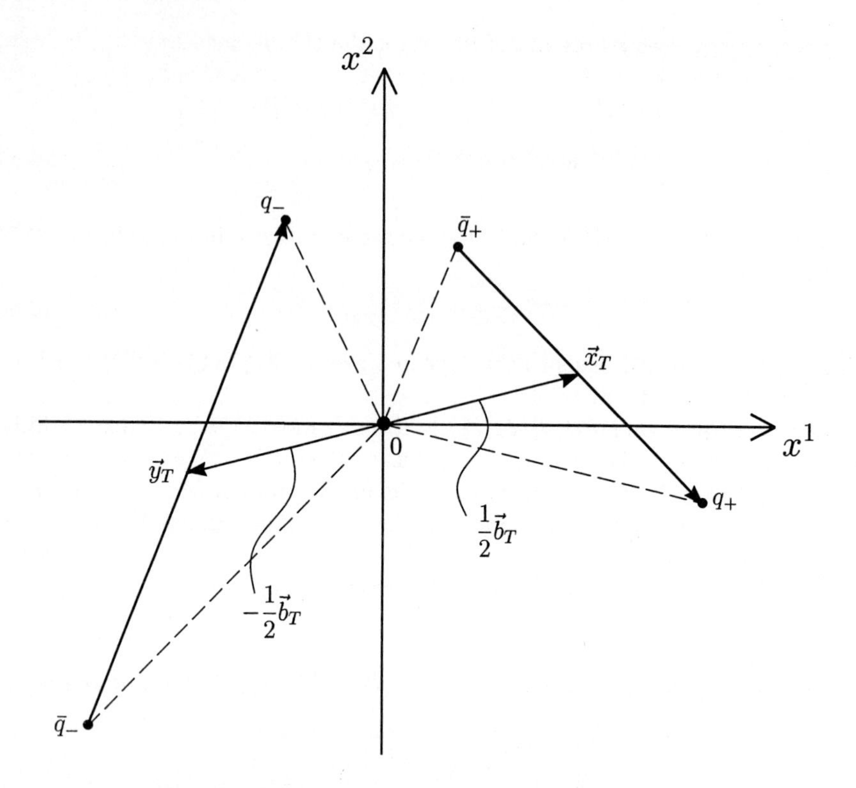

Fig. 16. *The projections of the two lightlike Wegner-Wilson loops C_+, C_- occurring in the definition of $W_+(\frac{1}{2}\mathbf{b}_T, \mathbf{x}_T), W_-(-\frac{1}{2}\mathbf{b}_T, \mathbf{y}_T)$ (cf. (3.87), (3.83)) into transverse space. The points marked $q_+, \bar{q}_+$ ($q_-, \bar{q}_-$) correspond to the projections of the quark and antiquark lines of C_+ (C_-).*

3.7 The Evaluation of Scattering Amplitudes in the Minkowskian Version of the Stochastic Vacuum Model

In the previous section we have derived expressions for the amplitudes of soft meson-meson scattering at high energies in terms of correlation functions of light-like Wegner-Wilson loops. The task is now to evaluate the corresponding functional integral $\langle\rangle_G$ in (3.87). Surely we do not want to make a perturbative expansion there, remembering our argument of Sect. 1 (cf. (1.14)). Instead, we will turn to the SVM which did quite well in its applications in Euclidean QCD (Sect. 2). Of course, the generalization of the SVM to Minkowski space-time is a bold step which was done in [59], [45]. The authors of these Refs. proposed to use in Minkowski space-time just the Assumptions 1-3 of the SVM (cf. Sect. 2.4), but after having made a suitable analytic continuation. In this way we obtain for instance the Minkowski version of Assumption 1 as (cf. (2.60), (2.61)):

Assumption 1: The correlator of two field strengths, shifted to a common reference point y, is independent of the connecting path and given by:

$$\langle\frac{g^2}{4\pi^2}\hat{G}^a_{\mu\nu}(y,x;C_x)\hat{G}^b_{\rho\sigma}(y,x';C_{x'})\rangle_G = \frac{1}{4}\delta^{ab}F_{\mu\nu\rho\sigma}(z),$$

$$F_{\mu\nu\rho\sigma}(z) = \frac{1}{24}G_2\{(g_{\mu\rho}g_{\nu\sigma} - g_{\mu\sigma}g_{\nu\rho})\kappa D(z^2) \tag{3.89}$$
$$+\frac{1}{2}\Big[\frac{\partial}{\partial z^\nu}(z_\sigma g_{\mu\rho} - z_\rho g_{\mu\sigma}) + \frac{\partial}{\partial z^\mu}(z_\rho g_{\nu\sigma} - z_\sigma g_{\nu\rho})\Big]\cdot(1-\kappa)D_1(z^2)\}.$$

Here $z = x - x'$ and $D(z^2)$ and $D_1(z^2)$ are defined as in (2.65), (2.67) for $z^2 \leq 0$ and by analytic continuation for $z^2 > 0$.

As a first application let us calculate the expectation value of one lightlike Wegner-Wilson loop. We have from (3.83) using the non-abelian Stokes theorem:

$$\langle\mathcal{W}_+(\mathbf{y}_T,\mathbf{z}_T)\rangle_G = \langle\frac{1}{3}\mathrm{TrP}\exp\big[-ig\int_{C_+}\mathrm{d}x_\mu G^\mu(x)\big]\rangle_G$$
$$= \langle\frac{1}{3}\mathrm{Tr\ P}\exp\big[-i\frac{g}{2}\int_{S_+}\mathrm{d}u\mathrm{d}v\frac{\partial(x^\mu,x^\nu)}{\partial(u,v)}\hat{G}_{\mu\nu}(R,x;C_x)\big]\rangle_G. \tag{3.90}$$

Here S_+ is the (planar) surface spanned into C_+ (Fig. 15) and parametrized by

$$x^\mu(u,v) = u\ n^\mu_+ + y^\mu - (v - \frac{1}{2})z^\mu,$$
$$-\infty < u < \infty,\ 0 \leq v \leq 1, \tag{3.91}$$

where n_+ is as in (3.42) and

$$y^\mu = \begin{pmatrix} 0 \\ \mathbf{y}_T \\ 0 \end{pmatrix}, \qquad z^\mu = \begin{pmatrix} 0 \\ \mathbf{z}_T \\ 0 \end{pmatrix}. \tag{3.92}$$

The reference point on the surface S_+ is denoted by R and C_x are straight lines running from x to R. From (3.91) we find

$$\frac{\partial(x^\mu,x^\nu)}{\partial(u,v)} = z^\mu n^\nu_+ - n^\mu_+ z^\nu. \tag{3.93}$$

Now we apply the cumulant expansion formulae (cf. Sect. 2.3) to (3.90) and use Assumptions 1-3 of the SVM. This is completely analogous to the calculations done in Sect. 2.5. We get:

$$\langle\mathcal{W}_+(\mathbf{y}_T,\mathbf{z}_T)\rangle_G = \exp\{-\frac{\pi^2}{6}\int_{S_+}\mathrm{d}u\mathrm{d}v\int_{S_+}\mathrm{d}u'\mathrm{d}v'K_2(x-x')$$
$$+\ \text{higher cumulant terms}\}, \tag{3.94}$$

where

$$K_2(x-x') = \frac{\partial(x^\mu, x^\nu)}{\partial(u,v)} \frac{\partial(x'^\rho, x'^\sigma)}{\partial(u',v')} F_{\mu\nu\rho\sigma}(x-x'),$$
$$x \equiv x(u,v), \quad x' \equiv x'(u',v'). \tag{3.95}$$

It is an easy exercise to evaluate $K_2(x-x')$ using (3.93) and $F_{\mu\nu\rho\sigma}$ from (3.89). The result is

$$K_2(x-x') = 0 \tag{3.96}$$

for all $x, x' \in S_+$. With Assumption 3 of the SVM (cf. (2.73)-(2.75)) all higher cumulants are related to the second one. Thus, (3.96) implies also the vanishing of all higher cumulant terms in (3.94) and we get

$$\langle \mathcal{W}_+(\mathbf{y}_T, \mathbf{z}_T) \rangle_G = 1. \tag{3.97}$$

A similar argument leads, of course, also to

$$\langle \mathcal{W}_-(\mathbf{y}_T, \mathbf{z}_T) \rangle_G = 1. \tag{3.98}$$

In Sect. 3.6 we have already used the result (3.97) in the discussion of the transition $M_1 \to M_3$ to obtain (3.86). We can also use it for the calculation of the wave function renormalization constant Z_ψ. The expression we obtained for Z_ψ in (3.48) can be interpreted as the expectation value of the non-abelian phase factor picked up by a very fast right-moving quark. Now, isolated quarks do not exist. The best approximation for it we can think of is a fast, right-moving quark-antiquark pair with the antiquark being very far away from the quark in transverse direction. In this way we obtain for Z_ψ instead of (3.48):

$$\begin{aligned} Z_\psi &= \lim_{|\mathbf{z}_T| \to \infty} \langle \frac{1}{3} \mathrm{Tr}[V_-(\infty,0,0) V_-^\dagger(\infty,0,\mathbf{z}_T)] \rangle_G \\ &= \lim_{|\mathbf{z}_T| \to \infty} \langle \mathcal{W}_+(-\frac{1}{2}\mathbf{z}_T, -\mathbf{z}_T) \rangle_G \\ &= 1. \end{aligned} \tag{3.99}$$

This result was already used in Sect. 3.5, (3.57)ff.

We come now to the evaluation of the loop-loop correlation function of (3.87):

$$\begin{aligned} &\langle \mathcal{W}_+(\frac{1}{2}\mathbf{b}_T, \mathbf{x}_T) \mathcal{W}_-(-\frac{1}{2}\mathbf{b}_T, \mathbf{y}_T) - 1 \rangle_G \\ &= \langle [\mathcal{W}_+(\frac{1}{2}\mathbf{b}_T, \mathbf{x}_T) - 1][\mathcal{W}_-(-\frac{1}{2}\mathbf{b}_T, \mathbf{y}_T) - 1] \rangle_G. \end{aligned} \tag{3.100}$$

Here we used (3.97), (3.98). The strategy is as before. We want to transform the line integrals of $\mathcal{W}_\pm$ (cf. (3.83)) into surface integrals using the non-abelian Stokes theorem of Sect. 2.2. Following the authors of [59], [45] we choose as surface with boundary C_+ and C_- a double pyramid with apex at the mid-point

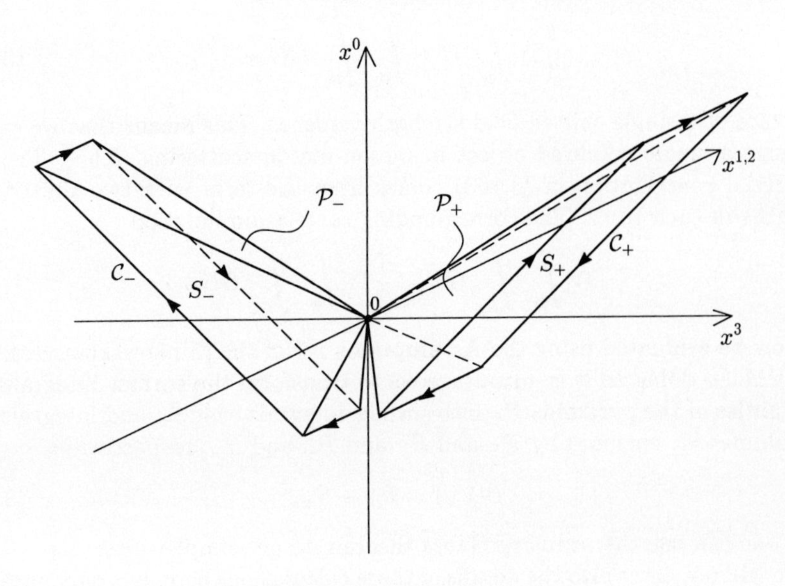

Fig. 17. *The curves C_+ and C_- along which the path integrals $\mathcal{W}_\pm$ in (3.100) are taken. The mantle of the pyramid with apex at the origin of the coordinate system and boundary C_+ (C_-) is $\mathcal{P}_+$ ($\mathcal{P}_-$), the basis surface S_+ (S_-).*

of C_+ and C_- which is the origin of our coordinate system (Fig. 17). The mantle of this pyramid is $\mathcal{P}_+ + \mathcal{P}_-$ and we have with suitable orientation

$$\partial(\mathcal{P}_+ + \mathcal{P}_-) = C_+ + C_-. \tag{3.101}$$

In the transverse projection of Fig. 16 the basis surface $S_+(S_-)$ appears as the line $\bar{q}_+ q_+$ ($\bar{q}_- q_-$) and the mantle surface $\mathcal{P}_+(\mathcal{P}_-)$ as the triangle $0\bar{q}_+q_+(0\bar{q}_-q_-)$. From the non-abelian Stokes theorem we obtain now for (3.100):

$$\begin{aligned} &\langle [\mathcal{W}_+(\tfrac{1}{2}\mathbf{b}_T, \mathbf{x}_T) - 1][\mathcal{W}_-(-\tfrac{1}{2}\mathbf{b}_T, \mathbf{y}_T) - 1]\rangle_G \\ &= \langle \{\tfrac{1}{3}\mathrm{TrPexp}[-i\tfrac{g}{2}\int_{\mathcal{P}_+} \mathrm{d}u\mathrm{d}v \frac{\partial(x^\mu, x^\nu)}{\partial(u,v)}\hat{G}_{\mu\nu}(0,x;C_x)] - 1\} \\ &\{\tfrac{1}{3}\mathrm{Tr\ P\exp}[-i\tfrac{g}{2}\int_{\mathcal{P}_-} \mathrm{d}u'\mathrm{d}v' \frac{\partial(x'^\rho, x'^\sigma)}{\partial(u',v')}\hat{G}_{\rho\sigma}(0,x';C_{x'})] - 1\}\rangle_G. \end{aligned} \tag{3.102}$$

So far, it has not been possible to use some version of the cumulant expansion for (3.102). Thus in [59], [45] the path-ordered exponentials on the r.h.s. of (3.102) were expanded directly. The structure of this expansion is as follows:

$$\begin{aligned} &\langle [\mathcal{W}_+ - 1][\mathcal{W}_- - 1]\rangle_G \\ &\sim \langle \tfrac{1}{3}\mathrm{Tr}[\int_{\mathcal{P}_+} \hat{G} + \int_{\mathcal{P}_+}\int_{\mathcal{P}_+} \hat{G}\hat{G} + ...] \end{aligned}$$

$$\frac{1}{3}\mathrm{Tr}[\int_{\mathcal{P}_-}\hat{G} + \int_{\mathcal{P}_-}\int_{\mathcal{P}_-}\hat{G}\hat{G} + ...]\rangle_G. \tag{3.103}$$

The trace of a single shifted field strength vanishes. This means that we cannot exchange a single coloured object in meson-meson scattering (3.65). The first non-trivial contribution in (3.103) comes from the term with two shifted field strengths in each trace. The corresponding correlation function

$$\langle\frac{1}{3}\mathrm{Tr}[\int_{\mathcal{P}_+}\int_{\mathcal{P}_+}\hat{G}\hat{G}]\ \frac{1}{3}\mathrm{Tr}[\int_{\mathcal{P}_-}\int_{\mathcal{P}_-}\hat{G}\hat{G}]\rangle_G$$

can now be evaluated using the Assumptions 1-3 of the Minkowskian version of the SVM. In doing so it is advantageous to transform the surface integrals over the mantles of the pyramids $\mathcal{P}_\pm$ into surface integrals over $S_\pm$ and integrals over the volumes $V_\pm$ enclosed by $\mathcal{P}_+$ and S_+ and $\mathcal{P}_-$ and S_-, respectively:

$$\partial(V_\pm) = \mathcal{P}_\pm - S_\pm. \tag{3.104}$$

Then one can use the ordinary Gauss theorem to get simpler integrals.

We will not enter into the details of these calculations here, but only note that the integrations along the directions x_+ and x_- can easily be done analytically and that one ends up with integrals over the projections of $S_\pm$ and $V_\pm$ into the transverse space. Thus, one finally needs the correlator (3.89) for space-like separations only:

$$\begin{aligned}&(x-x')^2 = z^2 \le 0,\\ &x = \begin{pmatrix}0\\ \mathbf{x}_T\\ 0\end{pmatrix},\quad x' = \begin{pmatrix}0\\ \mathbf{x}'_T\\ 0\end{pmatrix},\end{aligned} \tag{3.105}$$

where $\mathbf{x}_T$ runs over the triangle $\Delta_+ = 0\bar{q}_+q_+$ and $\mathbf{x'}_T$ over $\Delta_- = 0\bar{q}_-q_-$ in Fig. 16. For space-like separations the correlator functions $D(z^2)$ and $D_1(z^2)$ in (3.89) are as in Euclidean space time. The resulting expressions are then of the following structure

$$\begin{aligned}&\langle[\mathcal{W}_+ - 1][\mathcal{W}_- - 1]\rangle_G \sim \\ &\{\int_{\Delta_+} \mathrm{d}^2 z_T \int_{\Delta_-} \mathrm{d}^2 z'_T\, G_2[..D(-(\mathbf{z}_T - \mathbf{z}'_T)^2) + ...D_1(-(\mathbf{z}_T - \mathbf{z}'_T)^2)]\}^2.\end{aligned} \tag{3.106}$$

These integrals have to be evaluated numerically.

With the methods outlined above, we have obtained an (approximate) expression (3.106) for the functional integral $\langle\rangle_G$ governing the meson-meson scattering amplitude (3.87). Note that the nonperturbative gluon condensate parameter G_2 sets the scale in (3.106) and in the integrals to be performed there the vacuum correlation length a enters through the D and D_1 functions. Thus, on dimensional grounds, we must have in our approximation:

$$\langle(\mathcal{W}_+ - 1)(\mathcal{W}_- - 1)\rangle_G = G_2^2 a^8 f\left(\frac{\mathbf{b}_T}{a}, \frac{\mathbf{x}_T}{a}, \frac{\mathbf{y}_T}{a}\right), \tag{3.107}$$

where f is a dimensionless function. To obtain the meson-meson scattering amplitude (3.87) we still have to integrate over the profile functions $w_{3,1}(\mathbf{x}_T)$ and $w_{4,2}(\mathbf{y}_T)$. Here the transverse extensions of the mesons - i.e. of the wave packets representing them - enter in the results.

For a detailed exposition of the numerical results obtained in the way sketched above we refer to [45]. Here we only discuss the outcome for proton-proton scattering when treated in a similar way (cf. Appendix C). A fit to the numerical results gives for the total cross section and the slope parameter at $t = 0$ of elastic proton-proton scattering the following representation:

$$\sigma_{\rm tot}(pp) = 0.00881 \left(\frac{R_p}{a}\right)^{3.277} \cdot (3\pi^2 G_2)^2 \cdot a^{10}, \tag{3.108}$$

$$b_{pp} := \frac{\mathrm{d}}{\mathrm{d}t} \ln \frac{\mathrm{d}\sigma_{el}}{\mathrm{d}t}(pp)|_{t=0} = 1.558a^2 + 0.454R_p^2. \tag{3.109}$$

Here R_p is the proton radius and the formulae (3.108), (3.109) are valid for

$$1 \le R_p/a \le 3. \tag{3.110}$$

To compare (3.108), (3.109) with experimental results, we can, for instance, consider the c.m. energy $\sqrt{s} = 20$ GeV and take as input the following measured values (cf. [45]):

$$\begin{aligned} \sigma_{\rm tot}(pp)|_{\rm Pomeron\ part} &= 35\ \mathrm{mb}, \\ b_{pp} &= 12.5\ \mathrm{GeV}^{-2}, \\ R_p \equiv R_{p,\mathrm{elm}} &= 0.86\ \mathrm{fm}. \end{aligned} \tag{3.111}$$

We obtain then from (3.108) and (3.109):

$$\begin{aligned} a &= 0.31\ \mathrm{fm}, \\ G_2 &= (507\ \mathrm{MeV})^4. \end{aligned} \tag{3.112}$$

The values for the correlation length a and for the gluon condensate G_2 come out in surprisingly good agreement with the determination of these quantities from the fit to the lattice results (2.70).

But perhaps we were lucky in picking out the right c.m. energy $\sqrt{s}$ and radius for our comparison of theory and experiment. What about the s-dependence of the total cross section σ_{tot} and slope parameter b? The vacuum parameters G_2 and a should be independent of the energy $\sqrt{s}$. On the other hand, from the discussion in Sect. 3.1 leading to (3.2), it seems quite plausible to us that the effective strong interaction radii R of hadrons may depend on $\sqrt{s}$. Let us consider again pp (or $p\bar{p}$) elastic scattering. Once we have fixed G_2 and a from the data at $\sqrt{s} = 20$GeV (3.108) and (3.109) give us $\sigma_{\rm tot}(pp)$ and b_{pp} in terms of the single parameter R_p, i.e. we obtain as prediction of the model a curve in the plane b_{pp} versus $\sigma_{\rm tot}(pp)$. This is shown in Fig. 18. It is quite remarkable that the data from $\sqrt{s} = 20$GeV up to Tevatron energies, $\sqrt{s} = 1.8$ TeV follow this curve.

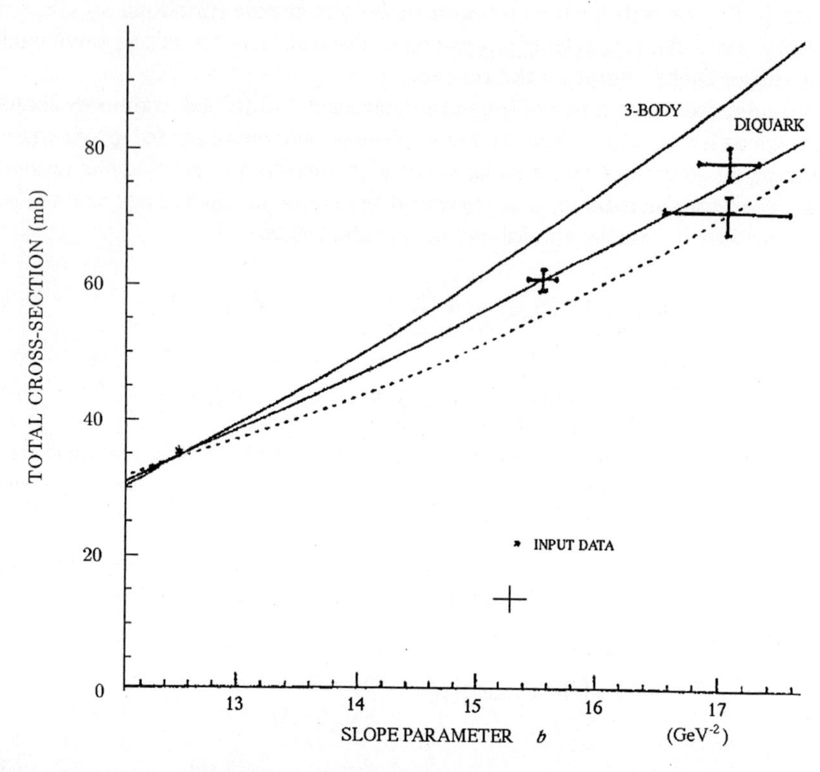

Fig. 18. *The relation between the total cross section σ_{tot} and the slope parameter b for proton-proton and proton-antiproton scattering. The dotted line is the prediction from Regge theory. The prediction of the calculation for soft high energy scattering in the stochastic vacuum model is that the data points should lie in the area between the full lines. In essence this is given by (3.108), (3.109) with an uncertainty estimate from different assumptions for the proton wave functions (cf. [45]).*

Summarizing this section we can say that explicit calculations for high energy-elastic hadron-hadron scattering near the forward direction have been performed combining the field-theoretic methods of [53] and the Minkowski version of the stochastic vacuum model of [59],[45]. The results are encouraging and support the idea that the vacuum structure of QCD plays an essential role in soft high-energy scattering. We want to point out that these calculations also resolve a possible paradoxon of QCD: On the one hand there are suggestions that the gluon propagator must be highly singular (probably $\propto (Q^2)^{-2}$, cf. e.g. [61]) for momentum transfers $Q^2 \to 0$ in order to produce confinement. On the other hand high energy scattering amplitudes are completely regular for $t \to 0$. A singular gluon propagator will lead in the 2-gluon exchange model to a singularity for $t = 0$ not only for quark-quark scattering but also for hadron-hadron scattering if the latter are considered e.g. as colour dipoles. The resolution of the paradoxon which we can present is intimately connected with the confinement mechanism which we found in the SVM in Sect. 2. The short range correlation of the gluon field strengths governs the t-dependence of the hadronic scattering amplitudes and gives rise to their regularity for $t = 0$. The gluon propagator on the other hand can be singular for $Q^2 \to 0$, since the gluon potentials have a long range correlation as we have seen in Sect. 2. The result (3.108) for the total cross section depends also on the proton radius. This radius dependence does not saturate for large radii in the calculation with non-abelian gluons but does saturate in an abelian model [45]. Thus with non-abelian gluons we do not get the additive quark model result [17], and thus not the picture of Donnachie and Landshoff [23], where the "soft Pomeron" couples to individual quarks in the hadrons. The strong radius dependence in (3.108) is due to the D-term in the correlator (3.89) which is related to the effective chromomagnetic monopole condensate in the QCD vacuum and which gives rise to the linearly rising quark-antiquark potential, i.e. to string formation (see Sect. 2.5). The calculations reported above suggest that in high energy scattering this same term gives rise to a string-string interaction which leads to the radius dependence in (3.108). Note that one does not have to put in the strings by hand. They enter the picture automatically through our lightlike Wegner-Wilson loops. The radius dependence of the cross section occurs, of course, also for meson-baryon and meson-meson scattering and gives a quantitative understanding of the difference between the Kp and πp total cross sections and slope parameters at high energies. For this and for further results we refer to [45], [62]. The success of the calculation for pp ($p\bar{p}$) scattering describing correctly the relation of the total cross section versus the slope parameter from $\sqrt{s} \simeq 20$ GeV up to $\sqrt{s} = 1.8$ TeV suggests the following simple interpretation: In soft elastic scattering the hadrons act like effective "colour dipoles" with a radius increasing with c.m. energy. The dipole-dipole interaction is governed by the correlation function of two lightlike Wegner-Wilson loops which receives the dominant contribution from the same non-perturbative phenomenon - effective chromomagnetic monopole condensation - which leads to string formation and confinement.

4 "Synchrotron Radiation" from the Vacuum, Electromagnetic Form Factors of Hadrons, and Spin Correlations in the Drell-Yan Reaction

Let us consider for definiteness again a proton-proton collision at high c.m. energy $\sqrt{s} \gg m_p$. We look at this collision in the c.m. system and choose as x^3-axis the collision axis (Fig. 19). According to Feynman's parton dogma [63] the hadrons look like jets of almost non-interacting partons, i.e. quarks and gluons. Accepting our previous views of the QCD vacuum (Sect. 2), these partons travel in a background chromomagnetic field.

What sort of new effects might we expect to occur in this situation? Consider

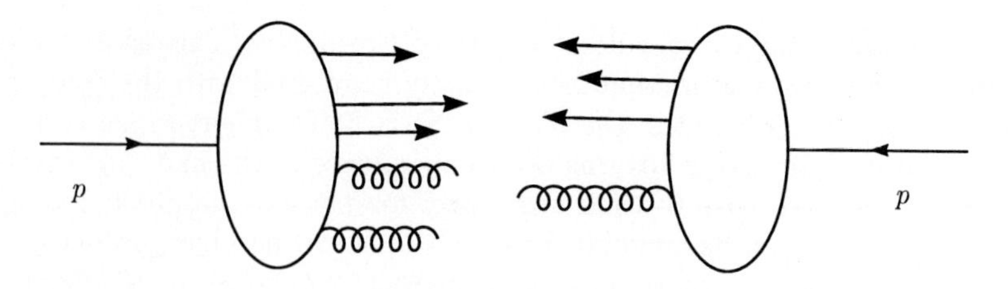

Fig. 19. *A proton-proton collision at high energies in the parton picture.*

for instance a quark-antiquark collision in a chromomagnetic field. In our picture this is very similar to an electron-positron collision in a storage ring (Fig. 20). We know that in a storage ring e^- and e^+ are deflected and emit synchrotron radiation. They also get a transverse polarization due to emission of spin-flip synchrotron radiation [64], [65]. Quite similarly we can expect the quark and antiquark to be deflected by the vacuum fields. Since quarks have electric and colour charge, they should then emit both photon and gluon "synchrotron radiation". Of course, as long as we have quarks within a single, isolated proton (or other hadron) travelling through the vacuum no emission of photons can occur, and we should consider such processes as contributing to the cloud of quasi-real photons surrounding a fast-moving proton. (This is similar in spirit to the well-known Weizsäcker-Williams approximation.) But in a collision process the parent quark or antiquark will be scattered away and the photons of the cloud can become real, manifesting themselves as prompt photons in hadron-hadron collisions.

In Ref. [14] we have given an estimate for the rate and the spectrum of such prompt photons using the classical formulae for synchrotron radiation [65]. A more detailed study of soft photon production in hadronic collisions was made in [66]. A sketch of our arguments and calculations is as follows.

In Sect. 2 we discussed the domain picture of the QCD vacuum. In Euclidean space time we have domains in the vacuum of linear size $\simeq a$. Inside one domain the colour fields are highly correlated. The colour field orientations and domain sizes fluctuate, i.e. have a certain distribution. If we translate this picture naively to Minkowski space, we arrive at colour correlations there being characterized by invariant distances of order a. Then the colour fields at the origin of Minkowski space, for instance, should be highly correlated with the fields in the region

$$|x^2| \lesssim a^2 \tag{4.1}$$

(cf. Fig. 21). Consider now a fast hadron passing by with one of its quarks going right through $x = 0$ on a nearly lightlike world line. It is clear that in such a situation the quark will, from the point of view of the observer, spend a long time in a correlated colour background field. An easy exercise shows furthermore that two quarks of the same hadron will have a very good chance to travel in two different colour domains. The argument is in essence as follows. The quarks have a transverse separation of the order of the hadron radius R whereas the transverse size of a domain is of order a and we have $a^2/R^2 \ll 1$ (cf. (2.72)).

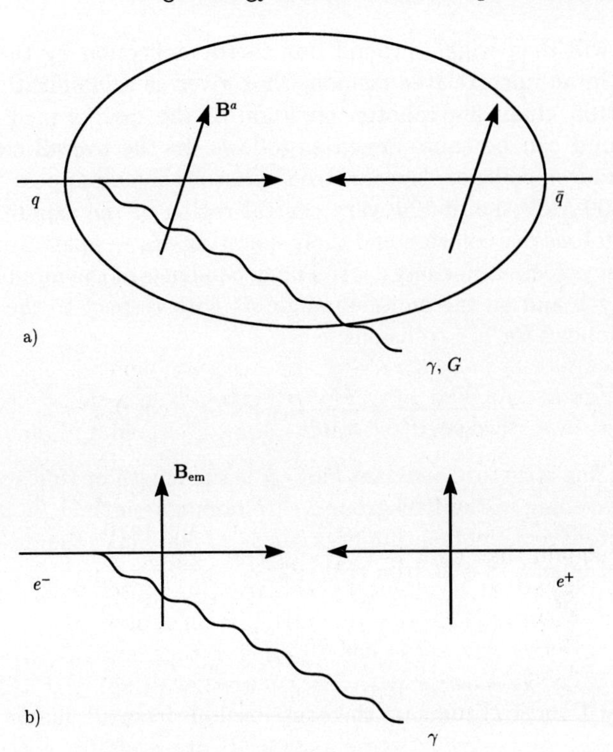

Fig. 20. *A quark and antiquark traversing a region of chromomagnetic field (a). An electron and a positron in a storage ring (b). In both cases we expect the emission of synchrotron radiation to occur.*

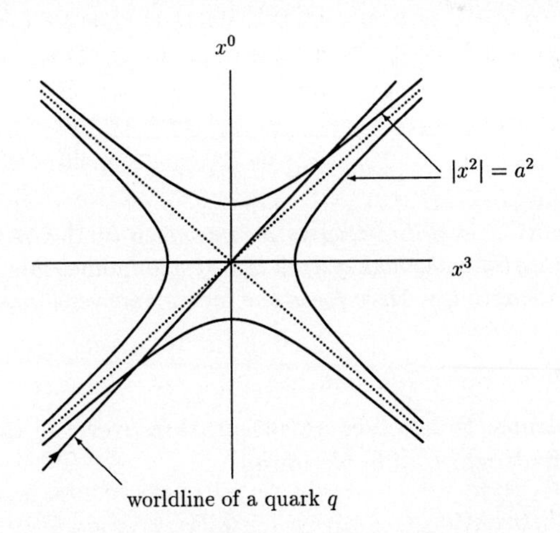

Fig. 21. *Sketch of a "colour domain" in Minkowski space and of the worldline of a quark from a fast hadron moving through it.*

Each quark will then wiggle around due to the deflection by the background colour fields in an uncorrelated fashion. This gives us a justification for adding the synchrotron gluon and photon emission of the quarks incoherently. The result we found can be summarized as follows: In the overall c.m. system of the hadron-hadron collision "synchrotron" photons should appear with energies $\omega < 300 - 500$ MeV, i.e. in the very central region of the rapidity space. The number of photons per collision and their spectrum are — apart from logarithms — independent of the c.m. energy $\sqrt{s}$. The dependence of the number of photons on the energy ω and on the emission angle ϑ^* with respect to the beam axis is obtained as follows for pN collisions:

$$\frac{dn_\gamma}{d\omega d\cos\vartheta^*} = \frac{2\pi\alpha}{\omega^{1/3}}(l_{\text{eff}})^{2/3} \cdot \Sigma(\cos\vartheta^*) \tag{4.2}$$

Here α is the fine structure constant and l_{eff} is the length or time over which the fast quark travelling in the background chromomagnetic field B_c obtains by its deflection a transverse momentum of order $\bar{p}_T \approx 300$ MeV, the mean transverse momentum of quarks in a hadron (Fig. 22):

$$l_{\text{eff}} = \frac{\bar{p}_T}{gB_c}. \tag{4.3}$$

The quantity Σ in (4.2) sums up the contributions from all quarks of the initial

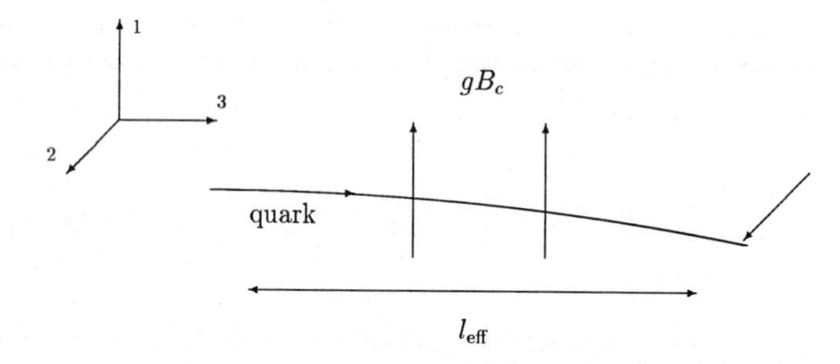

Fig. 22. *A quark moving in 3-direction in a transverse (in 1-direction) chromomagnetic field of strength gB_c and picking up a transverse momentum (in 2-direction) of magnitude $\bar{p}_T$ over a length l_{eff}. Here $\bar{p}_T$ is the mean transverse momentum of quarks in the hadron.*

and final state hadrons. It involves an integration over the quark distribution functions of these hadrons. In [66] we found

$$\Sigma(\cos\vartheta^*) \simeq \frac{0.21}{(\sin\vartheta^*)^{2/3}} \tag{4.4}$$

for pN collisions at $\sqrt{s} = 29$ GeV.

Our result (4.2) for synchrotron photons should be compared to the inner bremsstrahlung spectrum

$$\left.\frac{dn_\gamma}{d\omega d\cos\vartheta^*}\right|_{\text{bremsstr.}} \propto \frac{1}{\omega\sin^2\vartheta^*} \tag{4.5}$$

The "synchrotron" radiation from the quarks (4.2) is thus harder than the hadronic bremsstrahlung spectrum. This is welcome, since for $\omega \to 0$ bremsstrahlung should dominate according to Low's theorem [67].

It is amusing to note that in several experiments an excess of soft prompt photons over the bremsstrahlung calculation has been observed [68]-[72]. The gross features and the order of magnitude of this signal make it a candidate for our "synchrotron" process. A detailed comparison with our formulae has been made in [66] for the results from one experiment [72] with encouraging results. This is shown in Fig. 23 for the k_{T} spectrum of photons at $y = 0$, where

$$\begin{aligned} k_{\mathrm{T}} &= \omega\sin\vartheta^*, \\ y &= -\ln\tan(\vartheta^*/2). \end{aligned} \tag{4.6}$$

We see that the addition of synchrotron photons to the bremsstrahlung ones improves the agreement of theory with the data considerably. We deduce from Fig. 23 $l_{\text{eff}} \simeq 20 - 40$ fm. Taking $l_{\text{eff}} = 20$ fm and $\bar{p}_T = 300$ MeV, we find for the effective chromomagnetic deflection field from (4.3)

$$gB_c = \frac{\bar{p}_T}{l_{\text{eff}}} = (55 \text{ MeV})^2. \tag{4.7}$$

This is much smaller than the vacuum field strength (2.7). Our interpretation of this puzzle is as follows: The colour fields in a fast moving hadron must be shielded. Indeed, a chromomagnetic field of the strength (2.7) would lead to a ridiculously small value for the radius of cyclotron motion of a fast quark. The necessary shielding could be done by gluons in a fast hadron. We know from the deep inelastic lepton-nucleon scattering results that a fast nucleon contains many gluons. We may even be brave and turn the argument around: in order for a fast hadron to be able to move through the QCD vacuum, the very strong vacuum chromomagnetic fields must be shielded, making gluons in a fast hadron a <u>necessity</u>. Thus soft photon production in pp collisions may give us a quite unexpected insight into the quark and gluon structure of <u>fast</u> hadrons.

The next topic we want to discuss briefly concerns electromagnetic form factors of hadrons.

We have argued above that the colour fields in the vacuum should give a contribution to the virtual photon cloud of hadrons and we made an estimate of the distribution of these photons using the synchrotron radiation formulae. Consider now any reaction where a quasi-real photon is emitted from a hadron with the hadron staying intact and the photon interacting subsequently. In Fig. 24 we draw the corresponding diagram for a nucleon N:

$$N(p) \to N(p') + \gamma(q). \tag{4.8}$$

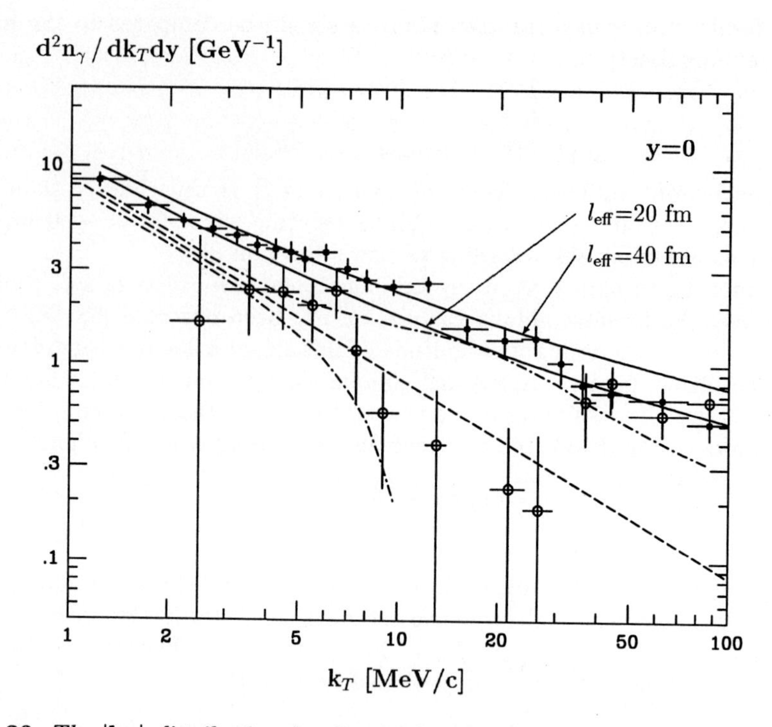

Fig. 23. *The $|\mathbf{k}_T|$ distribution for direct photons emitted at c.m. rapidity $y = 0$ in $p - Be$ collisions at 450 GeV incident proton momentum from [72]. The normalization is according to a private communication by H. J. Specht. The background of decay photons is subtracted. The dash-dotted line gives the expected yield of photons from hadronic bremsstrahlung, the dashed lines show the upper and lower limits including the systematic errors in the shape of the decay background and the bremsstrahlung calculation (cf. [72]). The lower (upper) solid line is the result of the calculation for synchrotron photons ((4.2ff.) with $l_{\text{eff}} = 20$ fm ($l_{\text{eff}} = 40$ fm) added to the spectrum of hadronic bremsstrahlung (cf. [66]).*

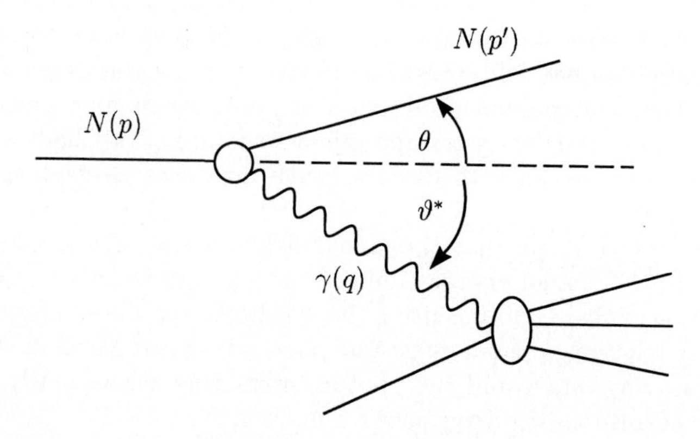

Fig. 24. *A nucleon interacting by emission of a quasi-real photon.*

The flux of these quasi-real photons is well known. The first calculations in this context are due to Fermi, Weizsäcker, and Williams [73]. For us the relevant formula is given in (D.4) of [74]. Let E be the energy of the initial nucleon, $G_E^N(Q^2)$ its electric Sachs formfactor, and let ω and $q^2 = -Q^2$ be the energy and mass of the virtual photon. Then the distribution of quasi-real photons in the fast-moving nucleon is given by

$$dn_\gamma^{(\text{excl})} = \frac{\alpha}{\pi}\frac{d\omega}{\omega}\frac{dQ^2}{Q^2}\left[G_E^N(Q^2)\right]^2 \tag{4.9}$$

where we neglect terms of order ω/E and Q^2/m_N^2 and assume

$$Q^2 \gg Q^2_{\min} \simeq \frac{m_N^2\omega^2}{E^2}. \tag{4.10}$$

We call (4.9) the exclusive flux since the nucleon stays intact. Now we want to translate (4.9) into a distribution in ω and the angle ϑ^* of emission of the γ (cf. Fig. 24). A simple calculation gives

$$Q^2 \simeq \omega^2 \sin^2\vartheta^*, \tag{4.11}$$

$$dn_\gamma^{(\text{excl})} \simeq \frac{2\alpha}{\pi}\frac{d\omega}{\omega}\frac{d\vartheta^*}{\sin\vartheta^*}\left[G_E^N(\omega^2\sin^2\vartheta^*)\right]^2. \tag{4.12}$$

Now we made an "exclusive-inclusive connection" argument in [66]: We require $dn_\gamma^{(\text{excl})}/d\omega d\cos\vartheta^*$ to behave as $\omega^{-1/3}$ for fixed ϑ^* as we found in (4.2). This implies for the form factor $G_E^N(Q^2)$ a behaviour as $(Q^2)^{1/6}$.

Thus we arrive at the following conclusion: The proton form factor G_E^p should contain in addition to a "normal" piece connected with the total charge and the hadronic bremsstrahlung in inelastic collisions a piece $\propto (Q^2)^{1/6}$ connected with "synchrotron" radiation from the QCD vacuum. For the neutron which has total charge zero we would expect the "normal" piece in G_E^n to be quite small and the "anomalous" piece to be quite important for not too large Q^2. Thus the neutron electric formfactor should be an interesting quantity to look for "anomalous" effects $\propto (Q^2)^{1/6}$.

In Fig. 25 we show the data on the electric formfactor of the neutron from [75, 76]. We superimpose the curve

$$G^n_{(\text{syn})}(Q^2) = 3.6\cdot 10^{-2}\left(\frac{Q^2}{5\,\text{fm}^{-2}}\right)^{1/6} \tag{4.13}$$

which is normalized to the data at $Q^2 = 5\,\text{fm}^{-2}$. We see that except in the very low Q^2 region we get a decent description of the data. For $Q^2 \to 0$ (4.13) has to break down since $G_E^n(Q^2)$ is regular at $Q^2 = 0$. Indeed one knows the slope of $G_E^n(Q^2)$ for $Q^2 = 0$ from the scattering of thermal neutrons on electrons ([77] and references cited therein):

$$\left.\frac{dG_E^n(Q^2)}{dQ^2}\right|_{Q^2=0} = 0.019\,\text{fm}^2. \tag{4.14}$$

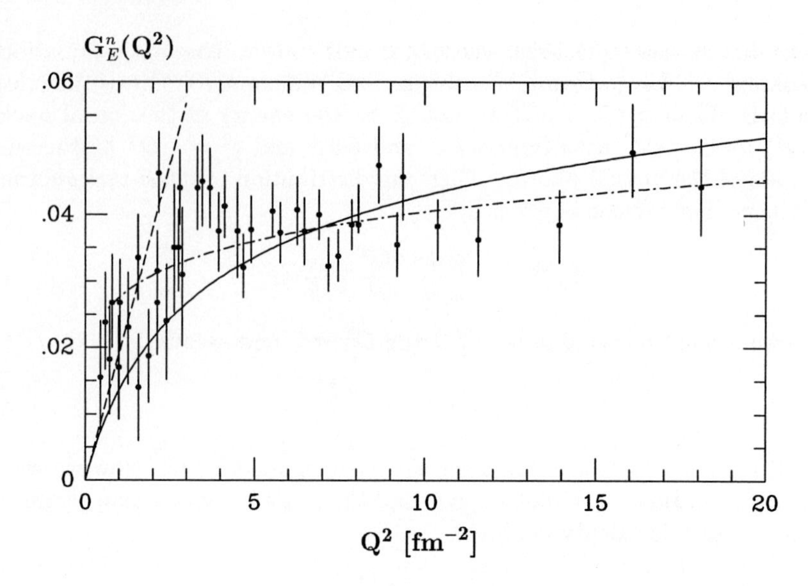

Fig. 25. *The data for the electric form factor of the neutron $G_E^n(Q^2)$ from Refs. [75, 76]. Dash-dotted line: our naive "synchrotron" prediction $\propto (Q^2)^{1/6}$ (4.13) normalized to the data at $Q^2 = 5$ fm^{-2}. Dashed line: the slope of $G_E^n(Q^2)$ at $Q^2 = 0$ as deduced from thermal neutron-electron scattering [77]. Full line: the ansatz (4.15).*

We see from Fig. 25 that the behaviour of $G_E^n(Q^2)$ has to change rather quickly as we go away from $Q^2 = 0$. We will now make a simple ansatz which takes into account that $G_E^n(Q^2)$ can have singularities in the complex Q^2-plane only for

$$-\infty < Q^2 \leq -4m_\pi^2,$$

where m_π is the pion mass. We require a $(Q^2)^{1/6}$ behaviour for positive Q^2 and take the slope of $G_E^n(Q^2)$ at $Q^2 = 0$ from experiment (4.14). This leads us to the following functional form for $G_E^n(Q^2)$:

$$G_E^n(Q^2) = 0.019 \text{ fm}^2 \cdot Q^2 \left[1 + \frac{Q^2}{4m_\pi^2}\right]^{-5/6} . \tag{4.15}$$

It is amusing to see that this gives a decent description of the data (Fig. 25).

What about the electric form factor of the proton $G_E^p(Q^2)$? Here, clearly, we have a dominant "normal" piece connected with the total charge. We will assume that this normal contribution is represented by the usual dipole formula

$$G_D(Q^2) = \left(1 + \frac{Q^2}{m_D^2}\right)^{-2} ,$$

$$m_D^2 = 18.23 \text{ fm}^{-2} \hat{=} 0.710 \text{ GeV}^2 \tag{4.16}$$

which gives a good representation of the data for $Q^2 = 2$–4 GeV^2 [78]. Let us add to this an anomalous piece for smaller Q^2, connected with synchrotron radiation, and let us assume that this is a purely isovector contribution, consistent with the singularity at $Q^2 = -4m_\pi^2$ in (4.15). We obtain then from (4.15) the following ansatz for $G_E^p(Q^2)$:

$$G_E^p(Q^2) = G_D(Q^2) - G_E^n(Q^2) = G_D(Q^2)(1 - \Delta(Q^2)), \tag{4.17}$$

where

$$\Delta(Q^2) = 0.019\ \mathrm{fm}^2 \cdot Q^2(1 + \frac{Q^2}{4m_\pi^2})^{-5/6} \cdot (1 + \frac{Q^2}{m_D^2})^2. \tag{4.18}$$

We predict a deviation of the ratio $G_E^p(Q^2)/G_D(Q^2)$ from unity for small Q^2. It is again amusing to note that such a deviation is indeed observed experimentally [47, 48]. Our ansatz does even quite well quantitatively (Fig. 26). For the electromagnetic radius of the proton we predict from (4.17)

$$\begin{aligned}\langle r_E^2 \rangle := -6 \frac{\mathrm{d}G_E^p(Q^2)}{\mathrm{d}Q^2}\Big|_{Q^2=0} &= \frac{12}{m_D^2} + 6 \cdot 0.019\ \mathrm{fm}^2 \\ &= (0.88\ \mathrm{fm})^2. \end{aligned} \tag{4.19}$$

This checks well with the experimental values quoted in [47]:

$$\begin{aligned}\langle r_E^2 \rangle^{1/2} &= 0.88 \pm 0.03\ \mathrm{fm},\ \text{or} \\ &\ 0.92 \pm 0.03\ \mathrm{fm}, \end{aligned} \tag{4.20}$$

depending on the fit used for $G_E^p(Q^2)$ at low Q^2.

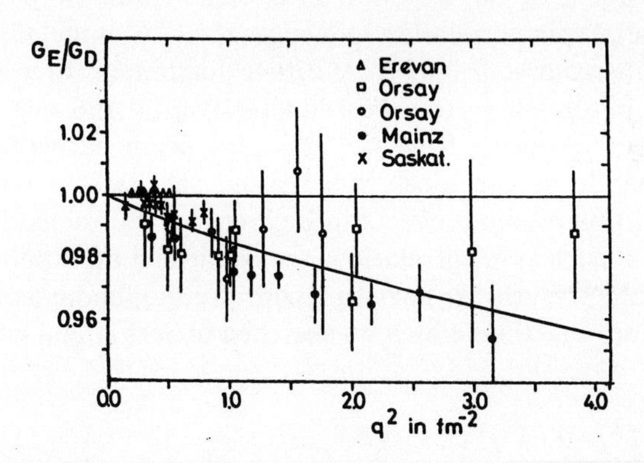

Fig. 26. *The ratio G_E^p/G_D of the electric form factor of the proton to the dipole fit versus Q^2. The data points are from various experiments as summarized in [48]. The solid line corresponds to the ansatz (4.17), (4.18).*

To conclude this brief discussion of nucleon form factors, we can summarize our picture as follows: The quarks in the nucleon make a cyclotron-type motion in the chromomagnetic vacuum field. This leads to a spreading out of the charge distribution of the neutron

$$\rho_n(\mathbf{x}) \propto |\mathbf{x}|^{-10/3} \tag{4.21}$$

corresponding to $G_E^n(Q^2) \propto (Q^2)^{1/6}$. The same effect should lead to a deviation of the proton form factor $G_E^p(Q^2)$ from the dipole formula for small Q^2. Concerning the sign of $G_E^n(Q^2)$ we can in essence follow the arguments put forward in [79].

One might think — maybe rightly — that these ideas are a little crazy. But we have also worked out some consequences of them for the Drell-Yan reaction (1.9), which make us optimistic. In the lowest order parton process contributing there, we have a quark-antiquark annihilation giving a virtual photon γ^*, which decays then into a lepton pair (Fig. 1):

$$q + \bar{q} \to \gamma^* \to \ell^+\ell^-. \tag{4.22}$$

In the usual theoretical framework q and $\bar{q}$ are assumed to be uncorrelated and unpolarized in spin and colour if the original hadrons are unpolarized. From our point of view we expect a different situation. Let us go back to Fig. 21 where we sketched the world line of a quark of one fast hadron in a colour domain of extension $|x^2| \lesssim a^2$. Let the quark q and antiquark $\bar{q}$ in (4.22) annihilate at the point $x = 0$. Here q and $\bar{q}$ come from two different hadrons h_1, h_2 moving with nearly light-like velocity in opposite directions. It is clear that in this situation q and $\bar{q}$ will spend a long time in a highly correlated colour background field (Fig. 27). In [14] we speculated that this may lead to a correlated transverse spin and colour spin polarization of q and $\bar{q}$ due to the chromomagnetic Sokolov-Ternov effect [64, 65]. In [80] we worked out this idea in more detail and found that a transverse $q\bar{q}$ spin correlation influences the $\ell^+\ell^-$ angular distribution in a profound way. Then our colleague H. J. Pirner pointed out to us that data which may be relevant in this connection existed already [81]. And very obligingly these data show a large deviation from the standard perturbative QCD prediction. On the other hand, we can nicely understand the data in terms of our spin correlations and thus vacuum effects in high energy collisions. For more details we refer to [80]. If such spin correlations are confirmed by experiments at higher energies, we would presumably have to reconsider the fundamental factorization hypothesis for hard reactions which we sketched in Sect. 1 and which is discussed in detail in [11].

5 Conclusions

In these lectures we have discussed various ideas connected with non-perturbative QCD and in particular with the QCD vacuum structure. In Sect. 2 we introduced connectors, the non-abelian Stokes theorem and the cumulant expansion. Then we presented the domain picture of the QCD vacuum and the stochastic vacuum model (SVM). The latter is consistent with the view of the QCD vacuum acting like a dual superconductor: We found that in the SVM the

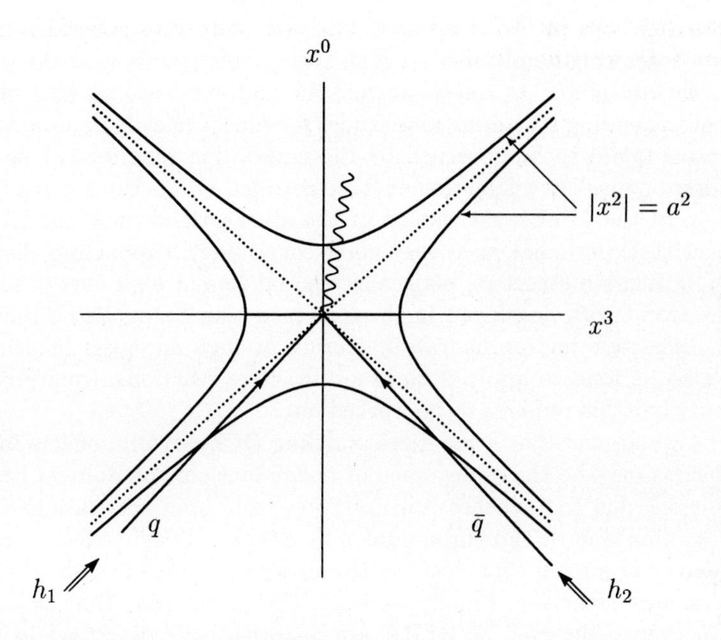

Fig. 27. *Annihilation of a $q\bar{q}$ pair with production of a virtual photon γ^* in a colour domain. Here q and $\bar{q}$ come from two different hadrons h_1 and h_2, respectively.*

vacuum contains an effective chromomagnetic monopole condensate, whose effect is parametrized by the D-term in the gluon field strength correlator (2.61). With these tools we could calculate the expectation value of the Wegner-Wilson loop in the SVM. We found a linearly rising potential between a heavy quark-antiquark pair, $Q\bar{Q}$. This is related to the formation of a "string", a chromoelectric flux tube between $Q\bar{Q}$ as can be seen explicitly in the SVM [52]. Thus in this framework confinement is an effect of the nontrivial QCD vacuum structure. All calculations in Sect. 2 were done in Euclidean space time.

In Appendix A we discuss some problems which arise when one considers higher cumulant terms in the SVM. We propose as remedy for these problems to do the calculation of the expectation value of the Wegner-Wilson loop in an iterative way, summing step by step over the contributions of various plaquettes. This also leads us to a proposal for including dynamical fermions in the calculation with the expected result that the linearly rising potential levels off and goes to a constant at large $Q\bar{Q}$ separation, where, of course, we have then mesons $Q\bar{q}, q\bar{Q}$ formed of a heavy quark Q (antiquark $\bar{Q}$) and a light antiquark $\bar{q}$ (quark q).

In Sects. 3, 4 we looked at various possible effects of the nontrivial vacuum structure in the Minkowski world. In Sect. 3 we gave a detailed account of a field-theoretic method for the calculation of scattering amplitudes of high-energy soft hadronic reactions. We started from the functional integral and made high energy approximations in the integrand. This led us to give general rules for writing down scattering amplitudes for high energy soft reactions at the parton level. Then we considered as representation for hadrons wave packets of partons. The corresponding scattering amplitudes for (quasi-)elastic hadron-hadron scattering were found to be governed by the correlation functions of lightlike Wegner-Wilson loops. The evaluation of these correlation functions was possible with the help of the Minkowski version of the stochastic vacuum model. The comparison with experiment gave very good consistency, supporting the view that the QCD vacuum structure plays an essential role in high energy soft reactions. The framework developed in these lectures can be applied directly to elastic and diffractive hadron-hadron scattering at high energies. In principle we should also be able to apply it to non-diffractive reactions, fragmentation processes etc., but this remains to be worked out.

In Sect. 4 we argued that some more startling QCD vacuum effects in high energy collisions may be the appearance of anomalous soft photons in hadron-hadron collisions due to "synchrotron radiation" and spin correlations in the Drell-Yan reaction due to the chromodynamic Sokolov-Ternov effect. Furthermore, we gave an argument that electromagnetic form factors at small Q^2 should reflect the vacuum structure. Finally we would like to mention that in [82] the rapidity gap events observed at HERA are quantitatively described in terms of the parton model but invoking again nonperturbative QCD effects, possibly connected with the vacuum structure. Another place where the QCD vacuum structure may show up is in certain correlations of hadrons in Z^0 decays to two jets [14, 83] for which there is also some experimental evidence [84].

We hope to have convinced the reader that the non-perturbative structure of the QCD vacuum is useful in order to understand confinement of heavy quarks. In our view this vacuum structure manifests itself also in high energy soft and hard reactions. We think it is very worth-while to study such effects both from the theoretical and the experimental point of view.

Acknowledgements

The author is grateful to the organizers of the Graduiertenkolleg Erlangen-Regensburg, especially F. Lenz, for the invitation to give lectures at Banz in 1993. Thanks are due to U. Grandel for collecting lecture notes then. The "35. Internationale Universitätswochen für Kern- und Teilchenphysik 1996" in Schladming, Austria, gave the author the opportunity to review and extend these lectures. The author thanks the organizers of that meeting headed by H. Latal for the invitation to lecture there. The present article is the contribution of the author to the proceedings of both meetings. For many fruitful discussions on topics of these lecture notes the author extends his gratitude to A. Brandenburg, W. Buchmüller, H. G. Dosch, U. Ellwanger, D. Gromes, P. Haberl, P. V. Landshoff, P. Lepage, H. Leutwyler, Th. Mannel, E. Mirkes, H. J. Pirner, M. Rueter, Yu. A. Simonov, G. Sterman, and W. Wetzel. Special thanks are due to H. G. Dosch for a critical reading of the manuscript and to E. Berger and P. Haberl for their help in the preparation of the manuscript. Finally the author thanks Mrs. U. Einecke for her excellent typing of the article.

Appendix A: Higher Cumulant Terms and Dynamic Fermions in the Calculation of the Wegner-Wilson Loop in the Stochastic Vacuum Model

In this appendix we discuss first some problems arising in the calculation of the Wegner-Wilson loop in the SVM (cf. Sect. 2.5) if higher cumulants are taken into account. Then we outline a possible remedy which may also point to a way of including the effects of dynamical fermions in the SVM. We start with the replacements (2.79) which allow us to use the cumulant expansion (2.41) for calculating $W(C)$. The second cumulant is given in (2.80):

$$K_2(1,2) = \frac{4}{3}\frac{\pi^2}{g^2}\mathcal{F}(1,2), \tag{A.1}$$

where we set with $F_{\mu\nu\rho\sigma}$ as defined in (2.61):

$$\mathcal{F}(i,j) \equiv F_{1414}(X^{(i)} - X^{(j)}), \tag{A.2}$$

$$\begin{aligned} F_{1414}(Z) &= \frac{G_2}{24}[\kappa D(-Z^2) + \frac{1}{2}\left(\frac{\partial}{\partial Z_1}Z_1 + \frac{\partial}{\partial Z_4}Z_4\right)(1-\kappa)D_1(-Z^2)], \\ Z^2 &= Z_1^2 + Z_4^2. \end{aligned} \tag{A.3}$$

With Assumptions 1-3 of the SVM (Sect. 2.4), the next nonvanishing cumulant is K_4, for which we obtain from (2.47), (2.60), (2.73), (2.75):

$$K_4(1,2,3,4) = -2\left(\frac{\pi^2}{g^2}\right)^2 [\mathcal{F}(1,3)\mathcal{F}(2,4)\Theta(1,2,3,4) \ + \ perm.]. \tag{A.4}$$

Here $\Theta(1,2,3,4) = 1$ if $X^{(1)} > X^{(2)} > X^{(3)} > X^{(4)}$ in the sense of the path-ordering on the surface S (cf. Figs. 8, 10) and $\Theta(1,2,3,4) = 0$ otherwise.

We start now again from the expression (2.78) for the expectation value of the Wegner-Wilson loop and use the cumulant expansion (2.41) with the identifications (2.79). We get then:

$$W(C) = \exp\left\{\sum_{n=1}^{\infty}\frac{(-ig)^n}{n!}\int_{S_1}...\int_{S_n}K_n(1,...,n)\right\}, \tag{A.5}$$

where

$$\int_{S_i} \equiv \int_S \mathrm{d}X_1^{(i)}\mathrm{d}X_4^{(i)}. \tag{A.6}$$

In the SVM as formulated in Sect. 2.4 the cumulants for odd n vanish (cf. Assumption 3). The lowest contribution in (A.5) is then from $n = 2$, the next from $n = 4$. Cutting off the infinite sum in (A.5) at $n = 4$, we get with (A.1) and (A.4):

$$\begin{aligned} W(C) &= \exp\left\{ -\frac{g^2}{2!} \int_{S_1} \int_{S_2} K_2(1,2) + \frac{(g^2)^2}{4!} \int_{S_1} \cdots \int_{S_4} K_4(1,2,3,4) \right\} \\ &= \exp\left\{ -\frac{1}{2!} \frac{4\pi^2}{3} \int_{S_1} \int_{S_2} \mathcal{F}(1,2) \right. \\ &\quad \left. -\frac{1}{4!} 2\pi^4 \int_{S_1} \cdots \int_{S_4} [\mathcal{F}(1,3)\mathcal{F}(2,4)\Theta(1,2,3,4) + \textit{perm.}] \right\} \end{aligned} \tag{A.7}$$

Consider now the contribution of the second cumulant in (A.7) for a large Wegner-Wilson loop (Fig. A1 with $R, T \gg a$):

$$I_2 = \int_{S_1} \int_{S_2} \mathcal{F}(1,2). \tag{A.8}$$

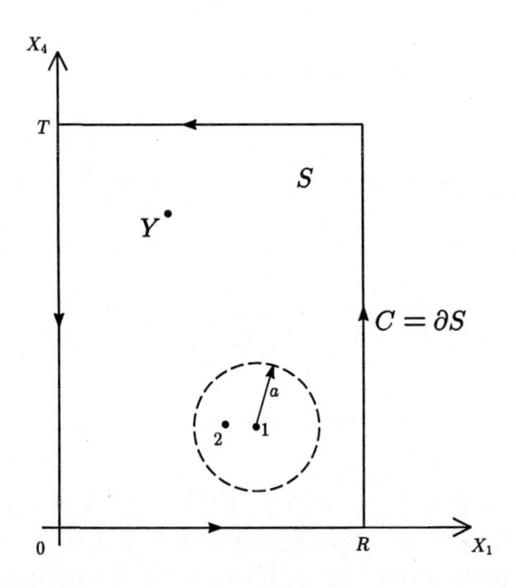

Fig. A1. *The relevant integration region for the integral (A.8). The point (1) runs freely over the surface S, the point (2) is constrained to a distance $\lesssim a$ from (1).*

For fixed integration point (1) on S the integration over (2) will give significant contributions only for a region of radius $\sim a$ around (1) since the functions $D(-Z^2)$ and $D_1(-Z^2)$, where $Z = (X^{(1)} - X^{(2)})$, are assumed to fall off rapidly with increasing Z^2. From (A.3) we see that the D_1-term contributes as a total divergence in $\mathcal{F}(1,2)$. Thus we can apply Gauss' law in 2 dimensions for it to transform it to an integral over the boundary $\partial S = C$. In this way we find a contribution from the D_1 term of order a^4/R^2, a^4/RT, a^4/T^2. From the D-term in (A.3) we get a factor $\propto a^2$ from the integration over (2) in (A.8). Then the integration over (1) is unconstrained and gives a factor proportional to the whole area of S:

$$I_2 \propto RTa^2 + O\left(a^4, a^4\frac{T}{R}, a^4\frac{R}{T}\right). \tag{A.9}$$

Putting in all factors from (A.3) and using the explicit form (2.65) for the function $D(-Z^2)$ gives for $T \to \infty$ and $R \gg a$ the result:

$$\frac{1}{2!}\frac{4\pi^2}{3}I_2 = RT\sigma \tag{A.10}$$

with σ as given in (2.83).

We turn next to the contribution of the 4th cumulant in (A.7) and study the integral

$$I_4 = \int_{S_1} \dots \int_{S_4} \mathcal{F}(1,3)\mathcal{F}(2,4)\Theta(1,2,3,4). \tag{A.11}$$

Consider again $R, T \gg a$ and fix the integration point (1) (Fig. A2). Then the short-range correlation of the field strengths requires (3) and (1) as well as (4) and (2) to be near to each other, i.e. at a distance $\lesssim a$. The function $\Theta(1,2,3,4)$ requires (1) > (2) > (3) > (4) in the sense of the path ordering on the surface S. Using the fan-type net as indicated in Fig. 8 for the application of the non-abelian Stokes theorem with straight line handles from the points $X^{(i)}$ in the surface to the reference point Y we see the following. The path-ordering function $\Theta(1,2,3,4)$ restricts (2) to be in the hatched sector of S in Fig. A2. In general this does not restrict (2) to a region close to (1). On the contrary, (2) can vary freely over a triangle of area $\gtrsim a.L$,where L is of order R or T or in between. Thus, the integration over (2) in (A.11) will give at least a factor $\propto (aT)$ if the hatched triangle has its long sides in X_4 direction (cf. Fig. A2). The integrations over (3) and (4) in (A.11) should give factors of a^2 each. The integration over (1) will give a factor RT. Thus we estimate

$$I_4 \propto RT.a^4.aT \propto T^2. \tag{A.12}$$

This is very unpleasant. The 4th cumulant gives a contribution which dominates over the one from the second cumulant for $T \to \infty$. There is no finite limit for $T \to \infty$ in the expression (2.77). The quark-antiquark potential comes out infinite.

This problem was recognized in Ref. [52] and eliminated by hand making an additional assumption: that all but the leading powers in the quotient of the correlation length to the extension parameters R, T of the loop could be neglected. Another simple cure of the problem would be to postulate instead of Assumption 3 (cf. (2.73)-(2.75)) a behaviour of the higher point correlation functions of the shifted gluon field strengths which gives precisely zero for the cumulants K_n with $n > 2$ in (A.5). In our opinion this would be an unsatisfactory

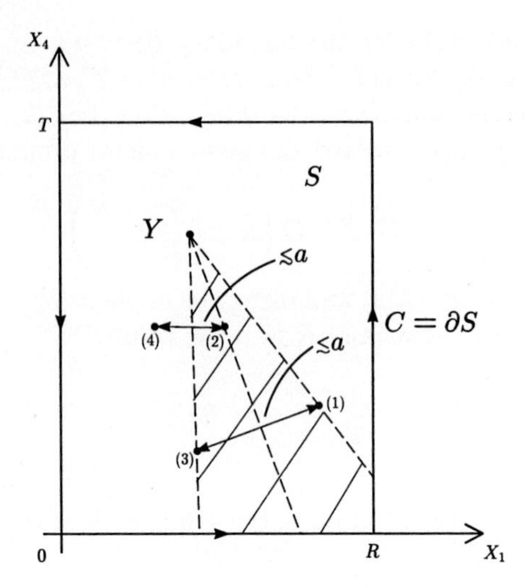

Fig. A2. *The relevant integration region for the integral (A.11). The point (1) runs freely over S. The points (1), (2), (3), (4) are ordered in the angle as seen from the reference point Y due to the path ordering function $\Theta(1,2,3,4)$. The points (1), (3) and (2), (4) must be at a distance $\lesssim a$ to each other.*

solution, since the model would then only work for a particular fine-tuned set of correlation functions whereas generically the above problem would remain.

We think that the origin of these difficulties in the SVM is Assumption 1, which states that the correlator should be independent of the reference point Y for arbitrary Y. But why should the field strength correlation (2.60) (cf. Fig. 5) be the same if we use a straight line path $C_X + \bar{C}_{X'}$ to connect X and X' or a path which runs on a loop behind the moon? We will replace Assumption 1 by a milder one:

- **Assumption 1':** $F_{\mu\nu\rho\sigma}$ in (2.60) is independent of the reference point Y and of the curves C_X and $C_{X'}$ if the reference point Y is close to X and X':

$$|X - Y| \lesssim a', \ |X' - Y| \lesssim a',$$

  where a' is of order of a.

Now we try to calculate the expectation value of the Wegner-Wilson loop (2.76) using only this weaker hypothesis. We start from the rectangle S of area RT and insert a smaller rectangular loop C_1 with sides $R - 2a', T - 2a'$. On C_1 we choose N_1 reference points $Y_1, ..., Y_{N_1}$ and we partition the area between C and C_1 in N_1 plaquettes $P_1, ..., P_{N_1}$ of size $\sim a'^2$ (Fig. A3).

We can now apply the non-abelian Stokes theorem. We start from Y_0 on C and construct a path equivalent to C in the following way: From Y_0 to Y_1, then in a fan-type net over the plaquette P_1 with Y_1 as reference point, etc., until we

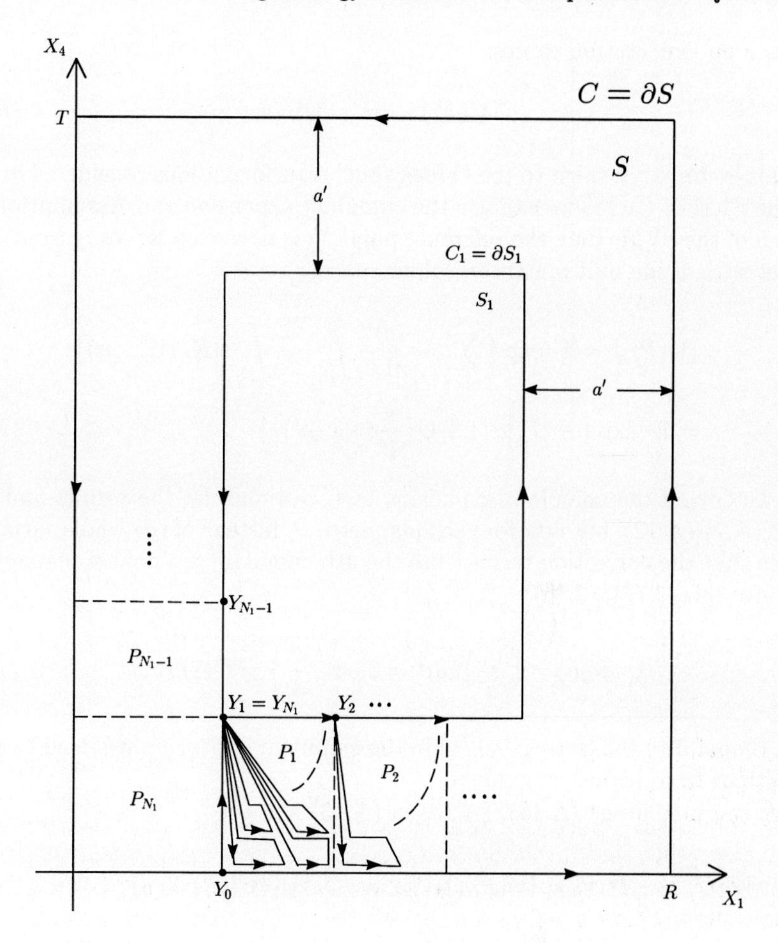

Fig. A3. *The rectangle S, $C = \partial S$ and N_1 reference points $Y_1, ..., Y_{N_1}$ on the curve C_1. The area between C_1 and C is partitioned in N_1 plaquettes $P_1, ..., P_{N_1}$.*

arrive at $Y_{N_1} \equiv Y_1$ from where the path runs back to Y_0. According to (2.36) we get then:

$$W(C) = \frac{1}{3}\mathrm{Tr}\langle V_{0,N_1} V(P_{N_1}) V_{N_1,N_1-1} ... V_{2,1} V(P_1) V_{1,0} \rangle, \tag{A.13}$$

where

$$V(P_j) = \mathrm{P}\ \exp\left[-ig \int_{P_j} \hat{G}_{14}(Y_j, X^{(j)}; C_{X^{(j)}}) \right] \tag{A.14}$$

and $V_{i,j}$ are the connectors from Y_j to Y_i on straight lines.

– **Assumption 4:** Now we will make a mean field-type approximation and replace the path-ordered integrals over the plaquettes P_j by the corresponding vacuum expectation values:

$$V(P_j) \to \langle V(P_j)\rangle \cdot 1\!\!1. \tag{A.15}$$

This is similar in spirit to the "block spin" transformations considered in [85]. For the r.h.s. of (A.15) we can use the cumulant expansion and Assumptions 1', 2 and 3 of the SVM. Here the reference point Y_j is never too far away from $X^{(j)}$. We get with $1\!\!1$ the unit matrix in colour space

$$\begin{aligned}\langle V(P_j)\rangle &= 1\!\!1 \cdot \exp\{\sum_{n=1}^{\infty} \frac{(-ig)^n}{n!} \int_{P_{j,1}} \cdots \int_{P_{j,n}} K_n(1,...,n)\} \\ &= 1\!\!1 \cdot \exp\{-|P_j|\sigma[1 + O(\frac{1}{4!}\sigma \cdot a \cdot a')]\}.\end{aligned} \tag{A.16}$$

Here we cut off the cumulant expansion at $n = 4$ and use the results and estimates (A.7)-(A.12), but now for each plaquette P_j instead of the whole surface S. We see that the correction terms from the 4th cumulant are now of manageable size since (cf. (2.70), (2.85))

$$\sigma a a' = \left(\frac{a'}{a}\right) \sigma a^2 = 0.56 \left(\frac{a'}{a}\right) = O(1). \tag{A.17}$$

Then (hopefully) the factorials $1/n!$ in the cumulant expansion will lead to small corrections from higher cumulants.

We can now insert (A.15), (A.16) in (A.13) and get

$$\begin{aligned}W(C) &\simeq \frac{1}{3}\mathrm{Tr}\langle V_{0,N_1}\langle V(P_{N_1})\rangle V_{N_1,N_1-1}...V_{2,1}\langle V(P_1)\rangle V_{1,0}\rangle \\ &= \frac{1}{3}\mathrm{Tr}\langle V_{0,N_1}V_{N_1,N_1-1}...V_{2,1}V_{1,0}\rangle \cdot \exp\{-\sum_{j=1}^{N_1}|P_j|\sigma[1+O(\frac{\sigma a a'}{4!})]\} \\ &= \frac{1}{3}\mathrm{Tr}\langle V_{N_1,N_1-1}...V_{2,1}\rangle \exp\{-(|S|-|S_1|)\sigma[1+O(\frac{\sigma a a'}{4!})]\}.\end{aligned} \tag{A.18}$$

Here we used also the cyclicity of the trace and $V_{1,0}V_{0,N_1} = 1\!\!1$ (cf. (2.24)). The product of the remaining connectors in (A.18) gives just the Wegner-Wilson loop of the curve $C_1 = \partial S_1$ in Fig. A3 and we obtain:

$$W(C) = W(C_1)\exp\left\{-(|S|-|S_1|)\sigma\left[1+O\left(\frac{\sigma a a'}{4!}\right)\right]\right\}. \tag{A.19}$$

The procedure can easily be repeated by inscribing a rectangle S_2 with boundary $C_2 = \partial S_2$ in S_1 etc. The final result of this iterative procedure obviously is:

$$W(C) = \exp\left\{-|S|\sigma\left[1+O\left(\frac{\sigma a a'}{4!}\right)\right]\right\}. \tag{A.20}$$

From this we deduce for the "true" string tension in the SVM:

$$\sigma_{\text{true}} = \sigma \left[1 + O\left(\frac{\sigma a a'}{4!} \right) \right] = \sigma \left[1 + \text{ terms of order } 5\% \right], \tag{A.21}$$

where σ is given in (2.83). Thus we have justified the result for the string tension in Sect. 2.5, where we used the second cumulant only. We have now relied on Assumption 1' and avoided transportation of field strengths to far away reference points Y. The fourth cumulant is estimated to give only a correction of a few percent and the contribution of the 6th, 8th, etc. cumulants can be estimated in an analogous way to be even smaller[1]. Most important, we have avoided the unpleasant result (A.12).

We will now discuss another question which can be raised in connection with the calculation of $W(C)$ in the SVM. Why should we span a minimal surface S into the loop C and not use some other, more complicated surface S' with the same boundary, $\partial S' = C$? From the point of view of applying the non-abelian Stokes theorem (2.37) any wiggly surface S' would be as good as the flat rectangle S which is the surface of minimal area. However, from the point of view of the iterative calculation with the plaquettes, as explained in this appendix, an arbitrary surface S' is clearly not equivalent to S. In (A.15) we made the approximation of replacing $V(P_j)$ $(j = 1, ..., N_1)$ by its vacuum expectation value. This should be a good approximation if the various plaquettes are well separated. Indeed, we would expect that then we can perform the functional integral over the variables related to the regions in Euclidean space time where the various plaquettes are located in a separate and independent way. If, however, two plaquettes overlap or are very near to each other, the above approximation will break down. The point is now that on the minimal surface S neighbouring plaquettes will be at maximal distance from each other. For some arbitrary surface S' with wiggles we will inevitably find plaquettes closer together which will make our approximation worse. This gives us some rationale to use a minimal surface S in the applications of the non-abelian Stokes theorem in the framework of the SVM.

So far our calculations should apply to QCD with dynamical gluons and static quarks only. The quark-antiquark potential $V(R)$ in (2.84) rises linearly for $R \to \infty$. In real life we have, of course, dynamic quarks. If we separate a heavy quark-antiquark pair $Q\bar{Q}$ starting from small R we will first see a linear potential as in (2.84) but at some point the heavy quark Q (antiquark $\bar{Q}$) will pick up a light antiquark $\bar{q}$ (quark q) from the vacuum and form a meson $Q\bar{q}$ $(\bar{Q}q)$. The two mesons can escape to infinity, i.e. the force between them vanishes, the potential $V(R)$ must go to a constant as $R \to \infty$:

$$V(R) \to V_\infty \quad \text{for} \quad R \to \infty. \tag{A.22}$$

[1] Numerical studies suggest that $a' \approx 3a$ should be large enough for obtaining the area law for the plaquettes P_j in (A.16) (H. G. Dosch, private communication). We obtain then $\sigma a a' \simeq 1.7$ and still correction terms $\lesssim 10$ % in (A.21).

Can we understand also this feature of nature in an extension of the SVM?

Let us go back to the approximation (A.15), where we have replaced the integral over the plaquette P_j by its vacuum expectation value. More generally we can argue that $V(P_j)$ as defined in (A.14) is an object transforming under a gauge transformation (2.17) as follows (cf. (2.20)):

$$V(P_j) \to U(Y_j)V(P_j)U^{-1}(Y_j). \tag{A.23}$$

Any approximation we make should respect this gauge property and indeed the replacement (A.15) does. Now we can ask for a generalisation of (A.15) in the presence of dynamic light quarks. We have then the quark and antiquark variables at the point Y_j at our disposal and can construct from them the object

$$q(Y_j)\bar{q}(Y_j) \tag{A.24}$$

which has the gauge transformation property (A.23) and is also rotationally invariant. Let us consider only u and d quarks. Then we suggest as generalization of (A.15) the following replacement:

$$V_{AB}(P_j) \to \langle V(P_j)\rangle_0 \delta_{AB} - w(P_j)q_{A,f,\alpha}(Y_j)\bar{q}_{B,f,\alpha}(Y_j), \tag{A.25}$$

where A, B are the colour indices, $f = u, d$ is the flavour index and α the Dirac index. Furthermore, $\langle V(P_j)\rangle_0$ is as in (A.15), (A.16), and $w(P_j)$ is a coefficient depending on the size of the plaquette P_j. It can be thought of as representing the chance of producing a $q\bar{q}$ pair from the vacuum gluon fields over the plaquette P_j. (This is inspired by the discussions of particle production in the LUND model [54]). On dimensional grounds $w(P_j)$ must be proportional to a volume, thus we will set

$$w(P_j) = c|P_j| \cdot a, \tag{A.26}$$

where c is a constant. The idea is that $q\bar{q}$ "production" should feel the gluon fields in a disc of area $|P_j|$ and thickness a.

We can now insert the ansatz (A.25) in (A.13). The resulting expression for $W(C)$ is of the form:

$$\begin{aligned} W(C) \cong \frac{1}{3}\mathrm{Tr}\langle V_{0,N_1}[\langle V(P_{N_1})\rangle - w(P_{N_1})q(Y_{N_1})\bar{q}(Y_{N_1})]V_{N_1,N_1-1} \\ ...V_{2,1}[\langle V(P_1)\rangle - w(P_1)q(Y_1)\bar{q}(Y_1)] \cdot V_{1,0}\rangle. \end{aligned} \tag{A.27}$$

If we mutliply out these brackets we get terms where we have again the Wegner-Wilson loop along C_1 (Fig. A3) but then also terms where quarks and antiquarks at different points Y_k, Y_l are connected. For two neighbouring points, for instance, $k = l + 1, l$ this would read:

$$...q(Y_{l+1})[-w(l)\bar{q}(Y_{l+1})V_{l+1,l}q(Y_l)]\bar{q}(Y_l)...$$

The importance of these terms will increase with increasing $w(P_j)$. Starting from (A.27) we can now inscribe plaquettes into the rectangle S_1 and transport everything, including the quark variables $q(Y_k)\bar{q}(Y_k)(k = 1, ..., N_1)$ to a curve C_2 etc.

To get an orientation we will now make a drastic approximation: For a loop with $R \ll R_c$, where R_c is some critical value to be determined, we neglect the dynamical quarks, saying that the $w(P_j)$ factors will still be too small. The result for $W(C)$ and $V(R)$ in the range $a \lesssim R \ll R_c$ will then be as in (A.20):

$$\begin{aligned} &W(C) = \exp(-RT\sigma), \\ &V(R) = \sigma R, \\ &(a \lesssim R \ll R_c), \end{aligned} \tag{A.28}$$

where we set $\sigma_{\text{true}} \cong \sigma$.

In the other limiting case, $R \gg R_c$, we can expect the $q\bar{q}$ terms in (A.27) to dominate. Indeed, let us start with the loop C with sides R and T of Fig. A3 and inscribe first plaquettes $P_{1,...}, P_{N_1}$ of size a'^2. With the replacement (A.25) this gives for $W(C)$ the expression (A.27). Now we inscribe plaquettes of size a'^2 in the loop C_1 and parallel transport everything to a smaller loop C_2. The new element is that we will now have to parallel-transport also the quark variables, but this poses no problem. We will again obtain an expression like the one shown in (A.27) but now for the curve C_2 and with increased weight factors w in front of the $q\bar{q}$ terms. We assume that we continue this procedure. Finally, we will obtain an expression like (A.27) but with the terms involving the quark variables at the reference points Y_j dominating the expression. Thus we set:

$$W(C) \cong \frac{1}{3}\text{Tr}\langle[-w(N)q(N)\bar{q}(N)]V_{N,N-1}...[-w(1)q(1)\bar{q}(1)]V_{1,N}\rangle, \tag{A.29}$$

where the points $1, ..., N$ are on a curve C' of sides R', T'. This summing up of quark contributions from various plaquettes of size a'^2 should be reasonable as long as the quarks are inside an area corresponding to their own correlation length, a_χ, the chiral correlation length. This can be estimated from the behaviour of the non-local $\bar{q}q$ condensate [86-88] for which one typically makes an ansatz of the form $(q = u, d)$:

$$\langle\bar{q}(X_2)V_{2,1}\, q(X_1)\rangle = \langle\bar{q}q\rangle e^{-|X_2-X_1|^2/a_\chi^2}. \tag{A.30}$$

Here $\langle\bar{q}q\rangle$ is the local quark condensate for which we set, neglecting isospin breaking (cf. [39]):

$$\langle\bar{q}q\rangle = \frac{1}{2}\langle\bar{u}u + \bar{d}d\rangle = -(0.23 \text{ GeV})^3. \tag{A.31}$$

For the correlation length a_χ one estimates (cf. [88]):

$$a_\chi \cong 1 \text{ fm}. \tag{A.32}$$

It will be reasonable to choose the reference points $1, \ldots, N$ on C' also such that the distance between neighbouring points is a_χ. Then

$$\begin{aligned} R' &= R - 2a_\chi, \\ T' &= T - 2a_\chi, \\ N &= \frac{2T' + 2R'}{a_\chi} = \frac{2(T+R)}{a_\chi} - 8. \end{aligned} \tag{A.33}$$

We will now estimate $W(C)$ from (A.29) as product of the expectation values of the nonlocal $\bar{q}q$ condensate. We set

$$w(1) = w(2) \ldots = w(N) = c \cdot a a_\chi^2, \tag{A.34}$$

where c should be a numerical constant of order 1. Furthermore we have from (A.30)

$$\begin{aligned} &\langle \bar{q}_{A,f,\alpha}(j)(V_{j,j-1})_{AB}\, q_{B,f',\alpha'}(j-1)\rangle \\ &= \frac{1}{4}\delta_{f,f'}\delta_{\alpha,\alpha'}\langle \bar{q}q\rangle \exp\left[-\frac{|X_j - X_{j-1}|^2}{a_\chi^2}\right] \\ &= \frac{1}{4e}\delta_{f,f'}\delta_{\alpha,\alpha'}\langle \bar{q}q\rangle \end{aligned} \tag{A.35}$$

for $|X_j - X_{j-1}| = a_\chi$. This gives:

$$\begin{aligned} W(C) &\cong -\frac{1}{3} \cdot \{[-w(N)\langle \bar{q}_{f_1,\alpha_1}(1)V_{1,N}\, q_{f_N,\alpha_N}(N)\rangle] \\ &\quad \ldots [-w(1)\langle \bar{q}_{f_2,\alpha_2}(2)V_{2,1}\, q_{f_1,\alpha_1}(1)\rangle]\} \\ &= -\frac{8}{3}\left[-\frac{w(1)\langle \bar{q}q\rangle}{4e}\right]^N \\ &= -\frac{8}{3}\left[-\frac{w(1)\langle \bar{q}q\rangle}{4e}\right]^{[\frac{2}{a_\chi}(T+R)-8]}. \end{aligned} \tag{A.36}$$

For the potential this leads to

$$V(R) = -\lim_{T\to\infty}\frac{1}{T}\ln W(C) = \frac{2}{a_\chi}\ln\frac{4e}{[-w(1)\langle \bar{q}q\rangle]} \equiv V_\infty. \tag{A.37}$$

Thus we find indeed a constant potential for $R \to \infty$.

For intermediate values of R the potential should change smoothly from the linear rise to the constant behaviour. As a crude approximation we will assume the potential to be continuous but turn abruptly from one to the other behaviour at a critical length $R = R_c$ which we define as

$$R_c = V_\infty/\sigma. \tag{A.38}$$

Our simple ansatz for $V(R)$ is then as follows (cf. Fig. A4):

$$V(R) = \begin{cases} \sigma R & \text{for } R \le R_c, \\ V_\infty = \sigma R_c & \text{for } R \ge R_c. \end{cases} \tag{A.39}$$

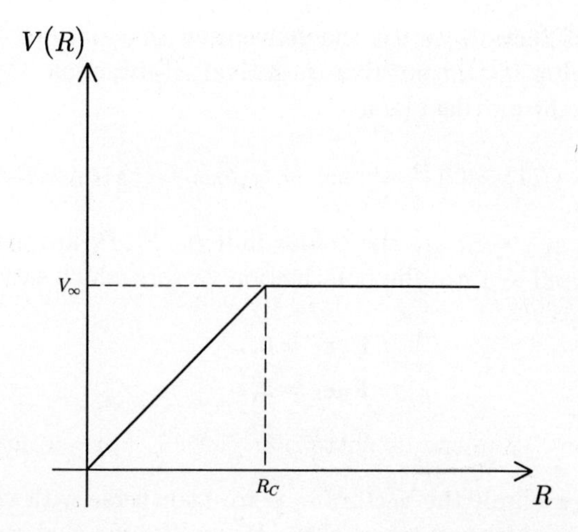

Fig. A4. *The potential* $V(R)$ *as defined in (A.39) with* R_c *the "string breaking" radius.*

How does the numerics work out taking (A.37) and (A.38) seriously? We can estimate the constant V_∞ crudely from the mass difference of two B-mesons and the Υ resonance

$$V_\infty \cong 2m_B - m_\Upsilon \cong 1.1 \text{ GeV}. \tag{A.40}$$

With σ from (2.85) we find for R_c (A.38):

$$R_c \cong 1,2 \text{ fm}. \tag{A.41}$$

A much more elaborate estimate of the string-breaking distance gives [89] $R_c \leq$ 1.6 to 2.1fm. The lattice calculations of [90] suggest an even larger value for R_c. From (A.40) and (A.37), (A.32) we get for $w(1)$ and c:

$$w(1) = -\frac{4\exp[1 - \frac{1}{2}a_\chi V_\infty]}{\langle \bar{q}q \rangle} = 0.42 \text{ fm}^3,$$

$$c = \frac{w(1)}{a_\chi^2 a} = 1.2. \tag{A.42}$$

Thus the probability factor $w(1)$ comes out as we estimated it on geometrical grounds in (A.26).

With these remarks we close this appendix. Of course, much more work is needed in order to decide if our simple ansatz for incorporating dynmical fermions in the stochastic vacuum model is viable or not.

Appendix B: The Scattering of Gluons

In this appendix we will discuss the contribution of gluons to the scattering amplitude of a general parton reaction, generalizing (3.50):

$$G(1)+G(2)+...+q(k)+\bar{q}(k')+... \to G(3)+G(4)+...+q(l)+\bar{q}(l')+.... \tag{B.1}$$

Here, as in all of Sect. 3, we use the convention that partons with odd (even) number are moving fast in positive (negative) x^3-direction. Consider thus the transition of a right-moving gluon:

$$G(1) \equiv G(P_1, \epsilon_1, a_1) \to G(3) \equiv G(P_3, \epsilon_3, a_3), \tag{B.2}$$

where $a_{1,3}$ $(1 \leq a_{1,3} \leq 8)$ are the colour indices, P_1, P_3 are the momenta with $P_{1+}, P_{3+} \to \infty$, and $\epsilon_{1,3}$ are the polarization vectors which satisfy:

$$\begin{aligned} \mathbf{P}_1 \epsilon_1 &= 0, \\ \mathbf{P}_3 \epsilon_3 &= 0, \\ |\epsilon_1| = |\epsilon_3| &= 1. \end{aligned} \tag{B.3}$$

In the high energy limit the vectors $\epsilon_{1,3}$ are transverse with corrections of order $|\mathbf{p}_{T1,3}|/P_{+1,3}$ which we can neglect. We will argue that such a gluon in a high energy soft reaction is equivalent to a quark-antiquark pair with the same quantum numbers in the limit of the q and $\bar{q}$ being so close to each other in position space that their separation cannot be resolved in the collision. From the contribution of such a $q\bar{q}$ pair to the scattering amplitude we will get in the above limit the gluon contribution.

We start by constructing wave packets similar to the mesonic wave packets of (3.66)

$$\begin{aligned} &|q\bar{q}; P_j, \epsilon_j, a_j\rangle = \int \mathrm{d}^2 p_T \int_0^1 \mathrm{d}\zeta \frac{1}{(2\pi)^{3/2}} h^j(\zeta, |\mathbf{p}_T|) \cdot \\ &\frac{1}{2}(\lambda_{a_j})_{A_j A'_j} (\epsilon_j \cdot \sigma\epsilon)_{s_j s'_j} |q(p_j, s_j, A_j) \bar{q}(P_j - p_j, s'_j, A'_j)\rangle \\ &(j = 1, 2, 3, 4), \end{aligned} \tag{B.4}$$

where p_j is as in (3.67), (3.68) and

$$\epsilon = \begin{pmatrix} 0 & 1 \\ -1 & 0 \end{pmatrix}. \tag{B.5}$$

We require the functions h^j to satisfy:

$$\begin{aligned} &(h^j(\zeta, |\mathbf{p}_T|))^* = h^j(\zeta, |\mathbf{p}_T|), \\ &h^j(\zeta, |\mathbf{p}_T|) = h^j(1 - \zeta, |\mathbf{p}_T|), \\ &\int \mathrm{d}^2 p_T \int_0^1 \mathrm{d}\zeta \; 2\zeta(1 - \zeta) |h^j(\zeta, |\mathbf{p}_T|)|^2 = 1. \end{aligned} \tag{B.6}$$

This gives us the normalization of the states (B.4) as

$$\langle q\bar{q}; P'_j, \epsilon'_j, a'_j | q\bar{q}, P_j, \epsilon_j, a_j\rangle = \epsilon'^*_j \cdot \epsilon_j \delta_{a'_j, a_j} (2\pi)^3 2P^0_j \delta^3(\mathbf{P}'_j - \mathbf{P}_j). \tag{B.7}$$

The $q\bar{q}$ states (B.4) have the same transformation properties under a parity (P) charge conjugation (C) and time reversal (T) transformation as a gluon state:

$$U(\mathrm{P})|q\bar{q}; P^0, \mathbf{P}, \epsilon, a\rangle = -|q\bar{q}; P^0, -\mathbf{P}, \epsilon, a\rangle, \tag{B.8}$$

$$U(\mathrm{C})|q\bar{q}; P^0, \mathbf{P}, \epsilon, a\rangle = -|q\bar{q}; P^0, \mathbf{P}, \epsilon, b\rangle \cdot \frac{1}{2}\mathrm{Tr}(\lambda_b \lambda_a^T), \tag{B.9}$$

$$V(\mathrm{T})|q\bar{q}; P^0, \mathbf{P}, \epsilon, a\rangle = -|q\bar{q}; P^0, -\mathbf{P}, \epsilon^*, b\rangle \frac{1}{2}\mathrm{Tr}(\lambda_b \lambda_a^T). \tag{B.10}$$

Here $U(\mathrm{P}), U(\mathrm{C})$ are the unitary operators, $V(\mathrm{T})$ is the antiunitary operator representing the P, C and T transformations, respectively, in the Fock space of parton states.

As in (3.71) we define the wave functions in transverse position space and longitudinal momentum fraction:

$$\varphi^j(\zeta, \mathbf{y}_T) := \sqrt{2\zeta(1-\zeta)}\frac{1}{2\pi}\int \mathrm{d}^2 p_T \exp(i\mathbf{p}_T \cdot \mathbf{y}_T) h^j(\zeta, |\mathbf{p}_T|). \tag{B.11}$$

Here we also define for the right (left) movers the wave functions in $y_-, \mathbf{y}_T$ space ($y_+, \mathbf{y}_T$ space) as:

$$\tilde{\varphi}^j(y_-, \mathbf{y}_T) := \frac{1}{\sqrt{4\pi}}\sqrt{P_{j+}}\int_0^1 \mathrm{d}\zeta\, \varphi^j(\zeta, \mathbf{y}_T) \exp[-\frac{i}{2}P_{j+}(\zeta - \frac{1}{2})y_-],$$
$$(j \text{ odd}), \tag{B.12}$$

$$\tilde{\varphi}^j(y_+, \mathbf{y}_T) := \frac{1}{\sqrt{4\pi}}\sqrt{P_{j-}}\int_0^1 \mathrm{d}\zeta\, \varphi^j(\zeta, \mathbf{y}_T) \exp[-\frac{i}{2}P_{j-}(\zeta - \frac{1}{2})y_+],$$
$$(j \text{ even}), \tag{B.13}$$

The normalization condition (B.6) implies:

$$\int \mathrm{d}^2 y_T \int_0^1 \mathrm{d}\zeta |\varphi^j(\zeta, \mathbf{y}_T)|^2 = 1,$$
$$\int \mathrm{d}y_- \int \mathrm{d}^2 y_T |\tilde{\varphi}^j(y_-, \mathbf{y}_T)|^2 = 1 \qquad (j \text{ odd}).$$
$$\int \mathrm{d}y_+ \int \mathrm{d}^2 y_T |\tilde{\varphi}^j(y_+, \mathbf{y}_T)|^2 = 1 \qquad (j \text{ even}). \tag{B.14}$$

To realize the condition that the $q\bar{q}$ pair acts like a gluon, we require for right (left) movers that they have similar longitudinal momenta and that their wave function in the relative q-$\bar{q}$ coordinates, $y_-, \mathbf{y}_T(y_+, \mathbf{y}_T)$ is nearly a δ function. To be concrete, we require:

$$\varphi^j(\zeta, \mathbf{y}_T) \neq 0 \quad \text{only for} \quad |\zeta - \frac{1}{2}| \leq \xi_0, \tag{B.15}$$

where

$$0 < \xi_0 \ll \frac{1}{2}, \tag{B.16}$$

and for j odd:

$$\tilde{\varphi}^j(y_-, \mathbf{y}_T) \neq 0 \quad \text{only if} \quad |y_-| \lesssim \frac{1}{P_+\xi_0} \quad \text{and} \quad |\mathbf{y}_T| \ll a. \tag{B.17}$$

For right movers we replace plus by minus signs in (B.17). Any $q\bar{q}$ wave packets with these properties should then look identical to a gluon for an observer in the femto universe with regard to "soft" scatterings.

Now we replace $G(1)$ and $G(3)$ in reaction (B.1) by the $q\bar{q}$ wave packets (B.4). According to the rules derived in Sect. 3.5 the scattering of the $q\bar{q}$ system

$$q(1)\bar{q}(1') \to q(3)\bar{q}(3') \tag{B.18}$$

gives the following factor in the S-matrix (cf. (3.58) (3.59)):

$$\mathcal{S}_{q+}(3,1)\mathcal{S}_{\bar{q}+}(3',1')$$

which still has to be integrated over the wave functions (B.4). In this way we obtain for the $q\bar{q}$-pair:

$$\begin{aligned}\mathcal{S}_{q\bar{q}}(3,1) = &\int d^2p'_T \int_0^1 d\zeta' \int d^2p_T \int_0^1 d\zeta \, \frac{1}{(2\pi)^3} h^3(\zeta', |\mathbf{p}'_T|) h^1(\zeta, |\mathbf{p}_T|)\\ &\frac{1}{2}(\lambda_{a_3})_{A'_3,A_3}(\epsilon_3{}^* \cdot \epsilon^T \boldsymbol{\sigma})_{s'_3,s_3} \, \frac{1}{2}(\lambda_{a_1})_{A_1,A'_1}(\epsilon_1 \cdot \boldsymbol{\sigma}\epsilon)_{s_1,s'_1}\\ &\mathcal{S}_{q+}(3,1) \, \mathcal{S}_{\bar{q}+}(3',1'),\end{aligned} \tag{B.19}$$

$$\begin{aligned}\mathcal{S}_{q\bar{q}}(3,1) = &\int dz_- \int dy_- \int d^2z_T \int d^2y_T \sqrt{P_{3+}P_{1+}} \epsilon_3^* \cdot \epsilon_1\\ &\exp[\frac{i}{2}(P_{3+} - P_{1+})z_- - i(\mathbf{P}_3 - \mathbf{P}_1)_T \mathbf{z}_T] \; \tilde{\varphi}^{3*}(y_-, \mathbf{y}_T)\tilde{\varphi}^1(y_-, \mathbf{y}_T)\\ &\frac{1}{2}\mathrm{Tr}[\lambda_{a_3} V_-(\infty, z_- + \frac{1}{2}y_-, \mathbf{z}_T + \frac{1}{2}\mathbf{y}_T)\\ &\lambda_{a_1} V_-^\dagger(\infty, z_- - \frac{1}{2}y_-, \mathbf{z}_T - \frac{1}{2}\mathbf{y}_T)].\end{aligned} \tag{B.20}$$

With our assumptions (B.17) $\tilde{\varphi}^{3*}(y_-, \mathbf{y}_T)\tilde{\varphi}^1(y_-, \mathbf{y}_T)$ acts like a δ-function at $y_- = 0$, $\mathbf{y}_T = 0$ and we get:

$$\begin{aligned}\mathcal{S}_{q\bar{q}}(3,1) = &\sqrt{P_{3+}P_{1+}} \; \epsilon_3^* \cdot \epsilon_1\\ &\int dy_- \int d^2y_T \; \tilde{\varphi}^{3*}(y_-, \mathbf{y}_T)\tilde{\varphi}^1(y_-, \mathbf{y}_T)\\ &\int dz_- \int d^2z_T \; \exp[\frac{i}{2}(P_{3+} - P_{1+})z_- - i(\mathbf{P}_3 - \mathbf{P}_1)_T \cdot \mathbf{z}_T]\\ &\frac{1}{2}\mathrm{Tr}[\lambda_{a_3} V_-(\infty, z_-, \mathbf{z}_T)\lambda_{a_1} V_-^\dagger(\infty, z_-, \mathbf{z}_T)].\end{aligned} \tag{B.21}$$

An easy exercise shows that

$$\frac{1}{2}\mathrm{Tr}[\lambda_{a_3}V_-(\infty,z_-,\mathbf{z}_T)\lambda_{a_1}V_-^\dagger(\infty,z_-,\mathbf{z}_T)] = \mathcal{V}_-(\infty,z_-,\mathbf{z}_T)_{a_3,a_1}, \quad \text{(B.22)}$$

where $\mathcal{V}_-$ is the connector analogous to (3.35) but for the adjoint representation (cf. (2.22)):

$$\mathcal{V}_-(\infty,z_-,\mathbf{z}_T) = \mathrm{P}\{\exp\big[-\frac{i}{2}g\int_{-\infty}^{\infty} dz_+ G_-^a(z_+,z_-,\mathbf{z}_T)T_a\big]\},$$
$$(T_a)_{bc} = \frac{1}{i}f_{abc}. \quad \text{(B.23)}$$

The f_{abc} are the structure constants of $SU(3)_c$. Inserting (B.22) in (B.21) gives

$$\mathcal{S}_{q\bar{q}}(3,1) = \int dy_- \int d^2y_T\ \tilde{\varphi}^{3*}(y_-,\mathbf{y}_T)\tilde{\varphi}^1(y_-,\mathbf{y}_T)\epsilon^*_{3j_3}\epsilon_{1j_1}\mathcal{S}_{G+}(3,1), \quad \text{(B.24)}$$

where

$$\mathcal{S}_{G+}(3,1) = \sqrt{P_{3+}P_{1+}}\ \delta_{j_3,j_1}\int dz_- \int d^2z_T$$
$$\exp\big[\frac{i}{2}(P_{3+}-P_{1+})z_- - i(\mathbf{P}_{3T}-\mathbf{P}_{1T})\cdot\mathbf{z}_T\big]$$
$$\mathcal{V}_-(\infty,z_-,\mathbf{z}_T)_{a_3,a_1}. \quad \text{(B.25)}$$

In a general scattering reaction (B.1) the factor $\mathcal{S}_{q\bar{q}}(3,1)$ (B.24) has to be inserted with other factors $\mathcal{S}_q, \mathcal{S}_{\bar{q}}, \ldots$ and then integrated over all gluon potentials as explained in Sect. 3.5. We note that $\mathcal{S}_{q\bar{q}}(3,1)$ in (B.24) factorizes into $\mathcal{S}_{G+}(3,1)$ (B.25) times the overlap of the internal wave functions of the incoming and outgoing $q\bar{q}$ pairs:

$$\int dy_- \int d^2y_T\ \tilde{\varphi}^{3*}(y_-,\mathbf{y}_T)\tilde{\varphi}^1(y_-,\mathbf{y}_T). \quad \text{(B.26)}$$

This means that the $q\bar{q}$ pair will come out with some distribution in total momentum P_3 and polarization vector $\epsilon_3 = \epsilon_1$ but always with internal wave function $\tilde{\varphi}^1(y_-,\mathbf{y}_T)$. With the conditions (B.15)-(B.17) $\tilde{\varphi}^1(y_-,\mathbf{y}_T)$ leads to a "permissible" internal wave function for a $q\bar{q}$ pair of momentum P_3, to be regarded as a gluon of momentum P_3 by our observer in the femto universe. Indeed we have from (B.12):

$$\tilde{\varphi}^1(y_-,\mathbf{y}_T) = \sqrt{\frac{P_{1+}}{4\pi}}\int_0^1 d\zeta\ \varphi^1(\zeta,\mathbf{y}_T)\exp[-\frac{i}{2}P_{1+}(\zeta-\frac{1}{2})y_-]$$
$$= \sqrt{\frac{P_{1+}}{4\pi}}\int_{-1/2}^{1/2} d\xi\ \varphi^1(\frac{1}{2}+\xi,\mathbf{y}_T)\exp[-\frac{i}{2}P_{1+}\xi\cdot y_-]$$
$$= \sqrt{\frac{P_{3+}}{4\pi}}\int_{\xi_-}^{\xi_+} d\xi'\ \varphi'^3(\frac{1}{2}+\xi',\mathbf{y}_T)\exp[-\frac{i}{2}P_{3+}\xi'\cdot y_-] \quad \text{(B.27)}$$

Here we define the internal wave function of the outgoing $q\bar{q}$ pair as

$$\varphi'^3\left(\frac{1}{2}+\xi', \mathbf{y}_T\right) = \sqrt{\frac{P_{3+}}{P_{1+}}}\varphi^1\left(\frac{1}{2}+\frac{P_{3+}}{P_{1+}}\xi', \mathbf{y}_T\right) \tag{B.28}$$

$$\text{and } \xi_\pm = \pm\frac{P_{1+}}{P_{3+}}\cdot\frac{1}{2}. \tag{B.29}$$

The condition (B.15) for φ^1 guarantees a similar condition for φ'^3 if P_{3+}/P_{1+} is of order 1, as we will always assume. Thus the integration limits $\xi_\pm$ in (B.27) can be replaced by $\pm\frac{1}{2}$ and also the normalization conditions (B.14) can easily be checked for φ'^3.

To summarize: In this appendix we have shown that suitable $q\bar{q}$ pairs which are indistinguishable from gluons for our observer in the femto universe scatter as entities in a soft reaction. Their internal wave function in momentum space is modified, but in a way not observable in a soft reaction. Thus we can consider $\mathcal{S}_{G+}(3,1)$ in (B.25) as the scattering contribution of a right-moving gluon in the transition (B.2). We quoted this result already in (3.64) in Sect. 3.5. For left moving gluons we just have to exchange everywhere + and − components.

Appendix C: The Scattering of Baryons

In this appendix we discuss high energy soft reactions, in particular elastic reactions involving baryons and antibaryons. We represent baryons by qqq wave packets:

$$\begin{aligned}|B_j(P_j)\rangle = \frac{1}{6\cdot(2\pi)^3}\int d\mu\; h^j(f^i, s^i, \zeta^i, \mathbf{p}_T^i)\,\cdot \\ \epsilon_{A^1A^2A^3}|q(p_j^1, f^1, s^1, A^1), q(p_j^2, f^2, s^2, A^2), q(p_j^3, f^3, s^3, A^3)\rangle.\end{aligned} \tag{C.1}$$

Here f^i, s^i, A^i $(i = 1,2,3)$ are the flavour, spin, and colour indices of the quarks and ϵ_{ABC} $(\epsilon_{123} = 1)$ is the totally antisymmetric tensor. For the momenta p_j^i we set for right-moving baryons (j odd, $P_{j+}\to\infty$):

$$\begin{aligned}p_{j+}^i &= \zeta^i P_{j+}, \\ \mathbf{p}_{jT}^i &= \frac{1}{3}\mathbf{P}_{jT} + \mathbf{p}_T^i, \\ (i &= 1,2,3)\end{aligned} \tag{C.2}$$

and for left-moving baryons (j even, $P_{j-}\to\infty$):

$$\begin{aligned}p_{j-}^i &= \zeta^i P_{j-}, \\ \mathbf{p}_{jT}^i &= \frac{1}{3}\mathbf{P}_{jT} + \mathbf{p}_T^i, \\ (i &= 1,2,3).\end{aligned} \tag{C.3}$$

The integral with the measure $\mathrm{d}\mu$ stands for

$$\int \mathrm{d}\mu \equiv \int\int\int \prod_{i=1}^{3} \mathrm{d}^2 p_T^i \delta^2(\sum_{i=1}^{3} \mathbf{p}_T^i) \cdot \int_0^1 \int_0^1 \int_0^1 \prod_{i=1}^{3} \mathrm{d}\zeta^i \delta(1 - \sum_{i=1}^{3} \zeta^i). \quad \text{(C.4)}$$

The flavour and spin of the baryon states (C.1) is, of course, determined by the functions h^j which must be totally symmetric under simultaneous exchange of the arguments $(f^i, s^i, \zeta^i, \mathbf{p}_T^i)$ for $i = 1, 2, 3$. In the following we will collectively set $(f, s, \zeta, \mathbf{p}_T) \equiv \alpha$ and $q(\alpha, A) \equiv q(p_j, f, s, A)$. We have then

$$h^j(\alpha, \beta, \gamma) = h^j(\beta, \alpha\gamma) = h^j(\alpha, \gamma, \beta). \quad \text{(C.5)}$$

$$|B_j(P_j)\rangle = \frac{1}{6.(2\pi)^3} \int \mathrm{d}\mu \; h^j(\alpha, \beta, \gamma) \epsilon_{ABC} |q(\alpha, A) q(\beta, B) q(\gamma, C)\rangle. \quad \text{(C.6)}$$

The normalization condition

$$\langle B_j(P_j') | B_j(P_j)\rangle = (2\pi)^3 2P_j^0 \delta^3(\mathbf{P}_j' - \mathbf{P}_j) \quad \text{(C.7)}$$

requires h^j to satisfy:

$$\int \mathrm{d}\mu \; 4\zeta^1\zeta^2\zeta^3 \; h^{j*}(\alpha, \beta, \gamma) h^j(\alpha, \beta, \gamma) = 1$$
$$\text{(no summation over} j). \quad \text{(C.8)}$$

We define the wave functions φ^j in transverse position and longitudinal momentum space and the transition profile functions w_{kj}^B for $B_j \to B_k$ as:

$$\varphi^j(f^i, s^i, \zeta^i, \mathbf{x}_T^i) := \frac{1}{(2\pi)^2} \int \prod_{i=1}^{3} \mathrm{d}^2 p_T^i \cdot \delta^2(\sum_{i=1}^{3} \mathbf{p}_T^i) \cdot$$
$$6 \cdot (\zeta^1\zeta^2\zeta^3)^{1/2} \exp(i \sum_{i=1}^{3} \mathbf{p}_T^i \cdot \mathbf{x}_T^i) h^j(\alpha^1, \alpha^2, \alpha^3), \quad \text{(C.9)}$$

$$w_{k,j}^B(\mathbf{x}_T^1, \mathbf{x}_T^2, \mathbf{x}_T^3) := \int_0^1 \int_0^1 \int_0^1 \prod_{i=1}^{3} \mathrm{d}\zeta^i \; \delta(1 - \sum_{i=1}^{3} \zeta^i)$$
$$\sum_{f^i, s^i} \varphi^{3*}(f^1, s^1, \zeta^1, \mathbf{x}_T^1; f^2, s^2, \zeta^2, \mathbf{x}_T^2; f^3, s^3, \zeta^3, \mathbf{x}_T^3)$$
$$\varphi^1(f^1, s^1, \zeta^1, \mathbf{x}_T^1; f^2, s^2, \zeta^2, \mathbf{x}_T^2; f^3, s^3, \zeta^3, \mathbf{x}_T^3). \quad \text{(C.10)}$$

The symmetry relations for h (C.5) and the normalization condition (C.8) imply:

$$w_{k,j}^B(\mathbf{x}_T^1, \mathbf{x}_T^2, \mathbf{x}_T^3) = w_{k,j}^B(\mathbf{x}_T^2, \mathbf{x}_T^1, \mathbf{x}_T^3) = w_{k,j}^B(\mathbf{x}_T^1, \mathbf{x}_T^3, \mathbf{x}_T^2), \quad \text{(C.11)}$$

$$\int \prod_{i=1}^{3} \mathrm{d}^2 x_T^i \; \delta^2(\mathbf{x}_T^1 + \mathbf{x}_T^2 + \mathbf{x}_T^3) w_{jj}^B(\mathbf{x}_T^1, \mathbf{x}_T^2, \mathbf{x}_T^3) = 1$$
$$\text{(no summation over } j). \quad \text{(C.12)}$$

As a concrete scattering reaction let us consider meson-baryon scattering:

$$B_1(P_1) + M_2(P_2) \to B_3(P_3) + M_4(P_4). \tag{C.13}$$

From the rules of Sect. 3.5 we get

$$\begin{aligned}
&S_{fi} \equiv \langle B_3(P_3), M_4(P_4)|S|B_1(P_1)M_2(P_2)\rangle = \\
&\frac{1}{6(2\pi)^6}\int \mathrm{d}\mu'\ h^{3*}(\alpha',\beta',\gamma')\epsilon_{A'B'C'}\int \mathrm{d}\mu\ h^1(\alpha,\beta,\gamma)\epsilon_{ABC} \\
&\frac{1}{3(2\pi)^3}\int \mathrm{d}^2p'_T\int_0^1 \mathrm{d}\zeta'\int \mathrm{d}^2p_T\int_0^1 \mathrm{d}\zeta\ h^{4*}_{s_4s'_4}(\zeta',\mathbf{p}'_T)h^2_{s_2s'_2}(\zeta,\mathbf{p}_T) \\
&\langle \mathcal{S}_{q+}(\alpha',\alpha)_{A'A}\mathcal{S}_{q+}(\beta',\beta)_{B'B}\mathcal{S}_{q+}(\gamma',\gamma)_{C'C} \\
&\delta_{A'_2A_2}\delta_{A'_4A_4}\mathcal{S}_{q-}(4,2)_{A_4A_2}\mathcal{S}_{\bar{q}-}(4',2')_{A'_4A'_2}\rangle_G.
\end{aligned} \tag{C.14}$$

After some straightforward algebra we get:

$$\begin{aligned}
&S_{fi} = \delta_{fi} + i(2\pi)^4\delta(P_3+P_4-P_1-P_2)\mathcal{T}_{fi}, \\
&\mathcal{T}_{fi} = -\frac{i}{(2\pi)^6}\int \mathrm{d}\mu'\ h^{3*}(\alpha',\beta',\gamma') \\
&\int \mathrm{d}\mu\ h^1(\alpha,\beta,\gamma)\prod_{i=1}^3(\delta_{f'^i,f^i}\delta_{s'^i,s^i})(P_{1+})^3\left[\prod_{i=1}^3\zeta'^i\zeta^i\right]^{1/2} \\
&\frac{1}{(2\pi)^3}\int \mathrm{d}^2p'_T\int_0^1 \mathrm{d}\zeta'\int \mathrm{d}^2p_T\int_0^1 \mathrm{d}\zeta\ h^{4*}_{s,r}(\zeta',\mathbf{p}'_T)h^2_{s,r}(\zeta,\mathbf{p}_T) \\
&2(P_{2-})^2\left[\zeta'(1-\zeta')\zeta(1-\zeta)\right]^{1/2} \\
&3^3\cdot\int \mathrm{d}^2b_T\exp(i\mathbf{q}_T\cdot\mathbf{b}_T)\int\prod_{i=1}^3(\mathrm{d}x^i_-\mathrm{d}^2x^i_T)\delta\left(\sum_{i=1}^3 x^i_-\right)\ \delta^2\left(\sum_{i=1}^3\mathbf{x}^i_T\right) \\
&\int \mathrm{d}y_+\mathrm{d}^2y_T\exp\Big\{i\sum_{i=1}^3\left[\frac{1}{2}P_{1+}(\zeta'^i-\zeta^i)x^i_- - (\mathbf{p}'^i_T-\mathbf{p}^i_T)\cdot\mathbf{x}^i_T\right] \\
&+i\frac{1}{2}P_{2-}(\zeta'-\zeta)y_+ - i(\mathbf{p}'_T-\mathbf{p}_T)\cdot\mathbf{y}_T\Big\} \\
&\langle\{V_-(\infty,x^1_-,\frac{1}{2}\mathbf{b}_T+\mathbf{x}^1_T)_{A'A}V_-(\infty,x^2_-,\frac{1}{2}\mathbf{b}_T+\mathbf{x}^2_T)_{B'B} \\
&V_-(\infty,x^3_-,\frac{1}{2}\mathbf{b}_T+\mathbf{x}^3_T)_{C'C}\ \frac{1}{6}\varepsilon_{A'B'C'}\ \varepsilon_{ABC} \\
&\frac{1}{3}\mathrm{Tr}\big[V_+(y_+,\infty,-\frac{1}{2}\mathbf{b}_T+\frac{1}{2}\mathbf{y}_T)V_+^\dagger(0,\infty,-\frac{1}{2}\mathbf{b}_T-\frac{1}{2}\mathbf{y}_T)\big]-1\}\rangle_G,
\end{aligned} \tag{C.15}$$

where

$$\mathbf{q}_T = (\mathbf{P}_1 - \mathbf{P}_3)_T. \tag{C.16}$$

Now we make the transformation of variables

$$x_-^i \to \frac{2}{P_{1+}} x_-^i,$$
$$y_+ \to \frac{2}{P_{2-}} y_+ \tag{C.17}$$

and use $P_{1+} \to \infty$, $P_{2-} \to \infty$. With the same arguments which led us from (3.77) to (3.79) we get

$$\begin{aligned}
\mathcal{T}_{fi} = &-2is \int \mathrm{d}^2 b_T \exp(i\mathbf{q}_T \cdot \mathbf{b}_T) \int \prod_{i=1}^{3} \mathrm{d}^2 x_T^i \, \delta^2(\mathbf{x}_T^1 + \mathbf{x}_T^2 + \mathbf{x}_T^3) \\
&w_{3,1}^B(\mathbf{x}_T^1, \mathbf{x}_T^2, \mathbf{x}_T^3) \int \mathrm{d}^2 y_T w_{4,2}^M(\mathbf{y}_T) \\
&\langle \{ V_-(\infty, 0, \tfrac{1}{2}\mathbf{b}_T + \mathbf{x}_T^1)_{A'A} V_-(\infty, 0, \tfrac{1}{2}\mathbf{b}_T + \mathbf{x}_T^2)_{B'B} \\
&V_-(\infty, 0, \tfrac{1}{2}\mathbf{b}_T + \mathbf{x}_T^3)_{C'C} \tfrac{1}{6} \epsilon_{A'B'C'} \epsilon_{ABC} \\
&\tfrac{1}{3} \mathrm{Tr}\big[V_+(0, \infty, -\tfrac{1}{2}\mathbf{b}_T + \tfrac{1}{2}\mathbf{y}_T) V_+^\dagger(0, \infty, -\tfrac{1}{2}\mathbf{b}_T - \tfrac{1}{2}\mathbf{y}_T \big] - 1 \} \rangle_G .
\end{aligned} \tag{C.18}$$

Here $w_{4,2}^M(y_T)$ is the transition profile function for the mesons as defined in (3.72) and $w_{3,1}^B$ is the corresponding function for the baryons (cf. (C.10)).

In the next step we follow [59], [45] and use relations which are valid for any 3×3 matrix: $H = (H_{AB})$:

$$H_{A'A''} H_{B'B''} H_{C'C''} \cdot \epsilon_{A''B''C''} = \det H \cdot \epsilon_{A'B'C'}, \tag{C.19}$$

$$\begin{aligned}
&\det H \cdot \epsilon_{A'B'C'} \epsilon_{ABC} = \\
&H_{A'A} H_{B'B} H_{C'C} + H_{A'B} H_{B'C} H_{C'A} + H_{A'C} H_{B'A} H_{C'B} \\
&- H_{A'B} H_{B'A} H_{C'C} - H_{A'A} H_{B'C} H_{C'B} - H_{A'C} H_{B'B} H_{C'A} .
\end{aligned} \tag{C.20}$$

We take H equal to the antiquark line integral at the central point of the baryon in transverse space:

$$H = V_-^*(\infty, 0, \tfrac{1}{2}\mathbf{b}_T). \tag{C.21}$$

As a $SU(3)$-connector V_-^* satisfies

$$\det V_-^*(\infty, 0, \tfrac{1}{2}\mathbf{b}_T) = 1. \tag{C.22}$$

This is easy to prove. From (C.19) we see that $\det V_-^*$ is the connector in the singlet part of the product of three $SU(3)$ antiquark representations: $\bar{\mathbf{3}} \times \bar{\mathbf{3}} \times \bar{\mathbf{3}}$. But for the singlet representation the connector equals 1 since we have to set $T_a = 0$ in (2.22).

In the following we will use as shorthand notation

$$V(i) \equiv V_-(\infty, 0, \frac{1}{2}\mathbf{b}_T + \mathbf{x}_T^i),$$
$$(i = 1, 2, 3),$$
$$V(0) \equiv V_-(\infty, 0, \frac{1}{2}\mathbf{b}_T). \tag{C.23}$$

With this we get for the qqq-contribution to the integrand in the functional integral in (C.18) using (C.20) to (C.22):

$$\mathcal{W}_+^B(\frac{1}{2}\mathbf{b}_T, \mathbf{x}_T^1, \mathbf{x}_T^2, \mathbf{x}_T^3) := \frac{1}{6} V(1)_{A'A} V(2)_{B'B} V(3)_{C'C}\ \epsilon_{A'B'C'} \epsilon_{ABC}$$
$$= \frac{1}{6} \{ \mathrm{Tr}[V(1)V^\dagger(0)] \cdot \mathrm{Tr}[V(2)V^\dagger(0)] \cdot \mathrm{Tr}[V(3)V^\dagger(0)]$$
$$+ \mathrm{Tr}[V(1)V^\dagger(0)V(2)V^\dagger(0)V(3)V^\dagger(0)]$$
$$+ \mathrm{Tr}[V(1)V^\dagger(0)V(3)V^\dagger(0)V(2)V^\dagger(0)]$$
$$- \mathrm{Tr}[V(1)V^\dagger(0)V(2)V^\dagger(0)] \cdot \mathrm{Tr}[V(3)V^\dagger(0)]$$
$$- \mathrm{Tr}[V(2)V^\dagger(0)V(3)V^\dagger(0)] \cdot \mathrm{Tr}[V(1)V^\dagger(0)]$$
$$- \mathrm{Tr}[V(1)V^\dagger(0)V(3)V^\dagger(0)] \cdot \mathrm{Tr}[V(2)V^\dagger(0)] \} . \tag{C.24}$$

As for the meson case (cf. (3.82)) we will now add suitable connectors at $\pm\infty$. In this way all the traces in (C.24) become closed light-like Wegner-Wilson loops. To give an example: The term

$$\mathrm{Tr}[V(1)V^\dagger(0)V(2)V^\dagger(0)]$$

should then be read as the loop in the hyperplane $x_- = 0$ in the limit $T \to \infty$ which connects the following points $(x_+, x_-, \mathbf{x}_T)$:

$$(T, 0, \frac{1}{2}\mathbf{b}_T),$$
$$(-T, 0, \frac{1}{2}\mathbf{b}_T),$$
$$(-T, 0, \frac{1}{2}\mathbf{b}_T + \mathbf{x}_T^2),$$
$$(T, 0, \frac{1}{2}\mathbf{b}_T + \mathbf{x}_T^2),$$
$$(T, 0, \frac{1}{2}\mathbf{b}_T),$$
$$(-T, 0, \frac{1}{2}\mathbf{b}_T),$$
$$(-T, 0, \frac{1}{2}\mathbf{b}_T + \mathbf{x}_T^1),$$
$$(T, 0, \frac{1}{2}\mathbf{b}_T + \mathbf{x}_T^1),$$
$$(T, 0, \frac{1}{2}\mathbf{b}_T) \tag{C.25}$$

on straight lines in the order indicated.

Inserting (C.24) in (C.18) and denoting the mesonic Wegner-Wilson loop as defined in (3.83) by $\mathcal{W}_-^M$, we get finally for the T-matrix element of baryon-meson scattering:

$$\begin{aligned}\mathcal{T}_{fi} &= -2is \int \mathrm{d}^2 b_T \exp(i\mathbf{q}_T \cdot \mathbf{b}_T) \\ &\int \prod_{i=1}^{3} \mathrm{d}^2 x_T^i \; \delta^2(\mathbf{x}_T^1 + \mathbf{x}_T^2 + \mathbf{x}_T^3) w_{3,1}^B(\mathbf{x}_T^1, \mathbf{x}_T^2, \mathbf{x}_T^3) \; \int \mathrm{d}^2 y_T \; w_{4,2}^M(\mathbf{y}_T) \\ &\left\langle \mathcal{W}_+^B \left(\frac{1}{2}\mathbf{b}_T, \mathbf{x}_T^1, \mathbf{x}_T^2, \mathbf{x}_T^3\right) \mathcal{W}_-^M \left(-\frac{1}{2}\mathbf{b}_T, \mathbf{y}_T\right) - 1 \right\rangle_G . \qquad \text{(C.26)}\end{aligned}$$

This formula is the starting point for the evaluation of the baryon-meson elastic scattering amplitude: One can now apply the Minkowskian version of the SVM to calculate the functional integral $\langle \quad \rangle_G$ in an appropriate way. Then one has to fold the result with the profile functions of the mesonic and baryonic transitions. At the present state one has to make a suitable ansatz for these profile functions.

For the case of right-moving antibaryons in an elastic reaction we just have to substitute the loop factor $\mathcal{W}_+^B$ by $\mathcal{W}_+^{\bar{B}}$ which is obtained by replacing the quark connectors V_- by the antiquark connectors V_-^* and vice versa in (C.23), (C.24). In an equivalent way we can get $W_+^{\bar{B}}$ from $\mathcal{W}_+^B$ by reversing the directional arrows on all Wegner-Wilson loops obtained in the way discussed above from (C.24). For left-moving baryons and antibaryons we have to exchange + and − components.

For the further treatment of scattering amplitudes involving mesons, baryons and antibaryons, for many results and a comparison with experiments we refer to [59], [45], [62].

References

[1] H. Fritzsch, , M. Gell-Mann, H. Leutwyler: Phys. Lett. **B47**, 365 (1973); M. Gell-Mann: Acta Physica Austriaca, Suppl. 9, 733 (1972)

[2] G. 't Hooft: remarks at a conference in Marseille (1972); H. D. Politzer: Phys. Rev. Lett. **30**, 1346 (1973); D. Gross, F. Wilczek: Phys. Rev. Lett. **30**, 1343 (1973)

[3] K. G. Wilson: Phys. Rev. **D10**, 2445 (1974)

[4] P. Lepage: "Designer Field Theory-Redesigning Lattice QCD" in "Perturbative and Nonperturbative Aspects of Quantum Field Theory", eds. H. Latal, W. Schweiger, to be published by Springer, Heidelberg

[5] G. Altarelli: Phys. Rep. **81**, 1 (1982)

[6] J. L. Cardy, G. A. Winbow: Phys. Lett. **B52**, 95 (1974); C. E. DeTar, S. D. Ellis, P. V. Landshoff: Nucl. Phys. **B87**, 176 (1975)

[7] H. D. Politzer: Nucl. Phys. **B129**, 301 (1977);
C. T. Sachrajda: Phys. Lett. **B73**, 185 (1978);
D. Amati, R. Petronzio, G. Veneziano: Nucl. Phys. **B140**, 54 (1978); **B146**, 29 (1978);
R. K. Ellis, H. Georgi, M. Machacek, H. D. Politzer, G. G. Ross: Nucl. Phys. **B152**, 285 (1979);
S. B. Libby, G. Sterman: Phys. Rev. **D18**, 3252 (1978);
S. Gupta, A. H. Mueller: Phys. Rev. **D20**, 118 (1979)

[8] J. C. Collins, D. E. Soper, G. Sterman: Nucl. Phys. **B261**, 104 (1985); **B308**, 833 (1988);
G. Sterman: Phys. Lett. **B179**, 281 (1986); Nucl. Phys. **B281**, 310 (1987);
J. C. Collins, D. E. Soper: Ann. Rev. Nucl. Part. Sci. **37**, 383 (1987);
T. Matsuura, W. L. van Neerven: Z. Phys. **C38**, 623 (1988);
J. C. Collins, D. E. Soper, G. Sterman: "Factorization of Hard Processes in QCD", in "Perturbative QCD", A. H. Mueller, ed., World Scientific, Singapore 1990

[9] G. T. Bodwin, Phys. Rev. **D31**, 2616 (1985);
G. T. Bodwin, S. J. Brodsky, G. P. Lepage: Phys. Rev. **D39**, 3287 (1989)

[10] A. H. Mueller, Les Houches Lectures 1991

[11] G. Sterman; "Factorization and Resummation", in "Perturbative and Nonperturbative Aspects of Quantum Field Theory", eds. H. Latal, W. Schweiger, to be published by Springer, Heidelberg

[12] J. Ellis, M. K. Gaillard, W. J. Zakrzewski: Phys. Lett. **B81**, 224 (1979)

[13] R. Doria, J. Frenkel, J. C. Taylor: Nucl. Phys. **B168**, 93 (1980)

[14] O. Nachtmann, A. Reiter: Z. Phys. **C24**, 283 (1984)

[15] T. T. Wu, C. N. Yang: Phys. Rev. **B137**, 708 (1965);
T. T. Chou, C. N. Yang: Phys. Rev. **B170**, 1591 (1968); Phys. Rev. **D19**, 3268 (1979); Phys. Lett. **B128**, 457 (1983); Phys. Lett. **B244**, 113 (1990);
J. Dias de Deus, P. Kroll: Phys. Lett. **B60**, 375 (1976); Nuovo Cimento **A37**, 67 (1977);
P. Kroll: Z. Phys. **C15**, 67 (1982);
J. Hüfner, B. Povh: Phys. Rev. **D46**, 990 (1992)

[16] C. Bourrely, J. Soffer, T. T. Wu: Nucl. Phys. **B247**, 15 (1984); Phys. Rev. Lett. **54**, 757 (1985); Phys. Lett. **B196**, 237 (1987); Z. Phys. **C37**, 369 (1988);
R. J. Glauber, J. Velasco: Phys. Lett. **B147**, 380 (1984);
R. Henzi, P. Valin: Phys. Lett. **B149**, 239 (1984)

[17] E. M. Levin, L. L. Frankfurt: JETP Lett. **2**, 65 (1965);
H. J. Lipkin, F. Scheck: Phys. Rev. Lett. **16**, 71 (1966);
H. J. Lipkin: Phys. Rev. Lett. **16**, 1015 (1966);
J. J. J. Kokkedee, L. Van Hove: Nuovo Cimento **A42**, 711 (1966);
V. V. Anisovich et al.: "Quark Model and High Energy Collisions", World Scientific, Singapore 1985

[18] P. D. B. Collins: "An Introduction to Regge Theory", Cambridge University Press, Cambridge, U.K. 1977;

L. Caneschi, ed.: *Regge Theory of low p_T hadronic interaction*, North Holland, Amsterdam 1989

[19] G. 't Hooft: Nucl. Phys. **B72**, 461 (1974);
G. Veneziano: Nucl. Phys. **B74**, 365 (1974); Phys. Lett. **B52**, 220 (1974); Nucl. Phys. **B117**, 519 (1976);
M. Ciafaloni, G. Marchesini, G. Veneziano: Nucl. Phys. **B98**, 472, 493 (1975);
A. Capella, U. Sukhatme, Chung-I Tan, J. Tran Thanh Van: Phys. Lett. **B81**, 68 (1979);
A. Capella, U. Sukhatme, J. Tran Thanh Van: Z. Phys. **C3**, 329 (1980);
B. Andersson, G. Gustafson, G. Ingelman, T. Sjöstrand: Phys. Rep. **C97**, 31 (1983);
X. Artru: Phys. Rep. **C97**, 147 (1983);
B. Andersson, G. Gustafson, B. Nilsson-Almqvist: Nucl. Phys. **B281**, 289 (1987);
K. Werner: Phys. Rep. **C232**, 87 (1993)

[20] R. C. Hwa: Phys. Rev. **D22**, 759, 1593 (1980);
R. C. Hwa, M. Sajjad Zahir: Phys. Rev. **D23**, 2539 (1981);
R. C. Hwa: "Central production and small angle elastic scattering in the valon model", Proc. 12th Int. Symp. Multiparticle Dynamics, Notre Dame, 1981 (W. D. Shephard, V. P. Kenny, eds.) World Scientific, Singapore 1982

[21] E. A. Kuraev, L. N. Lipatov, V. S. Fadin: Sov. Phys. J.E.T.P. **44**, 443 (1976), **45**, 199 (1977);
L. N. Lipatov: Sov. J. Nucl. Phys. **23**, 338 (1976);
Ya. Ya. Balitskii, L. N. Lipatov: Sov. J. Nucl. Phys. **28**, 822 (1978);
L. N. Lipatov: Sov. Phys. J.E.T.P. **63**, 904 (1986);
L. N. Lipatov: "Pomeron in Quantum Chromodynamics", in "Perturbative Quantum Chromodynamics" (A. H. Mueller, Ed.), World Scientific, Singapore, 1989;
A. R. White: Int. J. Mod. Phys. **A6**, 1859 (1990);
J. Bartels: Z. Phys. **C60**, 471 (1993)

[22] F. E. Low: Phys. Rev. **D12**, 163 (1975);
S. Nussinov: Phys. Rev. Lett. **34**, 1286 (1975);
J. F. Gunion, D. E. Soper: Phys. Rev. **D15**, 2617 (1977)

[23] A. Donnachie, P. V. Landshoff: Nucl. Phys. **B244**, 322 (1984); Nucl. Phys. **B267**, 690 (1986); Phys. Lett. **B185**, 403 (1987)

[24] W. Heisenberg: Z. Phys. **133**, 65 (1952)

[25] H. Cheng, T. T. Wu: "Expanding Protons", MIT Press, Cambridge, Mass. 1987, and references cited therein.

[26] P. V. Landshoff, O. Nachtmann: Z. Phys. **C35**, 405 (1987)

[27] G. K. Savvidy: Phys. Lett. **71B**, 133 (1977)

[28] A. I. Vainshtein, V. I. Zakharov, M. A. Shifman: JETP Lett. **27**, 55 (1978)

[29] N.K. Nielsen, P. Olesen: Nucl. Phys. **B144**, 376 (1978)

[30] M. A. Shifman, A. I. Vainshtein, V. I. Zakharov: Nucl. Phys. **B147**, 385, 448, 519 (1979)

[31] G. 't Hooft: Cargèse Lectures, 1979, ed. G. 't Hooft et al., Plenum, New York, London 1980;
G. 't Hooft: Acta Phys. Austriaca, Suppl. **22**, 531 (1980)
[32] G. Mack: Acta Phys. Austriaca, Suppl. **22**, 509 (1980)
[33] J. Ambjørn, P. Olesen: Nucl. Phys. **B170**, 60, 265 (1980)
[34] H. Leutwyler: Phys. Lett. **96B**, 154 (1980)
[35] T. H. Hansson, K. Johnson, C. Peterson: Phys. Rev. **D26**, 2069 (1982)
[36] E. V. Shuryak: Phys. Rep. **C115**, 151 (1984), and references cited therein
[37] H. M. Fried, B. Müller (eds.): "QCD Vacuum Structure", World Scientific, Singapore 1993
[38] J. C. Maxwell: Philos. Mag. **21**, 281 (1861), reproduced in: *The Scientific Papers of James Clerk Maxwell*, Vol. 1, p. 488; W. D. Niven, ed., Cambridge Univ. Press 1890
[39] H. G. Dosch: "Nonperturbative Methods in Quantum Chromodynamics", Progr. in Part. and Nucl. Phys. **33**, 121 (1994)
[40] H. G. Dosch: Phys. Lett. **B190**, 177 (1987);
H. G. Dosch, Yu. A. Simonov: Phys. Lett. **B205**, 339 (1988);
Yu. A. Simonov: Nucl. Phys. **B307**, 512 (1988)
[41] J. Schwinger: Phys. Rev. **82**, 664 (1951)
[42] I. Ya. Aref'eva: Theor. Mat. Phys. **43**, 353 (1980);
N. E. Bralić: Phys. Rev. **D22**, 3090 (1980);
P. M. Fishbane, S. Gasioroviwcz, P. Kaus: Phys. Rev. **D24**, 2324 (1981);
L. Diosi: Phys. Rev. **D27**, 2552 (1983);
Yu. A. Simonov: Sov. J. Nucl. Phys. **48**, 878 (1988)
[43] N. G. Van Kampen: Physica **74**, 215, 239 (1974); Phys. Rep. **C24**, 172 (1976)
[44] A. DiGiacomo, H. Panagopoulos: Phys. Lett. **B285**, 133 (1992)
[45] H. G. Dosch, E. Ferreira, A. Krämer: Phys. Rev. **D50**, 1992 (1994)
[46] V. A. Novikow, M. A. Shifman, A. I. Vainshtein, V. I. Zakharov: Nucl. Phys. **B191**, 301 (1981)
[47] F. Borkowski et al., Nucl. Phys. **A222**, 269 (1974)
[48] F. Borkowski et al.: Nucl. Phys. **B93**, 461 (1975)
[49] S. R. Amendolia et al.: Nucl. Phys. **B277**, 168 (1986)
[50] B. Povh, J. Hüfner, Phys. Rev. Lett. **58**, 1612 (1987)
[51] E. Eichten, K. Gottfried, T. Kinoshita, K. D. Lane, T. M. Yan: Phys. Rev. **D21**, 203 (1980)
[52] M. Rueter, H. G. Dosch: Z. Phys. **C66**, 245 (1995)
[53] O. Nachtmann: Ann. Phys. **209**, 436 (1991)
[54] B. Andersson, G. Gustafson, G. Ingelman, T. Sjöstrand: Phys. Rep. **C97**, 33 (1983)
[55] W. Buchmüller, A. Hebecker: "Semiclassical approach to structure functions at small x", report DESY 95-208 (1995)
[56] H. Verlinde, E. Verlinde: "QCD at high energies and two-dimensional field theory" report PUPT-1319, IASSNS-HEP-92/30 (1993)
[57] L. Van Hove: Phys. Lett. **24B**, 183 (1967)

[58] L. Lukaszuk, B. Nicolescu: Nuovo Cimento Lett. **8**, 405 (1973); D. Bernard, P. Gauron, B. Nicolescu: Phys. Lett. **B199**, 125 (1987); E. Leader: Phys. Lett. **B253**, 457 (1991)

[59] A. Krämer, H. G. Dosch: Phys. Lett. **B252**, 669 (1990), **B272**, 114 (1991); H. G. Dosch, E. Ferreira, A. Krämer: Phys. Lett. **B289**, 153 (1992)

[60] E. Berger, "Nichtstörungstheoretische Methoden zur Beschreibung von Reaktionen mit kleinem Impulstransfer in einem abelschen Modell", diploma thesis, Univ. of Heidelberg (1996), unpublished

[61] G. B. West: Phys. Lett. **B115**, 468 (1982); C. D. Roberts, A. G. Williams: Progr. Part. and Nucl. Phys. **33**, 477 (1994), and references cited therein; K. Büttner, M. R. Pennington: "Infrared behaviour of the gluon propagator: Confining or confined?", Durham Univ. report DTP-95/32 (1995)

[62] M. Rueter, H. G. Dosch: "Nucleon Structure and High Energy Scattering", Univ. of Heidelberg report HD-THEP-96-04, hep-ph/9603214 (1996)

[63] R. P. Feynman: Phys. Rev. Lett. **23**, 1415 (1969); "Photon-Hadron Interactions", W. A. Benjamin, Reading, Mass. 1972

[64] I. M. Ternov, Yu. M. Loskutov, L. I. Korovina: Zh. Eskp. Teor. Fiz. **41**, 1294 (1961) (Sov. Phys. - JETP **14**, 921 (1962)); A. A. Sokolov, I. M. Ternov: Dokl. Akad. Nauk SSSR **153**, 1052 (1963) (Sov. Phys.-Dokl. **8**, 1203 (1964))

[65] J. D. Jackson: Rev. Mod. Phys. **48**, 417 (1976)

[66] G. W. Botz, P. Haberl, O. Nachtmann: Z. Phys. **C67**, 143 (1995)

[67] F. E. Low: Phys. Rev. **110**, 974 (1958)

[68] P. V. Chliapnikov et al.: Phys. Lett. **141B**, 276 (1984)

[69] F. Botterweck et al.: Z. Phys. **C51**, 541 (1991)

[70] S. Abatzis et al.: Nucl. Phys. **A525**, 487c (1991)

[71] S. Banerjee et al. (SOPHIE/WA83 coll.): Phys. Lett. **305B**, 182 (1993)

[72] J. Antos et al.: Z. Phys. **C59**, 547 (1993)

[73] E. Fermi: Z. Phys. **29**, 315 (1924); C. F. v. Weizsäcker: Z. Phys. **88**, 612 (1934); E. J. Williams: Kgl. Danske Videnskab. Selskab. Mat.-Fiz. Medd. **13**, No 4 (1935)

[74] V. M. Budnev et al.: Phys. Rep. **C15**, 181 (1975)

[75] S. Platchkov et al.: Nucl. Phys. **A510**, 740 (1990)

[76] M. Meyerhoff et al.: "First measurement of the electric form factor of the neutron in the exclusive quasielastic scattering of polarized electrons from polarized ^3He", Univ. Mainz Report (1994)

[77] H. Leeb, C. Teichtmeister: Phys. Rev. **C48**, 1719 (1993)

[78] P. E. Bosted et al., Phys. Rev. Lett. **68**, 3841 (1992)

[79] R. D. Carlitz, S. D. Ellis, R. Savit: Phys. Lett. **68B**, 443 (1977); N. Isgur, G. Karl, R. Koniuk: Phys. Rev. Lett. **41**, 1269 (1978); D. Gromes: "Ordinary Hadrons", in Proc. Yukon Adv. Study Inst.: *The Quark Structure of Matter*, eds. N. Isgur, G. Karl, P. J. O'Donnell (World Scientific, Singapore 1985)

[80] A. Brandenburg, E. Mirkes, O. Nachtmann: Z. Phys. **C60**, 697 (1993)
[81] S. Falciano et al., (NA10 coll.): Z. Phys. **C31**, 513 (1986);
M. Guanziroli et al., (NA10 coll.): Z. Phys. **C37**, 545 (1988)
[82] W. Buchmüller, A. Hebecker: Phys. Lett. **B355**, 573 (1995);
A. Edin, G. Ingelman, J. Rathsman: Phys. Lett. **B366**, 371 (1996)
[83] A. Efremov, D. Kharzeev: Phys. Lett. **B366**, 311 (1996)
[84] W. Bonivento et al. (DELPHI Coll.): "A measurement of quark spin correlations in hadronic Z^0 decays." Contribution to the EPS-HEP 95 conference, Brussels, August 1995
[85] H. B. Nielsen, A. Patkos: Nucl. Phys. **B195**, 137 (1982)
[86] S. V. Mikhailov, A. V. Radyushkin: Sov. J. Nucl. Phys. **49**, 494 (1989)
[87] E. V. Shuryak: Nucl. Phys. **B328**, 85 (1989)
[88] P. Ball, V. M. Braun, H. G. Dosch: Phys. Rev. **D44**, 3567 (1991)
[89] C. Alexandrou, S. Güsken, F. Jegerlehner, K. Schilling, R. Sommer: Nucl. Phys. **B414**, 815 (1994);
R. Sommer: report DESY 94-011 (1994), unpublished
[90] U. M. Heller, K. M. Bitar, R. G. Edwards, A. D. Kennedy: Phys. Lett. **B335**, 71 (1994)

Perturbative QCD (and Beyond)*

Yu. L. Dokshitzer[1];
Notes in collaboration with R. Scheibl and C. Slotta[2]

[1] St. Petersburg Nuclear Physics Institute, Gatchina, Russia
[2] Institute for Theoretical Physics, University of Regensburg, Universitätsstraße 31, 93040 Regensburg, Germany

QCD is a gauge theory based on symmetry with respect to colour, described by the $SU(3)$ group of colour transformations.

The aim of this lecture is to introduce, step by step, basic elements of QCD by following, as an Ariadne thread, the basics of the $SU(N)$ group. Each new element of the colour algebra will be illustrated by a lowest-order perturbative QCD example. Along these lines we will make the acquaintance with QCD bremsstrahlung, which is responsible for particle multiplication in hard processes. Physics of QED and QCD bremsstrahlung are quite similar but for one important difference: The gluon, contrary to the photon, carries "colour charge" by itself and therefore is bound to play a double rôle: that of "radiation" and, at the same time, that of "radiator" with respect to the next-generation (softer) gluons. Special emphasis shall be given to coherent effects in parton cascades. They are essential in understanding parton multiplication in jets and can be tested against experiment through detailed investigation of quark/gluon-initiated hadron jets.

The author hopes that the reader who takes on the burden of slowly reading this lecture, and solving the suggested exercises, will master the graphic colour-algebra technique, which provides a simple means for calculating QCD colour factors without performing laborious calculations. For more information on $SU(N)$ groups and colour algebra, see textbooks about gauge theories and [1]. Additional and more detailed information on bremsstrahlung and coherence effects can be found in the textbooks [2] and [3] as well as in their bibliographies. They are used as reference throughout this text.

1 Basics of Colour Algebra

Quarks in QCD may exist in three different states of colour charge (red, green, blue); we therefore define a quark-state vector as

$$\psi = \begin{pmatrix} \psi_r \\ \psi_g \\ \psi_b \end{pmatrix} ,$$

* Lectures at the workshop "Nonperturbative QCD" organised by the Graduiertenkolleg Erlangen-Regensburg, held on October 10th–12th, 1995 in Kloster Banz, Germany

where ψ_r, ψ_g and ψ_b are Dirac spinors. The gauge theory of QCD is based on the invariance of the QCD-Lagrangian under local rotations of ψ in colour space:

$$\psi'(x) = U(x)\,\psi(x) \qquad \text{and} \qquad \bar{\psi}'(x) = \bar{\psi}(x)\,U^\dagger(x) \ ;$$

the rotation matrix U belongs to the $SU(3)$ colour group.

$SU(N)$ is the group of special unitary $N \times N$ matrices U:

$$UU^\dagger = U^\dagger U = 1\,, \tag{1a}$$

$$\mathrm{Det}\,U = \mathrm{e}^{\mathrm{Tr}\ \ln\ \mathrm{U}} = 1\ . \tag{1b}$$

The N^2 matrix elements of U have in general complex values, which gives $2N^2$ real numbers. Unitarity (1a) provides N^2 and (1b) one condition; what remains are $N^2 - 1$ independent, real parameters to characterize U.

1.1 Parametrization. Infinitesimal Rotations. Generators

The elements U of the group $SU(N)$ may then be parametrized by (we always sum over repeated indices):

$$U(\omega) = \mathrm{e}^{\mathrm{i}\omega^a t^a} \qquad (a = 1, \ldots, N^2 - 1)\ , \tag{2}$$

where ω^a are real parameters and t^a are $N \times N$ matrices. From (1a) and (1b) it follows that t^a are Hermitian and traceless:

$$t^a = \left(t^\dagger\right)^a\,, \qquad \mathrm{Tr}\,(\mathrm{t^a}) = 0\ . \tag{3}$$

Thus we have parametrized a unitary $SU(N)$ transformation with N^2-1 (eight for $SU^c(3)$) "colour rotation angles" ω^a. Consider an infinitesimal colour rotation with the angles $\delta\omega^a$, $|\delta\omega^a| \ll 1$. U is then close to the unit matrix, and in the linear approximation reads

$$U = \mathbb{1} + \mathrm{i}\delta\omega^a\, t^a + \ldots \tag{4}$$

Under such transformation the quark wave function acquires a small increment

$$\psi \to \psi' = U\psi \simeq \psi + \delta\psi\,, \quad \delta\psi = \mathrm{i}\delta\omega^a\, t^a\, \psi\,.$$

Correspondingly, for an *antiquark* we have

$$\bar{\psi} \to \bar{\psi}' = \psi U^\dagger \simeq \bar{\psi} + \delta\bar{\psi}\,, \quad \delta\bar{\psi} = -i\delta\omega^a\left(\bar{\psi}t^a\right),$$

where the brackets are there to remind you the matrix multiplication rule: row $\times$ column.

In general, an operator T^a that determines an infinitesimal transformation of an object R is called *generator* (acting in a given representation R). Thus, we have found the generators acting in the fundamental **3** and the conjugated fundamental $\bar{\mathbf{3}}$ representations corresponding to the quark and antiquark states:

$$R^i = \psi^i\,,\ T^a(\mathbf{3}) = t^a\,; \quad (T^a(\mathbf{3}) : R)^i = (t^a)^i_k\, R^k\,; \tag{5a}$$

$$R_i = \bar{\psi}_i\,,\ T^a(\bar{\mathbf{3}}) = -t^a\,; \quad (T^a(\bar{\mathbf{3}}) : R)_i = -R_k\,(t^a)^k_i\,. \tag{5b}$$

The notion of generators plays an important rôle in the theory: It is T^a that determines the amplitude of radiation of a gluon with colour a off a given object R, that is the "colour charge" of the latter (generalization of the e.m. charge). Given this convention, one may say that q and $\bar{q}$ *colour* charges (5a) and (5b) are *opposite*, as well as their *e.m.* charges.

1.2 Colour Neutrality (Meson)

Combining a u-quark of a given colour with its antiquark, e.g.

$$u^1 \cdot \bar{u}_1 \equiv u^{\mathrm{red}} \cdot \bar{u}_{\mathrm{red}}\,, \tag{6}$$

we obviously get a system with a zero e.m. charge. However "colour charges" there are not one but eight, by the number of colour generators ($a = 1, 2, \ldots N^2 - 1 = 8$). Some of them do vanish on the state (6). Meanwhile, a truly colour-neutral system, such that $T^a : R = 0$ for all a, is a coherent mixture of all colours:

$$R_0 \equiv R_{\text{"white"}} = \bar{u}_1 u^1 + \bar{u}_2 u^2 + \bar{u}_3 u^3 = \sum_{i=1}^{N} \bar{u}_i u^i = (\bar{u}u)\,. \tag{7}$$

Indeed, adding together increments due to colour rotation of u and $\bar{u}$ components (5), we obtain

$$\delta R_0 = \mathrm{i}\delta\omega^a \bar{u}(t^a u) + \mathrm{i}\delta\omega^a\,(\bar{u}(-t^a))\,u = 0\,.$$

This shows that the colour mixture (convolution) (7) is invariant or, in other words, constitutes a trivial irreducible representation — the singlet $\mathbf{1}$, for which $T^a(1) \equiv 0$.

The colour convolution (7) determines the structure of $\pi^0, \rho^0, \ldots$ mesons in the constituent quark model; it enters in the kinetic, mass and photon radiation terms of the u-quark Lagrangian, etc.

Quarks of different flavours transform identically under colour $SU^c(3)$ rotations, therefore the "multiflavour" colour convolutions are "white" as well. For example:

$$R = \bar{d}_i u^i = (\bar{d}u)\,, \quad R = \bar{s}_i d^i = (\bar{s}d)$$

determine $\pi^+, K^0, \ldots$ wave functions, enter in the weak currents (charged and neutral, respectively), etc.

We have verified that the meson-type construction $\bar{q}q$ is colourless. We could have done it easily without invoking infinitesimal transformations (5), since for an arbitrary U it follows directly from the unitarity condition that

$$U : (\bar{q}q) \;=\; (\bar{q}U^{\dagger})(Uq) = (\bar{q}q)\,; \quad U^{\dagger}U = \mathbb{1}\,.$$

A little harder is to prove the colourlessness of a "baryon" qqq — one more (and only one, not counting $\bar{q}\bar{q}\bar{q}$) independent colour-singlet fermion construction. Here we shall use the infinitesimal transformations in full scale. To this end we should address the following questions.

1.3 Two-Quark Compound. How Come $\mathbf{3}\otimes\mathbf{3}=\mathbf{6}\oplus\bar{\mathbf{3}}$?

Let us choose for the sake of definiteness the product of two quark wave functions $u^i d^j$. This object has $N\cdot N = 9$ components which transform according to (5) as

$$\delta\left\{u^i d^j\right\} = (\delta u^i)d^j + u^i(\delta d^j) = \mathrm{i}\delta\omega^a\left\{(t^a)^i_k u^k d^j + u^i (t^a)^j_k d^k\right\}.$$

From these nine numbers we can construct two tensors: one symmetric and one antisymmetric with respect to colour indices:

$$u^i d^j \;=\; R^{\{ij\}} + R^{[ij]}\,, \tag{8}$$

with

$$R^{\{ij\}} = \tfrac{1}{2}(u^i d^j + u^j d^i)\,, \tag{9a}$$

$$R^{[ij]} = \tfrac{1}{2}(u^i d^j - u^j d^i)\,. \tag{9b}$$

$R^{\{ij\}}$ possesses $\frac{1}{2}N(N+1) = 6$ components (independent parameters), while $R^{[ij]}$ has $\frac{1}{2}N(N-1) = 3$. The symmetry of each of the two tensors (9) is preserved under colour rotations, that is the components of $R^{\{ij\}}$ do not mix with those of $R^{[ij]}$.

Exercise 1. Verify that

$$\delta R^{\{ij\}} = \delta R^{\{ji\}}\,; \quad \delta R^{[ij]} = -\delta R^{[ji]}\,.$$

This means, by definition, that $R^{\{ij\}}$ and $R^{[ij]}$ constitute *irreducible representations* of the group. Thus, we have constructed a new colour object — the sextet **6** — colour-symmetrized product of two quarks. Its generator is

$$T^a(6) \;=\; t^a\cdot\mathbb{1} + \mathbb{1}\cdot t^a\,, \tag{10a}$$

where the first (second) matrix in the dot-product acts upon the first (second) index of the state R^{ij} (9a). Written in full,

$$(T^a(6) : R)^{ij} \;=\; \left((t^a)^i_k\cdot\delta^j_\ell + \delta^i_k\cdot(t^a)^j_\ell\right) R^{k\ell}\,. \tag{10b}$$

The dimension of the second — antisymmetric — combination (9b) coincides, for the $SU^c(3)$ group of interest, with that of the fundamental representation, $\frac{1}{2}N(N-1) = N = 3$. Let us check that its transformation properties correspond to the *conjugated* fundamental representation $\mathbf{\bar{3}}$. To this end we observe that the antisymmetric tensor is equivalent to the 3-component "vector"

$$R_m = \epsilon_{mij} u^i d^j = \epsilon_{mij} R^{[ij]}\,, \quad (m = 1,2,3) \tag{11a}$$

with ϵ_{mij} the standard fully antisymmetric symbol, whose non-zero components are

$$\epsilon_{123} = \epsilon_{231} = \epsilon_{312} = -\epsilon_{213} = -\epsilon_{132} = -\epsilon_{321} = 1\,.$$

Thus,

$$\begin{aligned} R_1 &= u^2 d^3 - u^3 d^2\,, \\ R_2 &= u^3 d^1 - u^1 d^3\,, \\ R_3 &= u^1 d^2 - u^2 d^1\,. \end{aligned} \tag{11b}$$

The relation inverse to (11) looks as follows:

$$2\,R^{[ij]} = u^i d^j - u^j d^i = \epsilon^{\alpha ij} R_\alpha\,. \tag{12}$$

It is straightforward to verify (12) using the tensor identity

$$\epsilon^{\alpha ij}\epsilon_{\alpha k\ell} = \delta^i_k \delta^j_\ell - \delta^i_\ell \delta^j_k \tag{13}$$

(recall, sum over $\alpha = 1,2,3$ is implied). Now we take the state R_m and perform a small colour rotation:

$$\delta R_m = \delta(\epsilon_{mij} u^i d^j) = \mathrm{i}\delta\omega^a \epsilon_{mij}\left((t^a u)^i d^j + u^i (t^a d)^j\right).$$

Let us first rename the summation indices in the second term and then use antisymmetry of the ϵ-symbol to write

$$\epsilon_{mij} u^i (t^a d)^j = \epsilon_{mji} u^j (t^a d)^i = -\epsilon_{mij} u^j (t^a d)^i\,.$$

Now we can factor out the matrix t^a:

$$\delta R_m = \mathrm{i}\delta\omega^a \epsilon_{mij} (t^a)^i_k \left[u^k d^j - u^j d^k\right].$$

Continue equality invoking (12) for the expression in the square brackets:

$$\delta R_m = \mathrm{i}\delta\omega^a (t^a)^i_k\, \epsilon_{mij}\, \epsilon^{\alpha kj} R_\alpha\,;$$

then use (13):

$$\delta R_m = \mathrm{i}\delta\omega^a (t^a)^i_k \left(\delta^\alpha_m \delta^k_i - \delta^k_m \delta^\alpha_i\right) R_\alpha\,. \tag{14}$$

The first term in (14) vanishes due to $(t^a)^i_k \delta^k_i \equiv \mathrm{Tr}\,(\mathrm{t}^a) = 0$, and we finally arrive at

$$\delta R_m = -\mathrm{i}\delta\omega^a (t^a)^\alpha_m R_\alpha = -\mathrm{i}\delta\omega^a (R t^a)_m\,, \tag{15}$$

which is identical to the transformation of the wave function of an *antiquark*, $T^a(\mathbf{\bar{3}})$, cf. (5b).

We have decomposed the wave function of a compound 2-quark object into irreducible representations:

$$\mathbf{3} \otimes \mathbf{3} = \mathbf{6} \oplus \mathbf{\bar{3}}\,.$$

1.4 Colour Neutrality (Baryon)

The question about colourlessness of a "baryon" is now readily understood. Adding one more quark (s, for simplicity), we get

$$\delta(\epsilon_{mij}u^i d^j\, s^m) = (\delta R_m)s^m + R_m(\delta s^m) = i\delta\omega^a\left\{-(Rt^a)s + R(t^a s)\right\} = 0\,,$$

which in fact is a repetition of the proof that the "meson" $\bar{q}s$ is white, where the rôle of $\bar{q}$ is played by the ud pair in the $\bar{\mathbf{3}}$ state. Let us recall that it is the colour-antisymmetry between three quarks that has come to rescue the naturality of the baryon state description by the orthodox constituent quark model.

The lightest baryons fall into two positive-parity multiplets: the *flavour* octet $p + n + \Lambda + 3\Sigma(1193) + 2\Xi(1318)$ and the *flavour* decuplet $4\Delta + 3\Sigma(1385) + 2\Xi(1530) + \Omega$. The former, as you know, have spin $\frac{1}{2}$, the latter spin $\frac{3}{2}$. Together they combine into the so-called **56**-plet of the approximate $SU(6)$ spin-flavour symmetry group, where 56 accounts for the total number of spin$\times$flavour states: $2*8+4*10$.

Interchanging two quarks, one interchanges their flavours, spins, coordinates and colours. Consider the Δ^{++} baryon state with spin projection $+\frac{3}{2}$. It contains three constituent u-quarks, and its wave function is obviously *flavour*-symmetric (identical flavours). Its *coordinate* wave function is symmetric as well, since it is natural to expect that in the ground state the orbital movement is minimal, so that each quark pair sits in the S-state. In the absence of orbital momenta, quark spins must add up coherently, which corresponds to the *spin* wave function being symmetric as well. As a result, Fermi statistics would be violated if not for *colour* degrees of freedom:

$$\Delta^{++} = |\,u\!\uparrow\, u\!\uparrow\, u\!\uparrow\rangle \implies \Delta^{++} = \frac{1}{\sqrt{6}}\epsilon_{ijk}|\,u\!\uparrow^i u\!\uparrow^j u\!\uparrow^k\rangle\,.$$

To finalize the discussion of colourlessness, it is worth while to remark a difference between "white" mesons and "white" baryons. Namely, a "meson" $(\bar{q}q)$ is a singlet for any $SU(N)$ group, while the baryon structure essentially depends on N: The number of quarks in a "baryon" always equals N. This is easy to realize simply by counting the number of states. With the quark colour index i running from 1 up to N to organize a compound object with exactly N degrees of freedom, we will have to antisymmetrize $N-1$ quark states,

$$R_m = \epsilon_{i_1 i_2 \ldots i_{N-1}}\, q^{i_1} q^{i_2} \cdots q^{i_{N-1}}\,.$$

A well-known example is the isotopic-spin group $SU(2)$. Here the ϵ-symbol bears only two indices, and the "baryon" consists of two "quarks". In a more familiar language, one constructs the isotope-singlet ("white") states $I = 0$ from the isospin-$\frac{1}{2}$ nucleon, p and n ("quarks"), in two ways:

$$\begin{aligned} &(\text{"meson"}) & \bar{N}_i N^i &= \bar{p}p + \bar{n}n \\ &(\text{"baryon"}) & \epsilon_{ij}N^i N^j &= pn - np \end{aligned}$$

If nature would have chosen to have four colours, the product of two quark states would give

$$q^i q^j \;=\; \mathbf{4}\otimes\mathbf{4} \;=\; R^{\{ij\}} + R^{[ij]} \;=\; \mathbf{10}+\mathbf{6}\,.$$

Only after adding one more quark would one find the $\bar{\mathbf{4}}$ term,

$$q^i q^j q^k = \mathbf{4}\otimes\mathbf{4}\otimes\mathbf{4} \;=\; \ldots + \bar{\mathbf{4}}\,,$$

suitable for building a white 4-quark baryon,

$$\mathrm{SU(4)\text{-}``baryon"} \;=\; \epsilon_{ijk\ell}\, q^i q^j q^k q^\ell\,.$$

1.5 Commutator. Structure Constants

Studying infinitesimal transformations gives a complete information about the structure of the group. The most important characteristics of the group is the *commutator* of two transformations $U(\delta\omega_1)$ and $U(\delta\omega_2)$. Comparing two small rotations (4) performed in different order, we get a mismatch

$$[\,U_2 U_1\,] \equiv U_2U_1 - U_1U_2 \;=\; (\mathrm{i}\delta\omega_2^a)(\mathrm{i}\delta\omega_1^b)\;\left[\,t^a t^b\,\right] \;+\; \mathcal{O}(\delta\omega^3)\;.$$

Two matrices t^a and t^b and, therefore, two group transformations, generally speaking, do not commute. If this is the case, the group is called non-Abelian. (Familiar examples of Abelian groups — translations, phase transformations $U(1)$; non-Abelian — 3-dimensional rotations $O(3)$.)

The matrix $[t^a t^b]$ is obviously traceless ($\mathrm{Tr(t^a t^b)} = \mathrm{Tr(t^b t^a)}$). Therefore it may be represented as a linear combination of the same t^c-matrices:

$$[t^a t^b] \;=\; \mathrm{i} f_{abc} t^c\,; \qquad a,b,c = 1,2,\ldots,N^2-1\,. \tag{16}$$

The expansion coefficients f_{abc} are real (since $[t^a t^b]$ is anti-Hermitian, and we have explicitly extracted the imaginary unit in the r.h.s. of (16)). So defined, f_{abc} are called the structure constants of the group.

By their very nature, gluons are intimately related with infinitesimal $SU^c(3)$ rotations. To construct the QCD Lagrangian one invokes the heuristic principle of invariance under *local* colour transformations. This means that one is looking for the theory invariant with respect to colour rotations with parameters $\omega^a(x)$ depending on the space-time coordinate x^μ.

Changing the rotation angles from point to point introduces a mismatch between the colour rotation phases in the nearby points,

$$\frac{\delta}{\delta x^\mu}\omega^a(x)\, t^a \;=\; (\nabla_\mu \omega^a(x))\, t^a\quad.$$

This mismatch breaks colour invariance of the Dirac equation for free quarks. To compensate for this effect, one introduces the gluon fields $A^a_\mu(x)$ interacting with quarks. It is therefore not surprising that the matrix t^a that generates infinitesimal colour rotation describes, at the same time, the coupling between quarks and gluons: The emission amplitude of a gluon with colour a off a quark, $q^i \to q^j + g^a$, has the colour factor $(t^a)^j_i$.

The same rôle for the gluon emission off a *gluon* (for the gluon self-interaction, in other words), is played by the structure constants: The amplitude $g^a \to g^b + g^c$ has the colour factor $\mathrm{i}f_{abc}$. We need to make a closer acquaintance with the key objects t^a and f_{abc}.

1.6 Standardization of *t*-Matrices

We choose the normalization of the generators t as follows:

$$\mathrm{Tr}\left(\mathrm{t^a t^b}\right) \equiv \mathrm{T_R}\,\delta^{\mathrm{ab}} = \frac{1}{2}\,\delta^{\mathrm{ab}}\,. \tag{17}$$

This condition fixes an orthogonal basis in the space of traceless $N \times N$ matrices. The normalization coefficient $T_R = \frac{1}{2}$ is a matter of convenience. It is related with the choice of the "colour spin" unit.

As far as QCD in concerned, it is this normalization that corresponds to the quark-gluon interaction amplitude shown in Fig. 1.

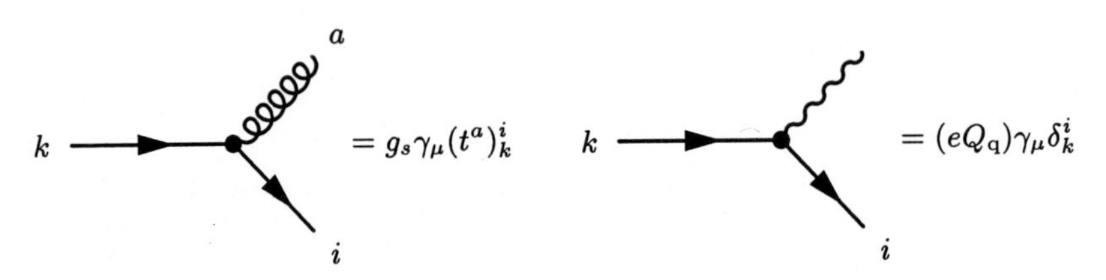

Fig. 1. *Quark-gluon vertex (left) and quark-photon vertex (right). k and i are quark colour indices; a the gluon colour. Gluon radiation affects the colour of the quark (coupling $(t^a)^i_k$); the colourless photon leaves it unchanged (coupling δ^i_k). Q_q is the fractional quark charge in electron units.*

Example: $\Gamma(q\bar{q} \to \gamma\gamma)/\Gamma(q\bar{q} \to gg)$. This knowledge is sufficient to consider an example of how to use relation (17).

Consider heavy onia, the mesons built of heavy quarks $c\bar{c}$ or $b\bar{b}$. Such states may have different spin J and different C- and P-parity, depending on the quark orbital momentum L and the total quark spin S.

The state is often represented in atomic spectroscopy standards as ${}^{(2S+1)}L_J$. For example, the famous J/ψ (and Υ) is the ${}^3\mathrm{S}_1$ state, that is the $S=1$, $L=0$ (S-wave, sorry for confusing S's) meson.

In general, a fermion-antifermion pair can have the quantum numbers

$$P = (-1)^{L+1}\,, \quad C = (-1)^{L+S}\,; \quad J = |L-S|, \ldots |L+S|\,. \tag{18}$$

This tells us that ψ/Υ is a vector C-odd state: $J^{PC} = 1^{--}$. Being C-odd, these mesons can be (and are) produced via one-photon e^+e^- annihilation,

$$e^+e^- \to \gamma^* \to J/\psi, \psi', \ldots; \Upsilon, \Upsilon', \ldots$$

Combining quark spins into $S=0$, instead of $S=1$ as for ψ/Υ, one gets pseudoscalar C-even (0^{-+}) mesons ${}^1S_0$ such as $\eta^c(2.98)$. Another possibility to construct C-even mesons is to take a P-wave ($L=1$) $q\bar{q}$ pair with the quark spin $S=1$. This gives three options for the spin (total momentum J) of χ-mesons: a scalar $\chi_0({}^3P_0)$, an axial-vector $\chi_1({}^3P_1)$, and a spin-2 tensor $\chi_2({}^3P_2)$.

C-even states cannot be produced via one (C-odd) photon, but they can decay into *two photons*, or into light (charmless, bottomless) hadrons via *two gluons* as shown in Fig. 2. (To be accurate, we should exclude the χ_1 mesons from this list: The Landau-Yang veto does not allow an axial state to decay into two on-mass-shell gauge bosons, be they photons or gluons.)

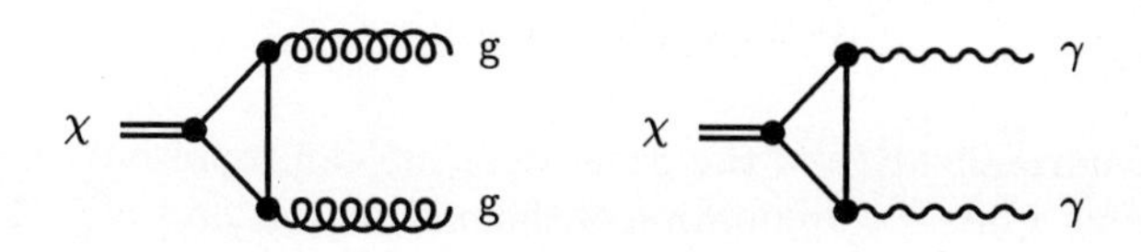

Fig. 2. *Decay of C-even quarkonia $\bar{Q}Q$ into two gluons or into two photons.*

Let us compare hadronic and 2-photon decay widths of C-even heavy onia. The matrix elements for the processes in Fig. 2 have identical Lorentz structures; they differ only in the strength of the coupling constants and in the colour structure. A diagrammatic representation of the squared matrix elements is shown in Fig. 3. The radiation of the two gluons or photons must not affect the net

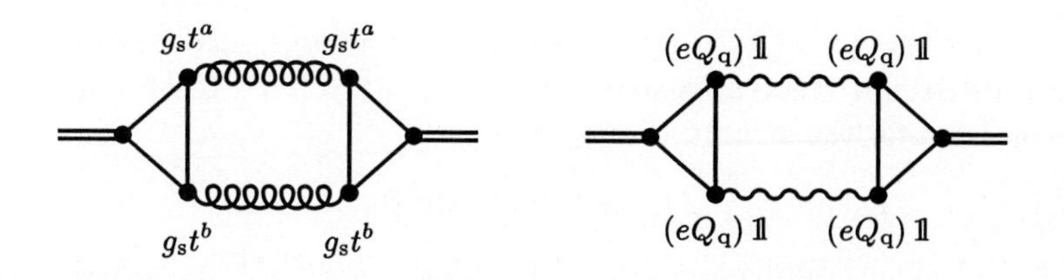

Fig. 3. *Squared matrix elements for the processes $\bar{Q}Q \to 2g,\, 2\gamma$.*

colour charge of the $\bar{c}c$ pair on either side of the diagrams in Fig. 3. To account for all possible combinations, we have to sum over the diagonal elements of $t^a t^b$ or $\mathbb{1}\mathbb{1}$. After squaring we obtain:

$$\left\{g_s^2 \,\mathrm{Tr}(\mathrm{t^a t^b})\right\}^2 = \left\{g_s^2 \,\tfrac{1}{2}\delta^{ab}\right\}^2 = g_s^4\,\frac{N^2-1}{4}\,; \tag{19}$$

$$\left\{e^2 Q_q^2 \,\mathrm{Tr}(\mathbb{1})\right\}^2 = \left\{e^2 Q_q^2\, N\right\}^2 = e^4 Q_q^4\, N^2\,. \tag{20}$$

Thus, for the charm mesons ($Q_c = 2/3$) we have the prediction

$$\begin{aligned}\frac{\Gamma_{gg}}{\Gamma_{\gamma\gamma}}\Big(\,\eta^c(2980)\Big) &= \frac{\Gamma_{gg}}{\Gamma_{\gamma\gamma}}\Big(\,\chi_0^c(3415)\Big) = \frac{\Gamma_{gg}}{\Gamma_{\gamma\gamma}}\Big(\,\chi_2^c(3555)\Big) \\ &= \left(\frac{\alpha_s}{\alpha_{em} Q_c^2}\right)^2 \frac{(N^2-1)}{4N^2} = \frac{9}{8}\left(\frac{\alpha_s}{\alpha_{em}}\right)^2 .\end{aligned} \tag{21}$$

In the beauty sector, the ratios similar to (21) should be 16 times larger ($Q_b = -1/3$), but neither η^b has been observed, nor 2-photon widths of $\chi_0^b(9860)$, $\chi_2^b(9915)$ have been measured.

1.7 Expansion in t^a

Having fixed the normalization of the generators, we can represent an arbitrary $N \times N$ matrix M as a linear combination of the unit matrix and $N^2 - 1$ matrices t^a :

$$M = n^0\,\mathbb{1} + n^a\, t^a\ . \tag{22a}$$

To find the coefficients n^0 and n^a we apply the trace operator to either side of the equation, once immediately, once after multiplying the equation by the matrix t^b:

$$\begin{aligned}\mathrm{Tr}(M) &= n^0 N + \ 0\,, \\ \mathrm{Tr}(M t^b) &= \ 0 \ \ + n^a\,\tfrac{1}{2}\delta^{ab}\ ,\end{aligned}$$

and end up with

$$M = \frac{1}{N}\,\mathrm{Tr}(M)\,\mathbb{1} + 2\,\mathrm{Tr}(M t^a)\, t^a\ . \tag{22b}$$

Example: Structure Constants Let us take $M = [t^a t^b]$ to obtain from (22) the expression for the structure constants:

$$i f_{abc} = 2\mathrm{Tr}\left(\mathrm{t^c}[\mathrm{t^a t^b}]\right). \tag{23}$$

The f_{abc} symbol is antisymmetric with respect to $a \to b$, as follows from its definition (16).

Exercise 2. Using cyclic permutation under the trace in (23), verify that f_{abc} is actually antisymmetric with respect to *any* pair of indices.

Example: $\mathbf{3} \otimes \bar{\mathbf{3}} = \mathbf{1} \oplus \mathbf{8}$. Consider a composite object $q_1\bar{q}_2$ as an $N \times N$ matrix given by the product of the quark (ψ) and antiquark ($\bar{\chi}$) wave functions:

$$\psi^i \bar{\chi}_k = \frac{1}{N}\delta^i_k(\bar{\chi}\psi) + R^i_k \,, \tag{24a}$$

where R^i_k is the traceless tensor

$$R^i_k = \psi^i\bar{\chi}_k - \frac{1}{N}\delta^i_k(\bar{\chi}\psi)\,, \qquad R^i_i = 0\,. \tag{24b}$$

Under the group transformation

$$\delta R^i_k = \mathrm{i}\delta\omega^a\,(t^a)^i_\ell R^\ell_k + \mathrm{i}\delta\omega^a\, R^i_\ell(-t^a)^\ell_k = \mathrm{i}\delta\omega^a([t^a R])^i_k\,, \tag{25}$$

the tracelessness stays intact:

$$\delta R^i_i \propto \mathrm{Tr}([\mathrm{t^a R}]) = 0 = \mathrm{R^i_i}\,.$$

Therefore, the N^2-1 parameters of R^i_k form a new (irreducible) representation — the $SU(3)$ octet.

According to (22), the traceless matrix R can be parametrized by a "vector" ϕ_a, $a = 1, 2, \ldots, N^2-1$, as

$$R^i_k = \phi_b(t^b)^i_k\,, \quad \phi_b = 2\mathrm{Tr}(\mathrm{Rt^b})\,. \tag{26}$$

1.8 Adjoint Representation. Generators $T^a(\mathbf{8})$

From the transformation law (25) we conclude that the generators of the matrix representation R of the octet are given by the commutation operation with t^a:

$$T^a(\mathbf{8}) : R = [t^a R]\,, \qquad T^a(\mathbf{8}) = [t^a \ldots]\,. \tag{27}$$

Substituting the representation (26), we define the generator in the "vector" representation by the relation

$$T^a(\mathbf{8}) : R = T^a(\mathbf{8}) : (\phi_c t^c) \equiv t^b\,(T^a : \phi)_b\,,$$

to obtain

$$\begin{aligned} t^b\,(T^a : \phi)_b &= [t^a t^c]\phi_c = \mathrm{i}f_{acb}t^b\,\phi_c = -\mathrm{i}f_{abc}t^b\,\phi_c \\ \Longrightarrow &\qquad (T^a : \phi)_b = -\mathrm{i}f_{abc}\phi_c\,. \end{aligned} \tag{28}$$

The representation whose generators coincide with the structure constants of the group is called the adjoint representation. The transformation law in the vector form,

$$\delta\phi = \mathrm{i}\delta\omega^a\, T^a : \phi\,, \tag{29a}$$

becomes

$$\delta\phi_b = \delta\omega^a\, f_{abc}\phi_c\,. \tag{29b}$$

Since both the rotation angles $\delta\omega^a$ and the elements f_{abc} are real numbers, the field ϕ can be chosen real as well. Example: an octet of real gluon potentials A_μ^a which transform according to (29) under the *global* (independent of the space-time point) colour rotations.

Treating f_{abc} as generator, $(\mathrm{i}T^a(\mathbf{8}))_{bc} = f_{abc}$, one should try to forget that the indices of the structure constant are "all equal" and look upon a as the label (number) of the generator and b, c as numerating rows and columns of the 8×8 matrix:

$$\delta\phi_b \propto (\mathrm{i}T^a)_{bc}\, \phi_c\,,$$

with c the index of the operand, b that of the result.

2 QCD and Conservation of Colour

2.1 Local Colour Invariance and Gluon Field Strength

Let us recall how the QCD Lagrangian emerges from the heuristic idea of the invariance of the theory under the *local* colour transformations.

One starts by considering the trivial Dirac fermionic Lagrange function describing free quarks. Its colour $SU(N)$ invariance gets lost due to the kinetic term with spatial derivative when the colour rotation parameters are taken to be x-dependent:

$$\bar\psi\, \partial_\mu \psi \;\to\; \bar\psi\, U^{-1} \partial_\mu (U\psi) = \bar\psi\, \partial_\mu \psi \;+\; \bar\psi\, U^{-1}(\partial_\mu U)\psi \;\neq\; \bar\psi\, \partial_\mu \psi\,.$$

To rescue invariance of the quark Lagrangian under local colour rotations one invokes the good old QED experience where the photon was known to act as a "compensating field" for the x-dependent Abelian phase shifts of the complex fermion (and/or scalar) fields. To this end we replace the usual derivative by the "covariant" one,

$$\mathrm{D}_\mu = \partial_\mu - \mathrm{i}gA_\mu\,, \tag{30a}$$

and design a proper inhomogeneous transformation for the compensating gluon field

$$A_\mu \to U A_\mu U^{-1} - \frac{\mathrm{i}}{g}(\partial_\mu U)U^{-1}\,. \tag{30b}$$

Here A is the matrix constructed of eight gluon-field potentials

$$A_\mu \equiv t^a A_\mu^a\,.$$

Within this convention, $\bar\psi\, \mathrm{D}_\mu \psi$ stays invariant since

$$\begin{aligned} \mathrm{D}_\mu\psi \;\to\; & U\partial_\mu\psi + (\partial_\mu U)\psi \;-\; \mathrm{i}g\left\{U A_\mu U^{-1} - \frac{\mathrm{i}}{g}(\partial_\mu U)U^{-1}\right\} U\psi \\ &= U\left(\partial_\mu\psi - \mathrm{i}gA_\mu\psi\right) \;=\; U(\mathrm{D}_\mu\psi). \end{aligned}$$

In other words the differential operator D_μ (unlike the potential A_μ, *sic*!) transforms as a proper adjoint-representation object: It "rotates" homogeneously under the local transformations according to

$$\begin{aligned} \bar\psi \mathrm{D}_\mu \psi &\to (\bar\psi U^{-1})\,(U\mathrm{D}_\mu U^{-1})\,(U\psi)\,, \\ \mathrm{D}_\mu &\to U\mathrm{D}_\mu U^{-1}\,. \end{aligned}$$

Obviously, so does the product of such objects and, therefore, the commutator of two covariant derivatives:

$$\frac{i}{g}\,[\mathrm{D}_\mu\,,\,\mathrm{D}_\nu] \;=\; \partial_\mu A_\nu - \partial_\nu A_\mu - \mathrm{i}g\,[A_\mu\,A_\nu] \;=\; t^a\,F^a_{\mu\nu}\,, \tag{31a}$$

$$F^a_{\mu\nu} \;\equiv\; \partial_\mu A^a_\nu - \partial_\nu A^a_\mu \,+\, g f_{abc} A^b_\mu A^c_\nu\,. \tag{31b}$$

This is nothing but the gluon field tensor that gives the *gluonic* piece of the QCD Lagrangian, $(F^a_{\mu\nu})^2$.

Restricting ourselves to infinitesimal colour rotations, $U \approx \mathbb{1} + \mathrm{i}\delta\omega^a(x)\,t^a$, from (30b) we get

$$A_\mu \;\to\; A_\mu \,+\, \mathrm{i}\delta\omega^a(x)\;[t^a\,A_\mu] \,+\, \frac{1}{g}\,t^a\,\partial_\mu\delta\omega^a(x)\,,$$

or in the "vector" form

$$A^b_\mu \;\to\; A^b_\mu \,+\, \delta\omega^a(x)\,f_{abc}A^c_\mu \,+\, \frac{1}{g}\,\partial_\mu\delta\omega^b(x)\,.$$

For the case of x-independent phases, $\partial_\mu\delta\omega^a(x) \equiv 0$, this leads us back to (27) and (29), showing that the gluon fields thus introduced indeed transform under the adjoint representation of the *global* $SU^c(N)$.

2.2 Jacobi Identity

Duality between the matrix and vector forms of the octet representation is practically very useful. As an example of its use we shall prove an important relation between the structure constants known as the Jacobi identity:

$$f_{abe}\,f_{cde} \,+ f_{bce}\,f_{ade} \,+ f_{cae}\,f_{bde} \;=\; 0\,, \tag{32}$$

with a, b, c, d arbitrary external indices (and e the summation index). A mnemonic rule for memorizing (32) is to think of the cyclic permutation of the triplet $\{a, b, c\}$. This relation looks quite cumbersome, but it becomes very transparent in the pictorial form (see below).

The Jacobi identity naturally appears in the problem of colour transformation properties of the gluon-field strength tensor (31b). Indeed, $F^a_{\mu\nu}$ contains the bilinear term which should behave, under the global rotations, exactly as the gluon potential itself (the linear term), that is according to the adjoint (octet) representation:

$$\mathrm{i}T^d(\mathbf{8}) \,:\; F^a \;=\; f_{dae}\,F^e\,.$$

For the bilinear term under interest the r.h.s. becomes

$$f_{dae}\, f_{ebc}\, A_\mu^b A_\nu^c\,. \tag{33a}$$

On the other hand, we can find the l.h.s. by explicitly acting on each of the two potentials:

$$\begin{aligned} \mathrm{i}T^d(8):\ \left\{f_{abc}A_\mu^b A_\nu^c\right\} &= f_{abc}\left\{(\mathrm{i}T^a : A_\mu)^b\, A_\nu^c \,+\, A_\mu^b(\mathrm{i}T^a : A_\nu)^c\right\} \\ = f_{abc}\left(f_{dbe}A_\mu^e A_\nu^c + f_{dce}A_\mu^b A_\nu^e\right) &= \left(f_{aec}\, f_{deb} + f_{abe}\, f_{dec}\right) A_\mu^b A_\nu^c\,, \end{aligned} \tag{33b}$$

where in the latter equality we have interchanged the summation indices $b \leftrightarrow e$ in the first term and $c \leftrightarrow e$ in the second one. From (33) we get

$$f_{dae}\, f_{ebc} \;=\; f_{aec}\, f_{deb} + f_{abe}\, f_{dec}\,,$$

which coincides with (32) after recalling the antisymmetry $f_{dae} = -f_{ade}$ and the cyclic permutation $f_{aec} = f_{cae}$, $f_{deb} = f_{bde}$ and $f_{dec} = f_{cde}$.

Thus we have checked that the Jacobi relation takes care of the proper colour transformation of the field strength. To prove (32), let us shift to the matrix language. We consider the matrix

$$\frac{\mathrm{i}}{g}F_{\mu\nu}^{(2)} \;\equiv\; \mathrm{i}f_{abc}A_\mu^b A_\nu^c \cdot t^a \;=\; [A_\mu\, A_\nu]$$

and apply the generator T^a to the commutator of two potentials,

$$\begin{aligned} T^a:\ \{[A_\mu\,,\, A_\nu]\} &= [(T^a : A_\mu)\,,\, A_\nu] + [A_\mu\,,\, (T^a : A_\nu)] \\ &= [\,[t^a, A_\mu],\, A_\nu] + [A_\mu,\, [t^a, A_\nu]\,] \;=\; [\,t^a\,,\, [A_\mu, A_\nu]]\,, \end{aligned}$$

which shows that $F^{(2)}$ indeed transforms as an octet, as it should. The second line results from the cyclic identity for commutators

$$\left[t^a,[t^b,t^c]\right] \,+\, \left[t^b,[t^c,t^a]\right] \,+\, \left[t^c,[t^a,t^b]\right] \;=\; 0\,. \tag{34}$$

As familiar analogue from vector algebra, there is a similar relation for the vector products:

$$[\mathbf{A}\times[\mathbf{B}\times\mathbf{C}]] \,+\, [\mathbf{B}\times[\mathbf{C}\times\mathbf{A}]] \,+\, [\mathbf{C}\times[\mathbf{A}\times\mathbf{B}]] \;=\; 0\,.$$

Exercise 3. Verify that expressing the commutators in (34) via structure constants directly leads to the Jacobi identity (32).

2.3 Universality of Commutation Relations

Let us convolute relation (34) with an octet field ϕ_c, $\phi = \phi_c t^c$ and rewrite the result as

$$[t^a\,,\,[t^b,\phi]] - [t^b\,,\,[t^a,\phi]] \;=\; [[t^a,t^b]\,,\,\phi] = \mathrm{i}f_{abc}[t^c\,,\,\phi]. \tag{35}$$

The commutator with t^a is the octet generator $T^a(\mathbf{8})$ (in the matrix language). Therefore (35) can be represented in the operator form as

$$\left[T^a(\mathbf{8})\,,\,T^b(\mathbf{8})\right] \;=\; \mathrm{i}f_{abc}\,T^c(\mathbf{8})\,. \tag{36a}$$

We have *defined* the structure constants by the relation

$$[t^a\,,\,t^b] \;=\; \mathrm{i}f_{abc}t^c\,.$$

Since t^a is the generator in the fundamental representation, this tells us

$$\left[T^a(\mathbf{3})\,,\,T^b(\mathbf{3})\right] \;=\; \mathrm{i}f_{abc}\,T^c(\mathbf{3})\,. \tag{36b}$$

We conclude that the generators of the adjoint representation commute with each other exactly as the generators in the fundamental one. This property is of the most general nature and holds for *all* irreducible representations R:

$$\left[T^a(\mathbf{R})\,,\,T^b(\mathbf{R})\right] \;=\; \mathrm{i}f_{abc}\,T^c(\mathbf{R})\,. \tag{37}$$

Exercise 4. Verify (37) for $T^a(\bar{\mathbf{3}})$ and $T^a(\mathbf{6})$. (Working with $T^a(\bar{\mathbf{3}}) = -t^a$, remember that the $\bar{\mathbf{3}}$ generators act on the antiquark state *from the right*).

2.4 Conservation of Colour Current

In physical terms, universality of the generator algebra is intimately related with *conservation of colour*. To illustrate this point let us consider production of a quark-gluon pair in some hard process and address the question of how this system radiates. Let p and k be the momenta of the quark and the gluon and b the octet colour index of the latter. For the sake of simplicity we concentrate on *soft* accompanying radiation, which determines the bulk of particle multiplicity inside jets, the structure of the hadronic plateau, etc. As far as emission of gluons with momenta $\ell \ll k, p$ is concerned, the so-called "soft insertion rules" apply, which tell us that the Feynman diagrams dominate where ℓ is radiated off external ("real") partons. The corresponding diagrams for our process are shown in Fig. 4.

Please notice that the colour factor in the 3-gluon vertex corresponds to the action of the generator $T^a(\mathbf{8})$ (emission of a gluon in the colour state a) on the octet A^b:

$$(T^a(\mathbf{8}) : A)_b \;=\; -\mathrm{i}f_{abc} = \mathrm{i}f_{bac}\,.$$

Do two emission amplitudes interfere with each other? It depends on the direction of the radiated gluon $\boldsymbol{\ell}$.

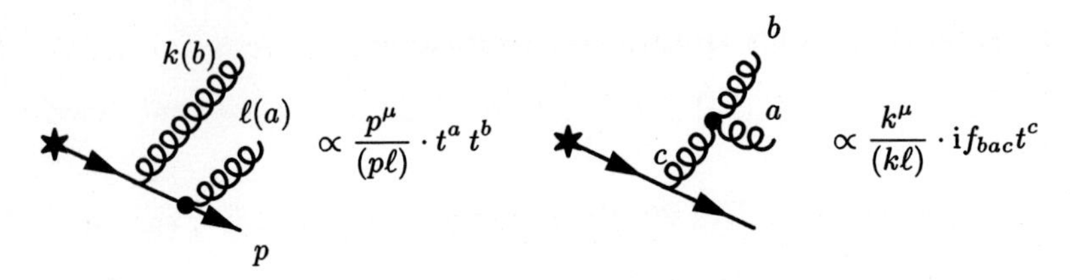

Fig. 4. *Feynman diagrams for radiation of the soft gluon with momentum ℓ and colour a off the qg system.*

In the first place, there are two bremsstrahlung cones centred around the directions of **p** and **k**:

$$\text{quark cone:} \quad \Theta_\ell \equiv \Theta_{\ell\mathbf{p}} \ll \Theta \approx \Theta_{\ell\mathbf{k}} \,,$$
$$\text{gluon cone:} \quad \Theta_\ell \equiv \Theta_{\ell\mathbf{k}} \ll \Theta \approx \Theta_{\ell\mathbf{p}} \,,$$

with $\Theta \equiv \Theta_{\mathbf{pk}}$ the aperture of the qg fork. In these regions one of the two amplitudes of Fig. 4 is much larger than the other, and the interference is negligible; the gluon ℓ is radiated independently and participates in the formation of the quark and gluon sub-jets. If Θ is sufficiently large and k sufficiently energetic (relatively hard, $k \sim p$), these two sub-jets can be distinguished in the final state. Generally speaking their properties should be remarkably different. In particular, the particle density in q and g jets (at least asymptotically) should be proportional to the probability of soft gluon radiation which, in turn, is proportional to the "squared colour charge" of a quark/gluon. As we shall shortly see, this results in the ratio

$$\left(\ell\frac{\mathrm{d}n}{\mathrm{d}\ell}\right)^{g}_{\Theta_\ell<\Theta} : \left(\ell\frac{\mathrm{d}n}{\mathrm{d}\ell}\right)^{q}_{\Theta_\ell<\Theta} = N : \frac{N^2-1}{2N} = 3 : \frac{4}{3} = \frac{9}{4}\,.$$

Multijet configurations are comparatively rare: Emission of an additional hard gluon $k \sim p$ at large angles $\Theta \sim 1$ constitutes a fraction $\alpha_s/\pi \lesssim 10\%$ of all events. Typically, k would prefer to belong to the quark bremsstrahlung cone itself, i.e. to have $\Theta \ll 1$. In such circumstances the question arises about the structure of the accompanying radiation at comparatively *large* angles

$$\Theta_\ell = \Theta_{\ell\mathbf{p}} \simeq \Theta_{\ell\mathbf{k}} \gg \Theta\,. \tag{38}$$

If the quark and the gluon were acting as independent emitters, we would expect the particle density to increase correspondingly and to overshoot the standard quark jet by a factor

$$\left(\ell\frac{\mathrm{d}n}{\mathrm{d}\ell}\right)^{g+q}_{\Theta_\ell>\Theta} : \left(\ell\frac{\mathrm{d}n}{\mathrm{d}\ell}\right)^{q}_{\Theta_\ell>\Theta} = N : \frac{N^2-1}{2N} + 1 = \frac{13}{4}\,. \tag{39}$$

However, in this angular region our amplitudes start to interfere significantly, so that radiation off the qg pair is no longer given by the *sum of probabilities* $q \to g$ plus $g \to g$. We have to square the *sum of amplitudes* instead.

This can easily be done by observing that in the large-angle kinematics (38) the angle Θ between $\mathbf{p}$ and $\mathbf{k}$ can be neglected, so that the accompanying soft radiation factors become indistinguishable,

$$\frac{k}{2(k\ell)} \simeq \frac{p}{2(p\ell)} = \frac{1}{2\ell(1-\cos\Theta_\ell)} \simeq \frac{1}{\ell\,\Theta_\ell^2} \simeq \frac{\ell}{\ell_\perp^2}\,.$$

Thus the Lorentz structure of the amplitudes becomes the same and it suffices to sum the colour factors:

$$t^a\,t^b + \mathrm{i}f_{bac}t^c = t^a\,t^b + \left[t^b\,,\,t^a\right] \equiv t^b\,t^a\,. \tag{40}$$

We conclude that the coherent sum of two amplitudes of Fig. 4 results in radiation at large angles *as if* off the initial quark, as shown in Fig. 5.

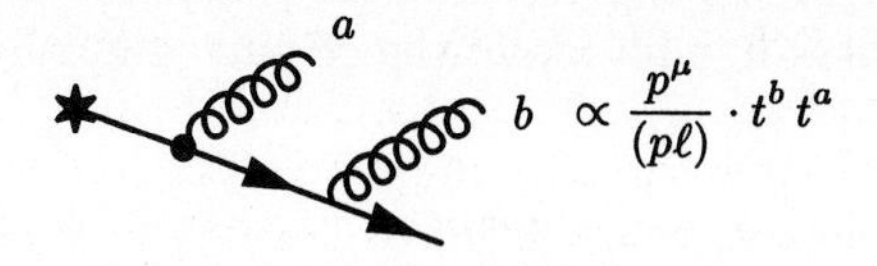

Fig. 5. *Radiation at large angles is determined by the total colour charge.*

This means that the naive probabilistic expectation of enhanced density (39) fails and the particle yield is equal to that for the quark-initiated jet instead: $13/4 \to 1$. It actually does not matter whether the gluon k was present at all, or whether there was a whole bunch of partons with small relative angles instead. Soft gluon radiation at large angles is sensitive only to the *total* colour charge of the final parton system, which equals the colour charge of the initial parton. This physically transparent statement holds not only for the quark as in Figs. 4, 5 but for an arbitrary object $\mathbf{R}$ (gluon, diquark, ..., you name it) as an initial object. In this case the matrices $t = T(\mathbf{3})$ should be replaced by the generators $T(\mathbf{R})$ corresponding to the colour representation $\mathbf{R}$, and (40) holds due to the universality of the generator algebra (37).

Exercise 5. Check the colour conservation for the case of an initial gluon. Consider the decay of a virtual gluon b into two "real" gluons $c + d$. Write down the colour factors for radiation of an extra gluon a off c and d. Use the Jacobi identity (32) to show that their sum (describing large-angle radiation) is equivalent to the radiation off the parent b.

Exercise 6. Produce a $q\bar{q}$ pair. Check that the coherent radiation off q and $\bar{q}$ cancels out when qq are produced by a colour singlet source (photon $\to q\bar{q}$) and adds up into radiation off a *gluon* in the case of $g \to q\bar{q}$ splitting process.

2.5 Angular Ordering in Parton Cascades

QCD is a *quantum* theory, no doubt about it. This very statement seems to make the problem of describing a parton system involving $n \gg n_0$ gluons and quarks (with the actual value of $n_0 \sim 1$ depending on your computer) look hopeless: Solving such a problem would call for sorting out and calculating $\mathcal{O}((n!)^2)$ Feynman diagrams.

Why should we worry about multiparton systems in the first place? Is it not true that the squared matrix element in the nth order of perturbation theory is proportional to $(\alpha_s/\pi)^n \lesssim (0.1)^n$ and, thus, vanishingly small for large n? The answer to this (as to many other questions, according to the celebrated Hegel dialectic wisdom) is: "Yes and No". Indeed,

Yes, it is very small, if we talk about a "multijet" configuration of 10 energetic quarks and gluons with large angles between them;

No, it is of order *unity*, if we address the *total* probability of having 10 extra gluons (and quarks) in addition to the $q\overline{q}$ pair originally produced in e^+e^- annihilation at LEP.

Allowing small relative angles between partons in a process with a large hardness Q^2 results in a logarithmic enhancement of the emission probability:

$$\alpha_s \Longrightarrow \alpha_s \frac{\mathrm{d}\Theta^2}{\Theta^2} \to \alpha_s \log Q^2\,, \tag{41a}$$

so that the total probability of one parton (E) turning into two ($E_1 \sim E_2 \sim \frac{1}{2}E$) may become of order 1, in spite of the smallness of the characteristic coupling, $\alpha_s(Q^2) \propto 1/\log Q^2$. A typical example of such a "collinear" enhancement — the splitting process $g \to q\overline{q}$.

Moreover, when we consider the *gluon* offspring, another "soft" enhancement enters the game, which is due to the fact that the gluon bremsstrahlung tends to populate the region of *relatively* small energies ($E \simeq E_1 \gg E_2 \equiv \omega$):

$$\alpha_s \Longrightarrow \alpha_s \frac{\mathrm{d}\omega}{\omega}\frac{\mathrm{d}\Theta^2}{\Theta^2} \to \alpha_s \log^2 Q^2\,. \tag{41b}$$

Thus the true perturbative "expansion parameter" responsible for parton multiplication via $q \to qg$ and $g \to gg$ may actually become *much larger* that 1!

In such circumstances we cannot trust the perturbative expansion in $\alpha_s \ll 1$ unless the logarithmically enhanced contributions (41) are taken full care of in all orders.

Fortunately, in spite of the complexity of high order Feynman diagrams, such a programme can be carried out. There is a physical reason for that: Large contributions (41) originate from a specific region of phase space, which can be viewed as a sequence of parton decays strongly ordered in fluctuation times. Given such a separation in time, successive parton splittings become independent, so that the emerging picture is essentially classical. This is how the parton cascades described by the classical equations of parton balance (evolution equations) come about.

However, in applying this picture some care should be exercised. As long as *soft* gluon radiation is concerned, being emitted *later* does not necessarily mean being emitted *independently*. As we have seen above, the quantum-mechanical coherence cuts off radiation at angles exceeding the angle between the emitters. As a result, the classical probabilistic picture of independent parton multiplication is applicable to *angular ordered* parton configurations with successively decreasing relative angles (the parton tree as a cypress rather than an oak).

It is universality of the algebra of generators (37), which is another way of pronouncing "conservation of colour", that makes the angular ordering prescription universal with respect to the nature (colour representation) of participating emitters.

3 Colour Charges

3.1 Casimir Operators and Gluon Radiation Intensity

Consider the "square" of the generator in some representation $\mathbf{R}$

$$T^2 \equiv T^a T^a .$$

This operator commutes with each of the generators. Indeed, such a commutator is identically zero as a convolution of an antisymmetric tensor with a symmetric one:

$$\begin{aligned}[T^2, T^b] &= T^a T^a T^b - T^b T^a T^a = T^a [T^a, T^b] + [T^a, T^b] T^a \\ &= i f_{abc} \left(T^a T^c + T^c T^a \right) = 0 .\end{aligned}$$

Therefore (Schur's Lemma) T^2 is proportional to the unit operator in a given representation. For example,

$$(T^a(\mathbf{3}))^i_j \, (T^a(\mathbf{3}))^j_k = (t^a)^i_j (t^a)^j_k = C_F \delta^i_k \equiv C_F \mathbb{1} , \quad \text{(42a)}$$

$$(T^a(\mathbf{8}))_{bd} (T^a(\mathbf{8}))_{dc} = (-\mathrm{i} f_{abd})(-\mathrm{i} f_{adc}) = f_{bad} f_{cad} = C_A \delta_{bc} \equiv C_A \mathbb{1} . \quad \text{(42b)}$$

The numbers C_F and C_A are known as "Casimir operators" of the fundamental (F) and adjoint (A) representations. On the other hand, these are the colour factors that determine intensity of gluon radiation off a quark and a gluon. Probability of gluon radiation is given by the squared amplitude summed over colours in the final state. Pictorially, it takes the form of (the imaginary part of) the loop diagram

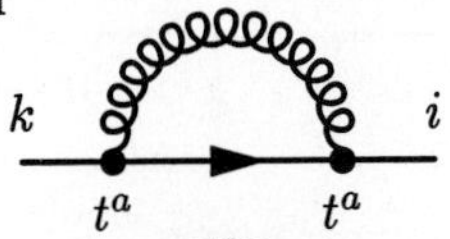

$$= C_F \mathbb{1} = \frac{N^2 - 1}{2N} \, \delta^i_k , \quad \text{(43a)}$$

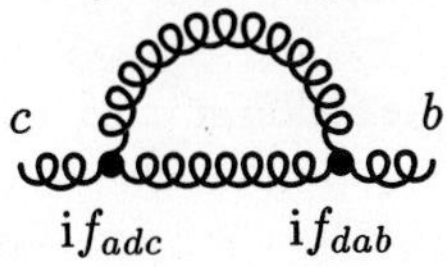

$$= C_A \mathbb{1} = N \, \delta_{bc} . \quad \text{(43b)}$$

From this point of view the statement of Schur's Lemma looks trivial: Emission and absorption of the same gluon leaves the colour state of the radiating object unchanged.

To find the squared "QCD charges" we first derive an extremely useful relation known as Fierz identity (completeness relation) for the $SU(N)$ group.

3.2 Fierz Identity

Let us again apply our knowledge of presenting an arbitrary $N \times N$ matrix M as a linear combination of generators. This time we choose for M an elementary matrix with only one non-zero component $M^i_k = \delta^{\,i}_{(j)}\,\delta^{(l)}_{\;k}$, where we assume j and l fixed. The general formula (22) then gives

$$\delta^i_j\,\delta^l_k = \frac{1}{N}\,\delta^i_k\,\delta^l_j + 2\,(t^a)^i_k\,(t^a)^l_j\;. \tag{44}$$

A pictorial representation of (44) is given in Fig. 6.

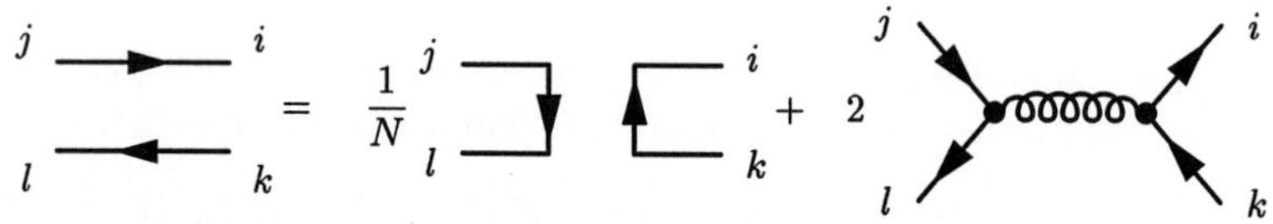

Fig. 6. *Pictorial representation of the Fierz identity (44).*

Considering i of M^i_k as the colour index of a quark and k as the colour index of an antiquark, the physical interpretation of this picture becomes clear: It is nothing but a decomposition of N^2 colour states of a $q\bar{q}$ system into the colour singlet and colour octet parts, which we have actually done before in the algebraic form, see (24) and (26).

It is convenient to introduce *colour projection operators* $P_{(0)}$ and $P_{(8)}$,

$$\mathbb{1}(\mathbf{3})\cdot\mathbb{1}(\bar{\mathbf{3}}) \;=\; P_{(0)} + P_{(8)}\,; \tag{45a}$$

$$P_{(0)} \;=\; \frac{1}{N} \tag{45b}$$

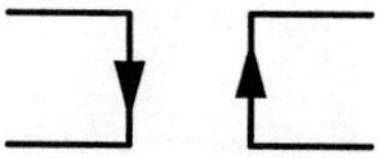

$$P_{(8)} \;=\; 2 \tag{45c}$$

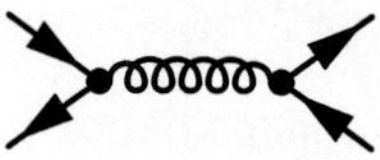

Exercise 7. Check *pictorially* that P are the true projectors: $P^2_{(0)} = P_{(0)}$, $P^2_{(8)} = P_{(8)}$, $P_{(0)}P_{(8)} = P_{(8)}P_{(0)} = 0$. Hint: use

$$= \mathrm{Tr}\,(\mathbb{1}) = \mathrm{N}\,; \qquad\qquad = \mathrm{Tr}\,(\mathrm{t^a}) = 0\,. \tag{46}$$

The Fierz identity is a powerful tool. An alternative way of looking at Fig. 6 is to express the colour structure of one-gluon exchange between two quarks (or a quark and an antiquark) in terms of plain quark lines. We have

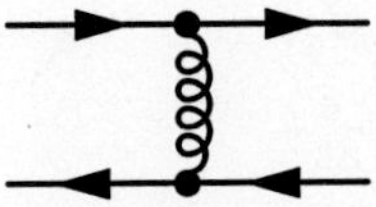

$$= \frac{1}{2} \qquad - \frac{1}{2N} \tag{47a}$$

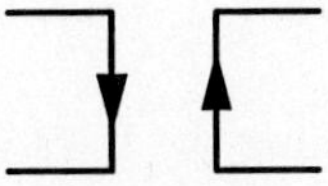

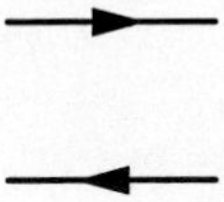

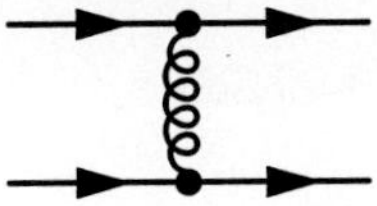

$$= \frac{1}{2} \qquad - \frac{1}{2N} \tag{47b}$$

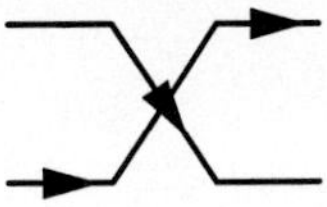

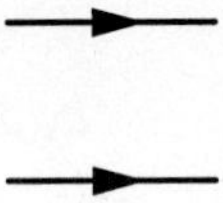

As for the 3-gluon vertex, it can be substituted (colourwise) by gluon-quark interactions according to (23), this relation having the following pictorial representation

$$\mathrm{i}f_{abc} \quad = \quad \overset{b}{\underset{a\qquad c}{}} \quad = \frac{1}{2}\left(\qquad - \qquad \right) \tag{48}$$

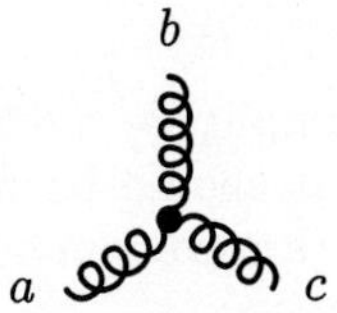

Hereafter we adopt the convention of labelling the 3-gluon f-vertices *clockwise*, as in (48).

Applying (47) and (48) allows one to get rid of gluon lines altogether and thus to trivialise the calculation of colour factors of arbitrarily complex QCD diagrams.

3.3 Quark Colour Charge

To find the squared colour charge of a quark, C_F, let us set and sum over $i = j$ in the Fierz identity (44). Pictorially this means joining the quark lines in Fig. 6 between the points i and j as shown below:

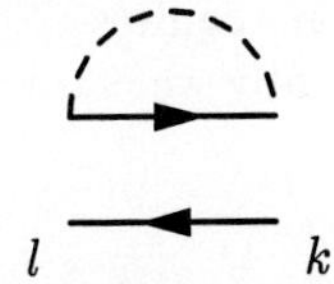

$$\underset{l\qquad\qquad k}{} = \frac{1}{N}\,\underset{l\qquad\qquad k}{} + \;2\;\underset{l\qquad\qquad k}{}$$

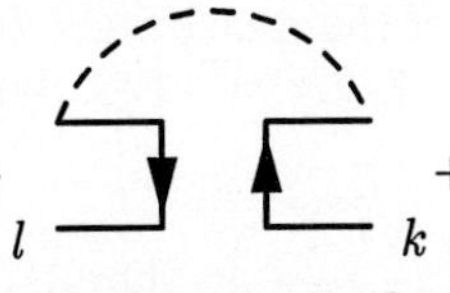

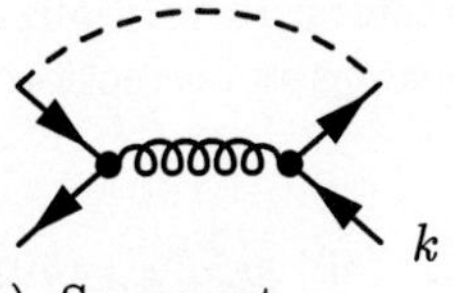

The last graph here is topologically identical to (43a). So we get

$$N\,\delta^l_k \;=\; \frac{1}{N}\,\delta^l_k \;+\; 2C_F\,\delta^l_k \qquad \Longrightarrow \qquad C_F = \frac{N^2-1}{2N} = \frac{4}{3}\,.$$

3.4 QCD Vertex Corrections

Let us now apply to Fig. 6 a similar trick of contracting a $q\bar{q}$ pair, but now sandwiching a gluon vertex between quarks:

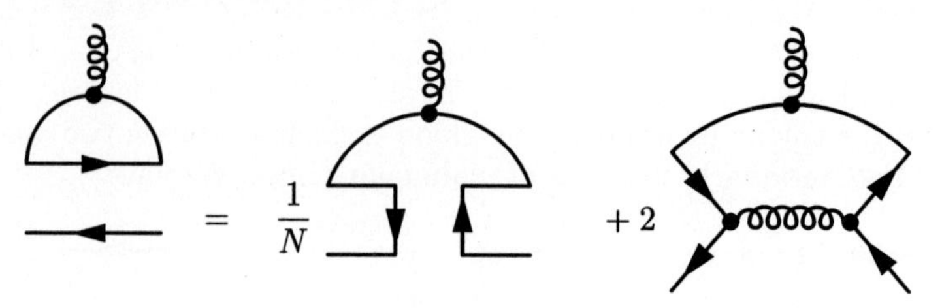

Making use of (46) and stretching the lines we arrive at

$$= -\frac{1}{2N} \tag{49a}$$

Algebraically,

$$t^a\, t^b\, t^a \;=\; -\frac{1}{2N}\, t^b\,. \tag{49b}$$

The graph in the l.h.s. of (49a) as a Feynman amplitude describes one-loop correction to the quark-gluon interaction vertex. It should be compared with the diagram for the one-loop correction to the quark wave function

$$= \frac{N^2-1}{2N} \tag{50}$$

These corrections are present in QED as well. Substituting electrons and photons for quarks and gluons we have exactly the same expressions, the only difference being the colour factors, which are obviously absent in the QED case.

Vertex and wave-function renormalization corrections possess logarithmic ultraviolet divergences. In QED they are known to cancel due to the Ward identity. In the QCD context, however, this cancellation is broken by mismatch in the colour factors: Renormalization of the quark wave function acquires $C_F = 4/3$, while the vertex correction only $-1/2N = -1/6$. Symbolically we can write the sum as

$$\delta Z_{\text{w.f.}}(50) + \delta Z_{\text{vert.}}(49) \;=\; \left(\frac{N^2-1}{2N}\right)\cdot A\ln\Lambda - \left(-\frac{1}{2N}\right)\cdot A\ln\Lambda\,. \tag{51}$$

The latter term is relatively suppressed by the factor $1/(N^2-1)$, which vanishes in the large-N limit. It looks as if the quark has become sterile in the vertex correction graph (49a). After virtually splitting q into a qg pair, we find the quark no longer capable of interacting with an external gluon as if it had lost its "colour charge". This is true in some sense. It had. But the colour did not vanish in thin air: It was transferred to the loop *gluon*, which is as legitimate a colour-bearer as a quark is. So we come to considering an additional vertex correction in which the loop gluon is being probed instead of the quark,

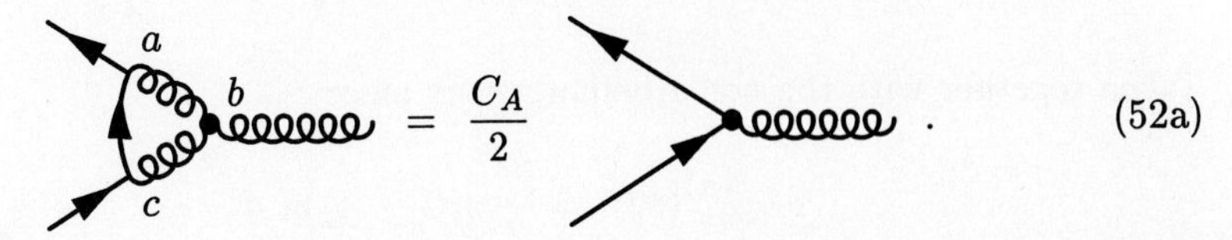

Algebraically,

$$t^a\, t^c\, \mathrm{i} f_{abc} \;=\; \frac{C_A}{2}\, t^b\,, \tag{52b}$$

which we immediately obtain by representing the product of generators as

$$t^a\, t^c \;=\; \tfrac{1}{2}[t^a,t^c] + \tfrac{1}{2}\{t^a,t^c\} \quad\Longrightarrow\quad \tfrac{1}{2}\,\mathrm{i} f_{acd}\, t^d + \dots\,,$$

dropping the symmetric piece ($\{t^a,t^c\}\, f_{abc} \equiv 0$) and invoking the definition of C_A in (42b).

This specific non-Abelian term in the $q\bar{q}g$ vertex function diverges logarithmically as well and works in a pool with the other two contributions to rescue colour conservation. Since we expect it to cover the mismatch between the already known colour factors in (51), this provides us with the means of finding the octet Casimir operator C_A.

3.5 Gluon Colour Charge

Construct the difference of the colour factors in (51) and trace the following chain of equalities:

$$\begin{aligned}\frac{N}{2}\, t^b &\equiv \left(\frac{N^2-1}{2N}\right) t^b - \left(-\frac{1}{2N}\right) t^b \\ &= t^a t^a\, t^b - t^a\, t^b\, t^a = t^a\,(\mathrm{i} f_{abc} t^c) = \left(\frac{\mathrm{i}}{2} f_{acd} t^d\right) \mathrm{i} f_{abc} = \tfrac{1}{2} f_{acd} f_{acb} t^d \equiv \frac{C_A}{2} t^b\,.\end{aligned}$$

The result is $C_A = N$, as was stated in (43b).

Another important message comes from the observation that we did not use, in the second line of the above equation, anything but the commutation relation (37), which is already known to us to be *universal.* Therefore, for an arbitrary object substituted for the quark we have

$$T^a(\mathbf{R})\, T^b(\mathbf{R})\, T^a(\mathbf{R}) \;=\; \left(C(\mathbf{R}) - \frac{N}{2}\right)\, T^b(\mathbf{R})\,. \tag{53}$$

$T^aT^bT^a$ is the colour factor for the pseudo-Abelian vertex correction; $C(\mathbf{R})$ that for the wave-function renormalization. Both depend on the colour representation of the scattered object. However, in their *difference*, which only enters into the expression for QCD coupling renormalization, the dependence on $\mathbf{R}$ cancels:

$$\delta Z^{(\mathbf{R})}_{\text{w.f.}}(50) + \delta Z^{(\mathbf{R})}_{\text{vert.}}(49) \;=\; C(\mathbf{R})\cdot A\ln\Lambda - \left(C(\mathbf{R}) - \frac{N}{2}\right)\cdot A\ln\Lambda\,. \quad (54a)$$

Taken together with the non-Abelian vertex piece,

$$\delta Z^{(\mathrm{NA})}_{\text{vert.}}(52) \;=\; \frac{N}{2}\cdot(G-A)\ln\Lambda\,, \quad (54b)$$

and (half of) the wave-function renormalization of the external gluon,

$$\tfrac{1}{2}\delta Z^{(\text{glue})}_{\text{w.f.}} \;=\; \left\{\frac{g_s^2}{16\pi^2}\left(\frac{11}{3}N - \frac{2}{3}n_f\right) - \frac{N}{2}\cdot G\right\}\ln\Lambda\,, \quad (54c)$$

this results in the *universal* renormalized coupling g_s. Remark that the pieces of (54) separately have little sense since the numbers A and G are gauge-dependent. Meanwhile, the physical coupling is not, and you recognize the famous one-loop QCD β-function in the coefficient of the ultraviolet log:

$$\begin{aligned}\delta g_s \;&=\; g_s\left(\delta Z^{(\mathbf{R})}_{\text{w.f.}} + \delta Z_{\text{vert.}} + \delta Z^{(\mathrm{NA})}_{\text{vert.}} + \tfrac{1}{2}\delta Z^{(\text{glue})}_{\text{w.f.}}\right)\\ &=\; \frac{g_s^3}{16\pi^2}\left(\frac{11}{3}N - \frac{2}{3}n_f\right)\ln\Lambda\,.\end{aligned} \quad (55)$$

3.6 Gluon Exchange Potential

The QED Feynman amplitude of scattering of two charges via photon exchange

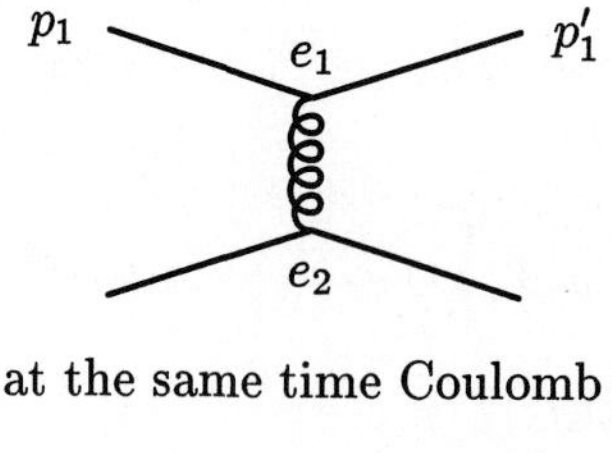

p_1 e_1 p_1'

e_2

$$\propto \frac{e_1\,e_2}{k^2}\,, \quad k = p_1' - p_1\,, \quad (56a)$$

describes at the same time Coulomb interaction energy

$$V(r) \;=\; \frac{e_1\,e_2}{r}\,, \quad (56b)$$

with $r = |\mathbf{r}_1 - \mathbf{r}_2|$ the distance between the charges. In QCD the *gluon* Green function coincides with that of the photon, at least in the region of large momentum transfer (virtuality) k^2. Therefore at sufficiently small distances $r \ll \Lambda^{-1} \sim 1\,\text{fm}$ we can talk about the Coulomb QCD interaction between quarks.

Turning to the QCD interaction picture, we have to replace the e.m. charges in (56) by the colour generators and sum over all colour states of the intermediate gluon:

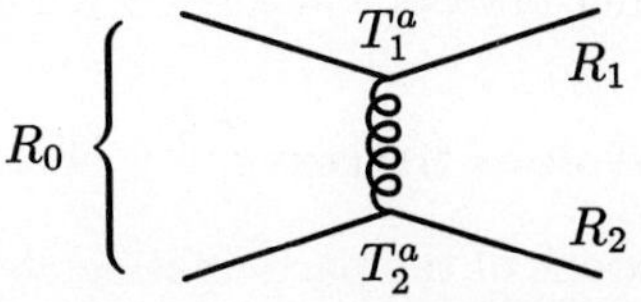

R_0 T_1^a R_1 T_2^a R_2

$$V(r) = g_s^2 \frac{T_1^a T_2^a}{r} \equiv \frac{g_s^2}{r} v_{12}\,, \tag{57}$$

Now $v_{12} = T_1^a T_2^a$ is an operator acting on the colour indices of the 2-particle state. As a result, the interaction energy (eigenvalue of the operator) will depend not only on the nature of participating objects ($\mathbf{R}_{1,2} = q, \bar{q}, g$, etc.) but on the overall colour state $\mathbf{R}_0$ of the pair as well. It is like the spin-spin interaction, whose magnitude surely depends on s_1, s_2 but also on the total spin of the system $s_0 = |\mathbf{s}_1 + \mathbf{s}_2|$, $|s_1 - s_2| \le s_0 \le s_1 + s_2$.

For example, for a quark-antiquark pair ($\mathbf{R}_1 = 3$, $\mathbf{R}_2 = \bar{3}$) we have

$$\begin{aligned} v_{\mathbf{3\bar{3}}} &= T^a(\mathbf{3})\, T^a(\mathbf{\bar{3}})\,; \\ v_{\mathbf{3\bar{3}}} : \psi^i \bar{\chi}_\ell &= (t^a)^i_k (-t^a)^j_\ell\, \psi^k \bar{\chi}_j\,. \end{aligned}$$

As we know, $q\bar{q}$ can be either in a singlet or an octet state. In the first case $\psi^k \bar{\chi}_j \propto \delta^k_j$ and we get

$$\begin{aligned} v_{\mathbf{3\bar{3}}} : \delta^i_\ell &= (t^a)^i_k (-t^a)^j_\ell\, \delta^k_j = -(t^a t^a)^i_\ell = -C_F \delta^i_\ell\,; \\ v_{\mathbf{3\bar{3}}}(\mathbf{1}) &= -C_F = -\frac{N^2-1}{2N} = -\frac{2}{3}\,. \end{aligned} \tag{58a}$$

For the octet state, $\psi^k \bar{\chi}_j (t^b)^j_\ell\, \phi_b$,

$$\begin{aligned} v_{\mathbf{3\bar{3}}} : (t^b)^i_\ell &= (t^a)^i_k (-t^a)^j_\ell\, (t^b)^k_j = -(t^a t^b t^a)^i_\ell = \frac{1}{2N}\,(t^b)^i_\ell\,; \\ v_{\mathbf{3\bar{3}}}(\mathbf{8}) &= \frac{1}{2N} = \frac{1}{6}\,, \end{aligned} \tag{58b}$$

where we have used (49). As expected, the operator v is diagonal in both cases: $\mathbf{1} \to \mathbf{1}$, $\mathbf{8} \to \mathbf{8}$.

Considering the interaction between two quarks we have

$$\begin{aligned} v_{\mathbf{3\bar{3}}} &= T^a(\mathbf{3})\, T^a(\mathbf{3})\,; \\ v_{\mathbf{3\bar{3}}} : \psi^i \chi^j &= (t^a)^i_k (t^a)^j_\ell\, \psi^k \chi^\ell\,. \end{aligned}$$

For the sextet we substitute the symmetric combination $R^{k\ell} = \psi^k \chi^\ell + \psi^\ell \chi^k$ to obtain

$$\begin{aligned} v_{\mathbf{33}} : R^{ij} &= (t^a)^i_k (t^a)^j_\ell\, R^{k\ell} = \left(\frac{1}{2}\delta^i_\ell \delta^j_k - \frac{1}{2N}\delta^i_k \delta^j_\ell\right) R^{k\ell} = \frac{1}{2} R^{ji} - \frac{1}{2N} R^{ij}\,; \\ v_{\mathbf{33}}(\mathbf{6}) &= \frac{N-1}{2N} = \frac{1}{3}\,. \end{aligned} \tag{59}$$

Here we employed the Fierz identity (44) and the symmetry of the R^{ij} state.

For the $\bar{\mathbf{3}}$ state of the qq pair, a straightforward calculation can similarly be performed. We shall use instead a flanking manoeuvre, which will teach us how to find interaction energy for *arbitrary* colour objects.

3.7 Interaction Between Arbitrary Colour States

Let $\mathbf{R}_1$ and $\mathbf{R}_2$ be colour group representations of the participating objects and R_0 their total colour state, as shown in (57). We may look upon it as a decay of $\mathbf{R}_0$ into $\mathbf{R}_1 + \mathbf{R}_2$. Since the colour current conserves in the course of "decay",

$$T^a(\mathbf{R}_0) \;=\; T^a(\mathbf{R}_1) + T^a(\mathbf{R}_2)\,. \tag{59a}$$

Squaring this equality we obtain

$$\begin{aligned} \left(T^a(\mathbf{R}_0)\right)^2 &= \left(T^a(\mathbf{R}_1)\right)^2 + \left(T^a(\mathbf{R}_2)\right)^2 + 2T^a(\mathbf{R}_1)T^a(\mathbf{R}_2)\,, \\ v_{\mathbf{R}_1\mathbf{R}_2}(R_0) &= T^a(\mathbf{R}_1)T^a(\mathbf{R}_2) \;=\; \tfrac{1}{2}\left(C(\mathbf{R}_0) - C(\mathbf{R}_1) - C(\mathbf{R}_2)\right). \end{aligned} \tag{59b}$$

Thus we have expressed the potential energy in terms of the Casimir operators (squared colour charges) of three representations.

For a pair of quarks in the antisymmetric state $\bar{\mathbf{3}}$, we have $C(\mathbf{R}_0) = C(\mathbf{R}_1) = C(\mathbf{R}_2) \equiv C_F$, which gives

$$v_{\mathbf{33}}(\bar{\mathbf{3}}) \;=\; -\tfrac{1}{2}C_F = -\frac{N^2-1}{4N} \;=\; -\frac{2}{3}\,. \tag{60}$$

Putting things together,

$$\begin{array}{lll} \text{Attraction:} & q\bar{q}(\text{singlet}) = -\dfrac{4}{3}\,; & qq(\overline{\text{triplet}}) \;=\; -\dfrac{2}{3}\,; \\ \text{Repulsion:} & q\bar{q}(\text{octet}) \;=\; +\dfrac{1}{6}\,; & qq(\text{sextet}) \;=\; +\dfrac{1}{3}\,. \end{array} \tag{61}$$

In reality one-gluon exchange may be too naive a picture to take responsibility for binding quarks into hadrons. In spite of this pessimistic remark, the very fact that the colour force between quarks according to (61) tends to attract $q\bar{q}$ into colourless mesons and qq pairs into baryonic $\bar{\mathbf{3}}$ compounds is supposed to give you that warm fuzzy feeling in your stomach.

What about interaction between two gluons or, say, between a quark and a gluon? For two gluons in colour *singlet* and *octet* states, we immediately derive from (59b)

$$v_{\mathbf{88}}(\mathbf{1}) \;=\; -N \;=\; -3\,; \qquad v_{\mathbf{88}}(\mathbf{8}) \;=\; -\tfrac{1}{2}N \;=\; -\frac{3}{2}\,. \tag{62}$$

This is not the end of the story, however, because the gg system has a richer "colour content":

$$\mathbf{8}\otimes\mathbf{8} \;=\; \mathbf{1}\oplus\mathbf{8}_a\oplus\mathbf{8}_s\oplus\mathbf{10}\oplus\overline{\mathbf{10}}\oplus\mathbf{27}\,. \tag{63}$$

As we shall see shortly there are two ways of constructing colour octets out of two gluons, namely antisymmetric (a) and symmetric (s) states with respect to the colours of the participants. Therefore two octets have appeared in the r.h.s. of (63), which have identical $SU^c(3)$ transformation properties.

To get any further, we need to learn about higher representations and their respective "charges" $C(\mathbf{R})$.

4 Beyond 3 and 8

4.1 High Irreducible Representations of $SU^c(3)$

The standard technique for constructing irreducible representations can be found in group theory textbooks. To give you a feeling, let us take qg system as an example. The product of quark and gluon wave functions bears three colour indices $\psi^i A^j_k$. With those we can do the following:
contract the upper and the lower index,

$$\psi^i A^j_i = R^j \Longrightarrow \mathbf{3} \tag{64a}$$

(Remember: Contraction of another pair $j = k$ would give zero because of irreducibility of $A(\mathbf{8})$),
antisymmetrize (pull down) two upper indices,

$$\psi^i A^j_k \cdot \epsilon_{ij\ell} = R_{k\ell} \Longrightarrow \bar{\mathbf{6}}\,, \tag{64b}$$

symmetrize the upper indices,

$$\psi^i A^j_k + \psi^i A^j_k = R^{ij}_k \Longrightarrow \mathbf{15}\,. \tag{64c}$$

The first two objects, we are familiar with; the latter (64c) is a new irreducible representation. It has dimension 15 and is described by a tensor with two upper and one lower quark indices (symbolically, $qq\bar{q}$).

In general, an irreducible representation $\mathbf{R}$ of the $SU(3)$ group can be described by a traceless tensor which is symmetric, separately, with respect to its p upper and q lower indices:

$$R^{\{i_1,i_2\ldots i_p\}}_{\{k_1,k_2\ldots k_q\}}\,; \qquad R^{\alpha,i_2\ldots i_p}_{\alpha,k_2\ldots k_q} = 0\,. \tag{65a}$$

Exercise 8. Prove that the tensor (65a) has

$$K(p,q) = \tfrac{1}{2}(p+1)(q+1)(p+q+2) \tag{65b}$$

degrees of freedom (dimension of the representation).

Given this technology, the multiplication table follows:

$$\mathbf{3}\otimes\mathbf{3} = \bar{\mathbf{3}}\oplus\mathbf{6}\,, \qquad \bar{\mathbf{3}}\otimes\mathbf{3} = \mathbf{1}\otimes\mathbf{8}$$
$$\mathbf{3}\otimes\mathbf{6} = \mathbf{8}\oplus\mathbf{10}\,, \qquad \bar{\mathbf{3}}\otimes\mathbf{6} = \mathbf{3}\oplus\mathbf{15}$$

$$\bar{\mathbf{3}} \otimes \mathbf{8} = \mathbf{3} \oplus \bar{\mathbf{6}} \oplus \mathbf{15}$$

$$\mathbf{3} \otimes \mathbf{10} = \mathbf{15} \oplus \mathbf{15'}\,, \qquad \bar{\mathbf{3}} \otimes \mathbf{10} = \mathbf{6} \oplus \mathbf{24}$$

$$\mathbf{6} \otimes \mathbf{6} = \bar{\mathbf{6}} \oplus \mathbf{15} \oplus \mathbf{15'} \qquad \bar{\mathbf{6}} \otimes \mathbf{6} = \mathbf{1} \oplus \mathbf{8} \oplus \mathbf{27}$$

$$\mathbf{6} \otimes \mathbf{8} = \bar{\mathbf{3}} \oplus \mathbf{6} \oplus \overline{\mathbf{15}} \oplus \mathbf{24}$$

$$\mathbf{6} \otimes \mathbf{10} = \overline{\mathbf{15}} \oplus \mathbf{24} \oplus \mathbf{21}\,, \qquad \bar{\mathbf{6}} \otimes \mathbf{10} = \mathbf{3} \oplus \mathbf{15} \oplus \mathbf{42}$$

$$\mathbf{8} \otimes \mathbf{8} = \mathbf{1} \oplus \mathbf{8_a} \oplus \mathbf{8_s} \oplus \mathbf{10} \oplus \overline{\mathbf{10}} \oplus \mathbf{27}$$

$$\mathbf{8} \otimes \mathbf{10} = \mathbf{8} \oplus \mathbf{10} \oplus \mathbf{27} \oplus \mathbf{35}$$

$$\mathbf{10} \otimes \mathbf{10} = \overline{\mathbf{10}} \oplus \mathbf{27} \oplus \mathbf{35} \oplus \mathbf{28}\,, \qquad \mathbf{10} \otimes \overline{\mathbf{10}} = \mathbf{1} \oplus \mathbf{8} \oplus \mathbf{27} \oplus \mathbf{64}$$

Each line can be "conjugated", for example,

$$\bar{\mathbf{3}} \otimes \bar{\mathbf{3}} = \mathbf{3} \oplus \bar{\mathbf{6}}\,, \qquad \bar{\mathbf{3}} \otimes \bar{\mathbf{6}} = \mathbf{8} \oplus \overline{\mathbf{10}}\,, \qquad \text{etc.}$$

Note that representations $\mathbf{8}, \mathbf{27}, \mathbf{64}, \ldots$, having $p = q$, are self-conjugated ("real").

Now we are in a position to calculate $C(R)$ for an arbitrary irreducible representation $R(p,q)$. We construct the corresponding composite generator

$$T^a = \sum_{\ell=1}^{p} T^a(\mathbf{3}) + \sum_{m=1}^{q} T^a(\bar{\mathbf{3}})$$

and square it to obtain the quadratic form

$$\begin{aligned} C(\mathbf{R}) &= (p+q)(T^a(\mathbf{3}))^2 + 2 \sum_{\ell'>\ell=1}^{p} T^a_\ell(\mathbf{3}) T^a_{\ell'}(\mathbf{3}) \\ &+ 2 \sum_{m'>m=1}^{q} T^a_m(\bar{\mathbf{3}}) T^a_{m'}(\bar{\mathbf{3}}) + 2 \sum_{\ell=1}^{p} \sum_{m=1}^{q} T^a_\ell(\mathbf{3}) T^a_m(\bar{\mathbf{3}})\,. \end{aligned} \tag{66a}$$

Here the first term accounts for the squared proper generators, while the rest is interference between p "quarks" and q "antiquarks". The latter contributions can be pinned down by observing that, due to the internal symmetry of the representation (65a), each couple of quarks in it is sitting in a sextet ($\bar{q}\bar{q}$ in $\bar{\mathbf{6}}$), while each $q\bar{q}$ pair sits in an octet (tracelessness). Therefore the products of two generators in (66a) give us colour potentials in the corresponding pair states:

$$C(\mathbf{R}) \;=\; (p+q)C_F \;+\; [\,p(p-1) + q(q-1)\,]\; v_{\mathbf{33}}(\mathbf{6}) \;+\; 2pq\, v_{\mathbf{3\bar{3}}}(\mathbf{8})\,. \tag{66b}$$

Substituting (59) and (58b) we arrive at

$$\begin{aligned} C(\mathbf{R}) &= (p+q)\frac{4}{3} \;+\; [\,p(p-1) + q(q-1)\,]\,\frac{1}{3} \;+\; 2pq\,\frac{1}{6} \\ &= \frac{1}{4}(p+q)(p+q+4) \;+\; \frac{1}{12}(p-q)^2\,. \end{aligned} \tag{67}$$

Using Table 1 and the general expression (59b) we can analyse the interaction strength (colour potential) between arbitrary objects.

Table 1. *Dimensions and charges of some representations of $SU^c(3)$.*

Composition	$K(\mathbf{R})$	$C(\mathbf{R})$	Comp.	$K(\mathbf{R})$	$C(\mathbf{R})$	Comp.	$K(\mathbf{R})$	$C(\mathbf{R})$
q	**3**	4/3	$qq\bar{q}\bar{q}$	**27**	8	q^5	**21**	40/3
$q\bar{q}$	**8**	3	$qqq\bar{q}$	**24**	25/3	$q^3\bar{q}^3$	**64**	15
qq	**6**	10/3	$qqqq$	**15'**	28/3	$q^4\bar{q}^2$	**60**	46/3
$qq\bar{q}$	**15**	16/3	$q^3\bar{q}^2$	**42**	34/3	$q^5\bar{q}$	**48**	49/3
qqq	**10**	6	$q^4\bar{q}$	**35**	12	q^6	**28**	18

Exercise 9. Turn back to the gg system (63). Verify that in the colour-tensor state **27** two gluons *repulse*, while in the **10** ($\overline{\mathbf{10}}$) state they *do not interact* via one-gluon exchange.

Exercise 10. Find the quark-gluon interaction energy in the state **15**. Do not confuse two different representations with the same dimension 15; see (64c) to fix a proper one.

4.2 The *d*–Symbol

To complete our excursion into basics of colour, one acquaintance still remains to be made. There is one more fundamental object of the $SU(N)$ group, the so-called d_{abc} symbol, which is analogous to the structure constant f_{abc}. It is related with the two ways of combining two gluons into an adjoint representation, namely antisymmetric and symmetric in gluon colours:

$$\phi^c_{(a)} \propto f_{abc}\, A^a_\mu(1) A^b_\nu(2)\,,$$
$$\phi^c_{(s)} \propto d_{abc}\, A^a_\mu(1) A^b_\nu(2)\,.$$

Correspondingly, there are two different colour singlets made up of three gluon fields:

$$S_{(a)} \propto f_{abc}\, A^a_\mu(1) A^b_\nu(2) A^c_\lambda(3)\,, \tag{68a}$$
$$S_{(s)} \propto d_{abc}\, A^a_\mu(1) A^b_\nu(2) A^c_\lambda(3)\,. \tag{68b}$$

Gluons are bosons. Therefore any physical state must be symmetric with respect to gluons, that is must stay invariant under an interchange of gluon momenta, polarization and *colour* indices. Since the state $S_{(s)}$ in (68b) is symmetric in colours, it is also symmetric with respect to transposition of gluon momenta and polarizations only. But this is exactly the symmetry of a 3-photon system. Therefore in the state (68b) we can ascribe to each gluon the negative charge parity of a photon, $C_g = -1 = C_\gamma$. Recall that the same analogy between gluons and photons was there for 2-boson (radiative and hadronic) decays of

C-even heavy mesons we discussed above. In that case symmetry with respect to gluon colours was trivial, since a colourless 2-gluon state was a convolution $A^a_\mu(1)A^a_\nu(2)$.

J/ψ and Υ meson families are C-odd. They are produced in e^+e^- annihilation via one photon and, therefore, cannot decay into two photons and/or via two gluons. On the other hand, decay into three C-odd bosons is allowed. Three-photon decay is of little interest. Much more interesting is 3-gluon decay, which determines the total hadronic width of such a meson. It is allowed as long as gluons constitute a colour-symmetric state (68b) based on the d-structure.

Consider the matrix

$$M \;=\; t^a t^b \,+\, t^b t^a \,-\, \frac{1}{N}\,\delta_{ab}\mathbb{1}\,. \tag{69a}$$

First notice that $\mathrm{Tr(M)} = \frac{1}{2} + \frac{1}{2} - \mathrm{N/N} = 0$. Therefore M can be expanded as

$$M \;=\; d_{abc}\, t^c\,. \tag{69b}$$

This defines the symbol d_{abc} in analogy with the structure constants f_{abc}, which have emerged as expansion parameters for the commutator.

Exercise 11. Derive

$$d_{abc} \;=\; 2\,\mathrm{Tr}\left(\,\mathrm{t^a t^b t^c} + \mathrm{t^b t^a t^c}\,\right) \tag{70a}$$

and verify its symmetry with respect to *all* three indices. Pictorially,

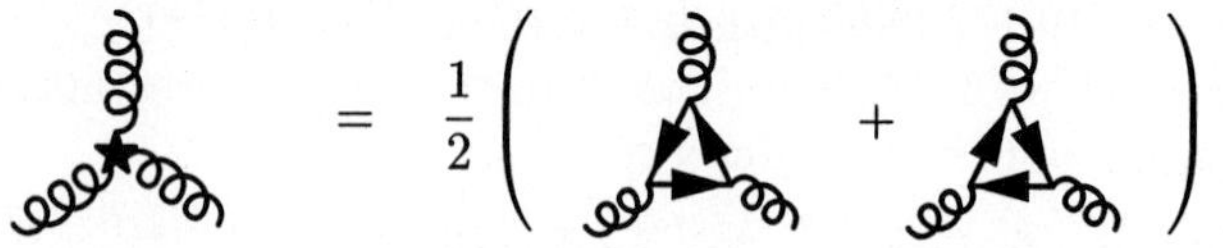

(70b)

Note that unlike f, the structure d_{abc} does not enter in the QCD Lagrangian, so that there is no dynamic 3-gluon interaction vertex with the d-coupling in the game. Therefore the picture (70b) is not a Feynman amplitude. However, as we shall see below, it is an important ingredient of the graphic technology for calculating colour factors of Feynman amplitudes.

a and b are parameters of the matrix (69a). Let us set $a = b$ and perform the standard summation $a = b = 1, 2, \ldots, (N^2 - 1)$. This gives

$$2C_F\mathbb{1} - \frac{N^2-1}{N}\mathbb{1} \;=\; d_{aac}t^c\,, \tag{71}$$

where we have recalled the definition of the quark Casimir operator, $t^a t^a = C_F\mathbb{1}$ and used $\delta_{aa} = N^2 - 1$. An expression proportional to the unit matrix equals another one expanded in t^c. Therefore both are zero. The l.h.s. $= 0$ reproduces the known result for C_F, the r.h.s. gives (remember, summation over repeated index a is always implied)

$$d_{aac} \;=\; 0\,. \tag{72}$$

This means that the d-symbol is *traceless* as a matrix in the adjoint representation space, i.e. viewed as N^2-1 matrices (label c) acting in the $(N^2-1)\times(N^2-1)$ space (indices a, b).

We now convolute the matrix M from (69a) with t^b:

$$t^b\left(t^a t^b + t^b t^a - \frac{1}{N}\,\delta_{ab}\mathbb{1}\right) = \left(-\frac{1}{2N} + \frac{N^2-1}{2N} - \frac{1}{N}\right) t^a = \frac{N^2-4}{2N}\, t^a\,.$$

On the other hand, from (69b) this very expression equals

$$d_{abc}\, t^b t^c = d_{abc}\cdot\tfrac{1}{2}\left(t^b t^c + t^c t^b - \frac{1}{N}\delta_{bc}\mathbb{1}\right) \equiv \tfrac{1}{2}\, d_{abc} d_{bce}\, t^e\,,$$

where we made use of $d_{abc} = d_{acb}$ and $d_{abb} = 0$. Thus we have arrived at the expression for the "square" of the d-symbol:

$$d_{abc}d_{ebc} = \frac{N^2-4}{N}\,\delta_{ae}\,; \qquad\qquad = \frac{N^2-4}{N} \qquad\qquad (73a)$$

This goes along with

$$f_{abc}f_{ebc} = N\,\delta_{ae}\,; \qquad\qquad = N \qquad\qquad (73b)$$

We are now ready to analyse J/ψ decays. Let us compare the amplitudes for $c\bar{c}$ annihilation into three gluons, which determine the hadronic width, and into two gluons and a photon (radiative width):

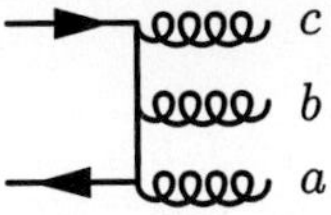

$$c,\ b,\ a \quad \propto g_s^3 \mathrm{Tr}\left(\mathrm{t^a t^b t^c}\right) = \frac{1}{4}\mathrm{g_s^3}\left(\mathrm{d_{abc}} + \mathrm{i f_{abc}}\right) \Longrightarrow \frac{1}{4}\mathrm{g_s^3}\,\mathrm{d_{abc}} \qquad (74a)$$

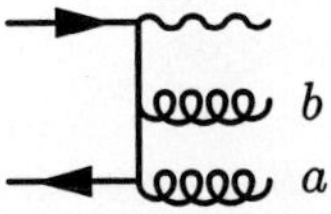

$$b,\ a \quad \propto eQ\, g_s^2 \mathrm{Tr}\left(\mathrm{t^a t^b}\right) = \frac{1}{2}\mathrm{eQ\, g_s^2}\,\delta_{\mathrm{ab}} \qquad (74b)$$

In the first amplitude we have kept the d_{abc} structure, which is the only one to survive in the C-odd $c\bar{c}$ state. In the radiative amplitude Q is the fractional e.m. charge of the quark ($Q_c = 2/3$).

The Lorentz structure of the amplitudes (74) is identical, and so are the factors coming from the $c\bar{c}$ wave function. Therefore, the *ratio* of the decay probabilities is determined by the colour factors (and the coupling constants). To calculate these probabilities we square the amplitudes and sum over colours in the final states (Fig. 7).

The only subtlety is the combinatorial factors we should supply the amplitudes with: $1/\sqrt{3!}$ (three identical gluons) in (74a) versus $1/\sqrt{2!}$ (two gluons) in the radiative decay amplitude (74b).

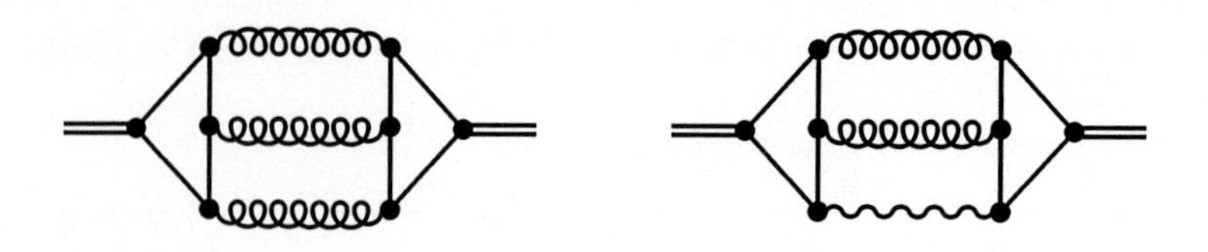

Fig. 7. *Squared matrix elements for onium decays $J/\psi \to 3g$ and $J/\psi \to 2g + \gamma$.*

We obtain

$$\frac{\Gamma_{ggg}}{\Gamma_{gg\gamma}} = \frac{2!}{3!} \frac{(g_s^3/4)^2\, d_{abc}d_{abc}}{(eQ\, g_s^2/2)^2\, \delta_{ab}\delta_{ab}} = \frac{\alpha_s}{12Q^2\,\alpha_{\rm em}} \frac{N^2-4}{N} = \frac{5}{36Q^2} \frac{\alpha_s}{\alpha_{\rm em}}\,. \tag{75}$$

This provides the means of a rough "measurement" of the QCD coupling at characteristic distances $r \sim M_c^{-1}$. Comparing the ratios of hadronic and radiative widths for J/ψ and Υ ($Q_b = -1/3$), one can even see that $\alpha_s(M_\Upsilon) < \alpha_s(J/\psi)$ in accord with the *asymptotic freedom*.

4.3 Successive 2-Gluon Radiation Off a Quark

Combining expressions for commutator and anticommutator of t matrices (16) and (69a), we obtain an important relation

$$t^a t^b = \frac{1}{2N}\,\delta_{ab}\mathbb{1} + \frac{\mathrm{i}}{2}\,f_{abc}\,t^c + \frac{1}{2}\,d_{abc}\,t^c\,, \tag{76a}$$

$$[a\ b] = \frac{1}{2N}\,[a\ b] + \frac{1}{2}\,[a\ b\ c]\;\mathrm{i}f_{abc} + \frac{1}{2}\,[a\ b\ c]\;d_{abc} \tag{76b}$$

It is necessary to bear in mind that the direction of the quark line, because of antisymmetry of the f-symbol, is essential for the graphic equation (76b). As long as we have chosen to label the f-vertex *clockwise*, the following rule applies:

> In graphic equations where a 3-gluon coupling appears (or disappears) as in (76), the f-vertex has to be positioned *to the right* from the direction of the fermion line.

Example: 2-Gluon Exchange. To illustrate the use of the graphic colour algebra let us analyse the colour structure of a 2-gluon exchange between quarks.

First, we invoke (76b) to draw

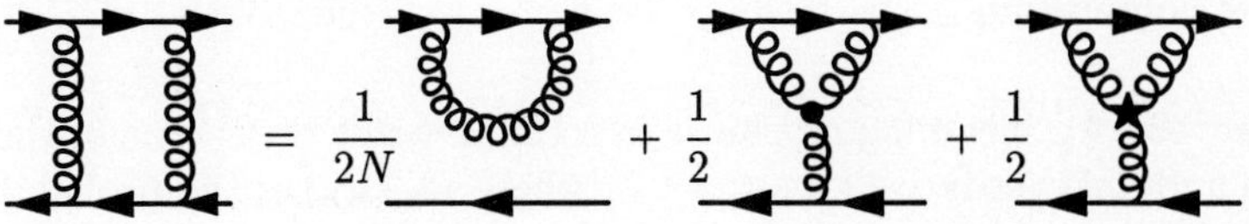

$$= \frac{1}{2N} \quad + \frac{1}{2} \quad + \frac{1}{2} \qquad (77a)$$

For the first two contributions in the r.h.s. we invoke the quark self-energy (43a) and the non-Abelian vertex (52) correspondingly. In the last term we apply the key equality (76b) once again to obtain

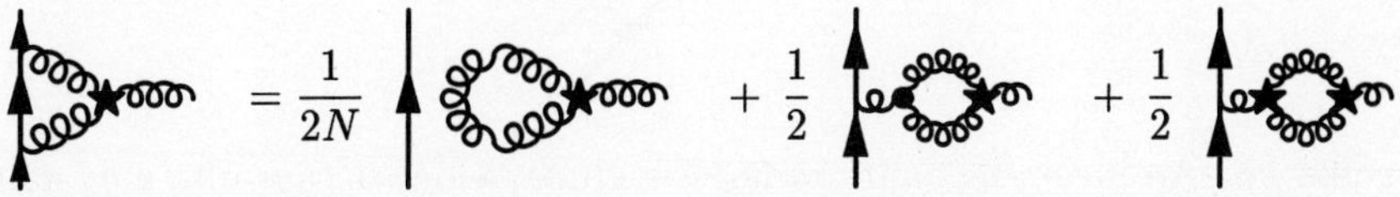

$$= \frac{1}{2N} \quad + \frac{1}{2} \quad + \frac{1}{2}$$

The first two tems here vanish ($d_{aac} = 0$, $f_{abc}d_{ebc} = 0$) and we arrive at

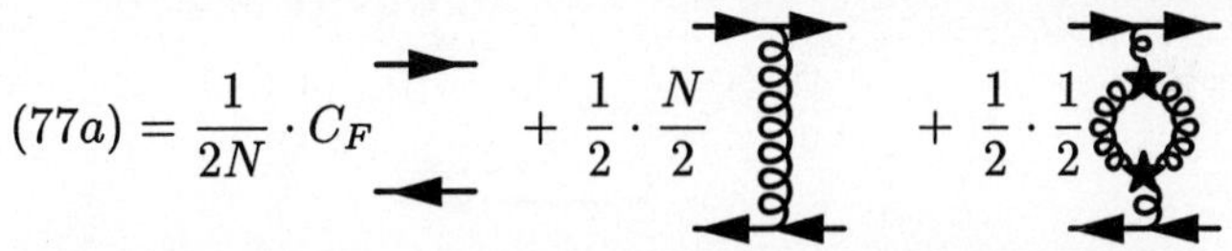

$$(77a) = \frac{1}{2N} \cdot C_F \quad + \frac{1}{2} \cdot \frac{N}{2} \quad + \frac{1}{2} \cdot \frac{1}{2} \qquad (77b)$$

Finally, fetching the value of $d^2 = (N^2 - 4)/N$ in (73a) and collecting terms we obtain

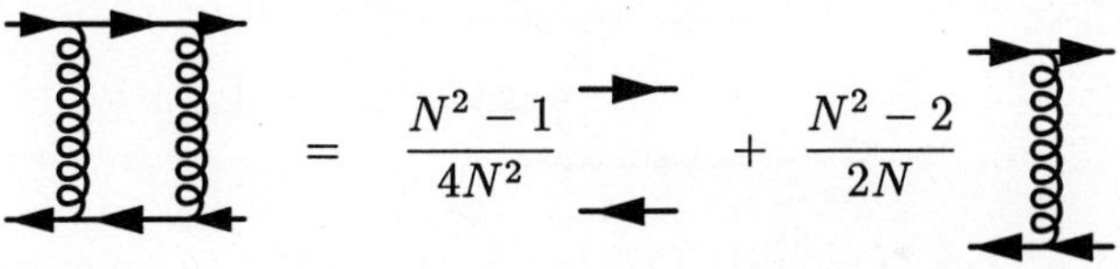

$$= \frac{N^2 - 1}{4N^2} \quad + \frac{N^2 - 2}{2N} \qquad (77c)$$

This is the colour structure of the 2-gluon exchange between a quark and an *antiquark*.

Exercise 12. For the double scattering of *two quarks*, show that

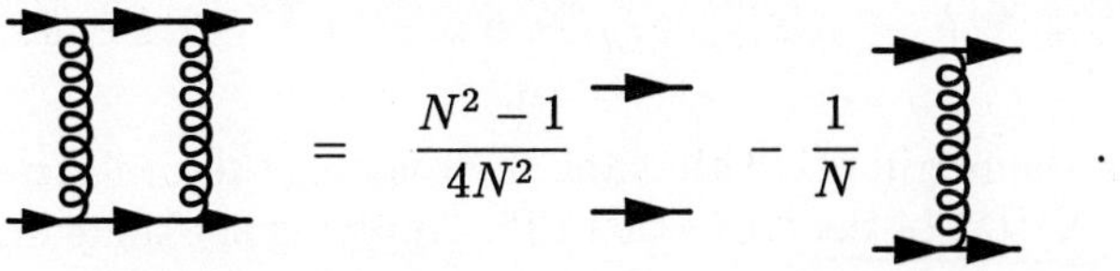

$$= \frac{N^2 - 1}{4N^2} \quad - \frac{1}{N} \quad . \qquad (78)$$

Equations (77) and (78) can be looked upon as a decomposition of the 2-gluon exchange into singlet and octet pieces from the t-channel point of view.

5 Practicum

Exercise 13. Derive (77) and (78) by applying singlet and octet state projectors (45) to $q\bar{q}$ pair in the "t-channel".

Exercise 14. Decompose $q\bar{q}$ scattering amplitude (77) into the <u>s-channel</u> singlet and octet components.

Exercise 15. An incoming $q\bar{q}$ pair is in a colour octet state. One of the quarks radiates a gluon. Verify $w_1 : w_8 = 1 : 7$, where w_1 (w_8) is the probability that the outgoing $q\bar{q}$ pair is in a colour singlet (octet) state.

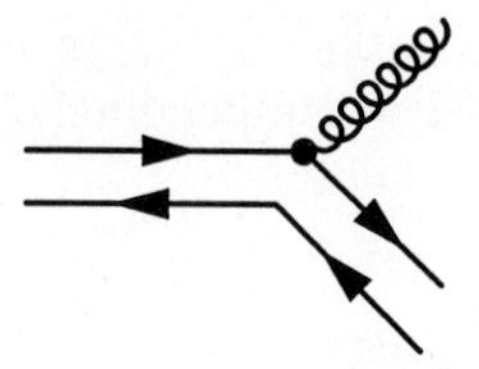

Exercise 16. An incoming quark radiates a gluon, which decays into a $q\bar{q}$ pair. The produced antiquark couples to the initial quark. Verify that this $q\bar{q}$ state is well prepared to form a "white" meson, namely, $w_1 : w_8 = 8 : 1$.

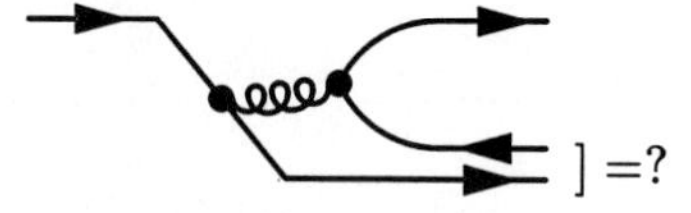

Exercise 17. Prove that an octet $q\bar{q}$ pair shaking off a gluon in a specific way displayed below *always* converts into a colour singlet:

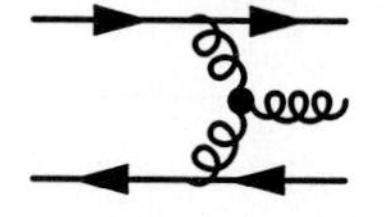

Exercise 18. The Jacobi identity (32) has the following nice pictorial representation

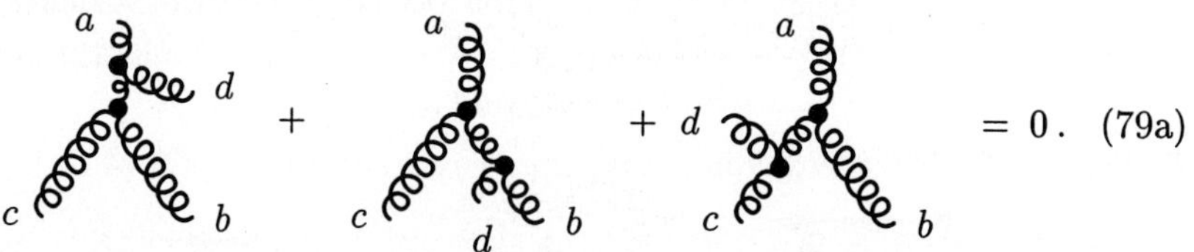

To understand the origin of (79a), think of "rotating" the *colourless* compound $\mathrm{i}f_{abc}A^a(1)A^b(2)A^c(3)$ in the "direction" T^d. Applying the same logic to another colouless object, namely, the *d*-coupled three gluons, results in

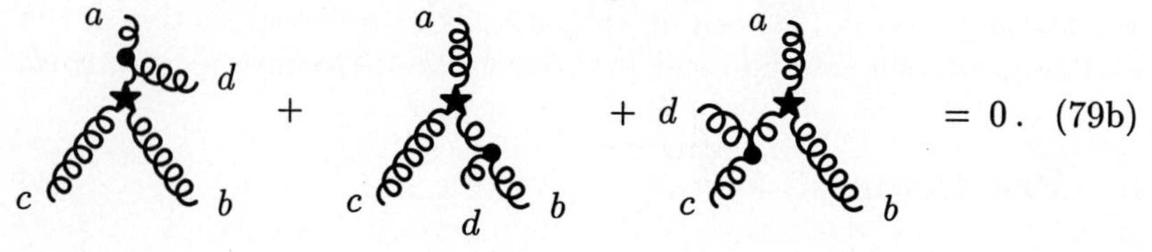

Suggest an algebraic proof of (79b), in analogy with (33) and (34).

Exercise 19. Let the gluon "d" be absorbed by one of the other three gluons in (79) and find graphically the gluon-loop correction to the 3-gluon vertices. (Use (43b) and be careful with the orientation of 3-gluon vertices; watch the signs!)

$$: \quad = \quad : \quad = \frac{N}{2}. \qquad (80)$$

Exercise 20. Make use of (79b) and (73a) to show

$$= \frac{N^2 - 4}{2N} \quad . \qquad (81a)$$

Masterpiece. This one won't be as easy:

$$= \frac{N^2 - 12}{2N} \quad . \qquad (81b)$$

Test. Giving a proof of the following identity by mere reflection, *without performing any calculations* (and even drawings), will show that you have mastered the graphic colour algebra:

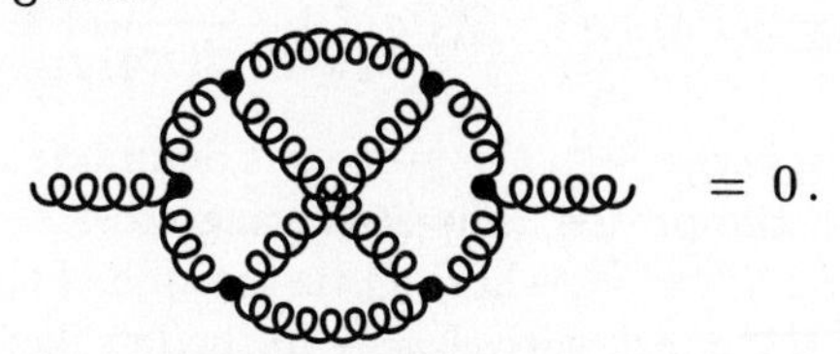

$$= 0 .$$

6 Colour Coherence in Scattering

We have considered an impact of coherence and colour conservation on internal time-like development of parton jets. In this section we shall discuss radiation accompanying space-like processes, namely scattering of charges. This gives us a reason to have a closer look at the physics of bremsstrahlung and to make an important application of the colour counting technique.

6.1 QCD DIS Minutes

Deep inelastic lepton-proton scattering (DIS) with large momentum transfer $-q^2 = Q^2 \gg \Lambda^2$ is a classical example of a hard QCD process, which is governed by the space-like evolution of the parton system. As is well known, the DIS cross section (structure functions) can be expressed in terms of parton distributions as a probability to find a quark with the longitudinal momentum fraction x inside a target hadron (proton). Typical graphs for DIS amplitudes are shown in Fig. 8.

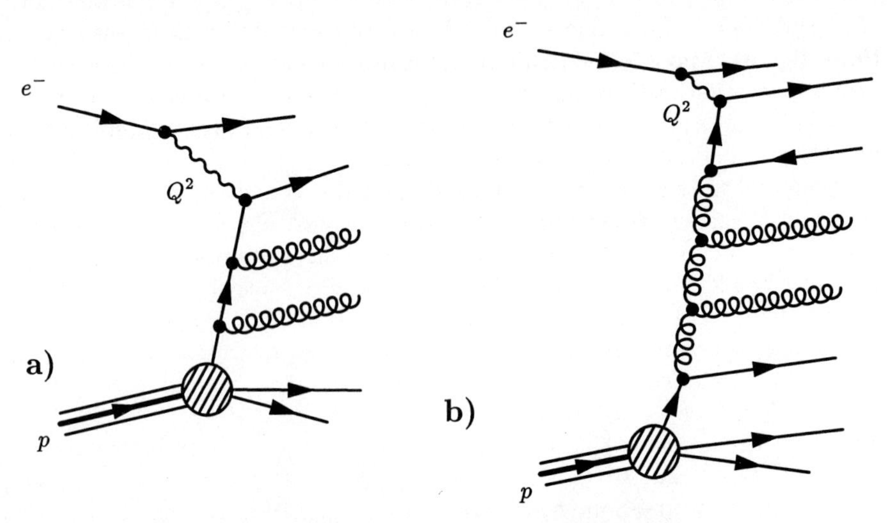

Fig. 8. *Valence (left) and Bethe-Heitler mechanism (right) of DIS.*

The value of Bjorken x, $x = -q^2/2(Pq) \leq 1$, determines the invariant mass of the produced final system, that is inelasticity of the process:

$$W^2 = (P+q)^2 - M_P^2 = 2(Pq) + q^2 = 2(Pq)\left(1 - \frac{-q^2}{2(Pq)}\right) \equiv s(1-x), \quad s = 2(Pq).$$

For moderate x-values, say, $x \sim \frac{1}{4}$, the process is dominated by lepton scattering off a valence quark in the proton. The scattering cross section has a standard energy behaviour $\sigma \propto x^{-2(J_{\mathrm{ex}}-1)}$, where J_{ex} is the spin of the exchanged particle in the t-channel. It is the quark with $J_{\mathrm{ex}} = \frac{1}{2}$ in the left picture of Fig. 8, so that the valence contribution to the cross section decreases with x as $\sigma \propto x$.

For high-energy scattering, $x \ll 1$, the Bethe-Heitler mechanism takes over which corresponds to the t-channel *gluon* exchange: $J_{\mathrm{ex}} = 1$, $\sigma \propto x^0$.

In the Leading Logarithmic Approximation (LLA) one insists on picking up, for each new parton taken into consideration, a logarithmic enhancement factor $\alpha_s \to \alpha_s \log Q^2$. In this approximation the scattering *probability* can be simply obtained by convoluting elementary probabilities of independent $1 \to 2$ parton splittings. To contribute to the LLA, the transverse momenta of produced partons should be strongly ordered, increasing up the "ladder": $k_{1\perp}^2 \ll \ldots \ll k_{n\perp}^2 \ll Q^2$. (At the level of Feynman *amplitudes* the ladder diagrams dominate, provided a special physical gauge is chosen for gluon fields.)

What can we say about the structure of the final states produced in DIS? The yield of final particles is driven by the accompanying emission of (relatively) soft gluons. Bearing this in mind, we notice that the valence mechanism is very much similar to $e^+e^- \to q\bar{q}$. In a reference frame where the struck quark (the top quark-line in Fig. 8) and the proton remnant move in the opposite directions (the Breit frame: $q_0 = 0$, $q_z = -Q$; $xP = |xP + q_z| = \frac{1}{2}Q$), the proton devoid of a triplet quark looks like a "hole" in a $\mathbf{\bar{3}}$ colour state, in close analogy with the antiquark from e^+e^- annihilation. In the e^+e^- case we break the vacuum to produce two relativistic colour charges $\mathbf{3}+\mathbf{\bar{3}}$, while in DIS we break up the proton. As long as soft accompaniment is concerned, this difference is insignificant since both the vacuum and the proton are "white".

However, for small x where the Bethe-Heitler contribution dominates, the situation is somewhat different. It is still true that the proton fragmentation region is in the $\mathbf{\bar{3}}$ state, as a whole. However, this antitriplet has now a non-trivial internal structure. Namely, it consists of an actual antiquark recoiling against the struck quark (upper quark-box in the right graph of Fig. 8) and the proton remnant *repainted* into an *octet*, due to t-channel gluon exchange; $\mathbf{\bar{3}} = \mathbf{\bar{3}} + \mathbf{8}$. As far as colour topology is concerned, this system rather reminds sort of a weird 3-jet e^+e^- annihilation event with two q, $\bar{q}$ jets, $E_q \simeq E_{\bar{q}} \simeq Q/2$, and an energetic "gluon jet" with $E_g \sim P \simeq Q/2x \gg E_{q,\bar{q}}$. In such circumstances we would expect radiation in the target fragmentation region to be enhanced as compared to the current fragmentation (struck quark), according to the ratio of the colour charges N/C_F.

The fact that the *current* quark jet is identical to "half of" e^+e^- annihilation is well established experimentally. Envisaged peculiarity of the *target* fragmentation at small x, which is due to the "colour-octet-proton" phenomenon, remains to be seen at HERA.

In the rest of this section we shall verify our qualitative expectation of the effect a colour exchange in a scattering process produces upon accompanying gluon radiation. To do so, we first recall the basics of QED bremsstrahlung.

6.2 Bremsstrahlung

Let us consider photon bremsstrahlung induced by a charged particle (electron) which scatters off an external field (e.g. a static electromagnetic field). The derivation is included in almost every textbook on QED, so we confine ourselves to the essential aspects.

The lowest order Feynman diagrams for photon radiation are depicted in Fig. 9, where p_1, p_2 are the momenta of the incoming and outgoing electron respectively and k represents the momentum of the emitted photon. The corresponding amplitudes, according to the Feynman rules, are given in momentum

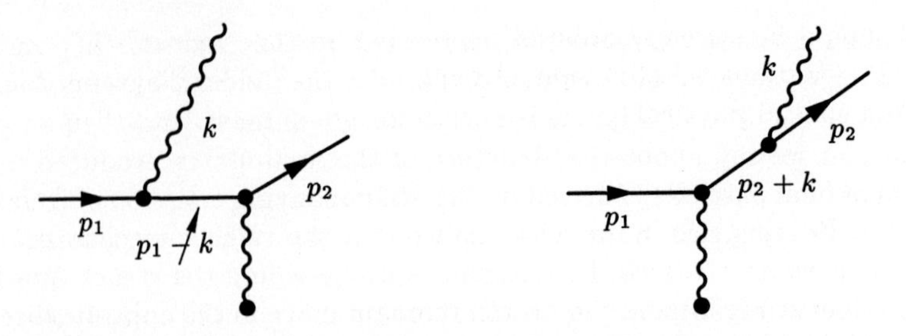

Fig. 9. Bremsstrahlung diagrams for scattering off an external field.

space by

$$M_{\rm i}^{\mu} = e\,\bar{u}(p_2,s_2)\,V(p_2+k-p_1)\,\frac{m+\not{p}_1-\not{k}}{m^2-(p_1-k)^2}\,\gamma^{\mu}\,u(p_1,s_1)\;, \tag{82a}$$

$$M_{\rm f}^{\mu} = e\,\bar{u}(p_2,s_2)\,\gamma^{\mu}\,\frac{m+\not{p}_2+\not{k}}{m^2-(p_2+k)^2}\,V(p_2+k-p_1)\;u(p_1,s_1)\;. \tag{82b}$$

Here V stands for the basic interaction amplitude which may depend in general on the momentum transfer (for the case of scattering off the static e.m. field, $V = \gamma^0$).

First we apply the soft-photon approximation, $\omega \ll p_1^0, p_2^0$, to neglect $\not{k}$ terms in the numerators. To deal with the remaining matrix structure in the numerators of (82) we use the identity $\not{p}\,\gamma^{\mu} = -\gamma^{\mu}\,\not{p} + 2\,p^{\mu}$ and the Dirac equation for the on-mass-shell electrons,

$$(m+\not{p}_1)\,\gamma^{\mu}\,u(p_1) = (2p_1^{\mu} + [\,(m-\not{p}_1\,])\,u(p_1) = 2p_1^{\mu}\,u(p_1)\,,$$
$$\bar{u}(p_2)\,\gamma^{\nu}\,(m+\not{p}_2) = \bar{u}(p_2)\,([\,(m-\not{p}_2\,] + 2p_1^{\nu}) = 2p_2^{\nu}\,\bar{u}(p_2)\,.$$

Denominators for real electrons ($p_i^2 = m^2$) and the photon ($k^2 = 0$) become $m^2 - (p_1-k)^2 = 2(p_1k)$ and $m^2 - (p_2+k)^2 = -2(p_2k)$, so that for the total amplitude we obtain the factorized expression

$$M^{\mu} = e\,j^{\mu}\,M_{\rm el}\,. \tag{83a}$$

Here $M_{\rm el}$ is the Born matrix element for non-radiative (elastic) scattering,

$$M_{\rm el} = \bar{u}(p_2,s_2)\,V(p_2-p_1)\,u(p_1,s_1) \tag{83b}$$

(in which the photon recoil effect has been neglected, $q = p_2 + k - p_1 \simeq p_2 - p_1$), and j^{μ} is the *soft accompanying radiation current*

$$j^{\mu}(k) \;=\; \frac{p_1^{\mu}}{(p_1k)} - \frac{p_2^{\mu}}{(p_2k)}\,. \tag{83c}$$

The factorization (83a) is of the most general nature. The form of j^{μ} does not depend on the details of the underlying process, neither on the nature of participating charges (electron spin, in particular). The only things which matter are the momenta and charges of incoming and outgoing particles. Generalization to an arbitrary process is straightforward and results in assembling the contributions due to all initial and final particles, weighted with their respective charges.

The soft current (83c) has a classical nature. It can be derived form the classical electrodynamics by considering the potential induced by change of the e.m. current due to scattering.

To calculate the radiation probability we square the amplitude projected onto a photon polarization state ε_μ^λ, sum over λ and supply the photon phase space factor to write down

$$dW = e^2 \sum_{\lambda=1,2} \left|\varepsilon_\mu^\lambda j^\mu\right|^2 \frac{\omega^2 \, d\omega \, d\Omega_\gamma}{2\omega\,(2\pi)^3} \, dW_{\rm el} \,. \tag{84}$$

The sum runs over two physical polarization states of the real photon, described by normalized polarization vectors orthogonal to its momentum:

$$\epsilon_\lambda^\mu(k)\cdot\epsilon^*_{\mu,\lambda'}(k) = -\delta_{\lambda\lambda'}\,, \quad \epsilon_\lambda^\mu(k)\cdot k_\mu = 0\,; \qquad \lambda\,,\lambda' = 1,2\,.$$

Within these conditions, the polarization vectors may be chosen differently. Due to gauge invariance such an uncertainty does not affect physical observables. Indeed, the polarization tensor may be represented as

$$\sum_{\lambda=1,2} \epsilon_\lambda^\mu \epsilon_\lambda^{*\,\nu} = -g^{\mu\nu} + \text{ tensor proportional to } k^\mu \text{ and/or } k^\nu\,. \tag{85}$$

The latter, however, can be dropped since the classical current (83c) is explicitly conserved, $j^\mu k_\mu = 0$. Therefore one may enjoy gauge invariance and employ an arbitrary gauge, instead of using the physical polarizations, to calculate accompanying photon production.

The Feynman gauge being the simplest choice,

$$\sum_{\lambda=1,2} \epsilon_\lambda^\mu \epsilon_\lambda^{*\,\nu} \Longrightarrow -g^{\mu\nu}\,,$$

we arrive at

$$\begin{aligned} dN \equiv \frac{dW}{dW_{\rm el}} &= -\frac{\alpha}{4\pi^2}\,(j^\mu)^2\,\omega\, d\omega\, d\Omega_\gamma \\ &\simeq \frac{\alpha}{\pi}\,\frac{d\omega}{\omega}\,\frac{d\Omega_\gamma}{2\pi}\,\frac{1-\cos\Theta_s}{(1-\cos\Theta_1)(1-\cos\Theta_2)}\,. \end{aligned}$$

The latter expression corresponds to the relativistic approximation $v_1, v_2 \approx 1$:

$$-(j^\mu)^2 = \frac{2(p_1p_2)}{(p_1k)(p_2k)} + \mathcal{O}\left(\frac{m^2}{p_0^2}\right) \simeq \frac{2}{\omega^2}\,\frac{(1-\mathbf{n}_1\cdot\mathbf{n}_2)}{(1-\mathbf{n}_1\cdot\mathbf{n})(1-\mathbf{n}_2\cdot\mathbf{n})}\,,$$

which disregards the contribution of *very* small emission angles $\Theta_i^2 \lesssim (1-v_i^2) = m^2/p_{0i}^2 \ll 1$, where the soft radiation vanishes (the so-called "Dead Cone" region). For the definition of angles see Fig. 10.

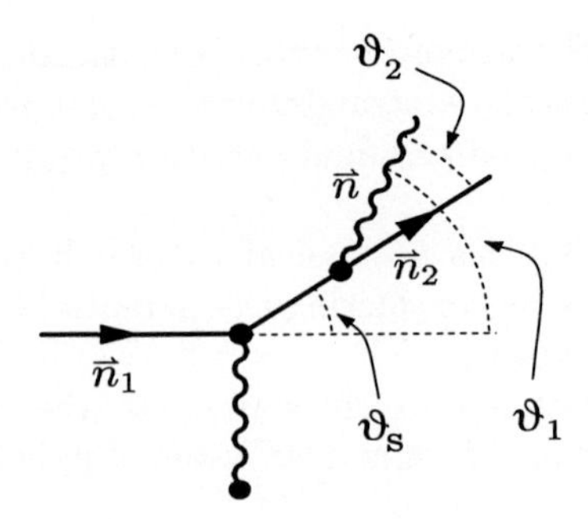

Fig. 10. Bremsstrahlung diagram containing the definitions of angles.

If the photon is emitted at a small angle with respect to, say, the incoming particle, i.e. $\Theta_1 \ll \Theta_2 \simeq \Theta_s$, the radiation spectrum (83) simplifies to

$$\mathrm{d}N \simeq \frac{\alpha}{\pi}\frac{\sin\Theta_1\,\mathrm{d}\Theta_1}{(1-\cos\Theta_1)}\frac{\mathrm{d}\omega}{\omega} \simeq \frac{\alpha}{\pi}\frac{\mathrm{d}\Theta_1^2}{\Theta_1^2}\frac{\mathrm{d}\omega}{\omega}\,.$$

Two bremsstrahlung cones appear, centred around incoming and outgoing electron momenta. Inside these cones the radiation has a *double-logarithmic* structure, exhibiting both the *soft* ($\mathrm{d}\omega/\omega$) and *collinear* enhancements ($\mathrm{d}\Theta^2/\Theta^2$).

6.3 Independent and Coherent Radiation

In the Feynman gauge, the result for the accompanying radiation factor $\mathrm{d}N$ is dominated by the *interference* between the two emitters:

$$\mathrm{d}N \propto -\left[\frac{p_1^\mu}{(p_1k)} - \frac{p_2^\mu}{(p_2k)}\right]^2 \approx \frac{2(p_1p_2)}{(p_1k)(p_2k)}\,.$$

Therefore it does not provide a satisfactory answer to the question, which part of radiation is due to the initial charge and which is due to the final one?

There is a way, however, to give a reasonable answer to this question. To do so, one has to sacrifice simplicity of the Feynman-gauge calculation and recall the original expression (84) for the cross section in terms of physical photon polarizations. It is natural to choose the so-called *radiative* (temporal) gauge based on the 3-vector potential $\mathbf{A}$, with the scalar component set to zero, $A_0 \equiv 0$. Our photon is then described by (real) 3-vectors orthogonal to one another and to its 3-momentum:

$$(\boldsymbol{\epsilon}_\lambda \cdot \boldsymbol{\epsilon}_{\lambda'}) = \delta_{\lambda\lambda'}\,, \qquad (\boldsymbol{\epsilon}_\lambda \cdot \mathbf{k}) = 0\,. \tag{86}$$

This explicitly leaves us with *two* physical polarization states. Summing over polarizations obviously results in

$$dN \propto \sum_{\lambda=1,2} \left|\mathbf{j}(k)\cdot\mathbf{e}_\lambda\right|^2 = \sum_{\alpha,\beta=1\ldots3} \mathbf{j}^\alpha(k)\cdot[\,\delta_{\alpha\beta} - \mathbf{n}_\alpha\mathbf{n}_\beta\,]\cdot\mathbf{j}^\beta(k)\,, \tag{87}$$

with α, β 3-dimensional indices. We now substitute the soft current (83c) in the 3-vector form, $p_i^\mu \to \mathbf{v}_i p_{0i}$, and make use of the relations

$$(\mathbf{v}_i)_\alpha \left[\delta_{\alpha\beta} - \frac{k_\alpha\, k_\beta}{\mathbf{k}^2}\right] (\mathbf{v}_i)_\beta = v_i^2 \sin^2 \Theta_i \,, \tag{88a}$$

$$(\mathbf{v}_1)_\alpha \left[\delta_{\alpha\beta} - \frac{k_\alpha\, k_\beta}{\mathbf{k}^2}\right] (\mathbf{v}_2)_\beta = v_1 v_2 (\cos\Theta_{12} - \cos\Theta_1 \cos\Theta_2) \,, \tag{88b}$$

to finally arrive at

$$dN = \frac{\alpha}{\pi} \left\{ \mathcal{R}_1 + \mathcal{R}_2 - 2\,\mathcal{J} \right\} \cdot \frac{d\omega}{\omega} \frac{d\Omega}{4\pi} \,. \tag{89a}$$

Here

$$\mathcal{R}_i = \frac{v_i^2 \sin^2 \Theta_i}{(1 - v_i \cos\Theta_i)^2} \,, \qquad i = 1, 2 \,, \tag{89b}$$

$$\mathcal{J} \equiv \frac{v_1 v_2 \,(\cos\Theta_{12} - \cos\Theta_1 \cos\Theta_2)}{(1 - v_1 \cos\Theta_1)(1 - v_2 \cos\Theta_2)} \,. \tag{89c}$$

The contributions $\mathcal{R}_{1,2}$ can be looked upon as being due to *independent radiation* off initial and final charges, while the $\mathcal{J}$-term accounts for *interference* between them. The independent and interference contribution, taken together, describe the *coherent* emission. It is straightforward to verify that (89) is identical to the Feynman-gauge result (83):

$$\mathcal{R}_{\text{coher.}} \equiv \mathcal{R}_{\text{indep.}} - 2\mathcal{J} \;=\; -\omega^2 (j^\mu)^2 \,, \qquad \mathcal{R}_{\text{indep.}} \equiv \mathcal{R}_1 + \mathcal{R}_2 \,. \tag{90}$$

6.4 The Rôle of Interference: *Strict* Angular Ordering

In the relativistic limit we have

$$\mathcal{R}_1 \simeq \frac{\sin^2 \Theta_1}{(1 - \cos\Theta_1)^2} = \frac{2}{a_1} - 1 \,, \tag{91a}$$

$$\mathcal{J} \simeq \frac{\cos\Theta_{12} - \cos\Theta_1 \cos\Theta_2}{(1 - \cos\Theta_1)(1 - \cos\Theta_2)} = \frac{a_1 + a_2 - a_{12}}{a_1 a_2} - 1 \,, \tag{91b}$$

where we introduced a convenient notation

$$a_1 = 1 - \mathbf{n}\mathbf{n}_1 = 1 - \cos\Theta_1 \,, \quad a_2 = 1 - \cos\Theta_2 \,,$$
$$a_{12} = 1 - \mathbf{n}_1 \mathbf{n}_2 = 1 - \cos\Theta_s \,.$$

The variables a are small when the agles are small: $a \simeq \frac{1}{2}\Theta^2$.

The independent radiation has a typical logarithmic behaviour up to large angles $a_1 \lesssim 1$:

$$\mathrm{d}N_1 \propto \mathcal{R}_1 \sin\Theta \mathrm{d}\Theta \propto \frac{\mathrm{d}a_1}{a_1} \,.$$

However, the interference effectively cuts off the radiation at angles exceeding the scattering angle:

$$dN \propto \mathcal{R}_{\text{coher.}} \sin\Theta d\Theta = 2a_{12}\frac{da}{a_1 a_2} \propto \frac{da}{a^2} \propto \frac{d\Theta^2}{\Theta^4}, \qquad a_1 \simeq a_2 \gg a_{12}\,.$$

To quantify this coherent effect, let us combine an independent contribution with half of the interference contribution to define

$$V_1 = \mathcal{R}_1 - \mathcal{J} = \frac{2}{a_1} - \frac{a_1 + a_2 - a_{12}}{a_1 a_2} = \frac{a_{12} + a_2 - a_1}{a_1 a_2}\,,$$
$$V_2 = \mathcal{R}_2 - \mathcal{J} = \frac{2}{a_2} - \frac{a_1 + a_2 - a_{12}}{a_1 a_2} = \frac{a_{12} + a_1 - a_2}{a_1 a_2}\,; \tag{92a}$$
$$\mathcal{R}_{\text{coher.}} = V_1 + V_2\,. \tag{92b}$$

The emission probability V_i can be still considered as "belonging" to the charge $\#i$ (V_1 is singular when $a_1 \to 0$, and vice versa). At the same time these are no longer *independent* probabilities, since V_1 explicitly depends on the direction of the partner-charge #2; *conditional* probabilities, so to say.

It is straightforward to verify the following remarkable property of the "conditional" distributions V: after *averaging* over the azimuthal angle of the radiated quantum, $\mathbf{n}$, with respect to the direction of the parent charge, $\mathbf{n}_1$, the probability $V_1(\mathbf{n},\mathbf{n}_1;\mathbf{n}_2)$ *vanishes* outside the Θ_s-cone, namely

$$\langle V_1 \rangle_{\text{azimuth}} \equiv \int_0^{2\pi} \frac{d\phi_{n,n_1}}{2\pi} V_1(\mathbf{n},\mathbf{n}_1;\mathbf{n}_2) = \frac{2}{a_1}\vartheta\left(a_{12} - a_1\right)\,. \tag{93}$$

It is only a_2 that changes under the integral (93), while a_1, and obviously a_{12}, stay fixed. The result follows from the angular integral

$$\int_0^{2\pi} \frac{d\phi_{n,n_1}}{2\pi}\frac{1}{a_2} = \frac{1}{|\cos\Theta_1 - \cos\Theta_s|} = \frac{1}{|a_{12} - a_1|}\,.$$

Naturally, a similar expression for V_2 emerges after the averaging over the azimuth around $\mathbf{n}_2$ is performed.

We conclude that as long as the *total* (angular-integrated) emission probability is concerned, the result can be expresses as a sum of two independent bremsstrahlung cones centred around $\mathbf{n}_1$ and $\mathbf{n}_2$, both having the finite opening half-angle Θ_s.

This nice property is known as a "strict angular ordering". It is an essential part of the so-called Modified Leading Log Approximation (MLLA), which describes the internal structure of parton jets with a single-logarithmic accuracy.

6.5 Angular Ordering on the Back of an Envelope

What is the reason for radiation at angles exceeding the scattering angle to be suppressed? Let us try our physical intuition and consider semi-classically how the radiation process really develops.

A physical electron is a charge surrounded by its proper Coulomb field. In the quantum language, the Lorentz-contracted Coulomb-disk attached to a relativistic particle may be treated as consisting of photons virtually emitted and, in due time, reabsorbed by the core charge. Such virtual emission and absorption processes form a coherent state which we call a physical electron ("dressed" particle).

This coherence is partially destroyed when the charge experiences an impact. As a result, a part of intrinsic field fluctuations gets released in the form of real photon radiation: The bremsstrahlung cone in the direction of the initial momentum develops. On the other hand, the deflected charge now leaves the interaction region as a "half-dressed" object with its proper field-coat lacking some field components (eventually those that were lost at the first stage). In the process of regenerating the new Coulomb-disk adjusted to the final-momentum direction, an extra radiation takes place giving rise to the second bremsstrahlung cone.

Now we need to be more specific to find out which momentum components of the electromagnetic coat do actually take leave of absence.

A typical time interval between emission and reabsorption of the photon k by the initial electron p_1 may be estimated as the Lorentz-dilated lifetime of the virtual intermediate electron state $(p_1 - k)$ (see the left graph in Fig.9),

$$t_{\text{fluct}} \sim \frac{E_1}{|m^2 - (p_1 - k)^2|} = \frac{E_1}{2p_1 k} \sim \frac{1}{\omega \Theta^2} = \frac{\omega}{k_\perp^2} \,. \tag{94}$$

Here we restricted ourselves, for simplicity, to small radiation angles, $k_\perp \approx \omega\Theta \ll k_{||} \approx \omega$. The fluctuation time (94) may become macroscopically large for small photon energies ω and enters as a characteristic parameter in a number of QED processes. As an example, let us mention the so called Landau-Pomeranchuk effect — suppression of soft radiation off a charge that experiences multiple scattering propagating through a medium. Quanta with too large a wavelength get not enough time to be properly formed before successive scattering occurs, so that the resulting bremsstrahlung spectrum behaves as $dN \propto d\omega/\sqrt{\omega}$ instead of the standard logarithmic $d\omega/\omega$ distribution.

The characteristic time scale (94) responsible for this and many other radiative phenomena is often referred to as the *formation time.*

Now imagine that within this interval the core charge was kicked by some external interaction and has changed direction by some Θ_s. Whether the photon will be reabsorbed or not depends on the position of the scattered charge with respect to the point where the photon was expecting to meet it "at the end of the day". That is, we need to compare the spatial displacement of the core charge

$\varDelta\mathbf{r}$ with the characteristic size of the photon field, $\lambda_{||} \sim \omega^{-1}$, $\lambda_\perp \sim {k_\perp}^{-1}$:

$$\begin{aligned} \varDelta r_{||} &\sim \left|v_{2||} - v_{1||}\right| \cdot t_{\text{fluct}} \sim \Theta_s^2 \cdot \frac{1}{\omega\Theta^2} = \left(\frac{\Theta_s}{\Theta}\right)^2 \lambda_{||} \Leftrightarrow \lambda_{||}\,; \\ \varDelta r_\perp &\sim \quad c\Theta_s \cdot t_{\text{fluct}} \sim \Theta_s \cdot \frac{1}{\omega\Theta^2} \quad = \left(\frac{\Theta_s}{\Theta}\right) \lambda_\perp \Leftrightarrow \lambda_\perp\,. \end{aligned} \tag{95}$$

For large scattering angles, $\Theta_s \sim 1$, the charge displacement exceeds the photon wavelength for arbitrary Θ, so that the two full-size bremsstrahlung cones are present. For numerically small $\Theta_s \ll 1$, however, it is only photons with $\Theta \lesssim \Theta_s$ that can notice the charge being displaced and thus the coherence of the state being disturbed. Therefore only the radiation at angles smaller than the scattering angle actually emerges. The other field components have too large a wavelength and are easily reabsorbed *as if* there were no scattering at all.

So what counts is a change in the current, which is sharp enough to be noticed by the "to-be-emitted" quantum within the characteristic formation/field-fluctuation time (94) of the latter.

Radiation at large angles has too short a formation time to become aware of the acceleration of the charge. No scattering — no radiation.

The same argument applies to the dual process of production of two opposite charges (decay of a neutral object, vacuum pair production, etc.). The only difference is that now one has to take for $\varDelta\mathbf{r}$ not a displacement between the initial and the final charges, but the actual distance between the produced particles (spatial size of a dipole), to be compared with the radiation wavelength.

6.6 Time Delay and Decoherence Effects

So far, we were dealing with particle scattering/production as *instant* processes. Such they usually are (compared to typical formation times). Nevertheless, let us imagine that our electron in Fig. 9 is delayed by some finite $\varDelta t = \tau$ "in the V-vertex". For example, as if some metastable state was formed with characteristic lifetime $\tau = \varGamma^{-1}$.

In such a case one would have to take into consideration an extra *longitudinal* charge displacement due to finite delay, and (95) would be modified as

$$\varDelta r_{||} \sim \left(\frac{\Theta_s}{\Theta}\right)^2 \lambda_{||} + c\tau \Leftrightarrow \lambda_{||} \sim \omega^{-1}\,.$$

Now the condition $\varDelta r_{||} < \lambda_{||}$ for radiation at $\Theta > \Theta_s$ to be coherently suppressed would imply an additional restriction $\tau < \omega^{-1}$ to be satisfied. For large enough values of the delay time, $\tau \gg E^{-1}$, this new condition seriously affects radiation with comparatively large energies $\omega > \tau^{-1}$ (but still *soft* in the overall energy scale, $\omega \ll E_i$). Such photons acquire sufficiently large resolutions for coherence to be completely destroyed by the delay. Therefore they are bound to form two independent bremsstrahlung cones even for $\Theta_s = 0$.

So we would expect the accompanying radiation pattern to be that of the coherent antenna $\mathcal{R}_{\text{coher.}}$ for softer radiation, $\tau^{-1} > \omega$, and, to the contrary, a sum of two independent sources $\mathcal{R}_1 + \mathcal{R}_2$ for relatively hard photons, $\tau^{-1} < \omega \ll E$. This qualitative expectation has a nice quantitative approval.

The initial and final electron currents in (83c) now acquire different phases due to difference between the Freeze! and Move! times t_{01} and t_{02}:

$$j^\mu \implies j^\mu_{\text{del.}}(k) = \frac{p_1^\mu}{(p_1 k)}\, e^{i\omega t_{01}} - \frac{p_2^\mu}{(p_2 k)}\, e^{i\omega t_{02}}\,. \tag{96}$$

We should be careful when calculating the radiation probability, since the new current (96) is no longer conserved: $(j^\mu{}_{\text{del.}}\, k_\mu) \neq 0$. In particular, we cannot use the Feynman-gauge square of this current. The conservation could be formally rescued by adding the term describing our charge being frozen within the time interval $t_{02} - t_{01}$, namely

$$\delta j^\mu_{\text{del.}}(k) \;=\; -\frac{\delta^{0\mu}}{\omega}\left\{ e^{i\omega t_{01}} - e^{i\omega t_{02}} \right\}.$$

However, we can still use the physical polarization method instead, which remains perfectly applicable. The relative phase enters in the interference term, so that the soft radiation pattern gets modified according to

$$dN \;=\; \frac{\alpha}{\pi}\left\{ \mathcal{R}_1 + \mathcal{R}_2 - 2\,\mathcal{J}\cdot \Re\left[e^{i\omega(t_{01}-t_{02})} \right] \right\} \frac{d\omega}{\omega}\frac{d\Omega}{4\pi}\,. \tag{97}$$

To make our pedagogical setup more realistic, imagine that it was the formation of a meta-stable (resonant) state that caused the delay. In such a case the delay-time $\tau \equiv t_{01} - t_{02}$ would be distributed according to the characteristic decay exponent

$$\left[\Gamma \int_0^\infty d\tau\, e^{-\Gamma\tau} \right].$$

Averaging (97) with this distribution immediately results in a Γ-dependent expression, namely,

$$dN \;=\; \frac{\alpha}{\pi}\left\{ \mathcal{R}_1 + \mathcal{R}_2 - 2\,\mathcal{J}\cdot \chi_\Gamma(\omega) \right\} \frac{d\omega}{\omega}\frac{d\Omega}{4\pi}$$

with the *profile factor*

$$\chi_\Gamma \equiv \Re\left[\Gamma \int_0^\infty d\tau\, e^{-\Gamma\tau}\cdot e^{i\omega\tau} \right] = \Re\left[\frac{\Gamma}{\Gamma - i\omega} \right] = \frac{\Gamma^2}{\Gamma^2 + \omega^2}\,.$$

The answer can be written as a *mixture* of independent and coherent patterns with the weights depending on the ratio ω/Γ via the profile function χ_Γ,

$$dN \;=\; \frac{\alpha}{\pi}\frac{d\omega}{\omega}\frac{d\Omega}{4\pi}\left\{ [1-\chi_\Gamma(\omega)]\cdot \mathcal{R}_{\text{indep.}}(\mathbf{n}) + \chi_\Gamma(\omega)\cdot \mathcal{R}_{\text{coher.}}(\mathbf{n}) \right\}. \tag{98}$$

$\chi(\omega)$ acts as a "switch": For long-wave radiation $\chi(\omega \ll \Gamma) \to 1$ and the standard coherent antenna pattern appears; vice versa, for large frequencies $\chi(\omega \gg \Gamma) \to 0$, and coherence between charges is dashed away, as we expected.

Example: Soft Photons and the W-Width. This simple phenomenon finds an intriguing practically important implication in the dual channel. Suppose that in the e^+e^- annihilation process a pair of non-relativistic W^+W^- is produced. An intermediate boson has a finite life-time, $\Gamma \simeq 2$ GeV, and decays either leptonically or into a quark pair producing two hadron jets at the end of the day. Thus, the decays of W^+ and W^- produce ultra-relativistic electromagnetic currents and occur independently from one another within a characteristic time interval $|\Delta t_0| \sim \Gamma^{-1}$. The process is displayed in Fig. 11.

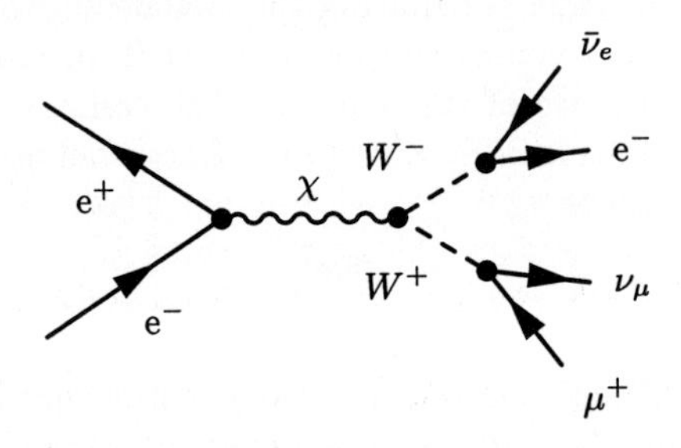

Fig. 11. Leptonic decay of a W^+W^- pair as an illustration of time-dependent decoherence effects.

Therefore one meets exactly the same "delayed acceleration" scenario as applied to the final-state currents. As a result, eq. (98) describes the photon radiation accompanying *leptonic* decays of non-relativistic W^+ and W^-.

A non-trivial ω-dependence of the profile function χ_Γ comes together with the functional dependence on the angle Θ_{12} between the leptons. This suggests a programme of measuring the W-width Γ_W by studying the variation of the radiation yield with Θ_{12}.

6.7 QCD Scattering and t-Channel Radiation

Both the qualitative arguments of the previous sections and the quantitative analysis of the two-particle antenna pattern apply to the QCD process of gluon emission in the course of quark scattering. So two gluon-bremsstrahlung cones with the opening angles restricted by the scattering angle Θ_s would be expected to appear.

There is an important subtlety, however. In the QED case it was deflection of an electron that changed the e.m. current and caused photon radiation. In QCD there is another option, namely to "repaint" the quark. Rotation of the *colour state* would affect the colour current as well and, therefore, must lead to gluon radiation irrespectively of whether the quark-momentum direction has changed or not.

This is what happens when a quark scatters off a *colour* field.

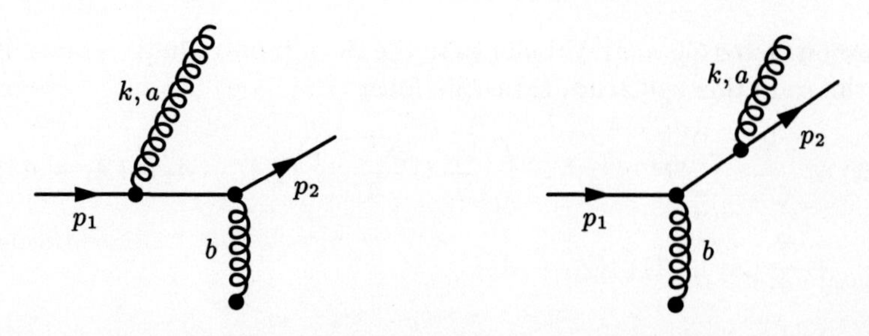

Fig. 12. Gluonic Bremsstrahlung diagrams for $k_\perp \ll q_\perp$. The characters a and b denote the colours of the radiated and exchanged gluons.

In Fig. 12 the amplitudes are shown for the radiation off the external quark lines. In principle, a graph with the gluon-gluon interaction vertex should also be considered. However, in the limit $k_\perp \ll q_\perp$, with $\mathbf{q}_\perp \approx \mathbf{p}_{2\perp} - \mathbf{p}_{1\perp}$ the momentum transfer in the scattering process, the first contributions dominate (recall the "soft insertion rules").

From the Feynman amplitudes of Fig. 12 the accompanying soft radiation current j^μ factors out, the only difference with the Abelian current (83c) being the colour generators:

$$j^\mu = \left[t^b\, t^a \left(\frac{p_1^\mu}{(p_1 k)} \right) - t^a\, t^b \left(\frac{p_2^\mu}{(p_2 k)} \right) \right] \,. \tag{99}$$

Introducing the abbreviation $A_i = \frac{p_i^\mu}{(p_i k)}$, we apply the standard decomposition of the product of two triplet colour generators (76),

$$t^a t^b = \frac{1}{2N} \delta_{ab} + \tfrac{1}{2} \left(d_{abc} + \mathrm{i} f_{abc} \right) t^c \,,$$

to rewrite (99) as

$$\begin{aligned} j^\mu &= \tfrac{1}{2}(A_1 - A_2) \left\{ t^b, t^a \right\} + \tfrac{1}{2}(A_1 + A_2) \left[t^b, t^a \right] \\ &= \tfrac{1}{2}(A_1 - A_2) \left(\frac{1}{N}\, \delta^{ab} + d^{abc}\, t^c \right) - \tfrac{1}{2}(A_1 + A_2)\, \mathrm{i} f^{abc}\, t^c \,. \end{aligned}$$

To find emission *probability* we need to sum the product of currents over colours. Three colour structures do not "interfere", so it suffices to evaluate the squares of the singlet, $\mathbf{8}_s$ and $\mathbf{8}_a$ structures:

$$\begin{aligned} \sum_{a,b} \left(\frac{1}{2N} \delta_{ab} \right)^2 &= \left(\frac{1}{2N} \right)^2 (N^2 - 1) = \frac{1}{2N} \cdot C_F \,; \\ \sum_{a,b} \left(\tfrac{1}{2}\, d_{abc}\, t^c \right)^2 &= \frac{1}{4} \frac{N^2 - 4}{N} \left(t^c \right)^2 = \frac{N^2 - 4}{4N} \cdot C_F \,; \\ \sum_{a,b} \left(\tfrac{1}{2}\, i f_{abc}\, t^c \right)^2 &= \frac{1}{4} N \left(t^c \right)^2 = \frac{N}{4} \cdot C_F \,. \end{aligned}$$

The common factor $C_F = \left(t^b\right)^2$ belongs to the Born (non-radiative) cross section, so that the radiation spectrum takes the form

$$\mathrm{d}N \propto \frac{1}{C_F}\sum_{\text{colour}} j^\mu \cdot (j_\mu)^* = \left(\frac{1}{2N} + \frac{N^2-4}{4N}\right)(A_1 - A_2)\cdot(A_1 - A_2)$$
$$+\frac{N}{4}(A_1 + A_2)\cdot(A_1 + A_2)\,.$$

A simple algebra leads to

$$\mathrm{d}N \propto C_F\,(A_1 - A_2)\cdot(A_1 - A_2) + N\,A_1\cdot A_2\,. \tag{100}$$

Dots here symbolize the sum over gluon polarization states. Similarly to the case of "delayed scattering" discussed above, the current (99) *does not conserve* because of non-commuting colour matrices. We would need to include gluon radiation from the exchange-gluon line *and* from the source, to be in a position to use an arbitrary gauge (e.g. the Feynman gauge) for the emitted gluon. Once again, the physical polarization technique (86) simplifies our task. To obtain the true accompanying radiation pattern (in the $k_\perp \ll q_\perp$ region) it suffices to use the projectors (88) for the dots in (100). In particular,

$$A_1\cdot A_2 \equiv \sum_{\lambda=1,2}\left(A_1 e^{(\lambda)}\right)\left(A_2 e^{(\lambda)}\right)^* = \mathcal{J} \qquad \{\neq -(A_1\,A_2)\ \textit{sic!}\,\}\,.$$

Accompanying radiation intensity finally takes the form

$$dN \propto C_F\,\mathcal{R}_{\text{coher.}} + N\,\mathcal{J}\,. \tag{101}$$

The first term proportional to the squared quark charge is responsible, as we already know, for two narrow bremsstrahlung cones around the incoming and outgoing quarks, $\Theta_1, \Theta_2 \le \Theta_s$. On top of that an additional, purely non-Abelian, contribution shows up, which is proportional to the *gluon* charge. It is given by the interference distribution (89c), (91b),

$$\mathcal{J} = \frac{a_1 + a_2 - a_{12}}{a_1 a_2} - 1\,,$$

which remains *non-singular* in the forward regions $\Theta_1 \ll \Theta_s$ and $\Theta_2 \ll \Theta_s$. At the same time, it populates large emission angles $\Theta = \Theta_1 \approx \Theta_2 \gg \Theta_s$ where

$$\mathrm{d}N \propto \mathrm{d}\Omega\,\mathcal{J} \propto \sin\Theta d\Theta\left(\frac{2}{a} - 1\right) \sim \frac{d\Theta^2}{\Theta^2}\,.$$

Indeed, evaluating the azimuthal average, say, around the *incoming* quark direction, we obtain

$$\int\frac{d\phi_1}{2\pi}\,\mathcal{J} = \frac{1}{a_1}\left(1 + \frac{a_1 - a_{12}}{|a_1 - a_{12}|}\right) - 1 = \frac{2}{a_1}\,\vartheta(\Theta_1 - \Theta_s) - 1\,.$$

Thus we conclude that the third complementary bremsstrahlung cone emerges. It basically corresponds to radiation at angles *larger* than the scattering angle, and its intensity is proportional to the colour charge of the t-channel exchange.

We could have guessed without actually performing the calculation that at large angles the gluon radiation is related to the *gluon* colour charge. As far as large emission angles $\Theta \gg \Theta_s$ are concerned, one may identify the directions of initial and final particles to simplify the total radiation amplitude as

$$j^\mu = T^b T^a \cdot \frac{p_1^\mu}{p_1 k} - T^a T^b \cdot \frac{p_2^\mu}{p_2 k} \approx \left(T^b T^a - T^a T^b\right) \cdot \frac{p^\mu}{pk}.$$

This, with account of the commutation relation (37), immediately leads to $N \propto (if_{abc})^2$ as a proper colour charge, *irrespective* of the nature of a scattered particle.

References

[1] Howard Georgi, *Lie Algebras in Particle Physics*, Benjamin & Cummings, Reading Massachusetts (1982).

[2] Richard D. Field, *Applications of Perturbative QCD*, ed. E. Pines, Addison-Wesley Publishing Company, Redwood City (1989).

[3] Yu. L. Dokshitzer, V. A. Khoze, A. H. Mueller and S. I. Troyan, *Basics of Perturbative QCD*, ed. J. Tran Thanh Van, Editions Frontières, Gif-sur-Yvette (1991).

[4] F. E. Close, *An Introduction to Quarks and Partons*, Academic Press, London (1979).

Quark Matter and High Energy Nuclear Collisions*

H. Satz[1];
Notes by S. Leupold[2]

[1] Theory Division, CERN, CH–1211 Geneva 23, Switzerland
and Fakultät für Physik, Universität Bielefeld, 33501 Bielefeld, Germany
[2] Institute for Theoretical Physics, University of Regensburg, Universitätsstraße 31, 93040 Regensburg, Germany

1 Introduction

The study of strongly interacting matter is a fascinating subject for a variety of reasons. It is needed to describe the first few microseconds in the evolution of our universe, and it provides the equation of state for stellar matter at high density. To study matter at extreme densities one has to develop the thermodynamics of strong interaction physics. But first of all we have to define what "extreme density" means. Let us have a look at the density of matter in the present world. It covers a range from 10^{-6} up to 10^{38} nucleons/cm^3. But what is beyond this

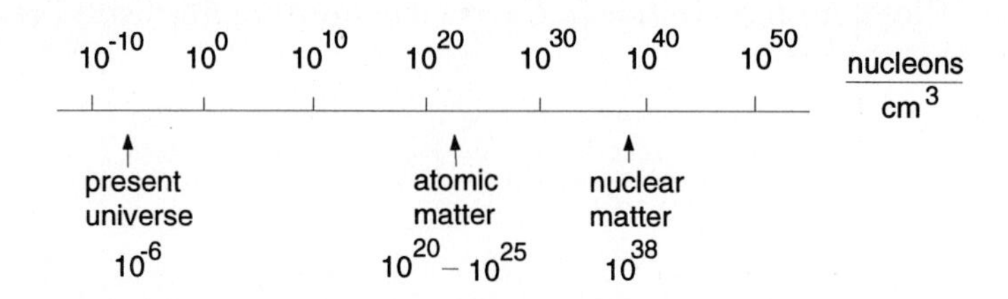

Fig. 1. *Range of observable matter densities.*

upper bound given by the density of nuclear matter?

To get a feeling for the new things which might appear we think about a gas of nucleons which is compressed more and more. Of course the nucleons have an intrinsic structure, the quarks. But in the beginning when the distance between the nucleons is much bigger than their extension it is well justified to speak about nucleons as separate particles. However when the density increases the concept of particles called "nucleons" becomes meaningless since one cannot

---

* Lectures presented at the workshop "Lattice QCD and Dense Matter" organised by the Graduiertenkolleg Erlangen–Regensburg, held on October 11th–13th, 1994 in Kloster Banz, Germany

isolate composite particles in the dense "quark soup". Thus the constituents of this new kind of matter are quarks and not hadrons anymore. At some point during the compression we expect a transition from hadronic matter to quark matter and the interesting questions are of course: What properties does quark matter have and can we create such a quark phase in the laboratory?

It is very instructive to compare strongly interacting matter described by QCD with the case of electrons and nuclei interacting via electromagnetic forces and forming a solid. In the same way as the electromagnetic Coulomb potential is screened in a dense medium (Debye screening) colour screening of the confining potential is expected in dense matter of quarks and gluons called the quark-gluon plasma. The confining potential which in a simple model is given by

$$V_{\mathrm{conf}}(r) = \sigma r \tag{1}$$

is replaced by [1]

$$V_{\mathrm{scr}}(r) = \frac{1 - e^{-\mu r}}{\mu r} \sigma r \tag{2}$$

which allows for asymptotic coloured states (cf. Fig. 2). This yields the following picture: At low density strongly interacting particles form a colour insulator, namely hadronic matter, whereas at high density due to screening of the potential matter becomes a colour conductor, the quark-gluon plasma. A second feature of

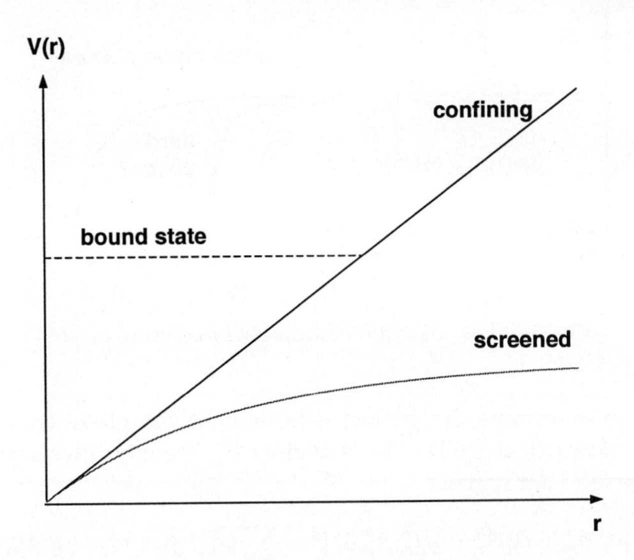

Fig. 2. *Colour confining and colour screened potential.*

quark matter is the restoration of chiral symmetry. Let us again compare QED and QCD. In metals the free electron mass becomes an effective mass due to interactions of the electron with lattice vibrations. In QCD it is just the other way round. Due to a non-trivial vacuum structure of QCD quarks in the nucleon have an effective mass (constituent mass) of about one third of the nucleon mass, whereas in the quark-gluon plasma the masses of quarks nearly vanish since we have entered the regime of perturbative QCD. Thus the chiral symmetry of the QCD Lagrangian which is spontaneously broken in hadrons is restored in the quark-gluon plasma.

These considerations yield the phase diagram shown in Fig. 3. At small temperatures T and baryochemical potential μ_B strongly interacting matter is in the hadronic phase. If both T and μ_B are very high the regime of the quark-gluon plasma is reached. However, there might be a third phase in between where quarks are deconfined but carry a constituent mass already. Again we can find an analogous case in solid state physics: Insulator, superconductor and conductor phases. To get a qualitative picture of the phases let us briefly discuss the two limiting cases of vanishing temperature or vanishing baryochemical potential, respectively. In the latter case the critical temperature of deconfinement and chiral symmetry restoration is expected to be the same, whereas at $T = 0$ the critical baryochemical potential of deconfinement might be smaller than the one of chiral symmetry restoration. To find out whether there are two or three

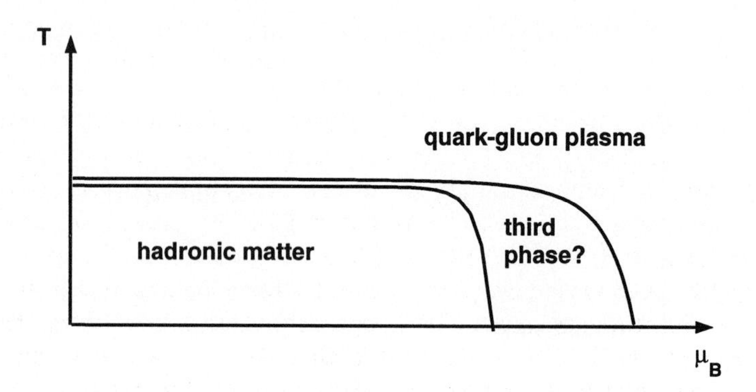

Fig. 3. *Phase diagram of strongly interacting matter.*

phases and to determine the critical behaviour of the phase transition(s) one has to study statistical QCD. The dynamics of strongly interacting matter is described by the Lagrangian

$$\mathcal{L} = -\frac{1}{4}[\partial_\mu A_\nu^a - \partial_\nu A_\mu^a - g f_{bc}^a A_\mu^b A_\nu^c]^2 - \sum_f \bar{\psi}_f^\alpha [i \not\partial + m_f - g \not{A}]_{\alpha\beta} \psi_f^\beta . \quad (3)$$

In contradistinction to QED gauge bosons (gluons) carry colour charges and thus interact directly with each other. Thermodynamic properties can be calculated from the QCD partition function

$$Z(T, V) = \mathrm{Tr}\left\{e^{-H(\mathcal{L})/T}\right\}$$

$$= \int \mathcal{D}\psi \, \mathcal{D}\bar{\psi} \, \mathcal{D}A \, e^{-S(A,\psi,\bar{\psi})} \tag{4}$$

with the action

$$S(A,\psi,\bar{\psi}) = \int_0^{1/T} d\tau \int_V d^3x \, \mathcal{L}(x, \tau = it) \tag{5}$$

where an imaginary time τ is introduced. Thus one has to work in Euclidean space with bosons (fermions) obeying (anti)periodic boundary conditions in the τ direction. Pressure and energy density of the system are given by

$$P = T \left(\frac{\partial \ln Z}{\partial V} \right)_T , \tag{6}$$

$$\epsilon = \frac{T^2}{V} \left(\frac{\partial \ln Z}{\partial T} \right)_V . \tag{7}$$

Obviously the evaluation of the path integral (4) is a highly non-trivial challenge as soon as one is leaving the perturbative regime. Indeed to describe the hadronic matter phase and the phase transition one has to deal with a large coupling constant which requires a non-perturbative treatment. Up to now the only known rigorous method to calculate the QCD partition function is to formulate the problem on a lattice [2]. This yields a generalized spin problem which can be solved by computer simulation [3]. Of course the computational evaluation of path integrals is also very much involved giving rise to a new branch of physics called computational particle physics. These computer simulations aim at the calculation of thermodynamic quantities and also of hadron masses in order to set the scale for other quantities. This approach enables us to study the critical behaviour of strongly interacting matter ab initio. For a discussion of these techniques we refer the reader to the literature and to the contributions by Negele and Hasenfratz within these proceedings [4]. In the following we rather present main results of lattice calculations obtained so far concerning thermodynamic quantities and especially the phase transition.

Figure 4 shows energy density ϵ and pressure p in the neighborhood of the phase transition. At small temperature these quantities are governed by the lightest hadrons, the pions. For a free pion gas the energy density is given by three times the pressure. The influence of all other hadrons is suppressed due to their masses being much bigger than the pion mass. At very high temperatures we expect to enter the perturbative regime of QCD since the running coupling constant decreases with increasing momenta and thus with increasing temperature. This means that for $T \to \infty$ the values of ϵ and p should be given by the energy density and pressure of a gas of free quarks and gluons. Calculations shown in Fig. 4 support the assumption that quarks and gluons are the relevant degrees of freedom at high temperatures.

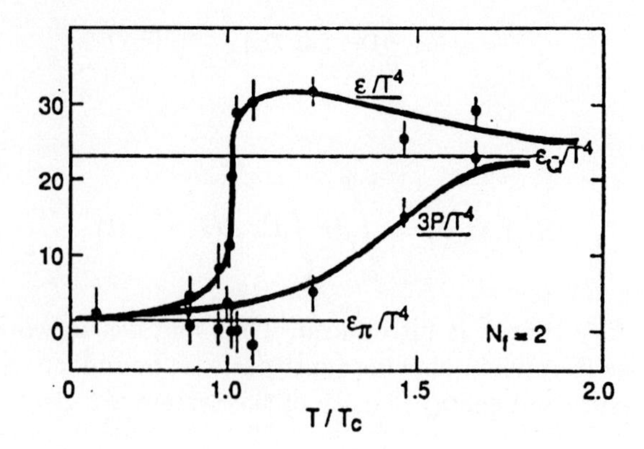

Fig. 4. *Energy density and pressure as functions of the temperature around the phase transition.*

In Fig. 5 two quantities are shown which parametrize deconfinement and chiral symmetry restoration, respectively. The Polyakov loop[1] [5]

$$\langle L \rangle = e^{-V(r=\infty)/T} \tag{8}$$

vanishes as long as the quarks are confined but gets a positive value in the deconfined phase. The chiral condensate [6]

$$\chi = \langle \bar{\psi} \psi \rangle \tag{9}$$

is proportional to the effective quark mass which is positive as long as one

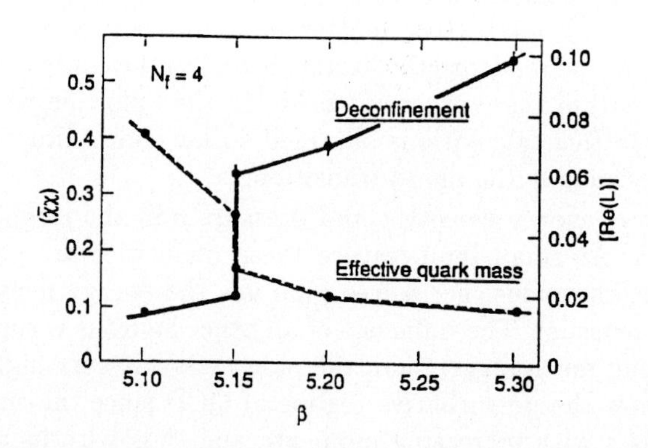

Fig. 5. *Polyakov line and chiral condensate as functions of the temperature around the phase transition.*

[1] $V(r)$ is the quark-quark potential (cf. Fig. 2).

is dealing with constituent quarks and vanishes when quarks become current quarks, i.e. when chiral symmetry is restored.[2] Both figures 4 and 5 seem to indicate that for $\mu_B = 0$ there is a single phase transition at about $T_c = 150\,\text{MeV}$ and $\epsilon_c = 1\,\text{GeV/fm}^3$ from hadronic matter to a plasma of deconfined and massless quarks and gluons. To get a feeling for the magnitude of the critical energy density it is instructive to compare with the typical density of nuclei $\epsilon_A = 0.15\,\text{GeV/fm}^3$ and the density of protons $\epsilon_p = 0.5\,\text{GeV/fm}^3$.

However all these considerations remain purely academic if there is no opportunity to observe such a phase transition. There might be or might have been such phase transitions in nature – probably the early universe was in a quark-gluon plasma phase – however the signals for the existence of this phase, if there are any, are hard to disentangle from other effects. Thus it would be a big achievement if one could create a quark-gluon plasma in the laboratory. Colliding two heavy nuclei at high momenta is a promising place to look for it since we expect the creation of a bubble of very dense matter at least in the center of the collision zone. If the collision energy is large enough the density of the bubble should be so high that the matter inside can no longer stay in the hadronic phase. Once the bubble is formed it expands, cools off, hadronizes again and finally emits free hadrons which can be detected. What do we need to perform such experiments? To get the nuclei accelerated one has to work with heavy ion beams. The demand of high collision energies requires powerful accelerators where ions can be injected. They are presently available at the AGS at Brookhaven National Laboratories (BNL) and at the SPS at CERN.

In the next section we will elaborate on the possibility of creating a quark-gluon plasma in high energy nuclear collisions. Section 3 is devoted to the hadron-quark transition whereas in the last section the question is discussed how to find unambiguous signals for the creation of a quark-matter phase.

2 Nuclear Collisions

Before talking about a theoretical description of heavy ion collisions it might be interesting to get some information about experiments which have been performed up to now and which are planned for the future. Status and plans of heavy ion experiments are briefly summarized in Table 1. The aims of heavy ion experiments can be divided in three (chronological) phases. In phase I the questions was whether the experiments are feasible, whether high densities are attainable and whether the observed particle behaviour shows new features. They have been answered positively. We will report on some of the new features in this section as well as in the last one. In phase II one has to find out whether the system thermalizes and whether it reaches the deconfined phase. If this is the case then finally the properties of the created quark-gluon plasma can be studied (phase III).

[2] Of course the quarks still carry their current quark mass which breaks chiral symmetry explicitly. However this becomes less important when the temperature increases.

Table 1. *Properties of heavy ion accelerators. Energies are given in GeV per nucleon.*

Time	Accelerator	Beam	CMS Energy (GeV/A)
1986 – 1993	BNL-AGS	$\to Si$	5
	CERN-SPS	$\to S$	20
1993 – 1999	AGS + booster	$\to U$	4
	SPS + Pb inj.	$\to U$	17
"2000"	BNL-RHIC	$\to U$	200
	CERN-LHC	$\to U$	5500

To decide whether it is possible to create a piece of quark matter inside a hot bubble originating from a heavy-ion collision we have to answer the following basic questions (see [7] and references therein): Is the bubble hot enough so that the system crosses the phase transition point? Is the bubble thermalized so that it reaches the deconfined phase? Is the bubble large enough so that we can study bulk properties and critical behaviour near the phase transition?

Let us start with an estimate of accessible energy densities using the multiplicity $(dN/dy)_{AB}$ of hadrons measured per unit central rapidity in A-B collisions. Assuming free flow the initial energy density is given by[3]

$$\epsilon_0 = \omega \frac{1}{V_0} \left(\frac{dN}{dy} \right)_{AB} \tag{10}$$

with the initial volume V_0 and the energy per hadron $\omega \simeq 0.5\,\text{GeV}$. The initial volume can be calculated via

$$V_0 \simeq \pi R_A^2 \tau \tag{11}$$

with the nuclear radius $R_A = 1.2\,A^{1/3}\,\text{fm}$ and the formation time $\tau \simeq 1\,\text{fm}$. Finally we have to calculate the multiplicity $(dN/dy)_{AB}$. It is related to the multiplicity $(dN/dy)_p$ of p-p and p-$\bar{p}$ collisions. A fit of all data from SPS to Tevatron energies suggests the form

$$(dN/dy)_p \simeq \ln[\sqrt{s}/2m_p]\,.\,. \tag{12}$$

This result can be extrapolated to A-B collisions using

$$\frac{(dN/dy)_{AB}}{(dN/dy)_p} \simeq \begin{cases} A^\alpha & \text{for } A = B \\ (A^{1/3}B^{2/3})^\alpha & \text{for } A \gg B \end{cases} \tag{13}$$

[3] The assumption of isentropic expansion instead of free flow gives approximately the same result in the studied energy range.

with $1 \leq \alpha \leq 4/3$. Collecting all formulae together we get

$$\epsilon_0^{AB} \simeq \left[\frac{(A^{1/3}B^{2/3})^\alpha}{2\pi(1.2\,B^{1/3})^2}\right] \ln[\sqrt{s}/2m_p]\,\text{GeV/fm}^3\,. \tag{14}$$

Using a value of $\alpha = 1.1$ as obtained from p-A collisions we can calculate the energy densities for *Pb-Pb* collisions for the various accelerators and find the results listed in Table 2. If more rescattering happens in *Pb-Pb* collisions com-

Table 2. *Attainable energy densities at various heavy ion colliders.*

Facility	$\sqrt{s}$ [GeV]	ϵ_0 [GeV/fm^3]
BNL-AGS	4	0.8
CERN-SPS	17	2.5
BNL-RHIC	200	5.2
CERN-LHC	6000	9.0

pared to *Pb-p* then α and thus ϵ would even be higher. By varying A, energy densities in the range from 0.5 to nearly $10\,\text{GeV/fm}^3$ can be covered by the aforementioned accelerators without any jumps or gaps as can be seen in Fig. 6. Thus the existing and planned facilities complement each other very well. Since the critical energy density of the phase transition is expected to be at about $1\,\text{GeV/fm}^3$ we conclude that the energy densities for a quark-gluon plasma can be attained in heavy ion collisions. To study the behaviour of bulk matter in

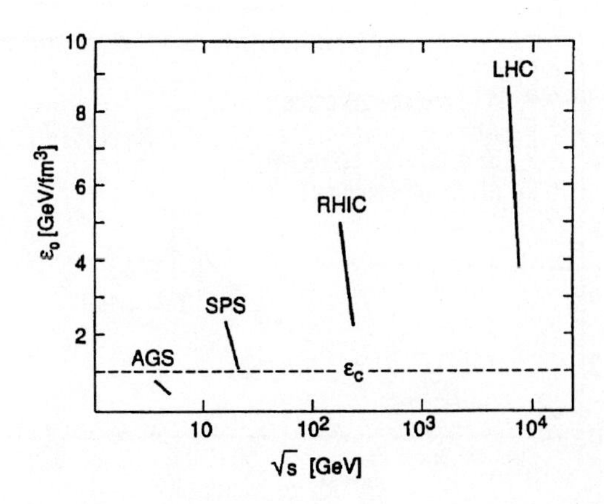

Fig. 6. *The estimated energy densities for AGS, SPS, RHIC and LHC as function of the incident energies.*

QCD, we need large systems; this is even more the case when we look for critical behaviour, since divergences or discontinuities become pronounced only in the large volume limit. The initial interaction volumes in A-A and p-p collisions are related by

$$V_0^{AA}/V_0^{pp} \simeq (R_A/R_p)^2 \simeq 2A^{2/3} . \tag{15}$$

This means that in a Pb-Pb collision the interaction volume is about 75 times that of p-p. But if the system has a sufficiently high initial density, it can expand considerably and still remain deconfinemed. The volume at freeze-out can be determined by requiring that at this point the mean free path λ of a hadron in the system should surpass the dimension of the system, i.e., the hadron "gets through" without interacting. λ can be related to the particle density n and the interaction cross section σ via

$$1/\lambda = n\sigma . \tag{16}$$

From

$$\lambda = (n\sigma)^{-1} = R_f , \tag{17}$$

we get with isentropic expansion and $\sigma \simeq \sigma_{\pi\pi} \simeq 20\,\text{mb}$ for the freeze-out of pions

$$R_f^\pi \simeq 0.7 \cdot (dN/dy)_{AA}^{1/2}\ \text{fm} \simeq 0.7 \cdot A^{0.55}[\ln\sqrt{s}/2m_p]^{1/2}\ \text{fm} \tag{18}$$

where we have used eqs.(12) and (13) and again a value of 1.1 for α. This shows that increasing A or $\sqrt{s}$ leads to higher ϵ_0, resulting in more expansion and hence a larger freeze-out radius – if the initial state indeed consisted of dense matter. The freeze-out radius can be measured via hadron interferometry [8]. There information about the size of a source emitting hadrons can be deduced from the fact that identical bosons/fermions attract/repulse each other. Figure 7

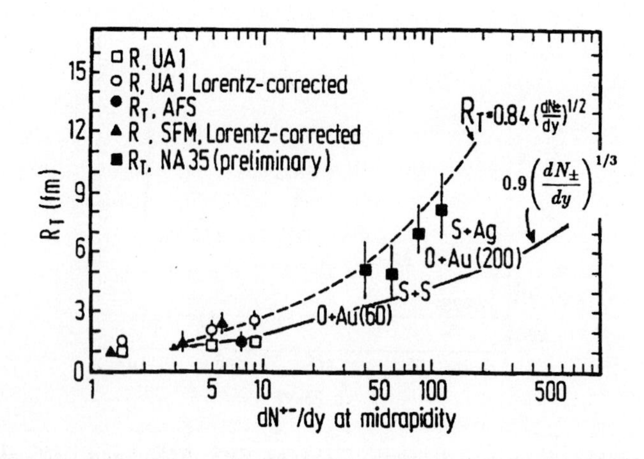

Fig. 7. *The freeze-out radius as a function of the multiplicity.*

shows that indeed the freeze-out radius grows with the multiplicity, i.e. there is expansion in the system. Using eq.(18) we can calculate the expected freeze-out volumes for *Pb-Pb* collisions listed in Table 3 These results foster our hope that

Table 3. *Estimated freeze–out volumes for Pb–Pb collisions at various accelerators.*

	SPS	RHIC	LHC
R_f/R_A	2.8	4.0	5.3
V_f [fm^3]	3×10^4	9×10^4	2×10^5

the systems are sufficiently macroscopic for large A and $\sqrt{s}$ to be amenable to studies in terms of QCD thermodynamics. However a caution is necessary. The volumes just referred to are calculated for the freeze-out of pions. But when the pions leave the hot bubble the system is (again) in the hadronic phase. Since we are interested in the deconfined phase we should instead determine the volume of the system when the transition from quark to hadronic matter is performed. This is clearly much more involved than the determination of the pion freeze-out.

Finally we discuss the question of thermalization. The evolution of the system towards equilibrium is driven by rescattering processes. Thus a necessary condition for the equilibration process is that a nucleus-nucleus collision is much more than a superposition of nucleon-nucleon collisions. To find some evidence for the appearance of a cascade of rescattering processes leading to equilibrium we can study the production ratios of hadrons and the strangeness evolution [9]. In general we expect that higher A and/or $\sqrt{s}$ increase the particle multiplicities dN/dy which is a hint for more rescattering. The ratio of kaons K^+ and pions π^+ might serve as an example. In *p-p* collisions this ratio takes a value of 0.06 ± 0.01 in the energy range of $5 \leq \sqrt{s} \leq 20\,\text{GeV}$. In contrast, the ratio for thermal emission is $0.20 - 0.25$ at temperatures between 150 MeV and 250 MeV and densities between one and four times nuclear density. Thus thermalization leads to an increase of the K^+/π^+ ratio of about four times the value obtained in nucleon-nucleon collisions. Figure 8 shows results from the BNL-E802 collaboration [10]. We see that the ratio indeed grows with increasing number of involved nucleons and reaches the regime predicted for thermal emission. This evolution towards thermalization can also be observed in many other ratios obtained in the *S-U* experiment at SPS. If there is a one-stage freeze-out of all thermal hadrons, then all production ratios are given in terms of two parameters T and μ_B [11]. In other words, we can take two measured ratios, such as $\Lambda/\bar{\Lambda}$ and $\Xi/\bar{\Xi}$, to fix the values of T and μ_B, and all other ratios are then predicted. Although previous analyses had been in accord for all strange baryon and meson ratios at $T \simeq 200\,\text{MeV}$ [12], $\mu_B \simeq 300\,\text{MeV}$ [13], newer and more precise data [14] no longer provide a common equilibrium freeze-out point even for strange hadrons.

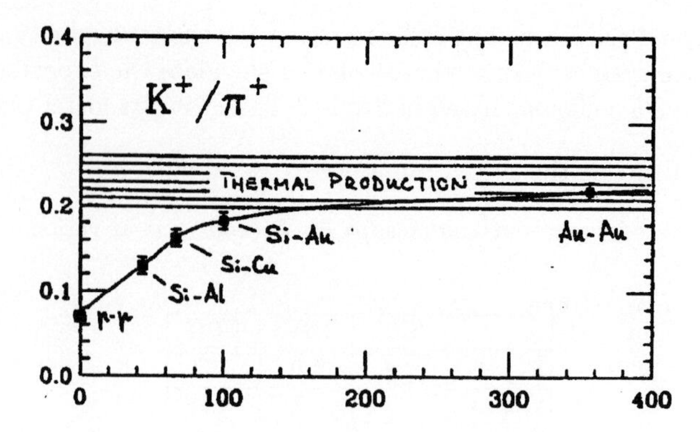

Fig. 8. *K^+/π^+ ratios vs. total number of participants at the AGS (central collisions with Au target).*

It therefore seems reasonable to consider when freeze-out will occur. Recalling our definition for freeze-out as given above (18), we get sequential freeze-out for the various particle species, since their freeze-out radius depends on their respective interaction cross section [15].

We conclude that the systems obtained in heavy-ion collisions are hot, large enough and thermalized so that we can hope to be able to study the properties of dense strongly interacting matter.

3 The Quark-Hadron Transition

In this section we will study the phase transition from hadronic to quark matter in greater detail. First we will discuss the thermodynamics of hadrons. Let us start with a hot ideal gas of pions. In the Boltzmann approximation the logarithm of the partition function is given by

$$\begin{aligned} \ln Z_0(T,V) &= \frac{V}{(2\pi)^3}\int d^3p\, e^{-\sqrt{m_0^2+p^2}/T} \\ &= \frac{VTm_0^2}{2\pi^2}K_2(m_0/T) \\ &= \frac{VT^3}{\pi^2}[1+o(m_0/T)]\,. \end{aligned} \tag{19}$$

Of course this is by no means an appropriate description of a system of dense hadrons since any kind of interaction is missing. To improve the model we include resonances by introducing a mass distribution $\rho(m)$ in the following way [16]:

$$\ln Z_R(T,V) = \frac{V}{(2\pi)^3}\int_{m_0}^{\infty} dm\, \rho(m)\int d^3p\, e^{-\sqrt{m^2+p^2}/T}$$

$$= \frac{VT}{2\pi^2} \int_{m_0}^{\infty} dm\, \rho(m) m^2 K_2(m/T) \ . \tag{20}$$

Note that one recovers the ideal pion gas we have started with in the case of

$$\rho(m) = \delta(m - m_0) \ . \tag{21}$$

The inclusion of resonances is attained by the ansatz [17]

$$\rho(m) \sim m^a e^{bm} \tag{22}$$

where a and b have to be determined from experimental data. In Hagedorn's statistical bootstrap model the partition function turns out to be

$$\begin{aligned} \ln Z_H(T,V) &\sim \frac{VT}{2\pi^2} \int_{m_0}^{\infty} dm\, m^{2+a} e^{bm} K_2(m/T) \\ &\approx V \left(\frac{T}{2\pi}\right)^{\frac{3}{2}} \int_{m_0}^{\infty} dm\, m^{\frac{3}{2}+a} e^{-m(\frac{1}{T}-b)} \ . \end{aligned} \tag{23}$$

Obviously the partition function diverges for temperatures larger than $T_H = 1/b$. While Hagedorn [17] argued that T_H is the ultimate temperature of matter, Cabibbo and Parisi [18] claimed that it marks the point where the transition to a new phase of strongly interacting matter takes place. Anyway it is an interesting observation that thermodynamics of hadrons defines its own limit. Of course the discussion about this limiting temperature got a new direction when the interpretation of hadrons as bound quark states was proposed by Gell-Man and Zweig.

Before discussing the phase transition and the transition temperature in the quark bag model we will present simple calculations concerning the second quantity which characterizes a state of matter, the density. To get insight in the typical magnitudes of the density for the three possible phases (cf. Fig. 3) we introduce a purely geometric model without any thermal features, the percolation model. Suppose putting spheres of volume V_0 into a box of volume V. At what density are there infinitely many clusters when V is taken to infinity? This question known as the percolation problem can be answered by a numerical calculation. Its solution is

$$n_p = 0.34/V_0 \ . \tag{24}$$

We want to apply this result to hadronic systems with hadrons described by an additive quark model [19]. There we can calculate the radii of the constituent quarks via the cross sections σ_{pp} and σ_{Qp} for proton-proton and quark-proton scattering, respectively:

$$\sigma_{pp}/\sigma_{Qp} = 3 \ . \tag{25}$$

From the collision geometry we get the obvious relation

$$\sigma_{pp}/\sigma_{Qp} = R_p^2/R_Q^2 \tag{26}$$

where $R_N \simeq 0.8\,\text{fm}$ is the radius of the nucleon and R_Q is the radius of the constituent quark. Comparing the last two equations we find

$$R_Q = R_N/\sqrt{3}\,. \tag{27}$$

The nucleons percolate at

$$n_H = 0.34/V_N \simeq 0.16\text{fm}^{-3} \tag{28}$$

which is about the same as the density n_0 of nuclear matter. Below n_H the hadrons form a gas, while above n_H they are in the hadronic matter phase. Using eq.(27) we find that the quarks themselves percolate at

$$n_Q = 0.34/V_Q \simeq 0.82\,\text{fm}^{-3} \simeq 5.1 \cdot n_0 \tag{29}$$

which gives the critical density for the phase transition from hadronic to quark matter.

Next we want to derive an estimate for the critical temperature exploiting the bag model [20]. To keep things as simple as possible we assume that there is only one phase transition and we neglect all baryons. We describe hadronic matter by an ideal gas of massless pions and quark matter by an ideal quark-gluon plasma. In both cases the partition function is given by [21]

$$\ln Z(T,V) = c\frac{\pi^2}{90}T^3V \tag{30}$$

where c denotes the number of degrees of freedom. For pions we have to take into account three isospin states, thus

$$c_\pi = 3\,, \tag{31}$$

whereas for the quark-gluon plasma we get

$$c_Q = 2 \times 8 + \frac{7}{8}(3 \times 2 \times 2 \times 2) = 37\,. \tag{32}$$

The first term is given by the number of possible polarizations and colours of gluons. The factor in front of the brackets accounts for the different statistics of fermions and bosons. Finally for quarks we have three colours, two spin states, particles and antiparticles and two light flavours. Calculating the pressure according to eq.(6) we find

$$P_\pi \approx \frac{1}{3}T^4 \tag{33}$$

and

$$P_Q \approx 4T^4\,. \tag{34}$$

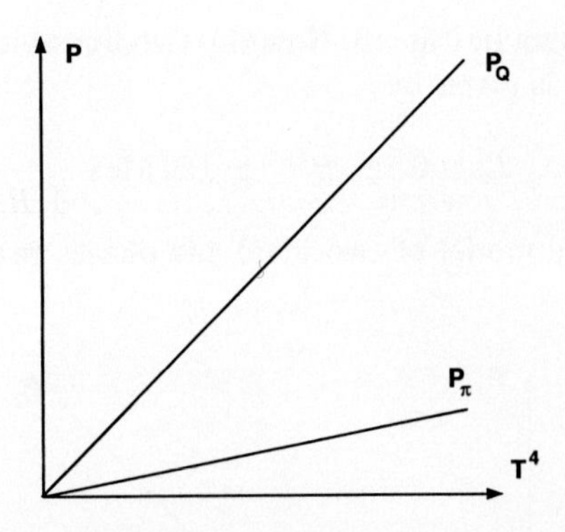

Fig. 9. *Pressure of pion gas and quark-gluon plasma, respectively, without bag pressure.*

From this calculation we would have to conclude that the quark-gluon plasma is the stable ground state of our world since a system always chooses the state of highest pressure (cf. Fig. 9). Thus our world of daily experience would be a metastable bubble only which would decay as soon as a quark-gluon plasma is created somewhere in the universe, e.g. in a heavy-ion collision [22]. Fortunately one feature is missing in this scenario preventing human civilization from being destroyed by its own enthusiasm about doing experiments. To get a more realistic picture we have to take into account the difference of pressure between the physical vacuum and the (perturbative) QCD vacuum, the bag pressure B. In the bag model the energy of a nucleon turns out to be

$$E = 3|\mathbf{p}| + BV \tag{35}$$

where $|\mathbf{p}|$ denotes the quark momentum and V the volume of the three-quark system. Due to the uncertainty relation we can relate $|\mathbf{p}|$ with the diameter $2R$ of the system:

$$|\mathbf{p}| \simeq 2/R\ . \tag{36}$$

Combining the last two equations we get the nucleonic energy as a function of the radius of the three-quark system. We expect that the energy takes its minimal value when the true nucleon radius R_N is inserted. From

$$\frac{dE}{dR} = 0 \tag{37}$$

we get an estimate for the bag pressure:

$$B^{1/4} = (3/2\pi)^{1/4} R_N^{-1} \simeq 210\,\mathrm{MeV}\ . \tag{38}$$

Subtraction of B from the pressure of quarks and gluons

$$P_Q \approx 4T^4 - B \tag{39}$$

yields the diagram shown in Fig. 10. Now the two lines intersect at some critical temperature T_c which is given by

$$T_c \simeq 0.72 \cdot B^{1/4} \simeq 150\,\mathrm{MeV}\ . \tag{40}$$

Thus even this simple model of two ideal gas phases predicts a quark-hadron

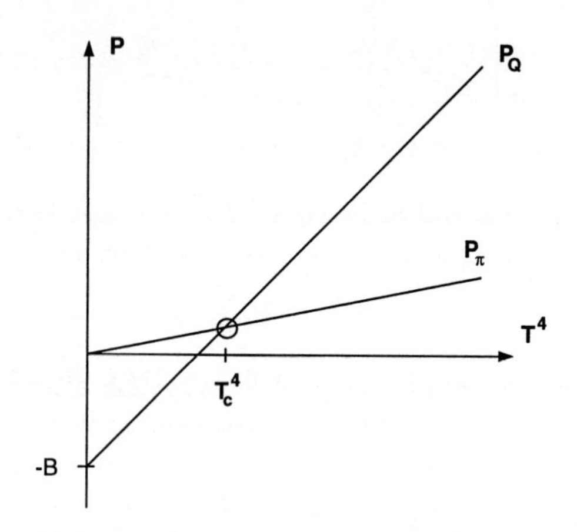

Fig. 10. *Pressures of pion gas and quark-gluon plasma with bag pressure included.*

transition in the expected range as soon as the bag pressure is taken into account. According to the model of quarks inside of a bag this transition can be interpreted as "bag fusion" [23]. Figure 11 shows the pressures as given in eqs.(33) and (39) and the energy densities

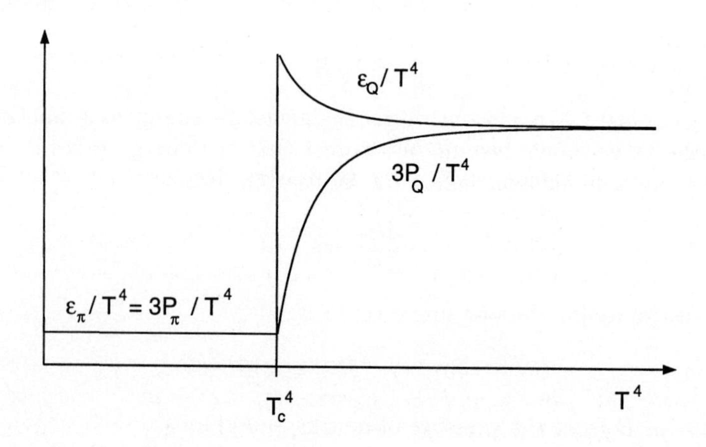

Fig. 11. *Energy density and pressure as functions of the temperature.*

$$\epsilon_\pi \simeq T^4 \tag{41}$$

and

$$\epsilon_Q \simeq 12T^4 + B . \tag{42}$$

Obviously the bag pressure becomes less important when the temperature increases. For $T \to \infty$ we recognize the relation

$$\epsilon_Q = 3P_Q \tag{43}$$

which holds for a quark-gluon plasma without any interactions. Finally in Fig. 12 the interaction measure Δ defined as

$$\Delta := (\epsilon - 3P)/T^4 \tag{44}$$

is plotted as a function of temperature. In the ideal pion gas phase Δ vanishes while in the quark-gluon plasma phase it turns out to be

$$\Delta = 4B/T^4 . \tag{45}$$

We conclude that the bag pressure parametrizes the (non-perturbative) inter-

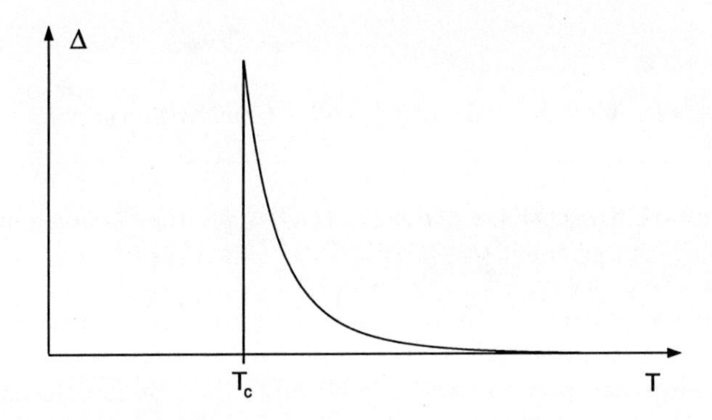

Fig. 12. *Interaction measure as a function of the temperature.*

action in the quark-gluon plasma phase.

To check the model independence of the results obtained so far we will present now the string model which also serves for the description of the quark structure of hadrons. There e.g. a meson consists of two massless point-like charges at distance $2r$ which are connected by a string of tension σ. The energy of the system is

$$E = 2|\mathbf{p}| + 2\sigma r \simeq \frac{2c}{r} + 2\sigma r \tag{46}$$

where we have used the uncertainty principle again and thus c is of the order of 1. The energy has to be minimized with respect to the distance. This yields

$$M_H = 4\sqrt{c\sigma} \tag{47}$$

for the hadron mass and

$$R_H = 2\sqrt{c/\sigma} \tag{48}$$

for the intrinsic size of the hadron. Now we consider a gas of such hadrons with interactions occurring only pairwise [24]. At low densities we get an ideal hadron gas. At high densities however the distance between the hadrons becomes comparable to the hadron size and thus the probability for bond rearrangement should increase (cf. Fig. 13). The occurrence of such "flip-flops" [25] enlarges the number of degrees of freedom since at high densities any charge can move anywhere via this flip-flop mechanism whereas at low densities all the charges are bound pairwise. Thus even for a string tension which is independent of temperature and density we get a two-phase structure of matter. In addition we get more degrees of freedom if we allow for different spin, colour and flavour states. The number of these intrinsic degrees of freedom is denoted by n_s. With

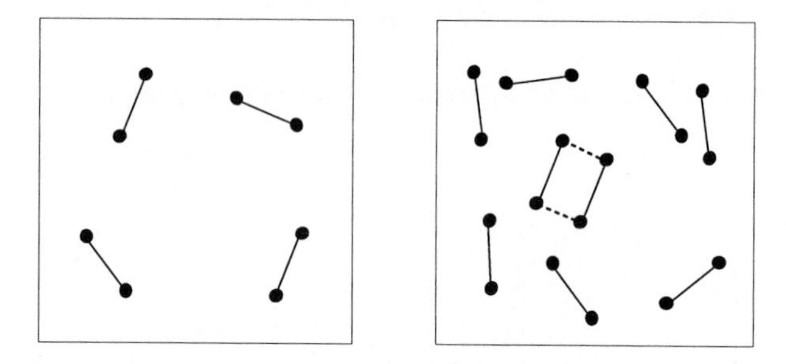

Fig. 13. *Quark strings at low and high densities and the flip-flop mechanism.*

these ingredients we are able to formulate the partition function and calculate thermodynamic quantities [25]. They are plotted as a function of temperature in Fig. 14. Energy and pressure are normalized to their Stefan-Boltzmann limits $\epsilon_{\rm SB}$ or $P_{\rm SB}$, respectively. The temperature is normalized to the string tension σ. We see that we get a phase transition even for a constant, i.e. temperature independent, value of σ parametrizing the non-perturbative quark-quark interaction. The nature of the phase transition depends on the number of intrinsic degrees of freedom n_s. Figure 15 shows the latent heat, i.e. the gap in the energy density at the critical temperature, as a function of n_s. For sufficiently large values of n_s the transition is of first order. Finally we turn to statistical QCD to obtain the critical behaviour and the phase structure from the fundamental theory of strongly interacting matter [26]. The equations (3)–(5) given in the first section constitute the complete basis of statistical QCD. We are left "only" with the question of how to carry out the calculation of the partition function eq.(4). So far, the computer simulation of the lattice formulation of QCD [2] is the only generally viable method for this, applicable in particular also in the critical region, where perturbative methods break down. Lattice QCD is obtained in three steps. First, we replace the space-time continuum $\tau = ix_0, \mathbf{x}$ in eq.(5) by

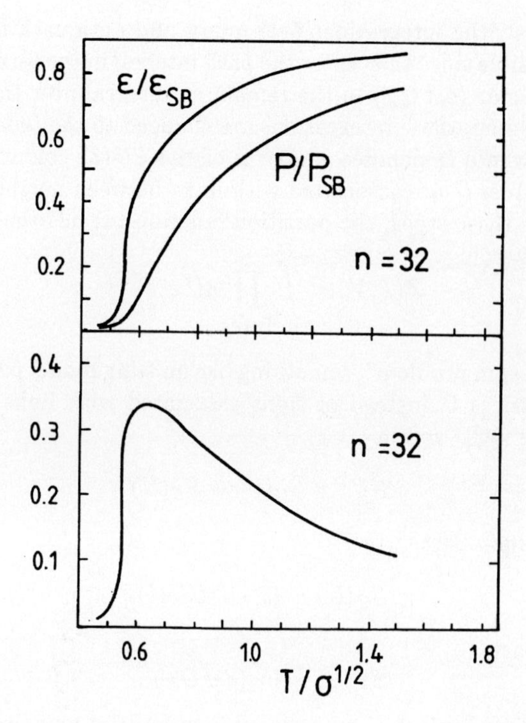

Fig. 14. *Energy density and pressure as functions of the temperature for $n_s = 32$ in the string model.*

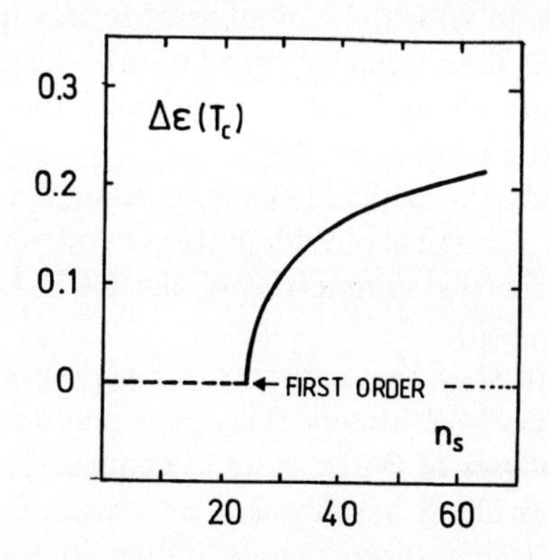

Fig. 15. *Latent heat as a function of the number of intrinsic degrees of freedom n_s in the string model.*

a lattice of $N_\tau \times N_\sigma{}^3$ discrete points, with $\{\tau = n_\tau a;\ x_i = n_{i\sigma} a,\ n_\tau = 1, .., N_\tau,\ n_{i\sigma} = 1, .., N_\sigma,\ i = 1, 2, 3\}$ and an isotropic lattice spacing a; hence $T = 1/(N_\tau a)$,

$V = (N_\sigma a)^3$. Next, the integrations over quark and antiquark fields are carried out, which is possible since they enter the path integral in the form $\exp(-\bar{\psi} Q_F \psi)$; this leads to a factor $(det\, Q_F)$ in the remaining integral over the gluon fields A [27]. In the third step, these integrations are changed to the (compact) variables $U = \exp(ixA)$, where U denotes a matrix of the $SU(3)$ colour gauge group of QCD; such variables U are associated with links between neighboring points on the lattice. After these steps, the partition function (4) has the structure

$$Z(T, V) = \int \prod_{\text{links}} dU\, e^{-S(U)} \tag{49}$$

of a "generalized spin problem", something like an Ising model partition function, with $SU(3)$ matrices U instead of spins associated with links instead of with lattice sites. The QCD action

$$S(U) = S_G(U) + S_F(U) \tag{50}$$

consists of the gluon part

$$S_G(U) \sim (1 - UUUU) \tag{51}$$

and the quark part

$$S_F(U) \sim \log(det\, Q_F)\ . \tag{52}$$

The product of four "spins" U on a closed loop built of four links ("plaquette") in the gluon action assures local gauge invariance [28], which an Ising like action with a product of only two links does not have. This completes the lattice formulation. The results which we want to obtain from it are supposed to hold in the continuum limit, in which the number of lattice points becomes infinite while the lattice spacing simultaneously goes to zero, keeping T and V constant (for a survey about methods and results concerning the simulation of strongly interacting matter on the lattice, see [29]).

Now we want to study the critical behaviour arising in statistical QCD. Since phase transitions are in statistical physics quite generally associated with changes in symmetry, we look for the symmetries of the QCD Lagrangian $\mathcal{L}$ or of the QCD action $S(U)$ in eq.(50).

Consider the limit of very heavy quarks: for $m_f \to \infty$, $S_F \to 0$, and we get thermodynamics of Yang–Mills theory. This pure gluon system already contains many of the crucial features of QCD, quite in contrast to QED, where the pure photon system forms an ideal boson gas. The reason for the difference is the term $gA_\mu A_\nu$ in eq.(3), which allows gluons, unlike photons, to interact directly, without intermediate fermions.

If we carry out the "spin flip"

$$U(n_0, \mathbf{n}) \to U'(n_0, \mathbf{n}) = zU(n_0, \mathbf{n}) \qquad \text{for all } \mathbf{n} \text{ at fixed } n_0, \tag{53}$$

where $z = \exp(i\pi r/N)$, $r = 0, .., N-1$, is an element of the center Z_N of the $SU(N)$ gauge group, then this leaves the action $S_G(U)$ invariant. Such an operation is just the analog of flipping all spins $s_i \to -s_i$ in the Ising model, which leaves the Ising action invariant. However, the "generalized spin"

$$L(U) \equiv \prod_{n_0=1}^{N_\tau} U(n_0, \mathbf{n}) \tag{54}$$

is "flipped" by this operation: $L(U') = zL(U)$. Its average,

$$\langle L \rangle = \int \prod_{\text{links}} dU\, L(U)\, e^{-S(U)}, \tag{55}$$

therefore measures a possible spontaneous breaking of the Z_N symmetry of the state of matter in which the system is, just as the average spin $\langle s \rangle$ in the Ising model indicates if the Ising system is in an ordered state with spontaneous magnetization or in a disordered state with $\langle s \rangle = 0$.

The transition from unbroken to broken Z_N symmetry corresponds in fact to deconfinement. The order parameter $\langle L \rangle$ (Polyakov loop) can be related to the potential $V(r)$ between static (infinitely heavy) quark and antiquark in the limit of infinite separation,

$$\langle L \rangle \sim \lim_{r \to \infty} e^{-V(r)/T}. \tag{56}$$

For a confined state, $V(r)$ diverges for $r \to \infty$, so that $\langle L \rangle = 0$; therefore Z_N symmetry signals confinement. In a deconfined state, $V(r)$ remains finite as $r \to \infty$, since colour screening cuts out the diverging (and hence confining) long distance part of the potential; thus the spontaneous breaking of the Z_N symmetry indicates deconfinement. Therefore the "generalized spin" $\langle L \rangle$ constitutes the theoretical probe for confinement or deconfinement. As noted, this is strictly true only in the limit of infinite quark masses, i.e. in pure $SU(N)$ gauge theory. For finite quark mass, thermal pair production leads to string breaking and hence to finite $\langle L \rangle$ for all $T \neq 0$. This is quite similar to the actual insulator-conductor transition, where thermal ionization also prevents the conductivity from vanishing in the insulator phase for $T \neq 0$. In both cases, however, the order parameter remains exponentially small in the "symmetric" phase: $\langle L \rangle \sim \exp(-m_h/T)$, with $m_h \sim m_\rho$ as typical hadron mass, corresponds to the conductivity $\sigma \sim \exp(-\Delta E/T)$, where ΔE denotes the ionization energy.

Figure 16 shows some lattice calculations of the Polyakov loop as a function of temperature[4] for different values of N_σ [30]. Obviously $\langle L \rangle$ rises with temperature. With increasing N_σ the shift from low to high values of $\langle L \rangle$ becomes more pronounced. Indeed the comparison with spin systems is not only stressed for didactical purposes. As shown in [31] the critical behaviour of a $SU(N)$ gauge theory in d+1 dimensions is the same as that of a Z_N spin system in d dimensions, since both are in the same universality class. Thus we have to compare

[4] In $SU(N)$ lattice gauge theory $2N/g^2$ is a dimensionless quantity which measures the temperature.

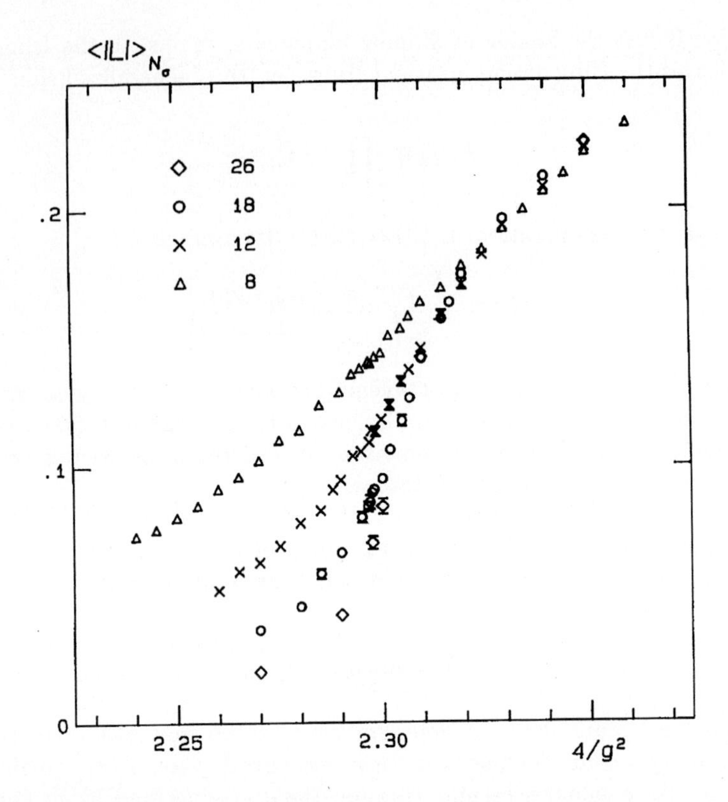

Fig. 16. *Polyakov loop for a $SU(2)$ gauge theory as a function of the coupling constant for different numbers of lattice points.*

e.g. a finite temperature $SU(2)$ gauge theory in $3+1$ dimensions with the Ising model in 3 dimensions. This statement is in full agreement with lattice calculations as can be read off from Table 4. where the critical exponents α, β, γ and ν are defined through the behaviour of "generalized spin" $\langle L\rangle$, susceptibility χ, correlation length ξ and specific heat C_V near the critical temperature T_c like[5]

$$\langle L\rangle \sim (T-T_c)^{\beta} \tag{57}$$

$$\chi \sim |T-T_c|^{-\gamma} \tag{58}$$

$$\xi \sim |T-T_c|^{-\nu} \tag{59}$$

$$C_V \sim |T-T_c|^{-\alpha} \; . \tag{60}$$

Using the theory of finite size scaling [30] one can extract some information about critical exponents from the values of $\langle L\rangle$ obtained for different values of N_σ. Plotting $\langle L\rangle N_\sigma^{\beta/\nu}$ against $xN_\sigma^{1/\nu}$ the result is expected to be independent

[5] Of course the first equation holds for $T > T_c$ only, since $\langle L\rangle$ vanishes in the symmetric phase.

Table 4. *Comparison between critical exponents in 3+1–dimensional SU(2) gauge theory and in the 3–dimensional Ising model.*

	$SU(2)$	$3d$ Ising
β/ν	0.545(30)	0.518(7)
γ/ν	1.931(15)	1.970(11)
$(\alpha-1)/\nu$	-1.36(4)	-1.41(?)
ν	0.65(4)	0.6289(8)
	$SU(3)$	$3d$ 1st order
γ/ν	3.02(14)	3
ν	0.339(13)	0.33

of N_σ for large enough values of N_σ. There the temperature is rescaled to

$$x = \frac{T - T_{c,\infty}}{T_{c,\infty}} \tag{61}$$

where $T_{c,\infty}$ is the critical temperature in the continuum limit. The critical exponents are taken from the Ising model. Figure 17 shows that lattice results agree very well with this expectation.

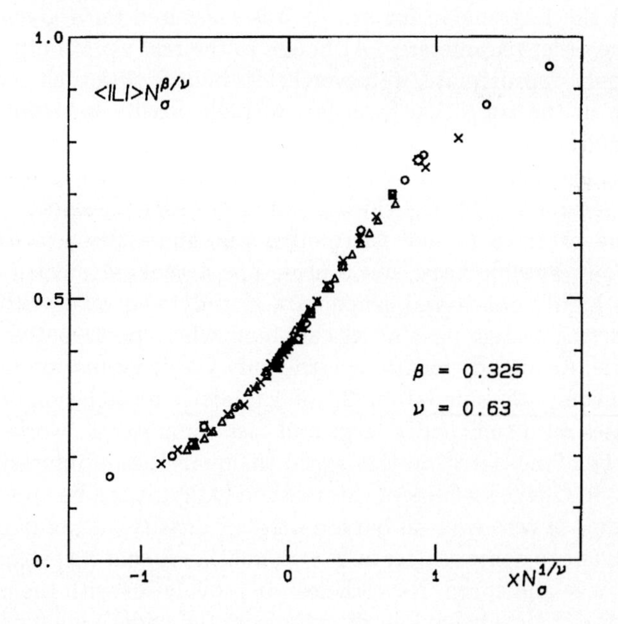

Fig. 17. *The N_σ independence of the rescaled Polyakov loop as a function of the rescaled temperature.*

We now turn to the already mentioned shift in mass of the constituents in connection with deconfinement and consider the Lagrangian in the limit of vanishing quark mass, $m_f = 0$. The four-spinor fields for quarks then decompose into a direct sum of a left-handed and a right-handed two-spinor, $\psi^{(4)} \to \psi_L^{(2)} \oplus \psi_R^{(2)}$. The chiral "rotation" between these,

$$\psi \to \psi' = e^{i\alpha\gamma_5}\psi\,, \tag{62}$$

leaves the $m_f = 0$ Lagrangian invariant, since $\bar{\psi}'\gamma_\mu\psi' = \bar{\psi}\gamma_\mu\psi$. However, the mass term in the original $m_f \neq 0$ Lagrangian would not remain invariant, since $\bar{\psi}'\psi' = \bar{\psi}e^{2i\alpha\gamma_5}\psi \neq \bar{\psi}\psi$. Hence $\chi \equiv \langle\bar{\psi}\psi\rangle$ measures if the system is in a chirally symmetric state or not. If $\chi \neq 0$, chiral symmetry is spontaneously broken, quarks have acquired a non-vanishing effective mass. For $\chi = 0$, chiral symmetry is restored and the constituents of the system are quarks with a vanishing effective mass, i.e., the current quarks of the massless Lagrangian. In the real world, the quarks in the Lagrangian cannot be massless, since the pion mass $m_\pi \sim m_q^2$ is not zero, so that we have approximate chiral symmetry only. In [32] it was argued that a system with chiral $SU_L(2) \times SU_R(2)$ symmetry is in the same universality class as the O(4) Heisenberg model, i.e. both show the same critical behaviour.

Statistical QCD thus leads to two critical phenomena, deconfinement and chiral symmetry restoration. Deconfinement is associated to a global Z_3 symmetry of the Lagrangian for $m_f \to \infty$, with $\langle L\rangle$ as order parameter; the chiral symmetry of the Lagrangian for $m_f \to 0$ is measured for a given state of the system by the order parameter χ. Although in the real world both of these symmetries are only approximate, we nevertheless believe that their remnant effects will show up in the transition from low to high density behaviour of strongly interacting matter.

So the basic quantities to be studied in statistical QCD are the deconfinement order parameter $\langle L\rangle$, the chiral symmetry order parameter χ, the energy density ϵ, the pressure P, and thermodynamic quantities derived from these, such as entropy, specific heat, and others. The actual calculation of these "observables" is highly non-trivial, since we are studying a relativistic field theory for an interacting system near a critical point, where perturbative methods are not applicable. As already mentioned, the only viable evaluation method in this region is the computer simulation [3] of the lattice formulation of the problem [2]: one creates on a sufficiently large and fast computer a "world according to QCD" and then "measures" in this world the quantities of interest; for a recent review, see [33]. One draw-back of this method is that it can be applied up to now only to systems of zero over-all baryon number density, i.e., only to hot matter with $n_B = 0$, not to dense matter at $T = 0$; another is that its precision is limited by computer size and speed. Nevertheless, it provides us with the unique chance to calculate the critical behaviour directly from the underlying fundamental theory, and the mentioned shortcomings will hopefully be removed or reduced in the future. The main results from computer simulation of lattice QCD at finite

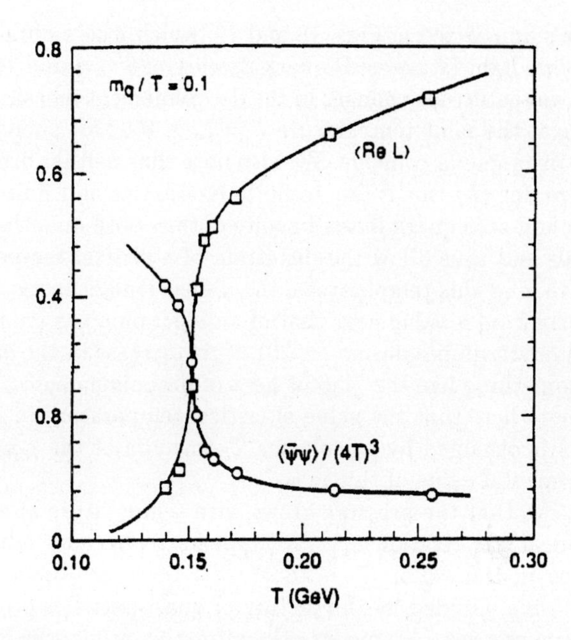

Fig. 18. *Temperature dependence of deconfinement measure* $\langle Re\ L\rangle$ *and chiral symmetry measure* χ.

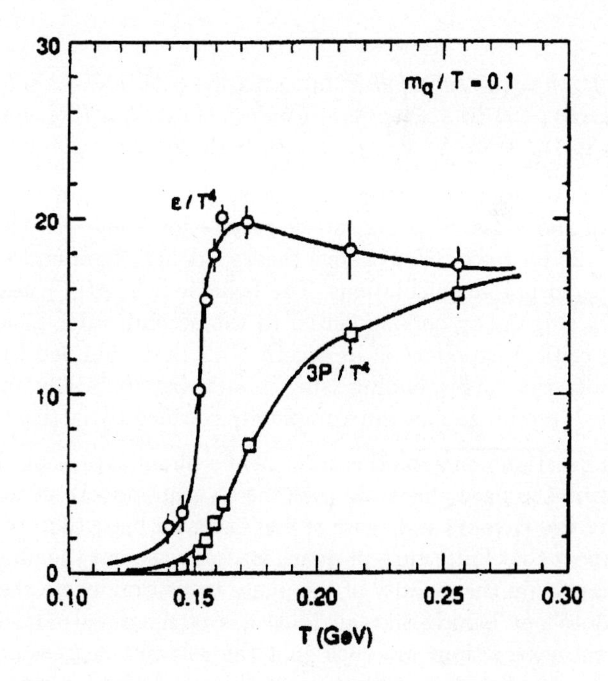

Fig. 19. *Temperature dependence of energy density* ϵ *and pressure* P.

temperature are summarized in Figs. 18 and 19, which where obtained in recent calculations for two light (staggered) quark species on $8^3 \times 4$ and $16^3 \times 4$ lattices [34]. In Fig. 18, we see abrupt changes in the deconfinement measure $\langle L \rangle$ and the chiral measure χ at the same temperature $T = T_c \simeq 150\,\mathrm{MeV}$; hence for $n_B = 0$, the two critical phenomena coincide. We also note that neither order parameter is ever really zero; for $\langle L \rangle$ this is due to finite lattice size and finite quark mass, for χ due to the non-zero quark mass. In spite of this, both quantities show clear transition signals and thus allow the definition of a critical temperature T_c. In Fig. 19, we see that at this temperature, the dimensionless energy density ϵ/T^4 increases abruptly from a value near that of an ideal pion gas ($\simeq 1$) to one near that of an ideal quark-gluon plasma ($\simeq 20$). This increase in the energy density $\Delta\epsilon$ represents something like the "latent heat of deconfinement".

We should note here that the value of critical temperature in physical units ($T_c \simeq 150\,\mathrm{MeV}$) is obtained by calculating T_c in units of the ρ-mass and then using the experimental value of this mass.

We also observe that the pressure grows with temperature at a much slower rate, and the ideal gas relation $\epsilon = 3P$ appears to become fulfilled only for $T \geq 2T_c$. Hence in the region $T_c \leq T \leq 2T_c$, there are definite interaction effects in the plasma. Guided by the picture of quasi-particles propagating in a refractive medium one can parametrize the refractive properties of the medium in terms of a temperature-dependent effective mass $M(T)$ of the excitations [35]. The partition function e.g. of an ideal gas of quasi-gluons of mass $M(T)$ is given by

$$-\ln Z(T,V) = (3V/\pi^2) \int_0^\infty dk\, k^2 \ln[1 - \exp(-\sqrt{M(T)^2 + k^2}/T)] \,. \qquad (63)$$

For given $M(T)$ one can extract the relations for energy density and pressure from this formula. Just the other way round we can obtain the temperature dependence of the mass M using lattice results for ϵ and P. The outcome is shown in Fig. 20 for pure $SU(2)$ gauge theory. At high temperatures the result agrees with perturbative calculations. The behaviour of $M(T)$ near the critical point shows a singularity corresponding to the second order phase transition; the resulting critical exponent is in accord with that obtained by universality arguments from the corresponding Ising model. Let us briefly summarize the questions which are up to now not completely clarified by lattice calculations:

1. The first question concerns the order of the phase transition: Is it really of first order in the pure glue state (like the Z_3 spin model), of second order in the case of two flavours and again of first order for more than two flavours? It might appear that finite current quark masses wash out the phase transition.
2. What happens in the vicinity of the phase transition where the quark-gluon plasma does not behave like an ideal gas of massless particles? It might be that the interactions are such that the relevant degrees of freedom are partons propagating in a refractive medium and thus carrying an effective mass which depends on temperature.

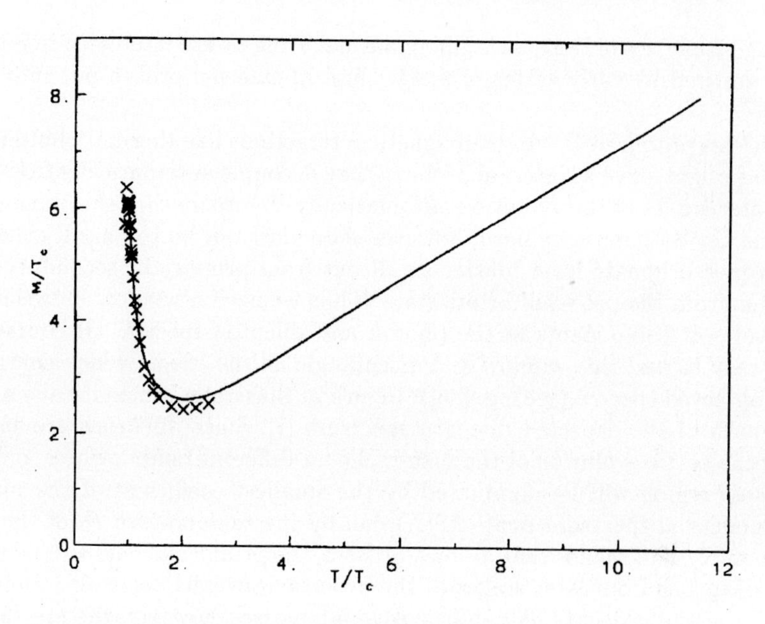

Fig. 20. *The effective gluon mass as a function of the temperature.*

3. Finally a serious problem is the introduction of non-vanishing baryon number n_B. Calculating average values like

$$\langle A \rangle \equiv \frac{\int dx\, A(x)\, e^{-S(x)}}{\int dx\, e^{-S(x)}} \tag{64}$$

via Monte-Carlo methods one needs a positive action: $S(x) > 0$. For $n_B \neq 0$ however this is no longer the case.

4 Probing the Quark-Gluon Plasma

Suppose now that we are able to create a quark-gluon plasma at the early stages of heavy-ion collisions. Then we have to find some observable quantities providing us with information about this dense system of deconfined quarks and gluons. This requires probes which exist at early times,which are hard enough to resolve short scales and which can distinguish between confined and deconfined quarks and gluons. Since thermal evolution destroys any memory of early stages of the system the probes we are looking for must not be in equilibrium with the later stages of matter, e.g. at pion freeze-out. The desired probes might decouple from the system before it reaches thermal equilibrium carrying information about the very early stages of its evolution; or the probes might be in thermal equilibrium with the deconfined phase without being influenced any more by rehadronization and the subsequent evolution in the hadronic phase. The latter are required for studying properties of the quark-gluon plasma. Thus we are interested in internal probes emitted by early states of matter and in external probes put into early matter.

Particles produced by electromagnetic interactions like thermal photons and dileptons might serve as internal probes. They decouple automatically from matter in later stages of the evolution automatically. To assure that they come from early matter they must be hard; however they must not be too hard, otherwise they do not originate from matter at all but from primary or secondary collisions, i.e. from the pre-equilibrium state. Thus we need a window between very hard and soft components in the photon and dilepton spectra. Of course it is not so easy to find this window and disentangle all the effects which contribute to the observed spectra (see e.g. [36]). Figure 21 illustrates schematically an idealized form of the expected dilepton spectrum [7]. Since dileptons are emitted during the entire evolution of the system, i.e. at different temperatures, only the high mass region will be dominated by the smallest coefficient of the mass in the exponential spectrum $\exp(-M/T)$, i.e. by the temperature T_0 of the early matter state. This means that at lower masses, the primordial part will be buried under dileptons from later stages of the evolution, in particular also under the tails of ρ, ω and ϕ decays. For sufficiently high masses, however, the pre-thermal Drell-Yan production will dominate, since it falls only as a power in the mass, in contrast to the exponential thermal spectrum. The Drell-Yan continuum is due to electromagnetic quark-antiquark annihilation into massive virtual photons, which decay into lepton pairs.

External probes are hadronic particles which are produced in primary collisions before matter is formed. Since they should retain memory of the deconfined phase they must not be in equilibrium with hadrons of the late states of matter. Of course these particles should behave differently when propagating through a confined or deconfined medium. Examples for hard probes are heavy quark pairs, e.g. charmonium or bottonium, and hard quarks and gluons observable as hard jets. External probes are not part of matter, i.e. they are not created in a thermal system, since they are suppressed by their masses or energies. For a given temperature of 400 MeV the charmonium state J/ψ is suppressed by $\exp(-M_\psi/T) \simeq 10^{-4}$, the bottonium state Υ by $\exp(-M_\Upsilon/T) \simeq 10^{-11}$ and hard jets by $\exp(-p_T^{\mathrm{jet}}/T) \simeq 10^{-11}$, where a transversal momentum of 10 GeV is assumed for a hard jet. Furthermore the behaviour of these probes in the vacuum should be well understood since we have to disentangle medium and vacuum properties. In the aforementioned cases the vacuum behaviour is given by perturbative QCD and in addition by universal non-perturbative phenomena which can be described by quark and gluon structure functions. For external probes produced in nuclear collisions one has to take into account the effects of the initial state, e.g. shadowing, and of the pre-equilibrium stage. The mentioned heavy quarkonium states mentioned before decay into $\mu^+\mu^-$ pairs and therefore also appear in the dilepton spectrum as shown in Fig. 21.

Besides these probes it is also useful to have a reference probe which remains unaffected by the medium produced in the collision. This enables us to com-

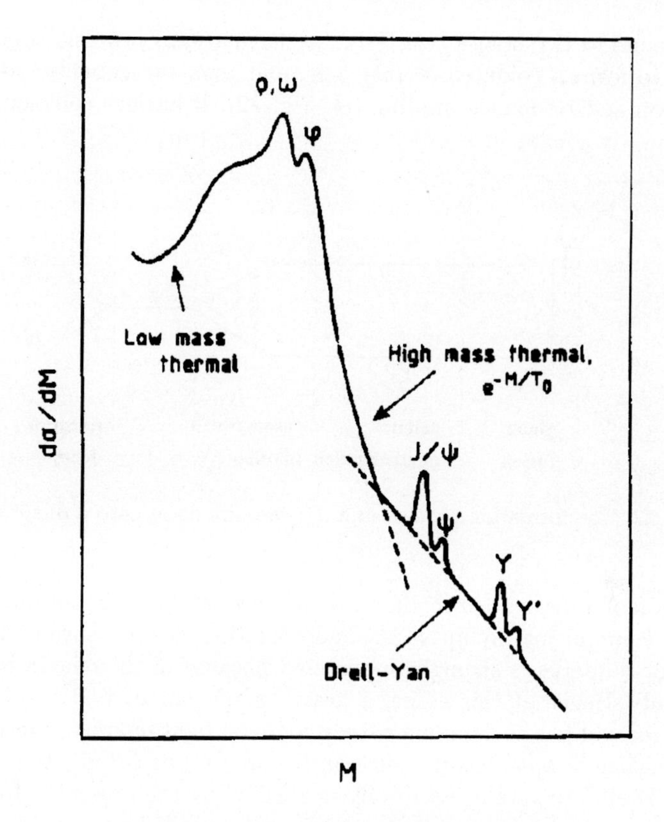

Fig. 21. *The ideal dilepton spectrum.*

pare the ratio of external and reference probe for various experiments, especially those where we expect the creation of a quark-gluon plasma with those where the system never leaves the hadronic phase. Examples for reference probes are hard direct photons, high mass Drell-Yan dileptons and mesons with open charm or beauty. These particles might also serve as a test for the pre-equilibrium state. They are influenced by initial state effects like nuclear shadowing or antishadowing and initial state parton scattering.

In the following we will concentrate on quarkonium as a candidate for indicating the appearance of a quark-gluon plasma [1]. The J/ψ is a bound state of the heavy "charmed" quark c and its antiquark $\bar{c}$; the quark mass m_c is about 1.5 GeV. It decays into a dilepton pair $\mu^+\mu^-$ which can be detected. Such quarks are not present in the colliding nucleons, and their creation in any thermal system is extremely unlikely, because of their large mass. For the Υ meson, consisting of a bottom quark b and its antiquark $\bar{b}$, with $m_b \simeq 3\,\mathrm{GeV}$, this is even more unlikely. Heavy quark-antiquark pairs ($c\bar{c}$, $b\bar{b}$) are therefore produced at a very early stage of the collision, by hard, pre-thermal interactions of quarks and gluons present in the collision partners. Thus charmonium production in hadronic collisions occurs in two stages [11]. First we have a hard process, the fusion of two gluons to form a coloured $c\bar{c}$ ($b\bar{b}$) pair, and then subsequently soft colour neutralization and resonance binding (cf. Fig. 22). If nuclear collisions lead to

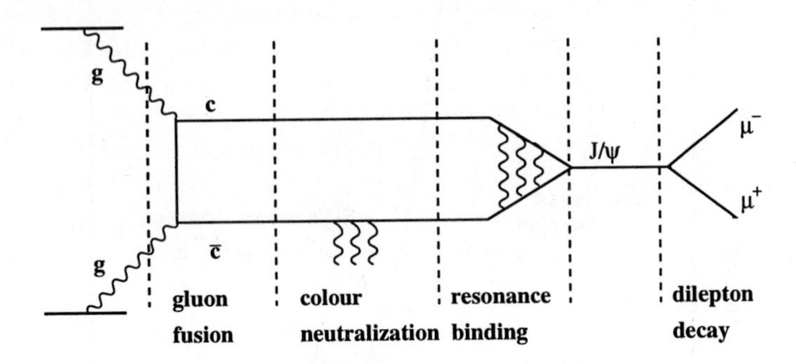

Fig. 22. The formation process of a J/ψ and its decay into a dilepton.

the formation of a deconfined medium, then the second step is inhibited so that c and $\bar{c}$ (or b and $\bar{b}$) just fly apart. At hadronization, the presence of additional thermal c or b quarks is strongly suppressed because of their large masses, as noted already. Hence at this stage, a heavy quark cannot find another heavy partner; it instead has to combine with a common light quark to form a meson with open charm or open beauty, such as D ($c\bar{u}$, etc.) or B ($b\bar{u}$, etc.). However, the overall Drell-Yan rate is essentially unaffected by the presence of a plasma. In the dilepton spectrum from high energy nuclear collisions, we thus expect a suppression of the J/ψ and Υ signals relative to the Drell-Yan continuum, if there was quark-gluon plasma formation. Obviously the melting of the heavy quarkonium states just described, is due to screening of the confining potential according to eq.(2). By solving the Schrödinger equation with the potential $V_{\rm scr}$ for a heavy quark pair of mass M we can determine the screening radius μ^{-1} for which binding is dissolved [1]. The screening radius can be related to tem-

Table 5. *Energy densities for melting of heavy quarkonium states*

State	J/ψ	χ_c	ψ'	Υ	χ_b	Υ'
M [GeV]	3.1	3.5	3.7	9.6	9.9	10.0
ϵ_d [GeV/fm^3]	1.9	1.0	1.0	47.1	1.0	1.6

perature and energy density via calculations within lattice QCD. This yields the respective energy density ϵ_d at which a charmonium or bottonium state melts. The results are given in Table 5 for various bound states of $c\bar{c}$ and $b\bar{b}$. Thus all mentioned quarkonium states except the Υ could melt at CERN-SPS energies. In the experiment NA38 [37], masses and momenta of the outgoing muon pairs are measured; in addition, the associated (neutral) energy E_T emitted in the plane orthogonal to the beam axis is determined by calorimeters. Since this transverse energy provides a measure for the associated hadron multiplicity and hence for the energy density ϵ_0, high E_T means central collisions with many hadronic secondaries, low E_T means peripheral events of lower multiplicity. In Fig. 23 the muon pair spectra from oxygen-uranium collisions for $E_T < 33\,\text{GeV}$ and $E_T > 82\,\text{GeV}$ are superimposed by matching the fitted Drell-Yan continua. We note that at high E_T and thus at high ϵ_0, the J/ψ signal has become considerably weaker; it has decreased by the shaded area. Any evidence for the ψ' has disappeared at high E_T. Indeed the ψ' is more strongly suppressed than

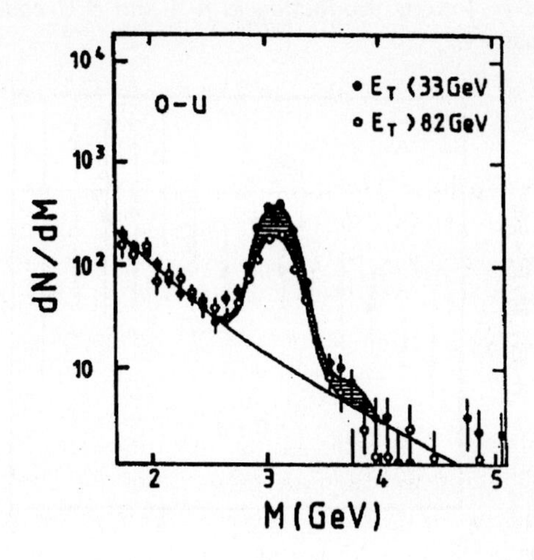

Fig. 23. *The dilepton spectrum from oxygen-uranium collisions at low and high E_T, with matched continua. The shaded area indicates the observed J/ψ and ψ' suppression.*

the J/ψ with growing ϵ_0, as can be seen in Fig. 24 [38].

Another interesting feature of J/ψ suppression can be seen in Fig. 25. We observe that the suppression depends on the transverse momentum P_T of the J/ψ; it vanishes for large P_T. To understand the aforementioned effects observed in A-B collisions we go one step back and study hadron-nucleus collisions first [11]. Analysing the data from $\pi^{\pm}$-A and p-A collisions at $17 \leq \sqrt{s} \leq 40\,\text{GeV}$ with respect to J/ψ, ψ' and Υ production it is found [39] that in such collisions charmonium and bottonium states are suppressed with increasing A, in comparison to the corresponding production on hadrons (cf. Fig. 26). In contrast there is no suppression of the Drell-Yan continuum. Figure 27 shows that the suppression also increases with increasing x_F. It is worth mentioning that all charmonium states are measured for $x_F \geq 0.1$, i.e. they move fast in the tar-

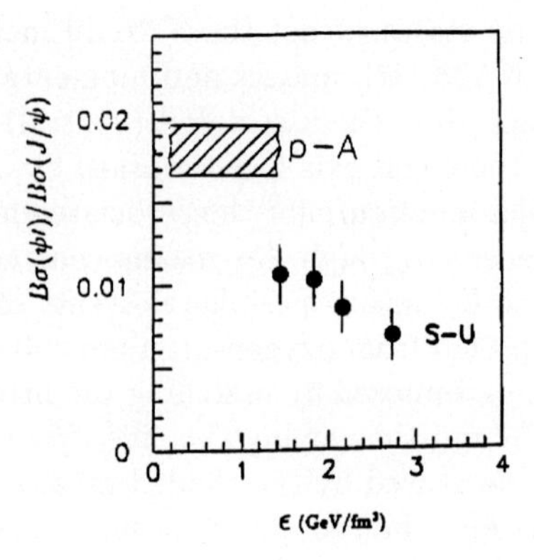

Fig. 24. *The ratio of ψ' to J/ψ production in p-A and S-U collisions as function of the initial energy density ϵ_0.*

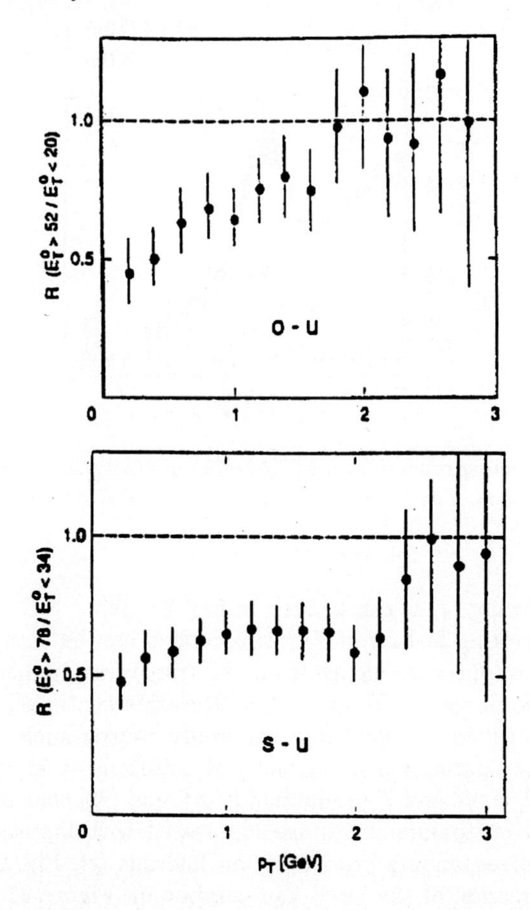

Fig. 25. *The P_T dependence of J/ψ suppression. The signals are normalized to the corresponding continuum.*

get system. This implies that the produced $c\bar{c}$ pairs leave the nucleus as a small colour octet state and reach physical resonance size only far outside the target; in A-B collisions, on the other hand, an interaction between resonances which are fully formed and partially stopped strongly interacting matter can take place. A specific consequence of the pre-hadronic character of the $c\bar{c}$ in h-A collisions is that the suppression of J/ψ and ψ' should be the same which is in fact observed (cf. Figs. 26 and 27). Thus the relevant mechanism for the suppression in h-A

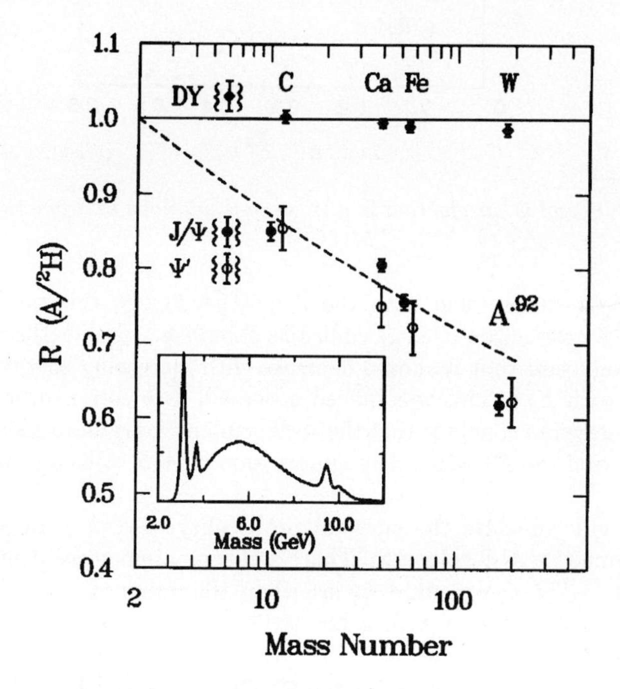

Fig. 26. *The ratios of heavy nucleus to deuterium integrated yields for the J/ψ and ψ' resonances and the Drell-Yan continuum. The insert shows the raw (no acceptance correction) dimuon invariant mass spectrum.*

collisions is not physical absorption of charmonium states but a combination of colour interactions [40] and nuclear modifications of the structure functions (gluon shadowing) [41].

We now return to the suppression in nucleus-nucleus collisions. As already noted, the accessible kinematic region ($0 < x_F < 0.2$) allows the formation of physical resonances in the medium. The observed suppression could thus be due to some type of absorption in confined matter at very high density [42] as well as to deconfinement, and we have to find a way to distinguish between the two mechanisms. But whatever origin of the suppression, the existence of physical resonance states predicts that the ψ', because of its larger geometric size, will

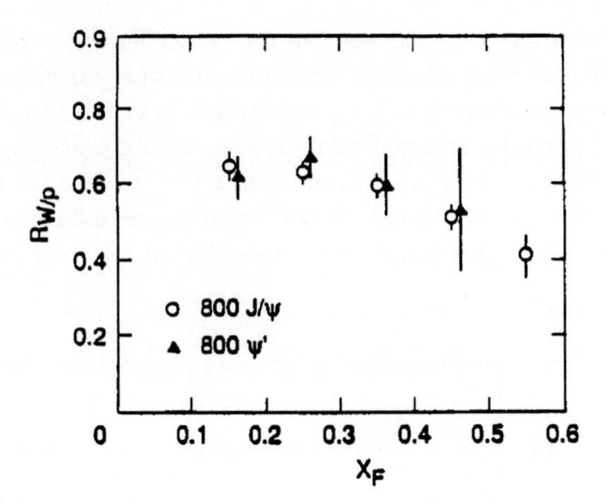

Fig. 27. *J/ψ and ψ' production in p-W vs. p-p collisions as a function of x_F.*

suffer a stronger suppression than the J/ψ. Specifically, this means that the ratio $\psi'/(J/\psi)$ production in A-B collisions should be smaller than that found in h-A collisions, and that it should decrease with increasing energy density (or equivalently, with E_T). This was indeed observed, as mentioned previously (cf. Fig. 24). We therefore conclude that the A-dependent charmonium suppression in h-A collisions and the E_T-dependent suppression in A-B collisions have different origins.

Next we will calculate the survival probability of a J/ψ in a deconfined medium obtained in A-B collisions. There a charmonium state x survives if the energy density ϵ_0 of the medium is less than the value ϵ_d^x needed to dissolve it. In the interaction region with a hot center and a cool surface the survival probability is given by

$$S_x = 1 - V_{\text{hot}}^x / V \tag{65}$$

where V_{hot}^x is that part of the interaction volume V in which $\epsilon_0 \geq \epsilon_d^x$. The profile of the initial energy density is assumed to be proportional to the density of the colliding nucleons

$$\epsilon_0(r) = \epsilon_0 \left[1 - (r/R_A)^2\right]^{2/3} . \tag{66}$$

With these ingredients the survival probability becomes

$$S_x = \Theta(\epsilon_d^x - \epsilon_0) + \left(\frac{\epsilon_d^x}{\epsilon_0}\right)^{9/4} \Theta(\epsilon_0 - \epsilon_d^x) . \tag{67}$$

Please note that there are no free parameters in this calculation. For the analysis of data it is important to take into account the J/ψ's which are produced via the decay of the excited χ_c state. The observed J/ψ's are about 70% directly produced and about 30% due to the decay $\chi_c \to J/\psi + \gamma$. Since the energy density needed to dissolve the χ_c differs from the one for J/ψ we get for the observable survival probability

$$S_{J/\psi}(\epsilon_0) = 0.7 \left[\Theta(\epsilon_d^\psi - \epsilon_0) + \left(\frac{\epsilon_d^\psi}{\epsilon_0}\right)^{9/4} \Theta(\epsilon_0 - \epsilon_d^\psi) \right] + 0.3 \left[\Theta(\epsilon_d^\chi - \epsilon_0) + \left(\frac{\epsilon_d^\chi}{\epsilon_0}\right)^{9/4} \Theta(\epsilon_0 - \epsilon_d^\chi) \right] . \quad (68)$$

Figure 28 shows that the result agrees very well with the data. In spite of this

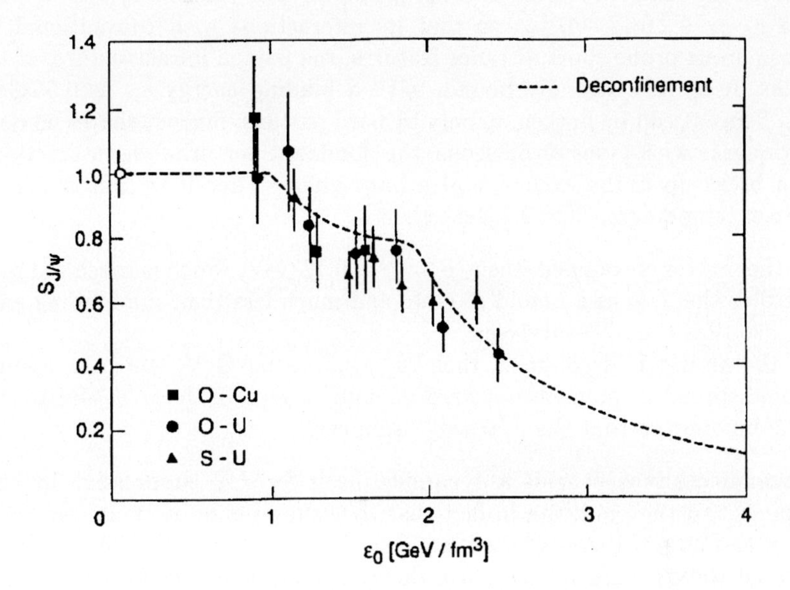

Fig. 28. *J/ψ suppression as function of the initial energy density ϵ_0 for deconfinement.*

good agreement we have to answer the question whether an absorption process in a confined medium can match the data similarily well. Thus we have to compare quarkonium interactions in confined and deconfined matter [43].

For matter in a confined state, the constituents will be hadrons; for simplicity, consider an ideal gas of pions [26]. Their momentum distribution will be thermal, i.e., for temperatures not too low it will be given by $\exp(-E_\pi/T) \simeq \exp(-p_\pi/T)$. Hence the average momentum of a pion in this medium is $\langle p_\pi \rangle = 3T$. The distribution of quarks and gluons within a pion is known from structure function studies; the gluon density is $g(x) \simeq 0.5\,(1-x)^3$ for large $x = p_g/p_\pi$. As a consequence, the average momentum of a gluon in confined matter is given by

$$\langle p_g \rangle_{\rm conf} = \frac{1}{5}\langle p_\pi \rangle = \frac{3}{5}T . \quad (69)$$

Hence in a medium of temperature $T \simeq 0.2\,\text{GeV}$, the average gluon momentum is around $0.12\,\text{GeV}$.

In contrast, the distribution of gluons in a deconfined medium is directly thermal, i.e., $\exp(-p_g/T)$, so that

$$\langle p_g \rangle_{\text{deconf}} = 3T \ . \tag{70}$$

Hence the average momentum of a gluon in a deconfined medium is five times larger than in a confined medium; for $T = 0.2\,\text{GeV}$, it becomes $0.6\,\text{GeV}$. We thus have to find a way to look for such a hardening of the gluon distribution in deconfined matter.

Indeed the J/ψ provides an ideal probe for this. It is very small, with a radius $r_\psi \simeq 0.2\,\text{fm} \ll \Lambda_{\text{QCD}}^{-1}$, so that its interactions with conventional light quark hadrons probe short distance features, the parton infrastructure, of these hadrons. It is very strongly bound, with a binding energy $\epsilon_\psi \simeq 0.65\,\text{GeV} \gg \Lambda_{\text{QCD}}$; hence it can be broken up only by hard partons. Since it shares no quarks or antiquarks with pions or nucleons, the dominant perturbative interaction for such a break-up is the exchange of a hard gluon. Thus if we put a J/ψ into matter of temperature $T = 0.2\,\text{GeV}$, then

- if the matter is confined, then $\langle p_g \rangle_{\text{conf}} \simeq 0.12\,\text{GeV}$, which is much too soft to resolve the J/ψ as a bound $c\bar{c}$ state and much less than the binding energy ϵ_ψ, so that the J/ψ survives;
- if the matter is deconfined, then $\langle p_g \rangle_{\text{deconf}} \simeq 0.6\,\text{GeV}$, which is, assuming some spread in momentum, hard enough to resolve the J/ψ and to break the binding, so that the J/ψ will disappear.

Our arguments thus provide a dynamical basis for J/ψ suppression by colour screening, and they indicate in fact that J/ψ suppression in dense matter will occur if and only if there is deconfinement.

To put these arguments on a firm theoretical basis, we need the cross section for the inelastic interaction of a J/ψ with a light hadron h, $\sigma_{\psi h}$, from which we can determine whether a J/ψ can be broken up on its passage through hadronic matter. Because of the small radius and large binding energy of the J/ψ, $\sigma_{\psi h}$ can be calculated by means of a short-distance analysis of QCD. The crucial feature for this calculation is the fact that heavy and tightly bound quarkonium states can be broken up by scattering on usual light hadrons only through the exchange of hard gluons. The momentum distribution of gluons within such a light hadron incident on a J/ψ, $g(x)$, with $x = p_g/p_h$, is a universal non-perturbative input, determined e.g. from parton counting rules or from deep inelastic processes. It has for large x in general the form $g(x) \simeq (1-x)^k$, with $k \simeq 3$ for mesons, $k \simeq 4$ for nucleons. The resulting (J/ψ)-h cross section then becomes [43]

$$\sigma_{\psi h}(s) \simeq 3\text{mb}(1 - \lambda_0/E_h)^{k+5/2} \ , \tag{71}$$

with

$$E_h = (m_h^2 + \mathbf{p}_h^2)^{1/2} \qquad \text{and} \qquad \lambda_0 = m_h + \epsilon_\psi \ . \tag{72}$$

For low collision energies $\sqrt{s}$, the cross section is thus determined by the behaviour of the gluon distribution at large x, and this leads to a very slow growth from threshold towards the asymptotic value of 3 mb. The functional form of this behaviour is shown in Fig. 29. Only in the limit of large quark mass the quarko-

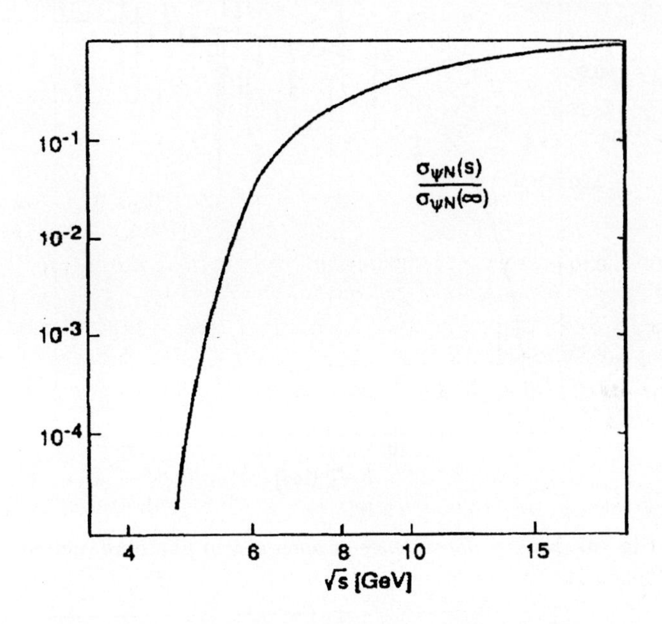

Fig. 29. *Energy dependence of the J/ψ-nucleon cross section.*

nium-hadron cross section can be calculated rigorously in short-distance QCD. We thus have to ask if the charm quark mass is sufficiently large to apply the results of heavy quark theory. For an empirical test, one can use the same approach to calculate the cross section for photoproduction of open charm, $\sigma_{\gamma h \to c\bar{c}}(s)$. The result is

$$\sigma_{\gamma h \to c\bar{c}}(s) \simeq 1.2\mu\mathrm{b}(1 - \nu_0/\nu)^k\,, \tag{73}$$

with

$$2\nu = s - m_h^2 \qquad \text{and} \qquad 2\nu_0 = M_\psi^2 + 2M_\psi m_h\,. \tag{74}$$

In Fig. 30 it is seen that the available data agree well with the heavy quark theory result eq.(73). A similar test is in principle provided by J/ψ-photoproduction, $\gamma N \to J/\psi\, N$. The forward scattering amplitude of this process will give the total charm photoproduction cross section $\sigma^{tot}_{\gamma N \to c\bar{c}}$, provided vector meson dominance holds and the real part of the amplitude can be neglected. Near threshold presumably neither is true, and so the result of our calculation falls there below the data (cf. Fig. 31).

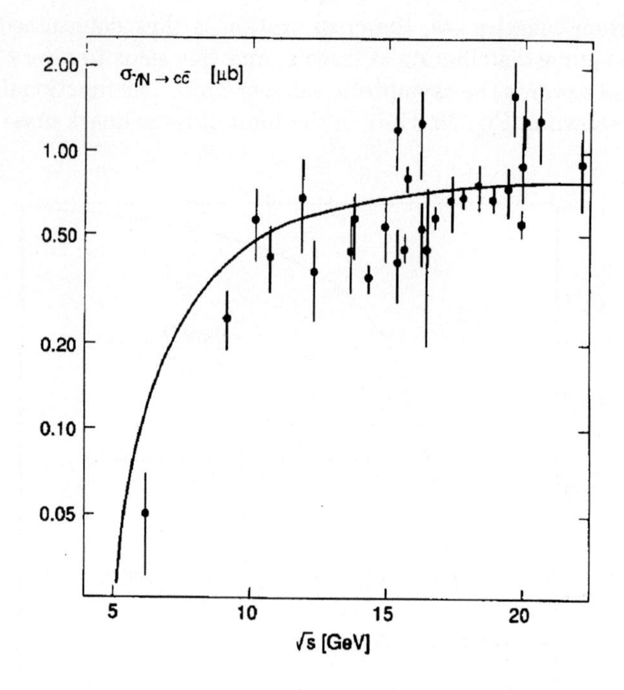

Fig. 30. *Energy dependence of open charm photoproduction.*

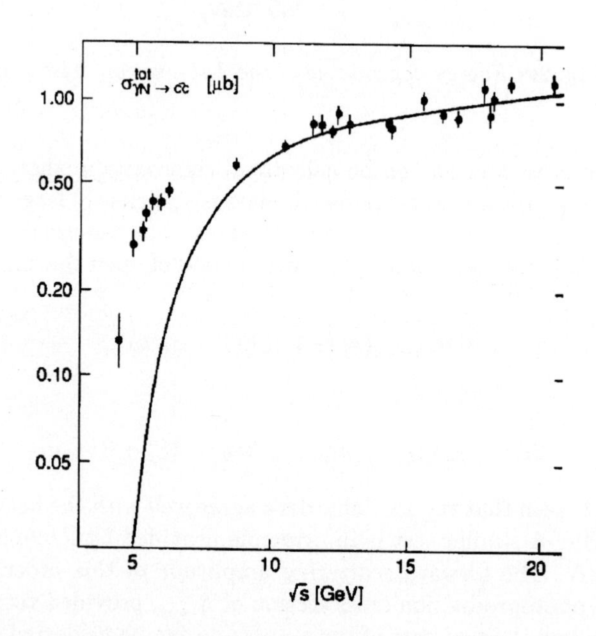

Fig. 31. *The total cross section for charm photoproduction, compared to the prediction of the short distance QCD analysis, with the real part of the amplitude neglected.*

To estimate the chance of J/ψ's surviving in confined hadronic matter of higher than standard nuclear density, we consider the survival probability

$$S = e^{-nL\sigma_{\psi h}} \tag{75}$$

where n denotes the density of the medium and L the radius of the bubble. The result for pionic matter (Table 6) shows essentially no suppression. The

Table 6. *Survival Probability for J/ψ in pionic matter at 5 times nuclear matter density and a bubble radius of 10fm for various temperatures.*

T [GeV]	0.2	0.3	0.4	0.5
S	1.00	1.00	1.00	0.98

calculation of the total $(J/\psi)h$ cross section as given above becomes exact in the heavy quark limit. It is interesting, however, to consider also non-perturbative mechanisms made possible by the finiteness of the charmed quark mass. Here, e.g., the process

$$J/\psi + \rho \to D + \bar{D} \tag{76}$$

near threshold might be worth studying. Imagine a J/ψ which is put "into" a ρ-meson. Apart from perturbative contributions, the amplitude for the reaction can acquire a non-perturbative piece, corresponding to the tunneling of the c-quark from the charmonium potential well into the potential well of the ρ-meson. The probability for the break-up of the J/ψ by tunneling can be estimated in a (non-relativistic) quantum-mechanical framework. In this case we can calculate the tunneling rate via

$$R = \omega P \tag{77}$$

where ω denotes the frequency of the J/ψ hitting the potential well and P the tunneling probability. An estimate for ω can be obtained by taking the difference of J/ψ and ψ' masses

$$\omega = M_{\psi'} - M_{\psi} \simeq 0.7\,\text{GeV} \tag{78}$$

while P is given by

$$P = e^{-2W} \tag{79}$$

with the Gamow factor

$$W = \int_{x_1}^{x_2} dx\,\sqrt{2m(V(x) - E)} \tag{80}$$

where $x_2 - x_1$ is the tunneling distance, i.e. between x_1 and x_2 the potential energy $V(x)$ is bigger than the total energy E of the particle with mass m.

Replacing $V(x) - E$ by the dissociation energy $\Delta E = 2M_D - M_\psi \simeq 0.7\,\text{GeV}$ we get

$$W \simeq (m_c \Delta E)^{1/2} R_t \tag{81}$$

with the tunneling distance R_t. For the J/ψ we have

$$R_t^\psi \simeq 0.7/\Lambda_{\text{QCD}} \,. \tag{82}$$

Thus the tunneling rate turns out to be

$$R \simeq 2.3 \cdot 10^{-3}\,\text{fm}^{-1} \,. \tag{83}$$

To get the survival probability for a J/ψ we need in addition the overlap time t, i.e. the time for J/ψ and ρ sharing the same spacial region. With the typical length scale of $d = 1/\Lambda_{\text{QCD}}$ we get

$$t = d/v_\rho \simeq \Lambda_{\text{QCD}}\, E_\rho/p_\rho \simeq 4\,\text{fm} \,. \tag{84}$$

This yields the survival probability

$$S_t^\psi \simeq e^{-Rt} \simeq e^{-(2.3 \cdot 10^{-3}) \cdot 4} \simeq 0.99 \,, \tag{85}$$

i.e. the J/ψ cannot break up non-perturbatively by tunneling. It is worth mentioning that this is due to its large dissociation energy. To illustrate this we calculate the respective survival probability for ψ'. There the dissociation energy is

$$\Delta E_{\psi'} = 2M_D - M_{\psi'} \simeq 0.05\,\text{GeV} \,. \tag{86}$$

Assuming a linear potential the tunneling distance becomes proportional to the dissociation energy (cf. Fig. 32). Thus

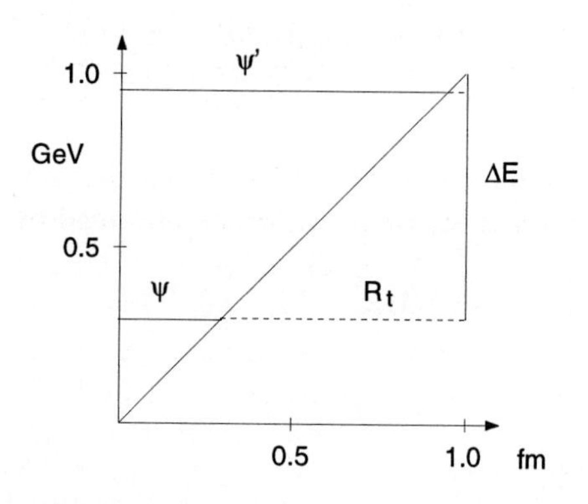

Fig. 32. *Dissociation energy and tunneling distance for J/ψ and ψ'.*

$$R_t^{\psi'} \simeq R_t^\psi \frac{\Delta E_{\psi'}}{\Delta E_\psi} \simeq 0.05/\Lambda_{\text{QCD}} \,. \tag{87}$$

With a frequency of

$$\omega_{\psi'} \simeq 0.08\,\mathrm{GeV} \tag{88}$$

the tunneling rate for ψ' becomes

$$R_{\psi'} \simeq 3.5 \cdot 10^{-1}\,\mathrm{fm}^{-1} \tag{89}$$

yielding the survival probability

$$S_t^{\psi'} \simeq 0.25\,, \tag{90}$$

i.e. ψ' can break up by tunneling in contrast to J/ψ.

A direct test of J/ψ absorption in confined matter will in fact soon become possible [26]. As mentioned above, the ideal way to study the fate of J/ψ's in confined matter is in principle provided by hadron-nucleus collisions, where the J/ψ is then observed by measuring its decay products, $J/\psi \to \mu^+\mu^-$. In practice, however, the muons are identified by passing through an absorber, and for this they have to be sufficiently energetic. As a consequence, experiments in which a hadron beam is incident on a nuclear target lead to J/ψ's which are very fast ($x_F \geq 0$) in the rest frame of the nucleus and hence leave the nuclear medium long before ever becoming fully formed physical resonances. Such experiments thus mainly study the behaviour of a coloured $c\bar{c}$ pair in the medium, and the behaviour of this system is quite different from that of a colour-singlet J/ψ. To test the interaction of the latter in the nuclear medium, we need J/ψ's in the momentum range $-1 \leq x_F \leq -0.3$, and as noted, these lead to dimuons too slow to be observable in the usual pA experiments. However, with the advent of the *Pb*-beam at CERN-SPS, it will become possible to study J/ψ-production from a nuclear beam incident on a hydrogen target. Fast J/ψ's in the lab system will then be slow in the nuclear rest frame and thus pass through the confined nuclear medium as fully formed physical resonances. The short-distance QCD analysis presented here predicts essentially no absorption for this passage, in contrast to a suppression of 25% or more if the asymptotic break-up cross section is used instead of eq.(71).

According to our QCD-based perturbative as well as non-perturbative calculations, we conclude that confined matter is transparent for J/ψ's while deconfined matter is opaque. Thus J/ψ probes colour confinement and therefore the existence of a quark-gluon plasma.

In closing, we comment briefly on the interpretation of the observed 50% J/ψ suppression in *O-U* and *S-U* collisions at CERN [37]. At this time, there remain at least two open questions which prevent us from concluding that this suppression is due to deconfinement. Our present considerations exclude J/ψ absorption in confined *matter*. The temporal sequence of J/ψ formation and equilibration in nuclear collision is not yet clear, however, and so energetic pre-equilibrium hadrons could have broken up the suppressed J/ψ's, either as fully formed resonances or in a nascent state. Such break-up processes can be studied theoretically, and it should be possible to determine their effect. Another uncertainty arises from J/ψ production through χ_c decay, which leads to a sizeable fraction of the overall production rate (30 – 40%). Since the χ's are much less tightly bound, their break-up into $D\bar{D}$ pairs is easier, and hence much of the observed suppression could in fact be χ suppression. In p-Li and $\pi^{\pm}$-Li interactions at 300 GeV beam momentum [44], the production rates of the different hidden charm final states (J/ψ, χ_1, χ_2, ψ') were measured, and if this would be done also in nuclear collisions, the effect of χ suppression could be determined. Thus both theorists and experimentalists have some work left to do before the relation between the observed J/ψ suppression and colour deconfinement is fully clarified.

References

[1] H. Satz, in: Quark-Gluon Plasma, p. 593, R. C. Hwa (ed.), Singapore (1990).
[2] K. G. Wilson, Phys. Rev. D10, 2445 (1974).
[3] M. Creutz, Phys. Rev. D21, 2308 (1980).
[4] P.Hasenfratz, in Lectures on QCD, Vol.1 Foundations, Lecture Notes in Physics No. 481, Springer Verlag, Heidelberg (1997);
J.Negele, ibid.
[5] B. Svetitsky, Phys. Rep. 132, 1 (1986).
[6] M. Shifman, A. Vainshtein, and V. Zakharov, Nucl. Phys. B147, 385 (1979).
[7] H. Satz, Nucl. Phys. A544, 371c (1992).
[8] D. Boal, C. K. Gelbke, and B. Jennings, Rev. Mod. Phys. 62, 553 (1990);
S. Pratt, T. Csörgö, and J. Zimányi, Phys. Rev. C42, 2646 (1990);
M. Gyulassy, S. K. Kauffmann, and L. W. Wilson, Phys. Rev. C20, 2267 (1979);
S. Chapman, P. Scotto, and U. Heinz, Regensburg preprint TPR-94-29, hep-ph/9409349, Heavy Ion Physics (Acta Phys. Hung., New Series) 1 (1995), in press.
[9] J. Rafelski and B. Müller, Phys. Rev. Lett. 48, 1066 (1982);
J. Cleymans and H. Satz, Z. Phys. C 57, 135 (1993);
U. Heinz, Nucl. Phys. A566, 205c (1994).
[10] M. Gonin, in: Heavy Ion Physics at the AGS '93, p. 184, MITLNS-2158.
[11] H. Satz, Z. Phys. C 62, 683 (1994).
[12] S. Abatzis et al. (WA85), Phys. Lett. B259, 508 (1991);
M. A. Mazzoni (NA34), Nucl. Phys. A544, 623c (1992);
R. Stock et al. (NA35), Nucl. Phys. A525, 497c (1992);
E. Andersen et al. (NA36), Phys. Lett. B294, 127 (1992);
C. Baglin et al., Phys. Lett. B272, 449 (1991).
[13] J. Cleymans et al., Z. Phys. C 58, 347 (1993).
[14] D. Evans, Nucl. Phys. A566, 225c (1994).
[15] S. Nagamiya, in: Quark Matter Formation and Heavy Ion Collisions, p. 281, M. Jacob and H. Satz (eds.), World Scientific, Singapore (1982).
[16] E. Beth and G. E. Uhlenbeck, Physica 4, 915 (1937).

[17] R. Hagedorn, Nuovo Cim. Suppl. 3, 147 (1965);
G. Veneziano, Nuovo Cim. 57A, 190 (1968).
[18] N. Cabibbo and G. Parisi, Phys. Lett. 59B, 67 (1975).
[19] G. Baym, Physica 96A, 131 (1979);
T. Çelik, F. Karsch, and H. Satz, Phys. Lett. 97B, 128 (1980).
[20] A. Chodos, R. L. Jaffe, K. Johnson, C. B. Thorn, and V. F. Weisskopf, Phys. Rev. D9, 3471 (1974).
[21] H. E. Haber and H. A. Weldon, J. Math. Phys. 23, 1852 (1982);
L. Dolan and R. Jackiw, Phys. Rev. D9, 3320 (1974).
[22] P. Hut, Nucl. Phys. A418, 301c (1983).
[23] J. Baacke, Acta Phys. Polonica B8, 625 (1977);
H. Satz, Phys. Rev. B113, 245 (1982).
[24] H. Miyazawa, Phys. Rev. D20, 2953 (1979).
[25] V.Goloviznin and H. Satz, CERN-preprint, CERN-TH 7268/94.
[26] H. Satz, CERN-preprint, CERN-TH 7410/94.
[27] T. Mathews and A. Salam, Nuovo Cim. 12, 563 (1954).
[28] F. J. Wegener, J. Math. Phys. 10, 2259 (1971).
[29] F. Karsch, in: Quark-Gluon Plasma, p. 61, R. C. Hwa (ed.), Singapore (1990).
[30] J. Engels, J. Fingberg, M. Weber, Nucl. Phys. B332, 737 (1990).
[31] B. Svetitsky and L. G. Yaffe, Nucl. Phys. 210B, 423 (1982).
[32] R. Pisarski and F. Wilczek, Phys. Rev. D 29, 338 (1984).
[33] F. Karsch and E. Laermann, Rep. Prog. Phys. 56, 1347 (1993).
[34] T. Blum, L. Kärkkäinen, D. Toussaint and S. Gottlieb, "The Equation of State for Two Flavor QCD", contribution to the XIIth International Symposium on Lattice Field Theory (Lattice '94), Bielefeld 1994.
[35] V. Goloviznin and H. Satz, Z. Phys. C 57, 671 (1993).
[36] P. V. Ruuskanen, in: Quark-Gluon Plasma, p. 519, R. C. Hwa (ed.), Singapore (1990).
[37] M. C. Abreu et al. (NA38), Z. Phys. C 38, 117 (1988);
L. Kluberg (NA38), Nucl. Phys. A488, 613c (1988);
J. Y. Grossiord (NA38), Nucl. Phys. A498, 249c (1989);
C. Baglin et al. (NA38), Phys. Lett. 220B, 471 (1989);
C. Baglin et al. (NA38), Phys. Lett. 255B, 459 (1991).
[38] B. Ronceux, Nucl. Phys. A566, 371c (1994).
[39] J. Badier et al., Z. Phys. C 20, 101 (1981);
D. M. Alde et al. (E772), Phys. Rev. Lett. 66, 133, 2285 (1991).
[40] D. Kharzeev and H. Satz, Z. Phys. C 60, 689 (1993);
G. Piller, J. Mutzbauer, and W. Weise, Z. Phys. A 343, 247 (1992).
[41] S. Gupta and H. Satz, Z. Phys. C 55, 391 (1992).
[42] See e.g.: J.-P. Blaizot, J.-Y. Ollitrault, in: Quark-Gluon Plasma, R. C. Hwa (ed.), Singapore (1990), and references given there.
[43] D. Kharzeev and H. Satz, Phys. Lett. B 334, 155 (1994).
[44] L. Antoniazzi et al. (E705), Phys. Rev. D46, 4828 (1992); Phys Rev. Lett. 70, 383 (1993).

Spin, Twist and Hadron Structure in Deep Inelastic Processes*

R. L. Jaffe[1,2];
Notes by H. Meyer and G. Piller[3]

[1] Center for Theoretical Physics, Massachusetts Institute of Technology, Cambridge, MA 02139 U.S.A.
[2] Department of Physics, Harvard University, Cambridge, MA 02138 U.S.A.
[3] Institute for Theoretical Physics, University of Regensburg, Universitätsstraße 31, 93040 Regensburg, Germany

Abstract. These notes provide an introduction to polarization effects in deep inelastic processes in QCD. We emphasize recent work on transverse asymmetries, subdominant effects, and the role of polarization in fragmentation and in purely hadronic processes. After a review of kinematics and some basic tools of short distance analysis, we study the twist, helicity, chirality and transversity dependence of a variety of high energy processes sensitive to the quark and gluon substructure of hadrons.

1 Introduction

In recent years hadron spin physics has emerged as one of the most dynamic areas of particle physics. During the same period the field has got considerably more complicated. In times past only longitudinal asymmetries, that have simple parton model interpretations, attracted much attention; only dominant effects, that scale in the Bjorken limit, were experimentally accessible; and only relatively crude experimental data were available. Now interest has spread to transverse polarization asymmetries, subdominant effects, polarization effects in fragmentation and in purely hadronic processes. The aim of these lectures is to present an introduction to spin dependent effects at dominant and subdominant order in deep inelastic processes including deep inelastic scattering of leptons, e^+e^- annihilation, and Drell-Yan processes. The methods can be extended relatively straightforwardly to other spin dependent effects in hard processes.

In a short set of lectures some detail and background must be sacrificed. As for background, I will assume that readers are familiar with the elementary parton model treatment of highly inelastic processes in the "infinite momentum frame". Anyone who is not familiar with basic parton model ideas should consult

* Lectures presented at the workshop "QCD and Hadron Structure" organised by the Graduiertenkolleg Erlangen–Regensburg, held on June 9th–11th, 1992 in Kloster Banz, Germany
This work is supported in part by funds provided by National Science Foundation (N.S.F.) grant #PHY 92-18167 and by the U.S. Department of Energy (D.O.E.) under contracts #DF-FC02-94ER40818 and #DF-FG02-92ER40702.

standard textbook presentations. [1, 2, 3] Although I will have a lot to say about the parton model, it may look poorly motivated to someone who has not seen the ideas presented in their simplest form first. As for detail, I will mostly ignore the complications of QCD radiative corrections, normally included via the renormalization group. There are many excellent treatments including books by Collins[4], Muta [5] and most recently in a context particularly well suited to these lectures, by Roberts. [6] Of course radiative corrections and the momentum scale dependence they generate are central to the understanding of QCD. Here we will be interested in the *classification* of scattering amplitudes in terms of helicity, chirality, twist, *etc.* – a classification which is largely (but not entirely) independent of radiative corrections. In many cases the soft, $\ln Q^2$ dependence they generate can be regarded as decorations of our primary results. Where this is not the case, I will try to warn the reader and refer to the appropriate literature.

The main question to be addressed here is: How can one classify and interpret the wide variety of spin dependent phenomena expected in hard processes? Which phenomena are displayed in which experiments? What are the selection rules enforced by the symmetries of QCD? Which phenomena dominate at large-Q^2, which are suppressed, and what is the physical origin of the suppression? In short, the object is to provide the background for both experimental and theoretical analysis of spin effects in hard processes. In contrast, I will resist almost entirely the temptation to speculate about the origins of spin effects based on models of hadron structure. These notes are not intended to be an introduction to the so-called "spin crisis" which grew out of the observation that only a small fraction of the nucleon's spin is carried by the spin of quarks. Theorists will not find their own or my own favorite explanation of the spin crisis in these lectures. That is a subject for another school.

Certain predictions of perturbative QCD are admired for being very general and independent of the difficult details of hadron structure. Examples include the cross section for $e^+e^- \to$ hadrons, event shapes in e^+e^- annihilation, the $\ln Q^2$ dependence of deep inelastic structure functions, and the Gross-Llewellyn Smith and Bjorken Sum Rules. Studies of these processes provide essential tests of QCD. These will not be major topics here. I will assume that perturbative QCD is correct and use it as a sophisticated probe of the poorly understood dynamics of confinement. As we shall see, perturbative QCD is by now so well understood that it is possible to "tune" the probe to measure the nucleon expectation values of a variety of quark and gluon distributions and correlations within hadrons. Probes can be selected for spin, twist and flavor quantum numbers, and can be used either to analyze the structure of hadronic targets or reaction fragments. No other approach yields such well defined information about hadronic bound states. This information may help guide us to a better understanding of confinement from first principles.

Many aspects of these lectures are based on work performed in collaboration with Xiangdong Ji. The reader who wishes to explore subjects in greater depth should look at refs. [7] – [12], as well as other references provided in the text. I would like to thank Xiangdong for the pleasure of this long collaboration. Thanks are also due to Matthias Burkardt, Gary Goldstein and Aneesh Manohar who collaborated on other projects related to this work. In addition I have benefited greatly from discussions with Guido Altarelli, Xavier Artru, Ian Balitsky, Vladimir Braun, Gerry Bunce, John Collins, Vernon Hughes, Gerd Mallot, Al Mueller, Richard Milner, John Ralston, Phil Ratcliffe, Klaus Rith, Jacques Soffer, and Linda Stuart.

These lecture notes grew out of talks at schools and conferences in the early 1990's. They began as a manuscript prepared by Drs. H. Meyer and G. Piller from their notes at the 1992 Graduiertenkolleg. I would like to thank them for the substantial work they undertook at that time. The present version was significantly edited, reformulated and expanded for the 1995 Erice School on the Internal Spin Structure of the Nucleon.

2 Kinematics and Other Generalities

The organizers of the school asked if I would briefly introduce the kinematic and dynamical variables common in the study of deep inelastic processes. So before getting down to the business of dynamics here is a short summary — the cogniscenti will certainly want to skip this section. Others, who may be familiar with less streamlined notation might wish at least to look at Eqs. (2.14), (2.15), (2.21), and (2.22). I hope students with less background in perturbative QCD will find this section useful.

2.1 Deep Inelastic Scattering

Basic Variables Deep inelastic scattering (DIS) is the archetype for hard processes in QCD: a lepton — in practice an electron, muon or neutrino — with high energy scatters off a target hadron — in practice a nucleon or nucleus, or perhaps a photon — transferring large quantities of both energy and invariant squared-four-momentum. For charged leptons the dominant reaction mechanism is electromagnetism and one photon exchange is a good approximation. For neutrinos either $W^{\pm}$ (charged current) or Z^0 (neutral current) exchange may occur. The weak interactions of electrons may also be studied either by means of small parity violating asymmetries originating in $\gamma - Z^0$ interference, or by means of the charged current reaction $e^- \to \nu_e$.

We are primarily interested in experiments performed with polarized targets. Neutrino scattering experiments require far too massive targets for polarization to be a practical option, so we will ignore them, although W-exchange has been observed in $e^- p \to \nu_e + X$, at HERA,[13] and could be extended to a polarized target, at least in principle. Thus we are mainly limited to charged lepton scattering by one photon exchange. The kinematics is shown in Fig. 1. The initial lepton with momentum k and energy E exchanges a photon of momentum q with a the target with momentum P. Only the outgoing electron with momentum k' and energy E' is detected. One can define the two invariants

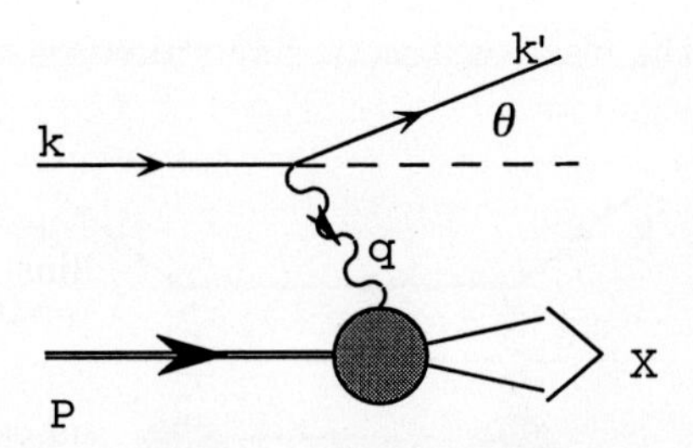

Fig. 1. *Kinematics of lepton-hadron scattering in the target rest frame.*

$$q^2 \equiv (k-k')^2 = q_0^2 - \mathbf{q}^2 = -4EE'\sin^2(\theta/2) = -Q^2 < 0 \tag{2.1}$$
$$\nu \equiv P\cdot q = M(E-E'), \tag{2.2}$$

where the lepton mass has been neglected (and will be neglected henceforth). The meaning of the scattering angle θ is clear from Fig. 1. Unless otherwise noted, E, E', θ and $q^0 \equiv E - E'$ refer to the target rest frame. The deep inelastic, or *Bjorken* limit is where Q^2 and ν both go to infinity with the ratio, $x \equiv Q^2/2\nu$ fixed. x is known as the Bjorken (scaling) variable.

Since the invariant mass of the hadronic final state is larger than or equal to the mass of the target, $(P+q)^2 \geq M^2$, one has $0 < x \leq 1$. It is convenient also to measure the energy loss using a dimensionless variable,

$$0 \leq y \equiv \frac{\nu}{ME} \leq 1. \tag{2.3}$$

We will find E, Q^2, x, and y to be a useful set of variables. Note that it is overcomplete since $xy = Q^2/2ME$, and note also that what we define as ν differs from common usage by a factor of M. The behavior of cross sections at large Q^2 is much more transparent using these variables than using the set (E, E', θ) favored by experimenters for the reason that $\theta \to 0$ as $Q^2 \to \infty$ at fixed x and y.

Cross Section and Structure Functions The differential cross-section for inclusive scattering ($eP \to e'X$) is given by:

$$d\sigma = \frac{1}{J}\frac{d^3k'}{2E'(2\pi)^3}\sum_X \prod_{i=1}^{n_X}\int \frac{d^3p_i}{(2\pi)^3 2p_{i0}}|\mathcal{A}|^2(2\pi)^4\delta^4(P+q-\sum_{i=1}^{n_X}p_i). \tag{2.4}$$

The flux factor for the incoming nucleon and electron is denoted by $J = 4P\cdot k$, which is equal to $J = 4ME$ in the rest frame of the nucleon. The sum runs over all hadronic final states X which are not observed. Each hadronic final state consists of n_X particles with momenta p_i ($\sum_{i=1}^{n_X} p_i \equiv p_X$). The squared-amplitude $|\mathcal{A}|^2$ can be separated into a leptonic ($l^{\mu\nu}$) and a hadronic ($W^{\mu\nu}$) tensor (see Fig. 2):

$$\left|\frac{\mathcal{A}}{4\pi}\right|^2 = \frac{\alpha^2}{Q^4} l^{\mu\nu} W_{\mu\nu}, \tag{2.5}$$

where $\alpha \sim 1/137$ is the electromagnetic fine structure constant. The leptonic

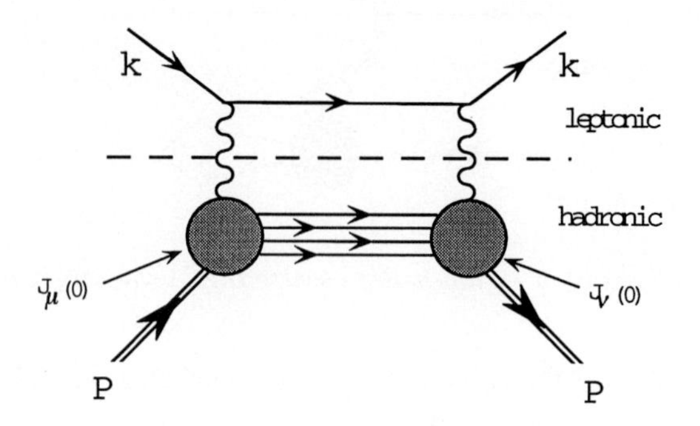

Fig. 2. *The squared amplitude $\mathcal{A}$ for electron-hadron scattering can be separated into a leptonic tensor $l^{\mu\nu}$ and a hadronic tensor $W^{\mu\nu}$.*

tensor $l^{\mu\nu}$ is given by the square of the elementary spin 1/2 current (summed over final spins):

$$\begin{aligned} l^{\mu\nu} &= \sum_{s'} \bar{u}(k,s)\gamma^\mu u(k',s')\bar{u}(k',s')\gamma^\nu u(k,s) \\ &= 2(k'^\mu k^\nu + k'^\nu k^\mu) - 2g^{\mu\nu} k\cdot k' + 2i\epsilon^{\mu\nu\lambda\sigma} q_\lambda s_\sigma, \end{aligned} \tag{2.6}$$

and consists of parts symmetric and antisymmetric in μ and ν. The antisymmetric part is linear in the spin vector s, which is normalized to $s^2 = -m^2$. While the leptonic tensor is known completely, $W^{\mu\nu}$, which describes the internal structure of the nucleon, depends on non-perturbative strong interaction dynamics. It is expressed in terms of the current J^μ as:

$$4\pi W^{\mu\nu} = \sum_X \langle PS|J^\mu|X\rangle\langle X|J^\nu|PS\rangle(2\pi)^4\delta(P+q-p_X) \tag{2.7}$$

$$= \int d^4\xi e^{iq\cdot\xi}\langle PS|[J^\mu(\xi), J^\nu(0)]|PS\rangle_c. \tag{2.8}$$

The steps leading from (2.7) to (2.8) include writing the δ function as an exponential,

$$(2\pi)^4\delta^4(K) = \int d^4\xi e^{i\xi\cdot K}, \tag{2.9}$$

translating the current, $e^{i\xi\cdot(P-p_X)}\langle PS|J^\mu(0)|X\rangle = \langle PS|J^\mu(\xi)|X\rangle$, and using completeness, $\sum_X |X\rangle\langle X| = 1$. Note that another term has been subtracted to convert the current product into a commutator. It is easy to check that the new term vanishes for $q^0 > 0$ which is the case for physical lepton scattering from a stable target. The subscript $_c$ means that the graphs associated with the matrix element must be connected. Finally, note that the states are covariantly normalized to:

$$\langle P|P'\rangle = 2E(2\pi)^3\delta^3(P-P'). \tag{2.10}$$

The optical theorem:

$$2\pi W^{\mu\nu} = \text{Im } T^{\mu\nu} \tag{2.11}$$

relates the hadronic tensor to the imaginary part of the forward virtual Compton scattering amplitude, T:

$$T_{\mu\nu} = i\int d^4\xi e^{iq\cdot\xi}\langle PS|T(J_\mu(\xi)J_\nu(0))|PS\rangle \tag{2.12}$$

as shown graphically in Fig. 3.

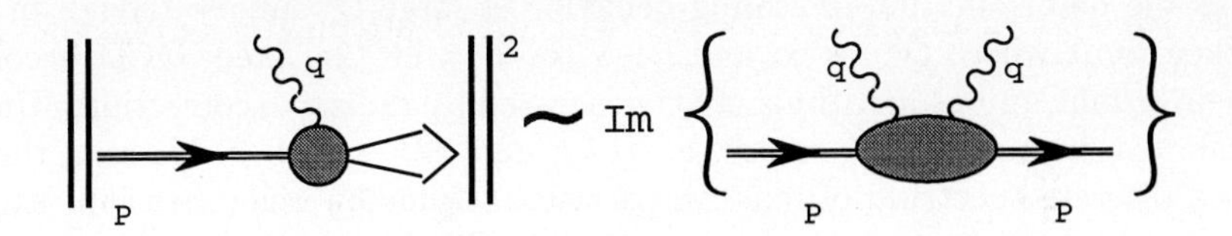

Fig. 3. *The optical theorem relates the hadronic structure tensor, $W^{\mu\nu}$ to the imaginary part of forward ($P = P'$, $q = q'$), virtual ($Q^2 < 0$) Compton scattering.*

Structure Functions Using Lorentz covariance, gauge invariance, parity conservation in electromagnetism and standard discrete symmetries of the strong interactions, $W^{\mu\nu}$ can be parametrized in terms of four scalar dimensionless structure functions $F_1(x,Q^2)$, $F_2(x,Q^2)$, $g_1(x,Q^2)$ and $g_2(x,Q^2)$. They depend only on the two invariants Q^2 and ν, or alternatively on Q^2 and the dimensionless Bjorken variable x. Splitting $W^{\mu\nu}$ into symmetric and anti-symmetric parts we have,

$$W^{\mu\nu} = W^{\{\mu\nu\}} + W^{[\mu\nu]}, \tag{2.13}$$

with

$$W^{\{\mu\nu\}} = \left(-g^{\mu\nu} + \frac{q^\mu q^\nu}{q^2}\right)F_1 + \left[\left(P^\mu - \frac{\nu}{q^2}q^\mu\right)\left(P^\nu - \frac{\nu}{q^2}q^\nu\right)\right]\frac{F_2}{\nu}, \tag{2.14}$$

$$W^{[\mu\nu]} = -i\epsilon^{\mu\nu\lambda\sigma}q_\lambda\left(\frac{S_\sigma}{\nu}(g_1+g_2) - \frac{q\cdot SP_\sigma}{\nu^2}g_2\right), \tag{2.15}$$

where S^σ is the polarization vector of the nucleon ($S^2 = -M^2$), $P\cdot S = 0$. S^σ is a pseudovector. Since $W^{[\mu\nu]}$ is a normal tensor, parity demands that the S^μ appear with another pseudotensor, and the only one available is the $\epsilon^{\mu\nu\sigma\lambda}$. Students often ask why $W^{\mu\nu}$ depends only linearly on S^μ – what is wrong with $S^\mu S^\nu$, for example? Lorentz invariance demands that $W^{\mu\nu}$, defined in (2.8) be linear in the initial and final nucleon spinors, $U(P,S)$ and $\bar{U}(P,S)$. Tensors constructed from these are either spin independent ($\bar{U}(P,S)\gamma^\mu U(P,S) = 2P^\mu$) or linear in S^μ (($\bar{U}(P,S)\gamma^\mu\gamma_5 U(P,S) = 2S^\mu$), but that is the end of it.

Note also that $W^{\mu\nu}$ is dimensionless (we shall have more to say about operator dimensions shortly). Factors of ν have been judiciously introduced into (2.15) and (2.14) so that the four structure functions, F_1, F_2, g_1, and g_2 are dimensionless. These structure functions are related to others in common use by:

$$W_1 = F_1, \quad W_2 = \frac{M^2}{\nu}F_2, \quad G_1 = \frac{M^2 g_1}{\nu}, \quad G_2 = \frac{M^4 g_2}{\nu^2}. \tag{2.16}$$

Scaling and Kinematic Domains Our choice of invariant structure functions makes the determination of scaling behavior at large Q^2 almost trivial. In the Bjorken limit where $Q^2 \to \infty$ and $\nu \to \infty$, $x = Q^2/2\nu$ fixed, QCD becomes scale invariant up to logarithms of Q^2 generated by radiative corrections. Under a scale transformation, $P \to \lambda P$, $q \to \lambda q$, and $M \to \lambda M \neq M$, so a theory with a discrete spectrum of massive particles cannot be scale invariant except in a limit in which all masses are negligible. Thus no masses can appear in $W^{\mu\nu}$ in the Bjorken limit; it must be a dimensionless function of P^μ, q^μ, S^μ, and the invariants Q^2 and ν. In particular, it cannot depend explicitly on the target mass, M. If, for example, a term like $\frac{P^\mu P^\nu}{M^2}W_2$ appeared in $W^{\{\mu\nu\}}$, it would violate scale invariance unless W_2 vanished like $\frac{M^2}{\nu}$ at large Q^2. Clearly, the way to avoid such pathological choices of structure functions is to write the dimensionless tensor $W^{\mu\nu}$ in terms of dimensionless invariant functions using ν (or Q^2) to supply dimensional factors as needed. The immediate conclusion is that the functions F_1, F_2, g_1, and g_2 defined in (2.14) and (2.15), become functions only of the dimensionless ratio $x = Q^2/2\nu$, modulo logarithms, in the Bjorken limit,

$$\begin{aligned} F_1(Q^2,\nu) &\to F_1(x, \ln Q^2), \quad F_2(Q^2,\nu) \to F_2(x, \ln Q^2) \\ g_1(Q^2,\nu) &\to g_1(x, \ln Q^2), \quad g_2(Q^2,\nu) \to g_2(x, \ln Q^2) \end{aligned} \tag{2.17}$$

as Q^2 and ν become large at fixed x. In practice it is observed that for $Q^2 > 1GeV^2$, the structure functions depend only very weakly on Q^2. Furthermore one observes an approximate relationship between F_1 and F_2, known as the Callan-Gross relation,[14]

$$F_1 - \frac{1}{2x}F_2 \sim \frac{1}{\ln Q^2}, \tag{2.18}$$

which indicates that the particles that carry electric charge (the quarks) have spin $\frac{1}{2}$. The different kinematic domains of interest in inelastic electron scattering are shown in Fig. 4.

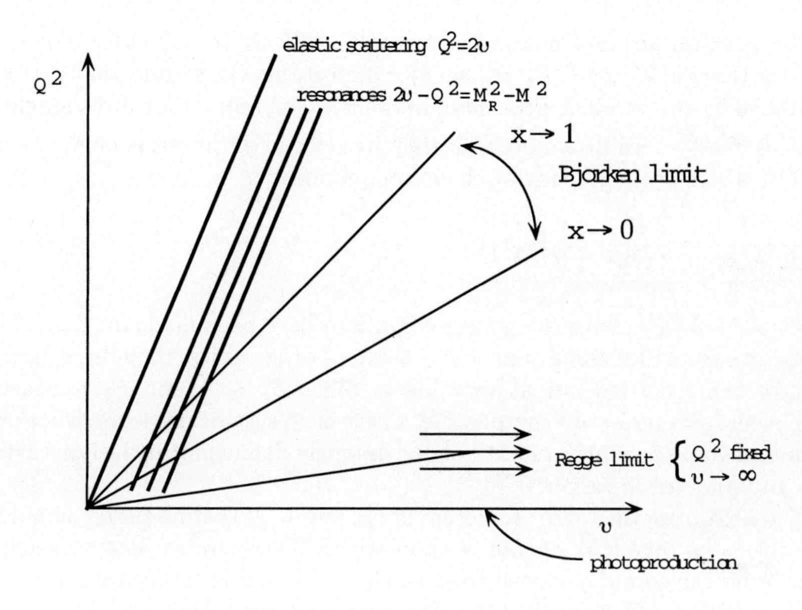

Fig. 4. *Kinematic domains in electron-nucleon scattering.*

Flavor Generalizations Only up, down and strange quarks appear to be important constituents of light hadrons. The processes of interest to us, therefore are mediated by currents lying in the $SU(3)_R \times SU(3)_L$ space of u, d, s vector and axial currents,

$$\begin{aligned} J^a_\mu &\equiv \tfrac{1}{2}\bar{\psi}\gamma_\mu\lambda^a\psi, \\ J^a_{\mu 5} &\equiv \tfrac{1}{2}\bar{\psi}\gamma_\mu\gamma_5\lambda^a\psi, \end{aligned} \tag{2.19}$$

where λ^a for $a = 1, 2 \ldots 8$ are the flavor $SU(3)$ matrices, which are normalized to $\mathrm{Tr}\lambda^a\lambda^b = 2\delta^{ab}$. Note, in particular, that $\lambda_3 = \mathrm{diag}(1, -1, 0)$ and $\lambda_8 = \mathrm{diag}\frac{1}{\sqrt{3}}(1, 1, -2)$. In addition one has the flavor singlet current $J^0_\mu = \sqrt{2/3}\bar{\psi}\gamma_\mu\psi$, acting like $\sqrt{\frac{2}{3}}\mathrm{diag}(1, 1, 1)$ in flavor space.

Cross Section for Electron-Hadron Scattering The differential cross section for unpolarized electron-hadron scattering can now be expanded in the Lorentz scalar structure functions by contracting the symmetric tensor (2.14) with the leptonic tensor, (2.6). Likewise the cross section for polarized scattering is obtained by contracting the antisymmetric tensor (2.15) with the same lepton tensor. The result is often quoted in terms of the experimenter's variables, Q^2, ν, θ, E and E', *e.g.* for the spin average case,

$$\frac{d^3\bar{\sigma}}{dE'd\Omega} = \frac{4\alpha^2}{MQ^4}E'^2\left\{2W_1(q^2,\nu)\sin^2\frac{\theta}{2} + W_2(q^2,\nu)\cos^2\frac{\theta}{2}\right\}. \tag{2.20}$$

The relative importance of the two terms is difficult to judge. Superficially it looks as though W_1 and W_2 are equally important. On second thought, W_1 is multiplied by $\sin^2 \frac{\theta}{2}$ which gets small in the Bjorken limit. On third thought, W_2 vanishes like $\frac{M^2}{\nu}$. To disentangle all this, we rewrite $d\bar{\sigma}$ in terms of F_1, F_2, x, y, and Q^2, where scaling behavior should be manifest,

$$\frac{d^3\bar{\sigma}}{dx\,dy\,d\phi} = \frac{e^4}{4\pi^2 Q^2}\left\{\frac{y}{2}F_1(x,Q^2) + \frac{1}{2xy}\left(1 - y - \frac{y^2}{4}(\kappa - 1)\right)F_2(x,Q^2)\right\} \tag{2.21}$$

with $\kappa \equiv 1 - \frac{4x^2M^2}{Q^2}$. No scaling approximations have been made in (2.21). Under typical experimental conditions y and x are of order unity, though experiments are now being carried out at very low-x. Since $F_1 \sim \frac{1}{x}$ and $F_2 \sim$ const. for small x, the two terms are comparable. There is no significant dependence on the azimuthal angle ϕ, which cannot even be uniquely defined for inclusive scattering with an unpolarized target.

It is clear from the tensor structure of $\ell_{\mu\nu}$ and $W_{\mu\nu}$ that no target spin dependent effects survive if the beam is unpolarized. Therefore we lose no generality by defining the spin dependent cross section, $\Delta\sigma$ as half the difference between right- and left-handed incident electron cross sections,[15]

$$\begin{aligned}\frac{d\Delta\sigma(\alpha)}{dx\,dy\,d\phi} = \frac{e^4}{4\pi^2Q^2}\Bigg\{&\cos\alpha\left\{\left[1 - \frac{y}{2} - \frac{y^2}{4}(\kappa - 1)\right]g_1(x,Q^2)\right.\\ &\left.-\frac{y}{2}(\kappa - 1)g_2(x,Q^2)\right\} - \sin\alpha\cos\phi\sqrt{(\kappa - 1)\left(1 - y - \frac{y^2}{4}(\kappa - 1)\right)}\,\cdot\\ &\cdot\left(\frac{y}{2}g_1(x,Q^2) + g_2(x,Q^2)\right)\Bigg\}\end{aligned} \tag{2.22}$$

Now the azimuthal angle ϕ and the angle, α, between the target spin $\hat{S}$ and the incident electron momentum, $\hat{k}$, make non-trivial appearances. These and other

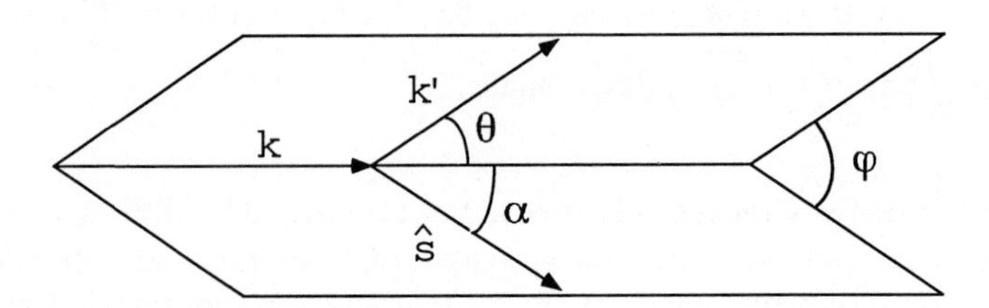

Fig. 5. *Kinematic variables in polarized lepton scattering from a polarized target.*

kinematic variables are defined in Fig. 5. Note the following:

- α is the angle between the spin vector of the target ($\hat{S}$) and the incident electron beam ($\hat{k}$), *not* the virtual photon direction ($\hat{q}$).

- ϕ is the azimuthal angle between the plane defined by $\mathbf{k}$ and $\mathbf{k}'$ and the plane defined by $\mathbf{k}$ and $\hat{S}$.
- Equations (2.21) and (2.22) are exact (except that lepton masses have been ignored): no scaling limit has been taken. $\kappa - 1 \equiv M^2Q^2/\nu^2 = 4M^2x^2/Q^2$ is a measure of the approach to the scaling limit, $Q^2 \to \infty$.
- To eliminate *spin-independent* effects one may either (i) subtract cross sections for different values of α; (ii) subtract cross sections for right- and left-handed leptons; or (iii) measure ϕ-dependence.

Notice that effects associated with $g_2(x, Q^2)$ are suppressed by a factor $\sqrt{\kappa - 1} = \frac{2Mx}{\sqrt{Q^2}}$ with respect to the dominant structure function $g_1(x, Q^2)$. In technical terms, this means that effects associated with g_2 are "higher twist" — suppressed by a power of Q relative to the leading phenomena in the Bjorken limit. However, at 90° the coefficient of the dominant term vanishes identically and allows the combination $\frac{y}{2}g_1 + g_2$ to be extracted cleanly at large Q^2. This is a unique feature of the spin-dependent scattering. Only very rarely, to my knowledge, can a higher twist effect be selected by an adroit kinematic arrangement, thereby avoiding the difficult process of fitting and subtracting away a leading twist effect to expose the higher twist correction underneath.

2.2 Other Basic Deep Inelastic Processes

Inclusive $e^+ e^-$ Annihilation In this process an electron with momentum k and a positron with momentum k' annihilate to form a massive time-like photon with momentum $q = k + k'$ ($Q^2 \equiv q^2 > 0$), which decays into an unobserved final state. Through the optical theorem, the total cross section is proportional to the imaginary part of the photon propagator (see Fig. 6),

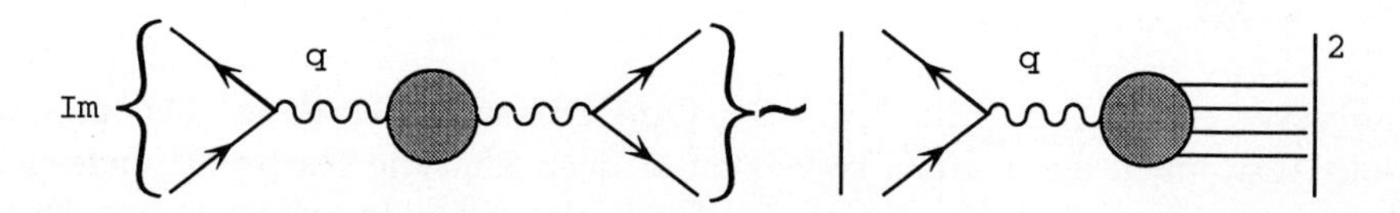

Fig. 6. *The optical theorem relates the total cross-section for e^+e^- annihilation to the imaginary part of the photon propagator.*

$$\sigma_{tot} = \frac{16\pi^2\alpha^2}{Q^2}\Pi(Q^2), \tag{2.23}$$

where $\Pi(Q^2)$ is the Lorentz scalar spectral function appearing in the photon propagator:

$$\Pi_{\mu\nu} = (q_\mu q_\nu - q^2 g_{\mu\nu})\Pi(Q^2), \tag{2.24}$$

and

$$\Pi(Q^2) = -\frac{1}{6Q^2}\int d^4\xi e^{iq\cdot\xi}\langle 0|\,[J_\mu(\xi), J^\mu(0)]\,|0\rangle. \tag{2.25}$$

Usually the data are expressed as a ratio to the pointlike annihilation cross section to muons (to lowest order in α_{EM}):

$$R(Q^2) \equiv \frac{\sigma_{tot}(e^+e^- \to hadrons)}{\sigma(e^+e^- \to \mu^+\mu^-)} = 12\pi\Pi(Q^2). \tag{2.26}$$

Since the hadronic process is initiated by the creation of a $q\bar{q}$ pair, R directly measures the number of colors. At large Q^2 it is modified only by perturbative QCD corrections:

$$R(Q^2) = \sum_q 3e_q^2\left\{1 + \frac{\alpha_{QCD}(Q^2)}{\pi} + 1.409\frac{\alpha_{QCD}(Q^2)}{\pi^2}\right. \tag{2.27}$$

$$\left. -12.805\frac{\alpha^3_{QCD}(Q^2)}{\pi^3} + \cdots\right\} + \text{quark mass corrections.}) \tag{2.28}$$

The coefficients in (2.28) are renormalization scheme dependent beyond lowest order. Those quoted in (2.28) were calculated in $\overline{MS}$ scheme with five flavors.[16] The formula for R does not depend on any details of hadronic structure, so it provides an important test of QCD (and measurement of α_{QCD}). Similar remarks apply to processes in which jets are observed in the final state of e^+e^- annihilation. Two jet events have the angular distribution that one expects for two spin 1/2 quarks; a third jet is associated with gluonic bremsstrahlung. These processes however are not sensitive to the structure of hadrons and we will not discuss them further here.

Inclusive e^+e^- Annihilation with One Observed Hadron This process looks very much like a timelike version of deep inelastic scattering. Indeed it shares many important characteristics, but it also differs in essential ways. From the point of view of a theorist interested in hadron structure, the opportunity to study unstable hadrons makes this process very attractive. Deep inelastic scattering from Λ-hyperons or π or ρ-mesons will never be more than a *gedanken* experiment. However, these and other unstable hadrons have already been studied in e^+e^--annihilation. The physical basis of "fragmentation" — the process by which a quark created by the current from the vacuum fragments into the observed hadron — is not as well understood as DIS, making this an area of considerable interest at the present time.

The kinematics for $e^+e^- \to P + X$ is illustrated in Fig. 7. Once again two kinematic invariants, Q^2 and $\nu = P\cdot q$, define the process. The limit of interest is $Q^2, \nu \to \infty$, with $z \equiv \frac{2\nu}{Q^2}$ fixed. The momentum of the virtual photon is time-

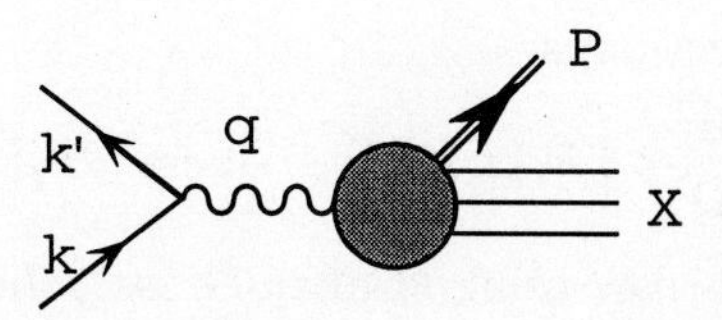

Fig. 7. *Kinematics for single particle inclusive annihilation* — $e^+e^- \to PX$.

like, and that makes a major difference as we shall see in Sect. 2. The invariants are often expressed in terms of quantities measured in the e^+e^- center of mass:

$$\begin{aligned} q^2 &\equiv Q^2 = 4E_e^2 > 0 \\ P \cdot q &\equiv \nu = E\sqrt{Q^2} \\ 0 < z &\equiv \frac{2P \cdot q}{q^2} = \frac{E}{E_e} \leq 1, \end{aligned} \tag{2.29}$$

where E is the energy of the observed hadron. We shall usually be interested the polarization dependence, but here we illustrate the kinematics for the simpler, spin-averaged case. The cross section can be written as the product of a leptonic $\hat{l}_{\mu\nu}$ and a hadronic tensor $\hat{W}^{\mu\nu}$:

$$d\sigma \sim \hat{l}_{\mu\nu} \hat{W}^{\mu\nu} \frac{d^3P}{(2\pi^3)2E} \tag{2.30}$$

The hadronic tensor is determined by the electromagnetic current and depends on two invariant "fragmentation functions" due to current conservation and C, P and T invariance:

$$\begin{aligned} \hat{W}^{\mu\nu} &= \frac{1}{4\pi} \sum_X (2\pi)^4 \delta^4(P + P_X - q) \langle 0|J_\mu|PX\rangle_{\text{out out}}\langle PX|J_\nu|0\rangle \\ &= \frac{1}{4\pi} \int d^4\xi e^{iq\cdot\xi} \sum_X \langle 0|J_\mu(\xi)|PX\rangle_{\text{out out}}\langle PX|J_\nu(0)|0\rangle \\ &= -\left(\frac{g_{\mu\nu} - q_\mu q_\nu}{q^2}\right) \hat{F}_1(z, q^2) + \frac{1}{\nu}\left(P_\mu - \frac{\nu}{q^2} q_\mu\right)\left(P_\nu - \frac{\nu}{q^2} q_\nu\right) \hat{F}_2(z, q^2). \end{aligned} \tag{2.31}$$

In contrast to DIS, the sum over unobserved hadrons X cannot be completed because the state $|PX\rangle_{\text{out}}$ depends non-trivially on the observed hadron. Even if P and X did not interact, Bose or Fermi statistics prevents the states X from being complete. In practice P and X interact dynamically, as indicated by the subscript "out". For simplicity we will generally suppress this subscript. Thus, $e^+e^- \to P + X$ is not controlled by the product of two operators (electroweak currents), a feature which complicates the study of $e^+e^- \to P + X$ significantly.

If $q^2 \to \infty$ at fixed z, the structure functions, $\hat{F}_1$ and $\hat{F}_2$ scale (up to logarithmic corrections) and obey a "Callan-Gross" relation, $\hat{F}_1 + (z/2)\hat{F}_2 \sim 1/\ln q^2$.

In this limit the cross section is:

$$\frac{d\sigma}{dz d\Omega} = \frac{\alpha^2}{Q^2} z \left\{ \hat{F}_1(z, \ln q^2) + \frac{1}{4} z \sin^2\theta \hat{F}_2(z, \ln q^2) \right\}. \tag{2.32}$$

In leading logarithmic order, using "Callan-Gross", the inclusive spectrum reduces to

$$\frac{1}{\sigma}\frac{d\sigma}{dz} \sim \frac{2}{R} z \hat{F}_1(z, q^2) + \cdots, \tag{2.33}$$

where R is defined by (2.26).

Lepton Pair Production The final process we will consider in detail is massive lepton-pair creation in hadron-hadron collisions – the so-called "Drell-Yan" process. The opportunities for study of novel aspects of hadron structure by means of polarized Drell-Yan experiments have motivated a major spin physics program at RHIC.[17] The kinematics of the lepton pair production are illustrated in Fig. 8. Two hadrons with momenta P and P' collide at a center of

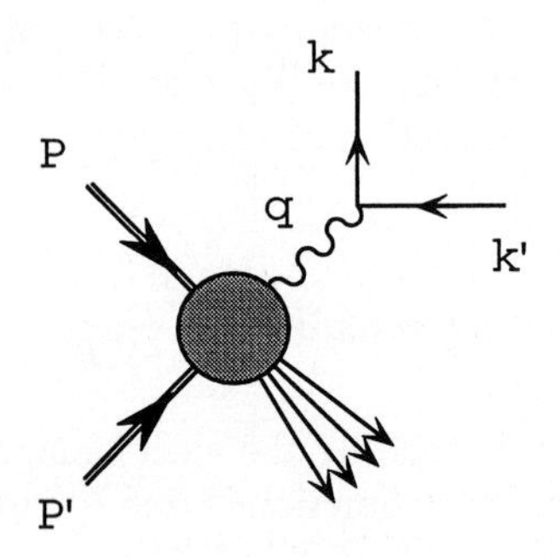

Fig. 8. *Kinematics of the Drell-Yan process.*

mass energy $s = (P + P')^2 = 4E_{CM}^2$. Two leptons with momenta k_1^μ and k_2^μ respectively are produced. They result from the decay of a timelike photon, $W^\pm$, or Z^0 carrying a momentum q, with $q^2 = Q^2 > 0$. Two dimensionless scaling variables are defined by $x = \frac{2P\cdot q}{q^2}$ and $y = \frac{2P'\cdot q}{q^2}$. It is easy to see that $xy = Q^2/s$. The differential cross section is

$$d\sigma \propto L_{\mu\nu} W^{\mu\nu} \frac{d^3 k_1}{(2\pi)^3 2k_1^0} \frac{d^3 k_2}{(2\pi)^3 2k_2^0}. \tag{2.34}$$

The decay of the virtual gauge boson is described by the leptonic tensor $L_{\mu\nu}$, whereas all information about the hadronic process are contained in $W^{\mu\nu}$:

$$\begin{aligned} W^{\mu\nu} &= \frac{1}{2} s \sum_X (2\pi)^4 \delta^4(P + P' - q - X) \ {}_{\text{in}}\langle PP'|J^\mu(0)|X\rangle \langle X|J^\nu(0)|PP'\rangle_{\text{in}} \\ &= \frac{1}{2} s \int d^4\xi e^{-iq\cdot\xi} \ {}_{\text{in}}\langle PP'|J^\mu(\xi)J^\nu(0)|PP'\rangle_{\text{in}}, \end{aligned} \tag{2.35}$$

where the in-state label on $|PP'\rangle$ will usually be suppressed. $W^{\mu\nu}$ contains many Lorentz invariant structure functions W_k. Depending on the experimental circumstances different combinations of the W_k and differential cross sections are of interest. As an example we consider the inclusive cross section where the lepton momenta have been integrated out, leaving $d\sigma/dq^4$,

$$\frac{d\sigma}{dq^4} = \frac{1}{6\pi^3}\frac{\alpha^2}{Q^2 s^2}\left(-W^\mu_\mu\right). \tag{2.36}$$

The scaling limit ($s, Q^2 \to \infty$ but $\tau \equiv Q^2/s$ fixed) once again yields a function of the dimensionless variables (x and y) modulo logarithms induced by QCD radiative corrections, and in this case W^μ_μ is of interest.

$$W^\mu_\mu \to W(x, y, \ln Q^2). \tag{2.37}$$

3 Deep Inelastic Processes from a Coordinate Space Viewpoint

Traditional introductions to the parton model stay fixed in momentum space, where they use the device of the "infinite momentum frame" to simplify dynamical arguments. More sophistication is necessary to handle the complexities introduced by spin dependence and the subdominant effects associated with transverse spin in DIS. It is particularly useful to employ coordinate space methods, mixing parton phenomenology with somewhat more formal methods of the operator product expansion.[18] Certainly, sophisticated momentum space methods[19] can achieve the same results. However, it is particularly easy to distinguish and catalogue dominant and sub-dominant contributions using the operator product expansion in coordinate space.

In this section we will explore the coordinate space structure of the hard processes introduced in Sect. 1. Much of this material is to be found in modern field theory texts,[20] however there is an advantage to providing a brief, self-contained introduction which stresses only those elementary aspects of the formalism that are useful in characterizing deep inelastic spin physics.

3.1 $e^+e^- \to$ hadrons – The Short-Distance Expansion

Inclusive e^+e^- annihilation into hadrons is the simplest process to analyze and illustrates the importance of Wilson's short distance expansion. As shown in Sect. 1, this process is described by the vacuum expectation value of a current commutator,

$$\Pi(Q^2) \propto \frac{1}{Q^2}\int d^4\xi\, e^{iq\cdot\xi}\, \langle 0|\, [J_\mu(\xi), J^\mu(0)]\, |0\rangle\,. \tag{3.1}$$

In the center of mass system we have $q = (\sqrt{Q^2}, \mathbf{0})$. Since the commutator is causal,

$$[J_\mu(\xi), J_\nu(0)] = 0 \quad \text{for} \quad \xi^2 < 0, \tag{3.2}$$

then $|\boldsymbol{\xi}| < \xi^0$ in the integral. Using the symmetry of the commutator one obtains:

$$\Pi(Q^2) \propto \int_0^\infty d\xi^0 \sin Q\xi^0 \int_{|\boldsymbol{\xi}|\leq\xi^0} d^3\xi \langle 0|\,[J_\mu(\xi), J^\mu(0)]\,|0\rangle \tag{3.3}$$

In the high energy limit, $Q \to \infty$, $\sin Q\xi^0$ oscillates rapidly, averaging out contributions except at the $\xi^0 = 0$ boundary of the integration region. This argument can be made more formal, leading to the conclusion that $\xi^0 \sim \frac{1}{q^0}$ gives the dominant contribution to the integral. Since $\xi^0 > |\boldsymbol{\xi}|$ we can conclude that e^+e^- annihilation into hadrons at high Q^2 is dominated by interactions at short distances, $\xi^\mu \to 0$. This is, of course, a Lorentz invariant condition.

The leading contribution to the annihilation process can now be found via the operator product expansion (OPE).[20] First postulated by Wilson, the existence of the OPE has been demonstrated to all orders in perturbation theory in renormalizable theories and also in various toy models which can be solved exactly. According to the OPE, a product of local operators $\hat{A}(\xi)$ and $\hat{B}(0)$ at short distances (here $\xi_\mu \to 0$) can be expanded in a series of *non-singular* local operators multiplying c-number *singular* functions,

$$\hat{A}(\xi)\hat{B}(0) \sim \sum_{[\alpha]} C_{[\alpha]}(\xi)\hat{\theta}_{[\alpha]}(0) \quad \text{as} \quad \xi^\mu \to 0. \tag{3.4}$$

In general the product $\hat{A}\hat{B}$ is singular as $\xi \to 0$. The substance of the expansion is that the singularities can be isolated in the c-number "Wilson coefficients", $C_{[\alpha]}$. The operators in (3.4) are cutoff independent renormalized operators and the Wilson coefficients are likewise cutoff independent.

The behavior of the Wilson coefficients at $\xi_\mu \to 0$ follows from dimensional analysis. In natural units, all quantities are measured in dimensions of mass to the appropriate power. For simplicity, if a quantity, θ has units m^{d_θ}, we write $d_\theta = [\theta]$. This is a simple concept, not to be confused with more subtle ones like anomalous dimensions or scale dimensions.[21] The dimension of all operators of interest to us can be deduced from the fact that charge and action are dimensionless. Thus $[J^\mu] = 3$ because $\int d^3x J^0(x) = Q$. For the quark field $[\psi] = \frac{3}{2}$ because the free Dirac action is $\int d^4x \bar{\psi} i\gamma\cdot\partial\psi + \ldots$, likewise for the gluon field strength $[G_{\mu\nu}] = 2$. Since we normalize our states covariantly, $\langle P|P'\rangle = 2E(2\pi)^3\delta^3(\mathbf{P} - \mathbf{P}')$, $[\langle P|P'\rangle] = -2$. For the vectors $[P^\mu] = [S^\mu] = 1$. We see that $W_{\mu\nu}$ is dimensionless, as reflected in the form of (2.14) and (2.15).

Dimensional consistency applied to the OPE requires,

$$[\hat{A}] + [\hat{B}] = [C_{[\alpha]}] + [\hat{\theta}_{[\alpha]}]. \tag{3.5}$$

What can account for the dimensions of the singular function $C_{[\alpha]}$? If the operators $\hat{A}$ and $\hat{B}$ are finite in the $m_{\text{quark}} \to 0$ limit, then powers of m_{quark} can only appear in the numerator of C. The renormalization scale, μ, necessary to render the theory finite can only appear in logarithms (of the form $\ln(\mu\xi)$) order by order in perturbation theory. This leaves the coordinate ξ itself to absorb the dimensions.

$$C_{[\alpha]}(\xi) \sim \frac{1}{\xi^{[\hat{A}]+[\hat{B}]-[\hat{\theta}_{[\alpha]}]}} \left(\ln^{\gamma_\theta}(\mu\xi) + \ldots\right) \quad \text{as} \quad \xi \to 0. \tag{3.6}$$

The exponent γ_θ is the "anomalous dimension" of the operator θ generated by radiative corrections. Without minimizing the importance of these logarithms, we will usually ignore them and focus on the gross, power law, behavior required by dimensional analysis. For given operators $\hat{A}$ and $\hat{B}$ the leading contribution at short distances comes from that term in the OPE having the lowest operator dimension $[\hat{\theta}_{[\alpha]}]$.

This can now easily be applied to e^+e^- annihilation. The dimension of the hadronic electromagnetic current is $[J_\mu] = 3$. No fields have negative dimensions, so the lowest dimension operator is the unit operator, $\hat{\theta}_0 \equiv \mathbf{1}$, with $[\mathbf{1}] = 0$. The $C_0(\xi) \sim 1/\xi^6$ and the dominant contribution in the OPE is,

$$\langle 0 | \left[J_\mu(\xi), J^\mu(0) \right] | 0 \rangle \sim \frac{1}{\xi^6}, \quad \text{modulo logarithms.} \tag{3.7}$$

Consequently the current correlation function scales like

$$\Pi(Q^2) \sim \frac{1}{Q^2} \int d^4\xi \, e^{iq\cdot\xi} \frac{1}{\xi^6} \sim 1, \tag{3.8}$$

again modulo logarithms, and the cross section (2.23) scales like:

$$\sigma\left(e^+e^- \to \text{hadrons}\right) \sim \frac{1}{Q^2}. \tag{3.9}$$

The logarithms can be gathered together into powers of $\frac{\alpha_s}{\pi}$ as anticipated in (2.28). Of course, having made no attempt to derive the OPE or to study the effects of radiative corrections and renormalization in detail, the example of the total e^+e^- annihilation cross section becomes rather trivial. Nevertheless it provides a useful introduction to the more complicated cases which follow.

3.2 $lp \to lX$ – The Light-Cone Expansion

Next we turn to deep inelastic scattering, which is characterized by two large invariants – Q^2 and ν. As we shall see, such processes are dominated by physics close to the light-cone.

Light-Cone Coordinates and Formulation of Deep Inelastic Scattering The four-momenta P^μ and q^μ can be used to define a frame and a spatial direction. Without loss of generality we can choose our frame such that P^μ and q^μ have components only in the time and $\hat{e}_3$ directions. It is helpful to introduce the light-like vectors

$$p^\mu = \frac{p}{\sqrt{2}}(1,0,0,1),$$
$$n^\mu = \frac{1}{\sqrt{2}p}(1,0,0,-1) \tag{3.10}$$

with $n^2 = p^2 = 0$ and $n \cdot p = 1$. Up to the scale factor p, the vectors p^μ and n^μ function as unit vectors along opposite tangents to the light-cone. They may be used to expand P^μ and q^μ,

$$q^\mu = \frac{1}{M^2}\left(\nu - \sqrt{\nu^2 + M^2Q^2}\right)p^\mu + \frac{1}{2}\left(\nu + \sqrt{\nu^2 + M^2Q^2}\right)n^\mu, \tag{3.11}$$
$$P^\mu = p^\mu + \frac{M^2}{2}n^\mu, \tag{3.12}$$

In the Bjorken limit q^μ simplifies to

$$\lim_{Bj} q^\mu \sim \left(\nu + \frac{1}{2}M^2x\right)n^\mu - xp^\mu + \mathcal{O}\left(\frac{1}{Q^2}\right). \tag{3.13}$$

p selects a specific frame. For example $p = M/\sqrt{2}$ yields the target rest frame, while $p \to \infty$ selects the infinite momentum frame. The decomposition along p^μ and n^μ is equivalent to the use of light-cone coordinates, which are defined as follows. An arbitrary four-vector $a^\mu = (a^0, a^1, a^2, a^3)$ can be rewritten in terms of the four components $a^\pm = \frac{1}{\sqrt{2}}(a^0 \pm a^3)$, and $\mathbf{a}^\perp = (a^1, a^2)$. In this basis, the metric $g_{\mu\nu}$ has non-zero components, $g_{+-} = g_{-+} = 1$ and $g_{ij} = -\delta_{ij}$, so $a \cdot b = a^+b^- + a^-b^+ - \mathbf{a}^\perp \cdot \mathbf{b}^\perp$. The transformation to light-cone components can be recast as an expansion in the basis vectors p^μ and n^μ,

$$a^\mu = \left(\frac{\sqrt{2}a^-}{p}\right)p^\mu + \left(\sqrt{2}a^+p\right)n^\mu + a^{\perp\mu}. \tag{3.14}$$

With these preliminaries it is easy to find the space-time region which dominates the DIS. Consider the hadronic tensor $W^{\mu\nu}$ defined in (2.8):

$$W^{\mu\nu} = \frac{1}{4\pi}\int d^4\xi e^{iq\cdot\xi}\langle P|\left[J^\mu(\xi), J^\nu(0)\right]|P\rangle, \tag{3.15}$$

Take the Bjorken limit by keeping P fixed and $q \to \infty$. Define

$$\xi^\mu \equiv \eta p^\mu + \lambda n^\mu + \xi^{\perp\mu}, \tag{3.16}$$

we find in the Bjorken limit:

$$\lim_{Bj} q \cdot \xi = \eta\nu - x\lambda. \tag{3.17}$$

Arguments similar to those used in the previous section show that the integral in (3.15) is dominated by $\eta \sim 1/\nu \sim 0$ and $\lambda \sim 1/x$, which is equivalent to $\xi^- \sim 0$ and $\xi^+ \sim 1/xp$ respectively.[22, 23] As in the previous case the commutator in (3.15) vanishes unless $\xi^2 = 2\lambda\eta - \boldsymbol{\xi}_\perp^2 \geq 0$ because of causality. Combining these results we find that the Bjorken limit of DIS probes a current correlation function near the light-cone $\xi^2 = 0$, extending out to distances (ξ^3 and ξ^0) of order $\frac{1}{xp}$.

Deep Inelastic Scattering and the Short Distance Expansion QCD simplifies at short distances on account of asymptotic freedom. The analysis of $e^+e^- \to$ hadrons simplifies greatly for this reason. Deep inelastic scattering is not a short distance process; it is light-cone dominated. Nevertheless it can be related to the OPE and to short distances with considerable resulting simplification.

To show this we consider the so-called the Bjorken-Johnson-Low limit ($\lim_{BJL}$) [24],[25]. This is a somewhat old fashioned method, mostly supplanted by Wilson's operator product expansion. It has the virtue that the connection between measurable structure functions and local operators is extremely clear (via dispersion relations). Use of the BJL limit prevents one making mistakes in subtle cases.[26, 27] In the BJL limit one takes $\mathbf{q} = 0$ and $q^0 \to i\infty$, which yields $q^2 \to -\infty$ and $x \to -i\infty$. In the physical region x is restricted to be real and between 0 and 1. So the hadronic tensor $W_{\mu\nu}$ cannot be measured in the BJL limit. It is useful because 1) it *is* dominated by short distances, and 2) it can be related to $W_{\mu\nu}$ in the physical region through dispersion relations. Remember that $W_{\mu\nu}$ is the imaginary part of the forward, virtual Compton amplitude, $T_{\mu\nu}$, by the optical theorem,

$$T_{\mu\nu}(q^2,\nu) = i\int d^4\xi\, e^{iq\cdot\xi} \langle P|T\left(J_\mu(\xi)J_\nu(0)\right)|P\rangle\,, \tag{3.18}$$

For simplicity we suppress Lorentz indices and spin degrees of freedom for a while. Standard dispersion theory arguments show that $T(q^2,\nu)$ is an analytic function of ν at fixed spacelike q^2 with branch points on the real–ν axis at $\nu = \pm\frac{Q^2}{2M}$, the threshold for the elastic process $\gamma^* p \to \gamma^* p$. In Fig. 9 one can see the physical region of this process and the area of the BJL limit in the complex $\omega = 1/x$ plane. The physical cuts lie on the real axis from $\frac{1}{x} = \pm 1$ to $\pm\infty$. This means that $T(q^2, \frac{1}{x})$ is analytic within the unit circle about the origin. The BJL limit takes $\frac{1}{x}$ to zero along the imaginary axis. Thus $T(q^2,\nu)$ can be expanded in a Taylor series in $(\frac{1}{x})$ about the origin in the BJL-limit. The coefficients in the Taylor expansion can be obtained from the dispersion relation obeyed by T. First remember that the optical theorem relates the imaginary part of T to the hadronic tensor $W(q^2,\nu)$ in the physical region,

$$\mathrm{Im}T(q^2,\nu) = 4\pi W(q^2,\nu). \tag{3.19}$$

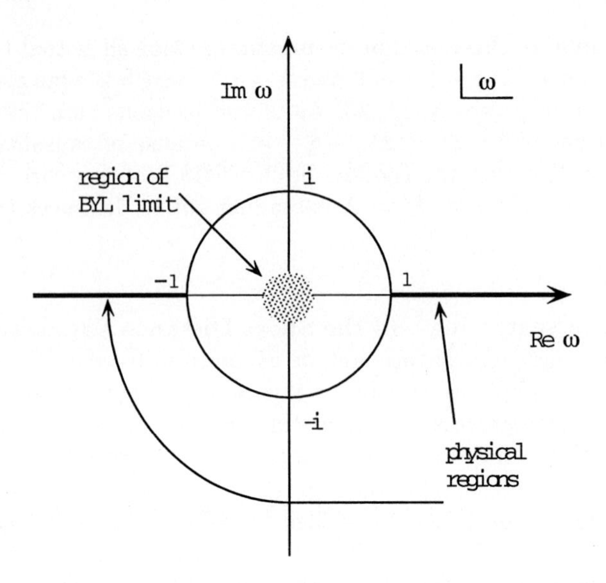

Fig. 9. *Physical region of the forward Compton amplitude and the BJL limit in the complex ω plane.*

Dispersion theory tells us that an analytic function can be represented in terms of its singularities in the complex plane,[28] in this case the physical cut on the real axis,

$$T(q^2,\omega) = 4\int_1^\infty d\omega'\,\omega'\,\frac{W(q^2,\omega')}{\omega'^2-\omega^2}. \tag{3.20}$$

Crossing, *i.e.* $T(\omega) = T(-\omega)$, has been used. Since $T(q^2,x)$ is analytic for $|\frac{1}{x}| < 1$ it may be expanded in a Taylor series in powers of $\frac{1}{x}$:

$$\lim_{\mathrm{BJL}} T(q^2,x) = 4\sum_{\mathrm{n\,even}} M^n(q^2)\,\frac{1}{x^n}, \tag{3.21}$$

with

$$M^n(q^2) = \int_0^1 dx x^{n-1} W(q^2,x). \tag{3.22}$$

Now consider where the BJL limit leads us in coordinate space. With $\mathbf{q}=0$ and $q^0 \to i\infty$, the factor $e^{iq\xi}$ in (3.18) reduces to $e^{-|q^0|\xi^0}$ and forces ξ^0 to zero. Although the time ordered product does not vanish outside the light-cone, it can be exchanged for a "retarded commutator",[25, 26] which does. Thus $\xi^0 \to 0$ forces $\xi^\mu \to 0$ and we conclude that the BJL-limit takes us to short distances where Wilson's operator product expansion may be used. The OPE analysis of the product of currents yields a power series in $\frac{1}{x}$ multiplying the matrix elements of local operators. Identifying terms in this Taylor series with the terms in (3.21) we obtain the celebrated "moment sum rules" relating integrals over deep inelastic structure functions to target matrix elements of local operators. We will not pursue this direction further here — it is treated in standard references.[29, 20]

3.3 $e^+e^- \to h\,X$– Once Again, the Light-Cone

Like deep inelastic scattering, single particle inclusive production in e^+e^- annihilation is dominated by the light-cone. However, the operator product expansion does not apply and no short distance analysis exists. The process is described by the tensor introduced in Sect. 1,

$$\bar{W}_{\mu\nu} = \frac{1}{4\pi}\int d^4\xi\, e^{iq\cdot\xi}\sum_X \langle 0|J_\mu(\xi)|PX\rangle\,\langle PX|J(0)_\nu|0\rangle \tag{3.23}$$

Once again, the nucleon and photon momenta may be expanded in terms of the light-like vectors introduced in Sect. 3.2,

$$\begin{aligned} P^\mu &= p^\mu + \frac{M^2}{2}n^\mu, \\ q^\mu &= \frac{1}{M^2}\left(\nu - \sqrt{\nu^2 - M^2Q^2}\right)p^\mu + \frac{1}{2}\left(\nu + \sqrt{\nu^2 - M^2Q^2}\right)n^\mu, \end{aligned} \tag{3.24}$$

and in the Bjorken limit ($Q^2, \nu \to \infty$ with z finite),

$$\lim_{Bj} q^\mu = \left(\nu - \frac{1}{2}\frac{M^2}{z}\right)n^\mu + \frac{1}{z}p^\mu. \tag{3.25}$$

It is traditional to use the photon rest frame ($\mathbf{q} = 0, p \sim \sqrt{\nu}$) to analyze the process. However, the label on the state, (P^μ) changes as the limit $Q^2 \to \infty$ is taken in this frame, making it difficult to sort out the important regions of the ξ-integration. Things are simpler in a frame where P is fixed, *e.g.* the rest frame of the produced hadron, where $p = \frac{M}{\sqrt{2}}$. In such a frame, the analysis of the fourier integral in (3.23) proceeds exactly in the same way as for the electroproduction process of Sect. 3.2. With

$$\xi^\mu = \eta p^\mu + \lambda n^\mu + \xi^{\mu\perp} \tag{3.26}$$

we find in the Bjorken limit

$$\lim_{Bj} q\cdot\xi = \eta\nu - \frac{\lambda}{z}. \tag{3.27}$$

So, $\nu \to \infty$ implies $\eta \to 0$ and $\lambda \sim z$, since z is finite. So light-like separations $\xi^\mu\xi_\mu \sim 0$ dominate again unless unusual variations occur in the matrix elements

$$\sum_X \langle 0|J(\xi)|PX\rangle\,\langle PX|J(0)|0\rangle\,, \tag{3.28}$$

which will not happen in the frame where P is independent of Q^2 and ν. Also the frequencies associated with the states in the sum $\sum_X |X\rangle\langle X|$ know nothing about Q^2 and ν, and will not spoil the argument. For a contrasting situation see the discussion of Drell-Yan in the following sub-section.

One can thus conclude that light-cone distances dominate fragmentation. However, in contrast to DIS the OPE cannot be applied here since the observed hadron state, $|P\rangle$ interferes with the attempt to complete the sum on X. Nevertheless nearly all of the QCD phenomenology developed for DIS can be carried over to this case, primarily using momentum space methods we will not discuss here.[30] In Sect. 6 we will see that the limitations on the coordinate space analysis do not prevent us from analyzing spin, twist and chirality in fragmentation.

3.4 $PP \to l^+l^-X$ – The Drell-Yan Process

Finally we consider the Drell-Yan process. Here the relevant hadronic tensor is (see Sect. 1):

$$W^{\mu\nu} = \frac{1}{2}s \int d^4\xi\, e^{iq\cdot\xi}\, \langle PP'|J^\mu(\xi)J^\nu(0)|PP'\rangle\,. \tag{3.29}$$

It is simplest to consider the case where only the dilepton invariant mass distribution $d\sigma/dq^2$ is measured, though other observables behave similarly. $\frac{d\sigma}{dq^2}$ depends only on W^μ_μ. We define a function, $W(s,Q^2)$ by integrating W^μ_μ over all q^μ with $q^2 = Q^2$ and $q^0 > 0$,

$$W(s,Q^2) = \frac{1}{(2\pi)^4}\int_R d^4q\, \delta(q^2 - Q^2)(-g^{\mu\nu}W_{\mu\nu})\theta(q^0), \tag{3.30}$$

$$= -4\pi^2 s \int_R d^4q\, \delta(q^2 - Q^2) \sum_X (2\pi^4)\delta^4(P + P' - q - X)$$

$$\times\, \langle PP'|J^\mu|X\rangle\, \langle X|J_\mu|PP'\rangle\,. \tag{3.31}$$

The virtual photon's momentum is integrated over all values consistent with the constraint $q^2 = Q^2$ and conservation of energy, $\sqrt{Q^2} < q_0 < (s + Q^2)/2\sqrt{s}$, which defines the region R. If we introduce the function

$$\Delta^R_+(\xi,Q^2) = \int_R \frac{d^4q}{(2\pi)^3}\, e^{-iq\cdot\xi}\delta(q^2 - Q^2)\theta(q_0), \tag{3.32}$$

then $W(s,Q^2)$ can be written as

$$W(s,Q^2) = -s \int d^4\xi\, \Delta^R_+(\xi,Q^2)\, \langle PP'|J^\mu(\xi)J_\mu(0)|PP'\rangle\,. \tag{3.33}$$

In the scaling limit ($Q^2, s \to \infty, \tau = Q^2/s$ fixed) $\Delta^R_+(\xi,Q^2)$ approaches a well-studied function of quantum field theory, the free field singular function, $\Delta_+(\xi,Q^2)$,[31],

$$\Delta_+(\xi,Q^2) = \frac{1}{4\pi}\epsilon(\xi_0)\delta(\xi^2) - \frac{mi}{8\pi\sqrt{\xi^2}}\theta(\xi^2)\left(N_1\left(Q\sqrt{\xi^2}\right) - i\epsilon(\xi_0)J_1\left(Q\sqrt{\xi^2}\right)\right)$$

$$+ \frac{mi}{4\pi^2\sqrt{-\xi^2}}\theta(-\xi^2)K_1\left(Q\sqrt{\xi^2}\right). \tag{3.34}$$

$\Delta_+(\xi, Q^2)$ is singular on the light-cone and would select out light-cone contributions were it not for high frequency variations in the matrix elements. These can occur because the hadron momenta P and P' cannot be kept fixed as $Q^2 \to \infty$ ($s \approx 2P \cdot P' > Q^2$) in any frame. Even in free field theory or the parton model the matrix element behaves like

$$\langle PP'|J^\mu(\xi)J_\mu(0)|PP'\rangle \sim \int d\alpha\, d\beta\, e^{i\alpha P\cdot\xi + i\beta P'\cdot\xi} f(\alpha,\beta), \tag{3.35}$$

where $f(\alpha, \beta)$ labels the momentum components of the partons that contribute to the current. To see that such variation can lead to contributions off the light-cone, consider a frame defined through the two vectors

$$\begin{aligned} p^\mu &= p\,(1,0,0,1), \\ p'^\mu &= \frac{s}{4p}\,(1,0,0,-1). \end{aligned} \tag{3.36}$$

The hadron and photon momenta can be written as

$$\begin{aligned} P^\mu &= p^\mu + \frac{M^2}{s} p'^\mu, \\ P'^\mu &= p'^\mu + \frac{M^2}{s} p^\mu, \\ q^\mu &= y p^\mu + x p'^\mu, \quad \text{for } |q_T| \ll \sqrt{Q^2}. \end{aligned} \tag{3.37}$$

With $\xi^\mu = \eta p^\mu + \lambda p'^\mu + \xi^\mu_\perp$ the hadronic tensor (3.29) is then equal to

$$W^\mu_\mu = \int d\lambda\, d\eta\, d^2\xi_T \int d\alpha\, d\beta\, f(\alpha,\beta) e^{i\frac{s}{2}[(\alpha-y)\lambda + (\beta-x)\eta]}. \tag{3.38}$$

Therefore the phases will cancel and the Drell-Yan process will escape from the light-cone if $\alpha \approx y = 2q \cdot P'/s$ and $\beta \approx x = 2q \cdot P/s$. In Sect. 5 we will return to this process and see that such phases are generated in a natural way.

3.5 Dominant and Subdominant Diagrams

Guided by our understanding of the regions of coordinate space important for various deep inelastic processes, we can return to the more familiar world of Feynman graphs and learn which diagrams are likely to give dominant and subdominant contributions. The quarks that couple to electroweak currents propagate according to $S_F(\xi)$, the Feynman propagator. In coordinate space, $S_F(\xi)$ behaves like $\frac{1}{\xi^3}$ at short distances (note $S_F(\xi) \sim \int d^4p \frac{e^{ip\xi}}{\gamma\cdot p - m}$), Interactions will not increase the singularity. For example, coupling a gluon to the propagating quark gives, Generally speaking in renormalizable field theories, interactions on propagating lines do not increase the order of the short-distance or light-cone singularity by more than logarithmic terms beyond free field theory. This can be used as a guideline to estimate the importance of different perturbative diagrams for hard processes.

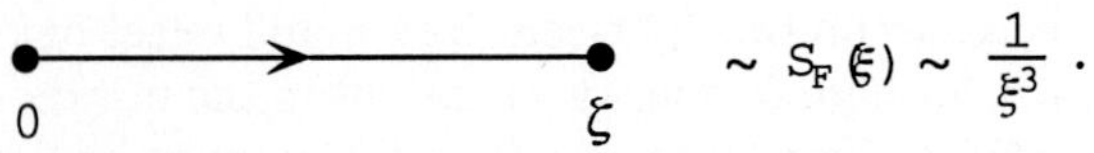

0 ζ $\sim S_F(\xi) \sim \frac{1}{\xi^3}\,.$

Fig. 10. *The Feynman Propagator.*

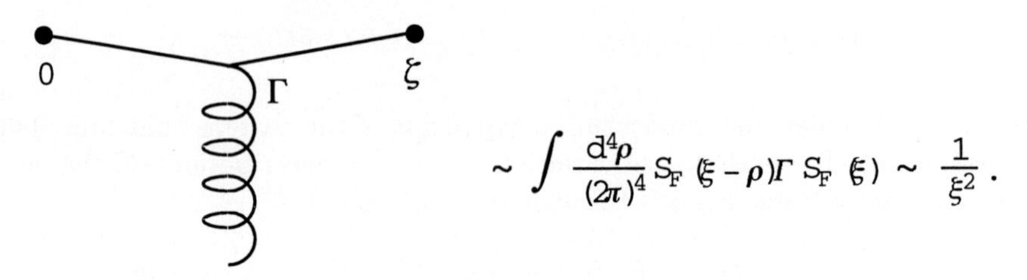

Fig. 11. *The coupling of a gluon to propagating quarks.*

As a first example, consider $e^+e^- \to$ hadrons. The total cross section is proportional to the vacuum polarization of the photon propagator, whose leading contribution results from the quark-antiquark loop fourier transformed, this

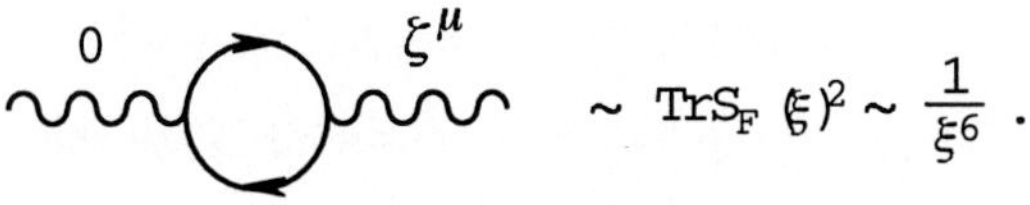

0 ζ^μ $\sim \mathrm{Tr} S_F(\xi)^2 \sim \frac{1}{\xi^6}\,.$

Fig. 12. *Quark–antiquark loop.*

$\mathcal{O}(\frac{1}{\xi^6})$ behavior generates a cross section which scales like $\frac{1}{Q^2}$. Radiative corrections introduce logarithmic dependence on ξ^2, $\ln \xi^2\mu^2$, where μ^2 is the renormalization point, but they do not change the power of the singularity in a renormalizable theory. The renormalization group may be used to sum classes of diagrams giving modifications of the $\frac{1}{\xi^6}$ behavior which go like powers of logarithms in an asymptotically free theory like QCD.

In deep inelastic scattering the leading contribution to the cross section or the forward Compton amplitude is shown in Fig. 13. It dominates because the free quark propagator has the greatest possible light-cone singularity. The modifications shown in Figs. 14-17 introduce only logarithmic modifications of the $\sim 1/\xi^3$ singularity. Renormalization group summation of leading $\ln \xi^2$ dependence leads to powers of $\ln \xi^2\mu^2$ but no change in the fundamental power singularity. All radiative corrections can be classified in the fashion outlined by Figs. 14-17. For single particle inclusive e^+e^- annihilation one finds in analogy to deep inelastic scattering the leading diagram Fig. 18 which has a singularity $1/\xi^3$ from a propagating quark. Radiative corrections can be treated as before. If light-cone dominance were the only consideration, the diagram in Fig. 19 would dominate

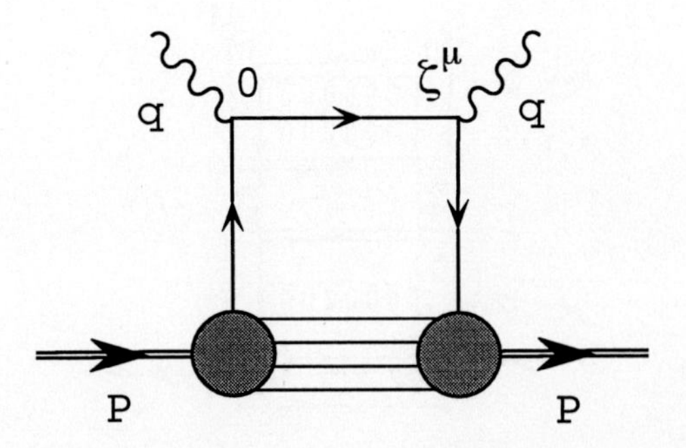

Fig. 13. *Leading diagram in deep inelastic scattering. The quark propagator between the two currents carries the large momentum* q^μ *and leads to a* $1/\xi^3$ *behavior at small distances.*

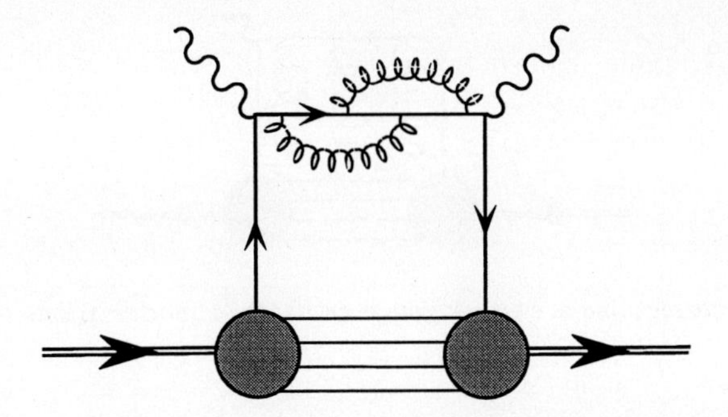

Fig. 14. *Radiative corrections to the quark propagator lead to* $O\left(\alpha_s(\xi^2)\right)$ *corrections to the coefficient of the leading* $1/\xi^3$ *term.*

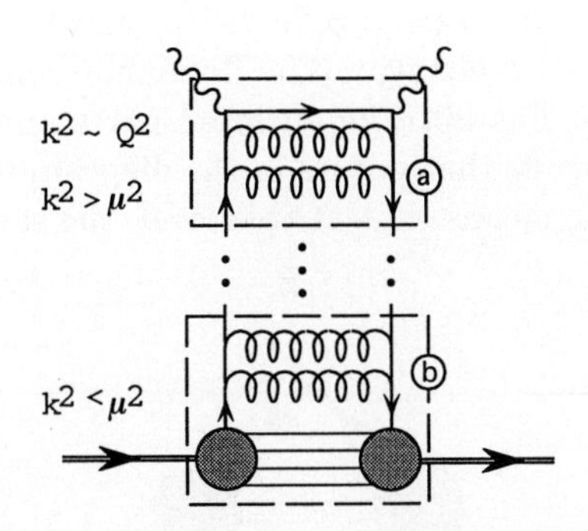

Fig. 15. *The higher order contributions in the upper part of the diagram (a), where the quark virtuality is greater than* μ^2 *lead to more* $O\left(\alpha_s(\xi^2)\ln\mu^2\xi^2\right)$ *corrections. The lower radiative corrections (b) can be absorbed into the quark-hadron amplitude which will then depend on the renormalization scale* μ^2.

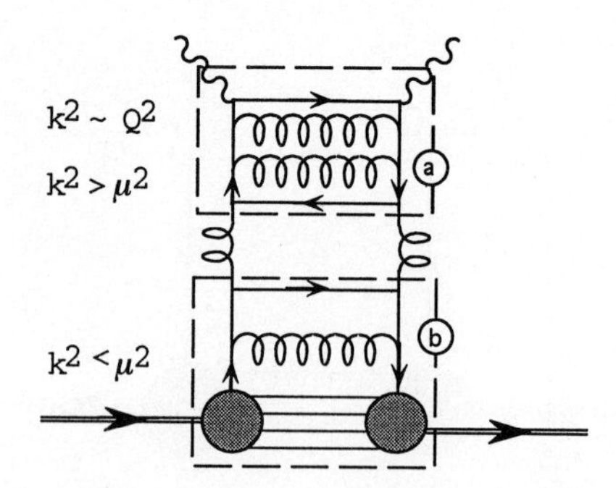

Fig. 16. *The upper part (a) here generates a c-number coefficient function for gluonic operators in the product of two currents, beginning at* $O\left(\alpha_s(\xi^2)\ln \mu^2\xi^2\right)$, *while the lower part (b) can be absorbed into a new gluon-hadron amplitude.*

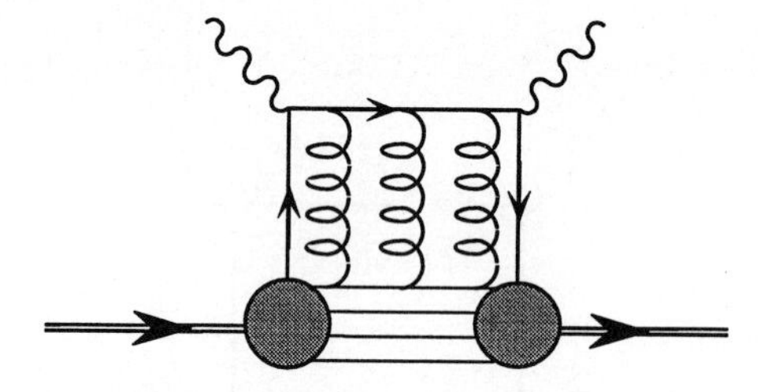

Fig. 17. *These corrections are either gauge artifacts or modifications of lower order in* $1/\xi$.

the Drell-Yan process. However, if one studies the flow of hard momentum, this diagram turns out to be suppressed. The quark which brehmsstrahlungs the massive photon must be far off-shell, which is unnatural in a hadron-hadron collision. In coordinate space this is reflected by the fact that no large phases are generated by the matrix element. The dominant contribution to the Drell-Yan process is shown in Fig. 20. The enclosed parts appear to be identical to the quark-hadron amplitude that occurs in the diagram that dominates in deep inelastic scattering. This means that at tree level, the same structure functions

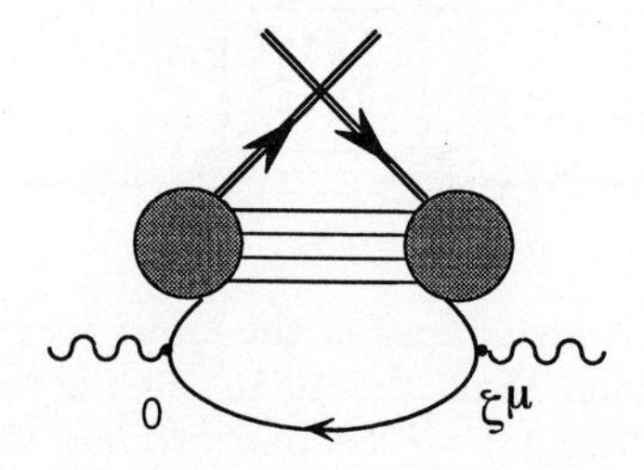

Fig. 18. *The dominant diagram for the single particle inclusive* e^+e^- *annihilation.*

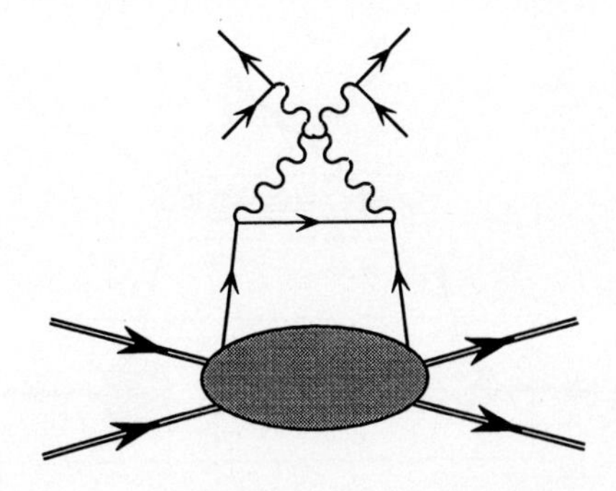

Fig. 19. *The diagram that has the leading light-cone singularity of the Drell-Yan process, but does not dominate.*

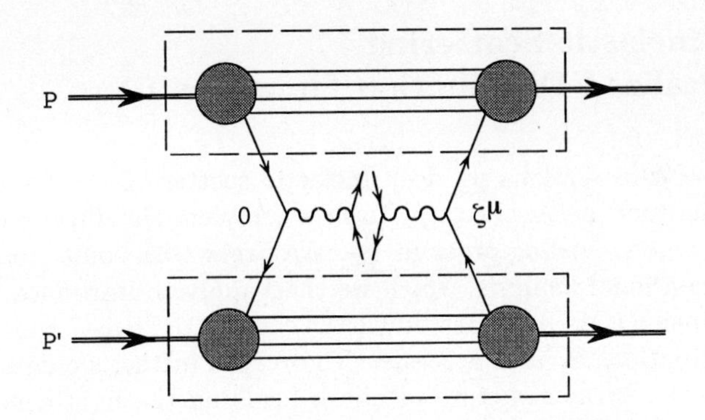

Fig. 20. *Dominant diagram of the Drell-Yan process.*

that appear in deep inelastic scattering also contribute to the Drell-Yan process. The diagram should still be dressed with QCD radiative corrections. The factorization theorem of QCD[32] says that this correspondence survives even in the presence of radiative corrections.

A subtlety of the Drell-Yan process is that the term most singular on the light-cone does not dominate, nevertheless the diagram gets its dominant contribution from $\xi^2 \approx 0$. Returning to the definition of $W(Q^2, s)$, we see that $\Delta_R(\xi, Q^2)$ forces ξ^2 to zero, but the phase factors generated by the two separate quark-hadron amplitudes select tangent planes to the light-cone that contribute to the ξ^+ and ξ^- integrals respectively.

One can now generalize these results to other processes, appealing to factorization. [32] For example Fig. 21 shows the dominant contribution to one particle inclusive deep inelastic scattering in the current fragmentation region. Building blocks of these calculations are (a) and (b), defined and measured in $e^+e^- \to$ hadrons and DIS respectively. Factorization allows them to be carried from one process to another. They are the fundamental objects of study in hard inclusive QCD and command our attention.

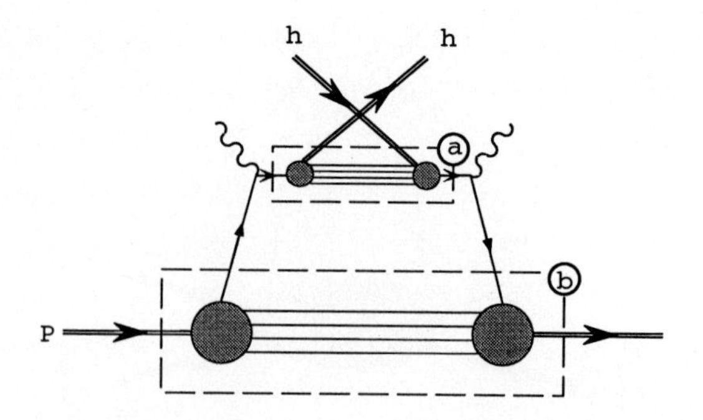

Fig. 21. *Dominant diagram for the one particle inclusive deep inelastic scattering.*

4 Deep Inelastic Scattering and Generalized Distribution Functions I

In this first of two sections on deep inelastic scattering the focus will be on developing the tools necessary to perform a complete classification of effects at leading and next-to-leading order in $\frac{1}{Q^2}$. We begin with some simple considerations of dimensional analysis, which we then apply to introduce the operator product expansion (OPE) and introduce the concept of "twist" which is useful to classify contributions to hard processes. To proceed further we must understand how to treat the Dirac structure of quark fields on the light-cone. This leads us briefly to explore light-cone quantization and introduce helicity, chirality and transversity as they apply to this problem. We will then look in some detail at a typical leading twist ($\mathcal{O}(1)$) and next-to-leading twist ($\mathcal{O}(\frac{1}{Q})$) distribution before attacking the complete problem in Sect. 4.

4.1 Twist

In Sect. 2 we introduced the operator product expansion (OPE) as a tool for analyzing $e^+e^- \to$ hadrons, where the operator with lowest dimension dominates. We also argued that light-like distances ($\xi^2 \sim 0$) dominate deep inelastic scattering. However, operators of high dimension can be important in this case. Instead a new quantum number, "twist", related to both the dimension and spin of an operator, orders the dominant effects.

Twist and the OPE As we learned in Sect. 2, the hadronic structure tensor of deep inelastic scattering,

$$4\pi W_{\mu\nu} = \int d^4\xi e^{iq\cdot\xi} \langle P, S|[J_\mu(\xi), J_\nu(0)]|P, S\rangle \tag{4.1}$$

is dominated by $\xi^2 \sim 0$ in the $Q^2 \to \infty$ limit. To make use of this, we expand the current commutator in terms of decreasing singularity around $\xi^2 = 0$,

$$[J(\xi), J(0)] \sim \sum_{[\theta]} K_{[\theta]}(\xi^2)\xi^{\mu_1} \dots \xi^{\mu_{n_\theta}} \theta_{\mu_1 \dots \mu_{n_\theta}}(0), \tag{4.2}$$

where $\theta_{\mu_1 \dots \mu_{n_\theta}}(0)$ are local operators, and $K_{[\theta]}(\xi^2)$ are singular c-number functions that can be ordered according to their degree of singularity at $\xi^2 = 0$. Operators of the same singularity at $\xi^2 = 0$ will be of the same importance as $\xi^\mu \xi_\mu \to 0$ even though numerator factors of ξ_μ render some less singular than others as $\xi_\mu \to 0$. For simplicity we have suppressed all labels, including spin, on the currents $J(\xi)$. It often convenient (and sometimes essential) to regroup the terms in (4.2) so that the operators $\theta_{\mu_1 \dots \mu_{n_\theta}}$ are traceless (*i.e.* $g^{\mu_1 \mu_2}\theta_{\mu_1 \dots \mu_{n_\theta}} = 0$, *etc.*) and symmetric in their Lorentz indices. We will assume this has been done.

Substituting the OPE into the definition of the structure function gives:

$$4\pi W = \int d^4\xi e^{iq\cdot\xi} \sum_{[\theta]} K_{[\theta]}(\xi^2)\xi^{\mu_1} \dots^{\mu_{n_\theta}} \langle P|\theta_{\mu_1 \dots \mu_{n_\theta}}(0)|P\rangle, \tag{4.3}$$

where the matrix elements have the form:

$$\langle P|\theta_{\mu_1 \dots \mu_{n_\theta}}(0)|P\rangle = P_{\mu_1} \dots P_{\mu_{n_\theta}} M^{d_\theta - n_\theta - 2} f_\theta + \dots . \tag{4.4}$$

The ... represent several types of terms which are less important in the Bjorken limit. We will return to them after looking at the dominant term.

Note that the power of a mass scale which appears in this expression is determined by dimensional analysis alone. We use the parameter M generically for a typical hadronic mass scale $M \sim \Lambda_{QCD} \sim R_{Bag}^{-1} \sim M_N/3$. The power with which M occurs defines the *twist* of the operator θ,

$$t_\theta \equiv d_\theta - n_\theta \tag{4.5}$$

The degree of the light-cone singularity of $K_{[\theta]} \sim \xi^{-6+t_\theta}$ is also determined by dimensional analysis and depends only on the twist, t_θ.

To carry out the fourier transformation make the substitution,

$$\xi^\mu \to -2iq^\mu \frac{\partial}{\partial q^2} \tag{4.6}$$

which yields,

$$4\pi W \sim \sum_{\{\theta\}} \left(\frac{M}{\sqrt{q^2}}\right)^{t_\theta - 2} \left(\frac{1}{x}\right)^{n_\theta} f_\theta \tag{4.7}$$

So the importance of an operator as $q^2 \to \infty$ is determined by its twist. As we shall see, it is typical for towers of operators with the same twist (and other quantum numbers such as flavor) and increasing spin to appear in the OPE. Then it is convenient to sum over spin — $\sum_{n_\theta} f_\theta \frac{1}{x}^{n_\theta} \equiv \tilde{f}_\theta(x)$ – where we now use the label θ to refer to the entire tower of operators.

The effect of radiative corrections is to introduce logarithmic dependence on Q^2 into the function $\tilde{f}_\theta$. Note however that the power law dependence on Q^2 is fixed by twist through dimensional analysis. Let us now return to the terms omitted in (4.4). These include terms like $P_{\mu_1} \ldots P_{\mu_{n-2}} g_{\mu_{n-1}\mu_n} M_N^2$ that make the expression traceless. It is easy to see that these contribute at most corrections of order $\frac{M_N^2}{Q^2}$ to the term we have kept. To carry through a complete analysis beyond order $\frac{1}{Q}$ it is necessary to keep careful track of these terms. This, and the fact that interesting spin effects appear at $\mathcal{O}(\frac{1}{Q})$, are the reasons we do not consider $\mathcal{O}(\frac{1}{Q^2})$ here.

The lowest twist operator towers in QCD have $t_\theta = 2$ and scale – modulo logarithms – in the Bjorken limit. This reflects the underlying scale invariance of the classical lagrangian. The matrix elements of higher twist operators, or the higher twist manifestations of twist-two operators are invariably signalled by the appearance of positive powers of mass in expressions analogous to (4.4). Dimensional analysis then forces compensating factors of large kinematic invariants in the denominator, suppressing the contribution. The simple conclusion is that *we can order the importance of effects in the deep inelastic limit simply by keeping track of masses we are forced to introduce into the numerators of parton-hadron amplitudes in order to maintain the correct dimensions.*

Examples and a Working Redefinition of Twist To make the preceding discussion clearer, here are some explicit examples from free field theories. These examples are not only pedagogical – the second one generates the leading twist effects in QCD up to logarithms. The light-cone singularities can be isolated easily. For the time ordered product of two scalar currents built from scalar fields, $J(\xi) =: \phi(\xi)\phi(\xi) :$, one can use Wick's theorem to show that,

$$T(J(\xi)J(0)) = -2\Delta_F^2(\xi) + 4i\Delta_F(\xi) : \phi(\xi)\phi(0) : + : \phi(\xi)\phi(\xi)\phi(0)\phi(0) :, \quad (4.8)$$

where the normal ordering operation is sufficient to render the operator products finite (in free field theory) as $\xi \to 0$, and

$$\Delta_F(\xi) = \frac{i}{4\pi^2}\frac{1}{\xi^2 - i\epsilon} \quad (4.9)$$

for a massless scalar field. To finally obtain the form of (4.2), simply Taylor expand the bilocal operators –

$$: \phi(\xi)\phi(0) := \sum_n \xi_{\mu_1} \ldots \xi_{\mu_n} \; : \{\partial^{\mu_1} \ldots \partial^{\mu_n} \phi(0)\}\phi(0) : \quad (4.10)$$

The current associated with a vector flavor symmetry of a fermion field is

$$J_\mu^a(\xi) =: \bar{\psi}(\xi)\frac{\lambda^a}{2}\gamma_\mu\psi(\xi) : . \quad (4.11)$$

Making use of the identity

$$[\bar{\psi}_1\psi_1, \bar{\psi}_2\psi_2] = \bar{\psi}_1\{\psi_1, \bar{\psi}_2\}\psi_2 - \bar{\psi}_2\{\psi_2, \bar{\psi}_1\}\psi_1 \tag{4.12}$$

(because $\{\psi_1, \psi_2\} = 0$ in free field theory), and, for a massless field,

$$\{\psi(\xi), \bar{\psi}(0)\} = \frac{1}{2\pi}\partial\!\!\!/ \epsilon(\xi_0)\delta(\xi^2), \tag{4.13}$$

one can now express the commutator of two currents in terms of bilocal operators [33]:

$$\begin{aligned}\left[J^{\mu a}(\xi), J^{\nu b}(0)\right] = -\frac{1}{4\pi}\left(\partial_\rho \epsilon(\xi_0)\delta(\xi^2)\right)\Big[&d^{abc}S^{\mu\rho\nu\alpha}A^c_\alpha(\xi,0) - id^{abc}\epsilon^{\mu\rho\nu\alpha}S^{5c}_\alpha(\xi,0)\\ &+ if^{abc}S^{\mu\rho\nu\alpha}S^c_\alpha(\xi,0) + f^{abc}\epsilon^{\mu\rho\nu\alpha}A^{5c}_\alpha(\xi,0)\Big],\end{aligned} \tag{4.14}$$

where the Lorentz structure is split into a symmetric and an antisymmetric part according to:

$$\begin{aligned}\gamma^\mu\gamma^\rho\gamma^\nu &= S^{\mu\rho\nu\alpha}\gamma_\alpha - i\epsilon^{\mu\rho\nu\alpha}\gamma_\alpha\gamma^5,\\ S_{\mu\rho\nu\alpha} &\equiv \frac{1}{4}\mathrm{Tr}\gamma_\mu\gamma_\rho\gamma_\nu\gamma_\alpha = g_{\mu\rho}g_{\nu\alpha} + g_{\mu\alpha}g_{\nu\rho} - g_{\mu\nu}g_{\alpha\rho},\end{aligned} \tag{4.15}$$

and the flavor structure is split in a similar way:

$$\lambda^a\lambda^b = \left(d^{abc}\lambda^c + if^{abc}\lambda^c\right). \tag{4.16}$$

The symmetric and anti-symmetric vector and axial *bilocal* currents are defined by,

$$S/A^{[5]c}_\alpha \equiv \bar{\psi}(\xi)\frac{\lambda^c}{2}\gamma_\alpha[\gamma^5]\psi(0) \pm \bar{\psi}(0)\frac{\lambda^c}{2}\gamma_\alpha[\gamma^5]\psi(\xi) \tag{4.17}$$

Once again the form of (4.2) is obtained by Taylor expanding the bilocal operators.

We have presented these formulas in their full complexity because they summarize the algebra of free quarks at short distances. All of the traditional results of the quark parton model applied to DIS (scaling relations, the Adler, Bjorken, Gross-Llewellyn Smith and other sum rules, the Callan Gross relation, *etc.*) can be obtained directly from these relations. [26]

The steps of first expanding the bilocal operators, then resumming the tower after fourier transformation are very inefficient. Clearly it should be possible to work directly with the bilocal operators. The twist content of a *bilocal operator* is somewhat more complicated than that of a local operator. Consider, for example, the bilocal current, $\bar{\psi}(0)\gamma^\mu\psi(\xi)$, which occurs in (4.14). The operator has dimension three and, were it a local operator, it would have spin-one. In fact it sums an infinite tower of operators of increasing spin and dimension, with $t \geq 2$. For example at short distance one can write:

$$\begin{aligned}\bar{\psi}(0)\gamma^\mu\psi(\xi) &= \bar{\psi}(0)\gamma^\mu\psi(0) + \xi_\nu\bar{\psi}(0)\gamma^\mu\partial^\nu\psi(0) + \dots\\ &\equiv J^\mu(0) + \xi_\nu\theta^{\mu\nu}(0) + \dots.\end{aligned} \tag{4.18}$$

$J^\mu(0)$ is traceless, symmetric and local and has twist-two. The operator $\theta^{\mu\nu}$ can be decomposed into a traceless operator and a "trace":

$$\theta^{\mu\nu} = \left\{\theta^{\mu\nu} - \frac{1}{4}g^{\mu\nu}\theta^\lambda_\lambda\right\} + \frac{1}{4}g^{\mu\nu}\theta^\lambda_\lambda. \tag{4.19}$$

The first term is traceless, symmetric, with twist-two. The second operator has spin-0, hence its twist is four. Further terms in the Taylor expansion of the *bilocal operator* each yield a tower of local operators beginning at twist-two and increasing in steps of two.

Up to now we have used *twist* only in the sense in which it was originally introduced — $t_\theta = d_\theta - n_\theta$. In practice, twist is used in a less formal way, to denote the order in $\frac{1}{Q^2}$ (modulo logarithms) at which a particular effect is seen in a particular experiment. If it behaves like $(1/Q^2)^p$, then the object of interest is said to have twist $t = 2 + 2p$. A traceless symmetric operator of twist t will generate contributions that go like $(1/Q^2)^{(2-t)}$, $(1/Q^2)^{(4-t)}\ldots$ as we saw explicitly for the operator $\theta_{\mu\nu}$. Although the two meanings of twist do not coincide perfectly, both are in common use.

We will make a definition of the *twist* of an *invariant matrix element* of a light-cone bilocal operators, that determines the scaling behavior of the matrix element. Matrix elements of operators like *e.g.* $\bar{\psi}(0)\gamma_\mu\psi(\lambda n)$ are the basic building blocks of the description of hard processes in QCD. So we will call "twist" the order in M/Q at which an operator matrix element contributes to deep inelastic processes. A few virtues of our working definition are a) that it is easily read off by inspection of matrix elements; b) that it directly corresponds to suppression in hard processes; and c) that effects we label twist-t never enter hard processes with suppression less than $(M/Q)^{t-2}$. The twist we associate with the invariant matrix element of a specific bilocal operator can be determined simply by considering the powers of mass which must be introduced to perform a Lorentz-tensor decomposition of the matrix element. The powers of mass carry through the entire calculation to the end where each power is compensated by a power of Q in the denominator. Twist-two results in no suppression, therefore $t-2$ is to be associated with the number of powers of mass introduced in the tensor decomposition of a matrix element.

The method is best explained by example. Consider the spin averaged matrix element of the bilocal current, $\bar{\psi}(\lambda n)\gamma^\mu\psi(0)$ on the light-cone,

$$\langle P|\bar{\psi}(\lambda n)\gamma^\mu\psi(0)|P\rangle = p^\mu f_1(\lambda) + n^\mu M^2 f_2(\lambda), \tag{4.20}$$

where the factor of M^2 must be introduced because $[n^\mu] = -1$. The twist of the first term is two but, due to the appearance of the factor M^2, the twist of the second term is four. In a physical application we assert that the factor of M^2 will survive all manipulations and appear in the result compensated dimensionally by a factor of $1/Q^2$. Note that it *is* possible for f_1 to pick up multiplicative factors of M^2/Q^2 during a calculation. Twist tells us the leading, not the exclusive, Q^2

dependence of an invariant piece of a light-cone bilocal operator. As a second example, consider

$$\langle P|\bar{\psi}(\lambda n)\psi(0)|P\rangle = Me(\lambda). \tag{4.21}$$

$e(\lambda)$ has twist-*three* due to the factor M which must be introduced to preserve dimensions. Finally, consider the matrix element of a gluonic operator

$$\langle P|G_{\mu}{}^{\alpha}(\lambda n)G_{\alpha\nu}(0)|P\rangle = p_\mu p_\nu f_1(\lambda) + (p_\mu n_\nu + p_\nu n_\mu)f_2(\lambda)M^2 + n_\mu n_\nu f_3(\lambda)M^4, \tag{4.22}$$

which has a twist content that can be worked out by the reader.

Spin and Twist Counting twist in the case of polarized targets (or fragments) has an added complication. The Lorentz tensors which describe a hadron's spin can appear in the Lorentz decomposition of matrix elements — their role in determining twist must be explained. The objects of interest in polarized scattering (or fragmentation) are forward scattering matrix elements on a null plane: $\langle P,\epsilon|\theta(\lambda n, 0)|P,\epsilon\rangle$. The matrix elements are bilinear in ϵ and ϵ^*, where ϵ and ϵ^* are the generalized spinors describing the target (Dirac spinors for spin 1/2, polarization vectors for spin 1, *etc.*). The matrix element is a tensor function of ϵ and ϵ^*. For spin-1/2 the only (non-trivial) tensors which can be built from $u \times \bar{u}$ are $\bar{u}\gamma^\mu u = 2P^\mu$, a vector, and $\bar{u}\gamma^\mu\gamma_5 u = 2S^\mu$, an axial vector. We have already analyzed P^μ (it gets decomposed into p^μ and n^μ). To expose the twist content of terms proportional to S^μ, express it in terms of p^μ and n^μ:

$$S^\mu = (S\cdot n)p^\mu + (S\cdot p)n^\mu + S_\perp^\mu. \tag{4.23}$$

Since $S\cdot p = -M^2 S\cdot n/2$, it is clear that the second term contributes at twist-four. The transverse spin term is more subtle. Because we have chosen to normalize $S^2 = -M^2$, $[S_\perp] = 1$ and because there are no transverse momenta in the problem, $S_\perp^\mu$ contains a hidden factor of the target mass. At the end of the day this factor will manifest itself in a suppression by $\frac{M}{Q}$. So we conclude that appearances of $S_\perp$ accompany *twist-three* distributions. An example is provided by:

$$\langle PS|\bar{\psi}(0)\gamma^\mu\gamma_5\psi(\lambda n)|PS\rangle = S\cdot n\; p^\mu g_1(\lambda) + S_\perp^\mu\; g_T(\lambda) + S\cdot p\; n^\mu g_3(\lambda). \tag{4.24}$$

According to dimensional analysis, g_1 is a twist-two object, g_T has twist-three and g_3 is a twist-four function. When combined with the analysis of the following section, one finds that the function we have labeled g_1 is the scaling limit of the "g_1" defined in Sect. 1. Similarly, g_T turns out to be $g_1 + g_2$. We discard g_3 because we are not concerned with twist-four.

The same method of analysis can be extended to higher spins. For a spin-1 target, all polarization information is contained in the spin-density matrix $\eta_{\mu\nu} \equiv \epsilon_\mu \epsilon_\nu^*$, which contains scalar $(\epsilon^*\cdot\epsilon)$, vector $(S_\mu \equiv \frac{i}{M^2}\epsilon_{\mu\nu\alpha\beta}P^\nu\epsilon^{*\alpha}\epsilon^\beta)$, and tensor $(\hat{\eta}_{\mu\nu} \equiv \epsilon_\mu^*\epsilon_\nu + \epsilon_\nu^*\epsilon_\mu + g_{\mu\nu}\frac{M^2}{2})$ polarization information. Note $[S^\mu] = 1$ and $[\hat{\eta}_{\mu\nu}] = 2$. To determine the twist of the associated distributions, η must be projected along p^μ, n^μ and transverse directions.[34] For even higher spins a multipole analysis is more streamlined.[35]

4.2 Dominant Diagram in Coordinate Space

As a final, and physically important example, we take the dominant diagram identified in Sect. 2 and use coordinate space methods to compute it. Since the quark that propagates between currents suffers no interactions (we are ignoring gluon radiative corrections here), we may use free field theory. Working out the commutator of free currents, we get

$$\begin{aligned} W^{ab}_{\mu\nu} &= \frac{1}{4\pi}\int d^4\xi e^{iq\cdot\xi}\langle P,S|[J^a_\mu(\xi),J^b_\nu(0)]|P,S\rangle \\ &= -\left(\frac{1}{4\pi}\right)^2\int d^4\xi e^{iq\cdot\xi}\partial^\rho(\delta(\xi^2)\epsilon(\xi^0))\left\{S_{\mu\rho\nu\alpha}d^{abc}\langle PS|A^{c\alpha}(\xi,0)|PS\rangle+\ldots\right\}, \end{aligned} \tag{4.25}$$

which corresponds to the handbag diagram of Fig. 13. The ... represent three more terms, given in (4.14). This simple free-field picture is modified by:

- vertex and self energy corrections, which modify the singular function (Fig. 14). They give rise to logarithmic corrections, as do the dominant parts of
- ladder graphs (Fig. 15), and
- box graphs, which mix in gluons at $\mathcal{O}(\alpha_s)$ (Fig. 16). Finally,
- in order to preserve color gauge invariance, one has to remember that the quark propagates in a gluon background (Fig. 17).

On account of the last point, the singular function of free field theory, $\{\psi(\xi),\bar\psi(0)\}=\frac{1}{2\pi}\partial\!\!\!/\epsilon(\xi_0)\delta(\xi^2)$, must be changed to

$$\{\psi(\xi),\bar\psi(0)\}\to\frac{1}{2\pi}\partial\!\!\!/\epsilon(\xi_0)\delta(\xi^2)\mathcal{P}\left(\exp i\int_0^\xi d\zeta^\mu A_\mu(\zeta)\right), \tag{4.26}$$

which is the quark propagator in a background gluon field. [The path ordering ($\mathcal{P}$) is necessary because $A_\mu(\zeta)$ is a matrix in color space.] The color field A^μ is that generated by remnants of the target nucleon and must be viewed as an operator sandwiched between the target hadron states. The bilocal operators in (4.25) are therefore replaced by,

$$\bar\psi(\xi)\Gamma\psi(0)\to\bar\psi(\xi)\mathcal{P}\left(\exp i\int_0^\xi d\zeta^\mu A_\mu(\zeta)\right)\Gamma\psi(0), \tag{4.27}$$

where Γ stands for whatever color/flavor/Dirac matrices appear between $\bar\psi$ and ψ. The δ-function in (4.25) selects the light-cone. If we expand ξ^μ about the null plane, $\xi^\mu=\lambda n^\mu+\hat\xi^\mu$, it is easy to see that the terms involving $\hat\xi^\mu$ are twist-four and higher. One therefore has:

$$\begin{aligned} &\langle PS|\bar\psi(0)\mathcal{P}\left(\exp -i\int_0^{\xi^\mu} d\zeta^\mu A_\mu(\zeta)\right)\psi(\xi)|PS\rangle = \\ &\langle PS|\bar\psi(0)\mathcal{P}\left(\exp -i\int_0^\lambda d\tau n\cdot A(\tau n)\right)\psi(\lambda n)|PS\rangle+\ldots, \end{aligned} \tag{4.28}$$

where the ... represent the parts that vanish on the light-cone and have a twist ≥ 4. In the light-cone gauge $n \cdot A = 0$, explicit reference to gluons disappears. However, the inclusion of the "Wilson link", $\mathcal{P}\left(\exp i \int_0^\xi d\zeta^\mu A_\mu(\zeta)\right)$, is essential in generating higher twist ($t \geq 4$) gluon corrections.

In the unpolarized case, the twist expansion of the bilocal operator matrix element gives

$$\int \frac{d\lambda}{2\pi} e^{i\lambda x} \langle P|\bar{\psi}_a(0)\gamma_\mu \psi_a(\lambda n)|P\rangle \equiv 2 f_{1a}(x) p_\mu + 2M^2 f_{4a}(x) n_\mu \tag{4.29}$$

and, carrying out the fourier transform in (4.25), we find

$$F_1(x) = \frac{1}{2} \sum_a e_a^2 (f_1^a(x) - f_1^a(-x)), \tag{4.30}$$

where $a = u, d, s, \ldots$ is the flavor index. The interpretation in terms of the parton model will be given later in this section.

A brief summary to this point is: Up to and including twist-three the basic objects of analysis in DIS are forward matrix elements of bilocal products of fields on the light-cone and in light-cone gauge,

$$\Gamma(x) = \int \frac{d\lambda}{2\pi} e^{i\lambda x} \langle PS|\bar{\psi}(0)\Gamma\psi(\lambda n)|PS\rangle. \tag{4.31}$$

Remember, that important $\ln Q^2$ radiative corrections have been ignored in pursuit of the twist and spin dependence.

4.3 Learning from Light-Cone Quantization

Since the dominant contribution to DIS comes from the light-cone, it is natural to consider a dynamical formulation in which the light-cone plays a special role. At the birth of deep inelastic physics it was recognized that field theories simplify in some important ways if they are quantized "on the light-cone" rather than at equal times.[36, 37] Unfortunately some features which are simple at equal times become difficult on the light-cone. Certainly, as we shall see, there is much insight to be gained by considering deep inelastic processes using light-cone quantization. The larger question – whether QCD simplifies in essential ways when quantized on the light-cone – will not be pursued here.

Field theories may be quantized by imposing canonical equal-time commutation (or anticommutation) relations on the dynamically independent fields.[20] Lorentz invariance requires that any other space-like hyperplane in Minkowski space would serve as well as $\xi^0 = 0$. A null-plane, such as $\xi \cdot n = 0$ is the limit of a sequence of space-like surfaces, and includes points that are causally connected. Although a field theory quantized on at $\xi \cdot n = 0$ could differ from one quantized at $\xi^0 = 0$, they coincide for all examples of which I am aware. Let us study what happens if we attempt to quantize fermions on the surface $\xi^+ = 0$.[38] First we must introduce and familiarize ourselves with the unusual kinematics of the light-cone.

Light-Cone Kinematics We have previously introduced light-cone coordinates $\xi^\pm = \frac{1}{\sqrt{2}}(\xi^0 \pm \xi^3)$ and $\boldsymbol{\xi}^\perp = (\xi^1, \xi^2)$, and the metric $g_{\mu\nu}$, with $g_{+-} = g_{-+} = 1$, and $g_{ij} = -\delta_{ij}$. The partially off-diagonal structure of g makes raising and lowering indices confusing, *viz.*, $a^+ = a_-$, and so forth. So we work with upper (contravariant) indices as much as possible.

Quantizing at (say) $\xi^+ = 0$, we are committed to ξ^+ as our evolution variable (just as quantization at $\xi^0 = 0$ fixes ξ^0 as the "time"). ξ^- and $\xi^\perp$ are therefore kinematic, not dynamical variables. The conjugate momenta p^+ and $\mathbf{p}^\perp$ parameterize the fourier decomposition of the independent light-cone fields, just like $\mathbf{p}$ in ordinary quantization. p^- is the "Hamiltonian" for light-cone dynamics.

Dirac Algebra on the Light-Cone The usual selection of $\gamma^0 = \text{diag}(1, 1, -1-1)$ is prejudiced toward equal time quantization. Then a (anti-) particle at rest has only ("lower") "upper" components in its Dirac spinor. Much of our analysis is simplified by choosing a representation for the Dirac matrices tailored to the light-cone.[38] To represent Dirac matrices compactly, we use the "bispinor" notation: let $(\sigma^1, \sigma^2, \sigma^3)$ and (ρ^1, ρ^2, ρ^3) be two copies of the standard (2×2) Pauli matrices. A 4×4 Dirac matrix can be represented as $\rho^i \otimes \sigma^j$. ρ controls the upper-versus-lower two-component space; σ controls the inner two-component space. An example will clarify the notation: the Dirac-Pauli representation used, for example, by Bjorken and Drell is,

$$\begin{aligned}
\gamma^0_{BD} &= \rho^3 \otimes 1 = \begin{pmatrix} 1 & 0 \\ 0 & -1 \end{pmatrix} \\
\gamma^j_{BD} &= i\rho^2 \otimes \sigma^j = \begin{pmatrix} 0 & \sigma^j \\ -\sigma^j & 0 \end{pmatrix} \\
\gamma^5_{BD} &= \rho^1 \otimes 1 = \begin{pmatrix} 0 & 1 \\ 1 & 0 \end{pmatrix}
\end{aligned} \tag{4.32}$$

The light-cone representation useful for us is instead,

$$\begin{aligned}
\gamma^0 &= \rho^1 \otimes \sigma^3 = \begin{pmatrix} 0 & \sigma^3 \\ \sigma^3 & 0 \end{pmatrix} \\
\boldsymbol{\gamma}^\perp &= 1 \otimes i\boldsymbol{\sigma}^\perp = \begin{pmatrix} i\boldsymbol{\sigma}^\perp & 0 \\ 0 & i\boldsymbol{\sigma}^\perp \end{pmatrix} \\
\gamma^3 &= -i\rho^2 \otimes \sigma^3 = \begin{pmatrix} 0 & -\sigma^3 \\ \sigma^3 & 0 \end{pmatrix} \\
\gamma^5 &= \rho^3 \otimes \sigma^3 = \begin{pmatrix} \sigma^3 & 0 \\ 0 & -\sigma^3 \end{pmatrix},
\end{aligned} \tag{4.33}$$

where $\perp = 1$ or 2. It is easy to check that (4.34) satisfy the usual algebra, $\{\gamma^\mu, \gamma^\nu\} = 2g^{\mu\nu}$, and $\gamma^5 = i\gamma^0\gamma^1\gamma^2\gamma^3$.

Operators which project on the upper and lower two component subspaces play a central role in light-cone dynamics. Define $\mathcal{P}_\pm$ by,

$$\mathcal{P}_\pm = \frac{1}{2}\gamma^\mp\gamma^\pm = \frac{1}{2}(1 \pm \alpha_3)$$
$$\gamma^\pm = \frac{1}{\sqrt{2}}(\gamma^0 \pm \gamma^3),$$

with the properties:

$$\mathcal{P}_-\mathcal{P}_+ = \mathcal{P}_+\mathcal{P}_- = 0$$
$$\mathcal{P}_\pm^2 = \mathcal{P}_\pm$$
$$\mathcal{P}_- + \mathcal{P}_+ = 1$$
$$\mathcal{P}_+ = \begin{pmatrix} 1 & 0 \\ 0 & 0 \end{pmatrix} \quad \mathcal{P}_- = \begin{pmatrix} 0 & 0 \\ 0 & 1 \end{pmatrix}$$

The "light-cone projections" of the Dirac field, $\psi_+ \equiv \mathcal{P}_+\psi$ and $\psi_- \equiv \mathcal{P}_-\psi$ are known as the "good" and "bad" light-cone components of ψ respectively. To save on subscripts we shall frequently replace $\psi_\pm$ as follows,

$$\psi_+ \Rightarrow \phi \quad \psi_- \Rightarrow \chi \tag{4.34}$$

Independent Degrees of Freedom The importance of $\mathcal{P}_\pm$ becomes clear when they are used to project the Dirac equation down to two two-component equations,

$$i\gamma^- D_- \chi = -\gamma^\perp \cdot \mathbf{D}^\perp \phi + m\phi$$
$$i\gamma^+ D_+ \phi = -\gamma^\perp \cdot \mathbf{D}^\perp \chi + m\chi, \tag{4.35}$$

where $D_\pm = \frac{\partial}{\partial \xi^\pm} - igA^\mp$. In the light-cone gauge $A^+ = 0$. ξ^+ is the evolution ("time") parameter, but the first of (4.35) only involves $\partial/\partial\xi^-$, so it appears that χ is not an independent dynamical field. Instead the Dirac equation constrains χ in terms of ϕ and $\mathbf{A}_\perp$ at fixed ξ^+,

$$i\gamma^- \frac{\partial}{\partial \xi^-}\chi = -\gamma^\perp \cdot \mathbf{D}^\perp \phi + m\phi \tag{4.36}$$

The longitudinal component of the electric field in electrodynamic is similarly constrained (*i.e.* determined at any time) by Gauss's Law in Coulomb gauge, $\boldsymbol{\nabla} \cdot \mathbf{E} = \rho$. Study of the gluon equations of motion indicates that A^- is also a constrained variable. The independent fields are therefore ϕ and $\mathbf{A}_\perp$. χ should be regarded as composite — as specified by (4.36) — $\chi = \mathcal{F}[\phi, \mathbf{A}_\perp]$.

By the way, the reduction of the four-component Dirac field to two propagating degrees of freedom is not unique to light-cone quantization. In the usual treatment of the Dirac equation one finds only two solutions for each energy and momentum, corresponding to the two spin states of a spin-1/2 particle. The two-degrees of freedom corresponding to the antiparticle are found in the solution with energy $-E$ and momentum $-\mathbf{p}$. In fact, the Dirac equation in momentum space is literally written in the form of a projection operation, $\Lambda_-\psi = 0$, where $\Lambda_\pm = \frac{1}{2m}(\not p \pm m)$ projects out two of the four components of the Dirac spinor.

Although the complete quantization of QCD requires much more work, the implication for the Dirac field is already clear: the *good* components should be regarded as independent propagating degrees of freedom; the *bad* components are dependent fields – actually quark-gluon composites.

The classification of quark spin states depends on the Dirac matrices which a) commute with $\mathcal{P}_\pm$ and b) commute with one-another. Returning to (4.34) we see that γ^1, γ^2, and γ_5 commute with $\mathcal{P}_\pm$. Furthermore, the component of the generator of spin-rotations along the $\hat{e}_3$-direction,

$$\Sigma^3 \equiv \frac{i}{2}[\gamma^1, \gamma^2] = \begin{pmatrix} \sigma^3 & 0 \\ 0 & \sigma^3 \end{pmatrix}, \tag{4.37}$$

also commutes with $\mathcal{P}_\pm$. Note that for a Dirac particle with momentum in the $\hat{e}_3$-direction, Σ^3 measures the *helicity*. This set of operators suggests two different maximal sets of commuting observables:

- Diagonalize γ_5 and Σ^3 — a *chirality* or *helicity basis*, or
- Diagonalize γ^1 (or equivalently, γ^2) – a *transversity basis*.

Let us consider these in turn –

Helicity Basis In the helicity basis, both the good and bad components of ψ carry *helicity* labels – the eigenvalues of Σ^3,

$$\psi = \begin{pmatrix} \phi_+ \\ \phi_- \\ \chi_+ \\ \chi_- \end{pmatrix}. \tag{4.38}$$

Note that upper and lower components of ψ correspond to *good* and *bad* light-cone components respectively. Referring back to the form of γ_5, (4.34), we see that helicity and chirality are identical for the good components of ψ but opposite for the bad components,

$$\gamma_5\psi = \begin{pmatrix} +\phi_+ \\ -\phi_- \\ -\chi_+ \\ +\chi_- \end{pmatrix}. \tag{4.39}$$

This may look strange at first, but it follows immediately from the composite nature of χ. A quantum of χ with positive helicity is actually a composite of a transverse gluon and a quantum of ϕ. Since the gluon carries helicity-one (but no chirality), angular momentum conservation requires that the ϕ-quantum have negative helicity and therefore negative chirality. Remembering this association will help sort out the chirality and helicity selection rules which appear in the following sections.

Transversity Basis Alternatively, we can diagonalize one of the transverse γ-matrices, to be specific, γ^1. We define eigenstates of the transverse spin-projection operators, $\mathcal{Q}_\pm = \frac{1}{2}(1 \mp \gamma_5\gamma^1)$, (which commute with $\mathcal{P}_\pm$),

$$\mathcal{Q}_+\phi \equiv \phi_\perp \tag{4.40}$$

$$\mathcal{Q}_-\phi \equiv \phi_\top, \tag{4.41}$$

and similarly for χ. Of course $\phi_{\top/\perp}$ are linear combinations of $\phi_\pm$. Note however that $\phi_{\top/\perp}$ are *not* eigenstates of the transverse spin operator $\Sigma^1 = \begin{pmatrix} 0 & i\sigma^1 \\ i\sigma^1 & 0 \end{pmatrix}$, which is not diagonal in the basis of good and bad components of ψ. So we have to be careful that we do not carelessly confuse *transversity*, the quantum number associated with $\mathcal{Q}_\pm$, which is simple in this picture, with *transverse spin*, which is not.

4.4 The Parton Model

Following the path we are on, the parton model is merely the light-cone Fock space decomposition of the matrix elements which control hard processes. Since we have both the matrix elements and the Fock space in hand, it is straightforward to construct the parton model. We will verify that the parton interpretation emerges as expected for twist-two and then explore twist-three. The reader should beware that twist-four is considerably more complicated. A parton model picture of twist-four does exist, however much work is required to make it obvious.[39, 19, 40]

The Fock space in the two bases can be constructed by defining operators that create the appropriate components of ϕ. In the helicity basis we define $R^\dagger(k^+, \mathbf{k}_\perp)$ to create a right-handed (positive *helicity*) component of ϕ and $L^\dagger(k^+, \mathbf{k}_\perp)$ to create a left-handed (negative helicity) component of ϕ, and $R_c^\dagger$ and $L_c^\dagger$, which do the same for the antiparticle field ϕ_c. In the transversity basis we define the operators $\alpha^\dagger(k^+, \mathbf{k}_\perp)$ and $\beta^\dagger(k^+, \mathbf{k}_\perp)$ that create the $\perp$ and $\top$ components of ϕ, respectively.

Twist-Two We begin with the simplest case – the spin average, twist-two deep inelastic scattering which is controlled by the bilocal operator defined in (4.29),

$$\int \frac{d\lambda}{2\pi} e^{i\lambda x} \langle PS|\bar{\psi}(0)\gamma_\mu\psi(\lambda n)|PS\rangle = 2\left\{f_1(x)p_\mu + f_4(x)n_\mu\right\}. \tag{4.42}$$

We project out the twist-two part, f_1, by contracting with n^ν.

$$\begin{aligned} f_1(x) &= \frac{1}{2}\int \frac{d\lambda}{2\pi} e^{i\lambda x} \langle PS|\bar{\psi}(0)\not{n}\psi(\lambda n)|PS\rangle \\ &= \frac{1}{\sqrt{2}p^+}\int \frac{d\lambda}{2\pi} e^{i\lambda x} \langle PS|\phi^\dagger(0)\phi(\lambda n)|PS\rangle. \end{aligned} \tag{4.43}$$

where we have used the Dirac algebra to express the quark field in terms of its light-cone components. Notice that only the dynamically independent "good" light-cone components occur. If we make a momentum $(k^+, \mathbf{k}_\perp)$ decomposition of ϕ and separate helicity states, we find,

$$f_1(x) = \frac{1}{x}\int d^2k_\perp \langle P|R^\dagger(xp, \mathbf{k}_\perp)R(xp, \mathbf{k}_\perp) + L^\dagger(xp, \mathbf{k}_\perp)L(xp, \mathbf{k}_\perp)|P\rangle, \quad (4.44)$$

This *is* the parton model as illustrated in Fig. 22: f_1 is expressed as a sum

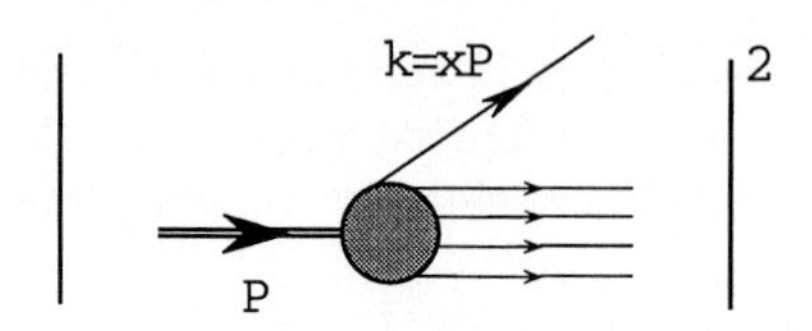

Fig. 22. *The classical parton model.*

of probabilities to find a (light-cone quantized) quark with $p^+ = xp$ and any transverse momentum, summed over helicities and weighted by the phase space factor $1/x$. Perhaps the reader is more familiar with the "infinite momentum frame" form of the model, where f_1 is written as the sum of probabilities to find an (equal-time quantized) quark with a fraction x of the target's (infinite) longitudinal momentum. The two formulations are equivalent since the boost to an infinite momentum frame is equivalent to a light-cone formulation. Since (4.44) is valid in any frame, it can be used in (*e.g.*) the lab frame to provide parton distributions which can be associated with quark models.[18] One must, however, be careful to remember that the fields in (4.44) are good light-cone Dirac components quantized at equal ξ^+, not equal ξ^0.

An identical calculation for $x < 0$ captures antiquark operators and leads to the standard crossing relation for f_1,

$$f_1(x) = -\bar{f}_1(-x) \quad (4.45)$$

where $\bar{f}_1$ is given by (4.44) with $R \to R_c$ and $L \to L_c$. The other (spin-dependent) quark distributions are explored in the following Section.

Twist-Three Now let us apply the same analysis to the simplest twist-three distribution function,

$$\int \frac{d\lambda}{2\pi} e^{i\lambda x}\langle P|\bar{\psi}(0)\psi(\lambda n)|P\rangle \equiv 2Me(x). \quad (4.46)$$

Decomposing in terms of ϕ and χ, we find

$$e(x) = \frac{1}{2M}\int \frac{d\lambda}{2\pi} e^{i\lambda x}\langle P|\phi^\dagger(0)\gamma^0\chi(\lambda n) + \chi^\dagger(0)\gamma^0\phi(\lambda n)|P\rangle, \quad (4.47)$$

which contains the dynamically dependent operator χ. If we use the constraint to eliminate χ we obtain

$$e(x) = -\frac{1}{4Mx}\int \frac{d\lambda}{2\pi} e^{i\lambda x} \langle P|\bar{\phi}(0) \not{n} \not{D}_\perp(\lambda n)\phi(\lambda n)|P\rangle + \text{h.c.} \tag{4.48}$$

So $e(x)$ is really a quark-*gluon* correlation function on the light-cone. It has no simple Fock-space interpretation in terms of quarks alone, despite the apparently simple form of (4.46).

We have happened upon a general (and very useful) result: Every factor of χ in the light-cone decomposition of a light-cone correlation function contributes an additional unit of twist to the associated distribution function,

$$\begin{aligned} \phi^\dagger\phi &\Leftrightarrow \text{Twist-2} \\ \phi^\dagger\chi &\Leftrightarrow \text{Twist-3} \\ \chi^\dagger\chi &\Leftrightarrow \text{Twist-4.} \end{aligned}$$

Likewise each unit of twist introduces an additional independent field in the null plane correlator:

$$\begin{aligned} &\text{Twist-2} \rightarrow \phi^\dagger(0)\phi(\lambda n) \\ &\text{Twist-3} \rightarrow \phi^\dagger(0)D_\perp(\lambda n)\phi(\mu n) \\ &\text{Twist-4} \rightarrow \phi^\dagger(0)D_\perp(\lambda n)D_\perp(\mu n)\phi(\nu n). \end{aligned}$$

It is as if ϕ had twist-one and χ had twist-two.

Twist-three is tractable, using the methods that have been developed in these lectures. Twist-four requires a more extensive analysis based on operator product expansion methods developed during the 1980's.[39, 19]. With these general tools in hand, we turn in the next section to the analysis of the specific distributions which appear in deep inelastic scattering of leptons.

5 Deep Inelastic Scattering and Generalized Distribution Functions II

In this section we use the tools developed in Sect. 3 to classify and interpret the quark distribution functions which appear in the analysis of DIS. The topics will include the classification and parton interpretation of the three leading twist quark distribution functions; a discussion of the physics of the less well known *transversity* distribution, h_1; a review of transverse spin in hard processes; a short digression on higher spin targets and gluon distribution functions; and a summary of the physics associated with the twist-three transverse spin distribution, g_2.

5.1 Helicity Amplitudes

Part of the task is simply to enumerate the independent distribution functions at twist-two and twist-three. This is simplified by viewing distribution functions as discontinuities in forward parton-(quark or gluon) hadron scattering. Suppressing all momentum indices, each quark distribution can be labeled by four helicities: a target of helicity Λ emits a parton of helicity λ which then participates in some hard scattering process. The resulting parton with helicity λ' is reabsorbed by a hadron of helicity Λ'. The process of interest to us is actually a *u-channel* discontinuity of the forward parton-hadron scattering amplitude $\mathcal{A}_{\Lambda\lambda,\Lambda'\lambda'}$ as shown in Fig. 23. Note the ordering of indices – although Λ

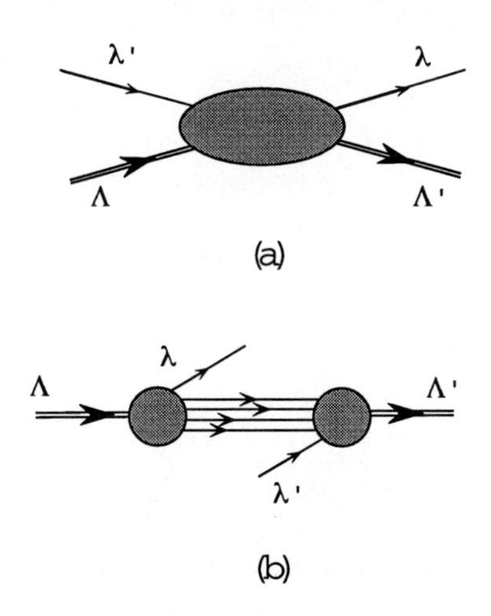

Fig. 23. *Helicity structure a) of the parton-hadron forward scattering amplitude; b) of the u-channel discontinuity which contributes to a parton distribution function.*

and λ' are the incoming helicities, it is convenient to label the amplitude in the sequence: initial hadron, struck quark, final hadron, returned quark.

Since the parton-hadron amplitude results from squaring something like $\langle X|\psi|PS\rangle$, the amplitude must be diagonal in the target spin. However spin eigenstates (in particular, *transverse* spin eigenstates) are linear superpositions of helicity eigenstates, so the $\{\mathcal{A}\}$ *do not* have to be diagonal in the target helicity. Only forward scattering is of interest, so the initial and final helicities must be the same,

$$\Lambda + \lambda' = \Lambda' + \lambda. \tag{5.1}$$

Also, the parity and time reversal invariance of the strong interactions place respective constraints on the $\{\mathcal{A}\}$,

$$\mathcal{A}_{\Lambda\lambda,\Lambda'\lambda'} = \mathcal{A}_{-\Lambda-\lambda,-\Lambda'-\lambda'} \tag{5.2}$$

$$\mathcal{A}_{\Lambda\lambda,\Lambda'\lambda'} = \mathcal{A}_{\Lambda'\lambda',\Lambda\lambda} \ . \tag{5.3}$$

Clearly the helicity counting outlined above applies equally well to good and bad light-cone components of quark or gluon fields. Therefore we can use it together with the methods of the previous section to enumerate quark distribution function through twist-three. To work through twist-three we will have to consider the case of one good and one bad light-cone component. We will identify any bad light-cone fields in helicity amplitudes by an asterix on the helicity label. Thus $\mathcal{A}_{0\frac{1}{2},0\frac{1}{2}^*}$ corresponds to emission of a good light-cone component and absorption of a bad one.

5.2 Quark Distributions in Targets with Spin-0, 1/2 and 1

Spin-0 Target Only $\Lambda = 0$ is available. Parity equates $\mathcal{A}_{0\frac{1}{2},0\frac{1}{2}}$ and $\mathcal{A}_{0-\frac{1}{2},0-\frac{1}{2}}$. Time reversal equates $\mathcal{A}_{0\frac{1}{2},0\frac{1}{2}^*}$ and $\mathcal{A}_{0\frac{1}{2}^*,0\frac{1}{2}}$. So there is only one distribution function at twist-two and one at twist-three. The twist-two function is none other than f_1 associated with the bilocal operator $\bar{\psi}(0)\not{n}\psi(\lambda n)$ and conserves quark chirality ("chiral even"). The twist-three function is e, associated with the scalar bilocal operator $\bar{\psi}(0)\psi(\lambda n)$ and flips quark chirality ("chiral odd"). These properties are summarized by Table 1.

Table 1. *Quark distributions in a spin-0 hadron through twist-three.*

Twist	Λ	λ	Λ'	λ'	Chirality
Two	0	1/2	0	1/2	Even
Three	0	1/2*	0	1/2	Odd

Spin-1/2 Target In the spin-1/2 case, for each twist assignment there are three independent helicity amplitudes. The reader may wish to verify that parity and time reversal invariance relate the many helicity amplitudes to the six listed in the table below (through twist-three). We leave the interpretation of these six

Table 2. *Quark distributions in a spin-1/2 hadron through twist-three.*

Twist	Λ	λ	Λ'	λ'	Chirality
Two	1/2	1/2	1/2	1/2	Even
Two	1/2	−1/2	1/2	−1/2	Even
Two	1/2	1/2	−1/2	−1/2	Odd
Three	1/2	1/2*	1/2	1/2	Odd
Three	1/2	−1/2*	1/2	−1/2	Odd
Three	1/2	1/2*	−1/2	−1/2	Even

distribution functions for the next section where they are discussed in detail.

Spin-1 Target A massive spin-one target has three independent helicity states. A new complication appears at twist-three: two helicity flip distributions arise which are not related by any of the symmetries of QCD. One can easily check that no such complication occurs for spin $\frac{1}{2}$. There is much interesting physics in these spin-one structure functions, however time will not permit us to work through it here. Instead we refer the interested reader to the original literature.[34, 35, 41, 42]

Table 3. *Quark distributions in a spin-1 hadron through twist-three.*

Twist	Λ	λ	Λ'	λ'	Chirality
Two	1	1/2	1	1/2	Even
Two	1	−1/2	1	−1/2	Even
Two	0	1/2	0	1/2	Even
Two	0	−1/2	1	1/2	Odd
Three	1	1/2*	1	1/2	Even
Three	1	−1/2*	1	−1/2	Even
Three	0	1/2*	0	1/2	Even
Three	0	−1/2*	1	1/2	Odd
Three	0	−1/2	1	1/2*	Odd

5.3 Quark Distribution Functions for the Nucleon

The distribution functions for a spin-$\frac{1}{2}$ target deserve special attention because protons and neutrons are the principal targets of interest. In Table 4 the quark distribution functions for a nucleon target are listed through twist-three. They are classified according to their twist (or light-cone projection) and their helicity.

The distribution functions f_1, g_1 and g_T are familiar because they can be measured in lepton scattering. The others are less well known, but are essential to understand the nucleon spin substructure in deep inelastic processes. All of them are defined by the matrix elements of quark bilocal operators,

$$\int \frac{d\lambda}{2\pi} e^{i\lambda x} \langle PS|\bar{\psi}(0)\gamma_\mu \psi(\lambda n)|PS\rangle = 2\left\{f_1(x)p_\mu + M^2 f_4(x) n_\mu\right\},$$

$$\int \frac{d\lambda}{2\pi} e^{i\lambda x} \langle PS|\bar{\psi}(0)\gamma_\mu\gamma_5 \psi(\lambda n)|PS\rangle = 2\{g_1(x)p_\mu S\cdot n + g_T(x) S_{\perp\mu} + M^2 g_3(x) n_\mu S\cdot n\},$$

$$\int \frac{d\lambda}{2\pi} e^{i\lambda x} \langle PS|\bar{\psi}(0)\psi(\lambda n)|PS\rangle = 2e(x),$$

Table 4. *Nucleon structure functions classified according to their twist and target helicities.*

twist $O(1/Q^{t-2})$	Name	Helicity Amplitude	Measurement	Chirality
Two	f_1	$\mathcal{A}_{1\frac{1}{2},1\frac{1}{2}} + \mathcal{A}_{1-\frac{1}{2},1-\frac{1}{2}}$	Spin average	Even
Two	g_1	$\mathcal{A}_{1\frac{1}{2},1\frac{1}{2}} - \mathcal{A}_{1-\frac{1}{2},1-\frac{1}{2}}$	Helicity difference	Even
Two	h_1	$\mathcal{A}_{0\frac{1}{2},1-\frac{1}{2}}$	Helicity flip	Odd
Three	e	$\mathcal{A}_{1\frac{1}{2}^*,1\frac{1}{2}} + \mathcal{A}_{1-\frac{1}{2},1-\frac{1}{2}}$	Spin average	Odd
Three	h_L	$\mathcal{A}_{1\frac{1}{2}^*,1\frac{1}{2}} - \mathcal{A}_{1-\frac{1}{2},1-\frac{1}{2}}$	Helicity difference	Odd
Three	g_T	$\mathcal{A}_{0\frac{1}{2}^*,1-\frac{1}{2}}$	Helicity flip	Even

$$\begin{aligned}\int \frac{d\lambda}{2\pi} e^{i\lambda x} \langle PS|\bar{\psi}(0) i\sigma_{\mu\nu}\gamma_5 \psi(\lambda n)|PS\rangle = 2\{&h_1(x)(S_{\perp\mu}p_\nu - S_{\perp\nu}p_\mu)/M \\ &+ h_L(x)M(p_\mu n_\nu - p_\nu n_\mu)S\cdot n \\ &+ h_3(x)M(S_{\perp\mu}n_\nu - S_{\perp\nu}n_\mu)\}, \quad (5.4)\end{aligned}$$

Some twist-four distributions (f_4, g_3, and h_3) appear in these matrix elements. However, they are joined by many other quark-quark and quark-gluon distributions from which they cannot be separated, so there is no point in keeping track of them in this analysis.

Nucleon Spin Structure at Twist-Two f_1, g_1 and h_1 are twist-two, *i.e.* they scale modulo logarithms. They can be projected out of the general decompositions, (5.4),

$$\begin{aligned} f_1(x) &= \int \frac{d\lambda}{4\pi} e^{i\lambda x} \langle P|\bar{\psi}(0) \not{n} \psi(\lambda n)|P\rangle \\ g_1(x) &= \int \frac{d\lambda}{4\pi} e^{i\lambda x} \langle PS_\parallel|\bar{\psi}(0) \not{n}\gamma_5 \psi(\lambda n)|PS_\parallel\rangle \\ h_1(x) &= \int \frac{d\lambda}{4\pi} e^{i\lambda x} \langle PS_\perp|\bar{\psi}(0) [\not{S}_\perp, \not{n}]\gamma_5 \psi(\lambda n)|PS_\perp\rangle \qquad (5.5)\end{aligned}$$

To understand their physical significance — in particular, to see why a third quark distribution in addition to f_1 and g_1 is necessary to describe the nucleon's quark spin substructure at leading twist in the parton model — it suffices to decompose them with respect to a light-cone Fock space basis. If we use the *helicity* basis, then

$$
\begin{aligned}
f_1(x) &= \frac{1}{x}\left\langle P|R^\dagger(xp)R(xp) + L^\dagger(xp)L(xp)|P\right\rangle, \\
g_1(x) &= \frac{1}{x}\left\langle P\hat{e}_3|R^\dagger(xp)R(xp) - L^\dagger(xp)L(xp)|P\hat{e}_3\right\rangle, \\
h_1(x) &= \frac{2}{x}Re\left\langle P\hat{e}_1|L^\dagger(xp)R(xp)|P\hat{e}_1\right\rangle,
\end{aligned} \tag{5.6}
$$

in analogy with (4.44), where we have integrated out the dependence on transverse momentum. Here $\hat{e}_3$ and $\hat{e}_1$ are unit vectors parallel and transverse, respectively, to the target nucleon's three-momentum. Clearly f_1 and g_1 can be interpreted in a probabilistic way: f_1 measures quarks independent of their helicity and g_1 measures the helicity asymmetry. But h_1 does not appear to have a probabilistic interpretation, instead it mixes right and left handed quarks.

If instead we use a *transversity* basis, diagonalizing γ^1, we find,

$$
\begin{aligned}
f_1(x) &= \frac{1}{x}\langle P|\alpha^\dagger(xp)\alpha(xp) + \beta^\dagger(xp)\beta(xp)|P\rangle, \\
g_1(x) &= \frac{2}{x}\mathrm{Re}\langle P\hat{e}_3|\alpha^\dagger(xp)\beta(xp)|P\hat{e}_3\rangle, \\
h_1(x) &= \frac{1}{x}\langle P\hat{e}_1|\alpha^\dagger(xp)\alpha(xp) - \beta^\dagger(xp)\beta(xp)|P\hat{e}_1\rangle.
\end{aligned} \tag{5.7}
$$

Clearly h_1 can be interpreted as the probability to find a quark with spin polarized along the transverse spin of a polarized nucleon minus the probability to find it polarized oppositely. f_1 still has the same interpretation, while now g_1 lacks a clear probabilistic interpretation. Of course the structure here is merely that of a $2 \otimes 2$ – spin density matrix, with the assignments $f_1 \leftrightarrow 1$, $g_1 \leftrightarrow \sigma_3$, and $h_1 \leftrightarrow \sigma_1$ in the basis of helicity eigenstates. The remaining element, σ_2, is related to h_1 by rotation about the $\hat{e}_3$-axis. In non-relativistic situations, spin and space operations (Euclidean boosts, *etc.*) commute and it is easy to show that $g_1 = h_1$, so h_1 is a measure of the relativistic nature of the quarks inside the nucleon.

The chirally odd structure functions like h_1 (Fig. 24a) are suppressed in DIS. The dominant handbag diagram Fig. 24b as well as the various decorations which generate $\log Q^2$ dependences, $\alpha_{QCD}(Q^2)$ corrections and higher twist corrections, examples of which are shown in Figs. 24c-e involve only chirally-even quark distributions because the quark couplings to the photon and gluon preserve chirality. Only the quark mass insertion, Fig. 24f, flips chirality. So up to corrections of order m_q/Q, $h_1(x, Q^2)$ decouples from electron scattering.

There is no analogous suppression of $h_1(x, Q^2)$ in deep inelastic processes with hadronic initial states such as Drell-Yan. The argument can be read from the standard parton diagram for Drell-Yan, Fig. 25. Although chirality is conserved on each quark line separately, the two quarks' chiralities are unrelated. It is not surprising, then, that Ralston and Soper [43] found that $h_1(x, Q^2)$ determines the transverse-target, transverse-beam asymmetry in Drell-Yan.

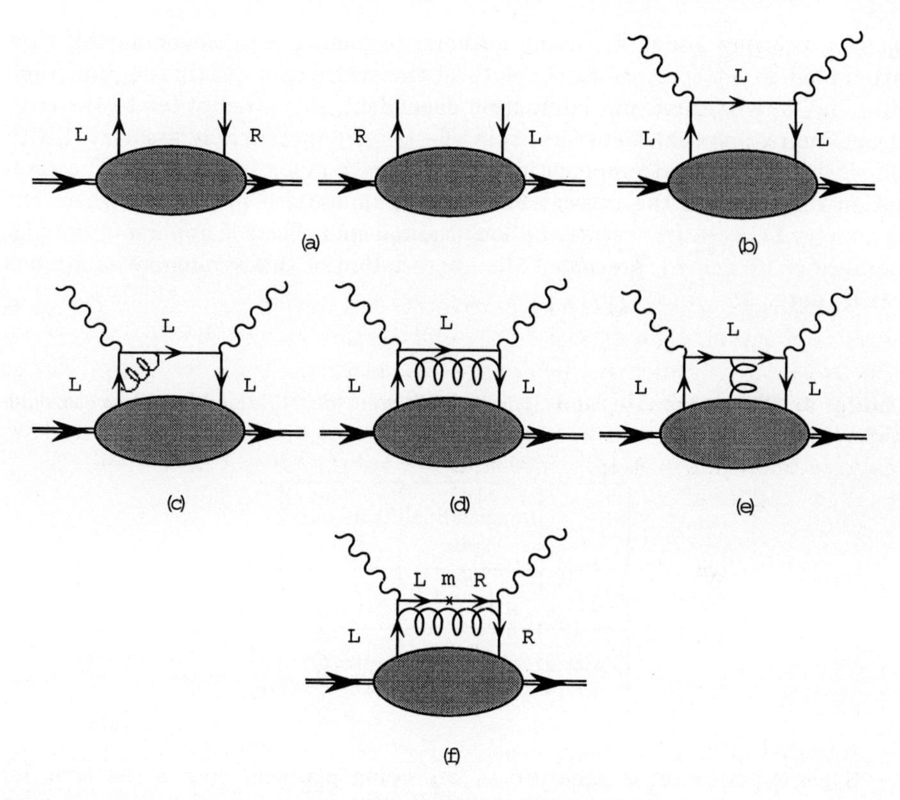

Fig. 24. *Chirality in deep inelastic scattering: a) Chirally odd contributions to $h_1(x)$; b)-e) Chirally even contributions to deep inelastic scattering (plus $L \leftrightarrow R$ for electromagnetic currents); f) Chirality flip by mass insertion.*

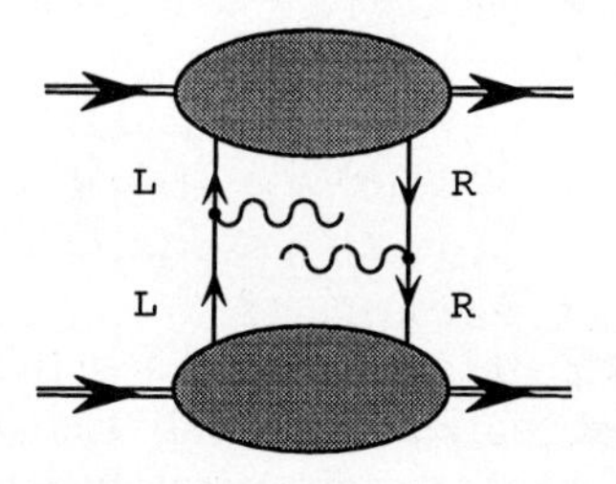

Fig. 25. *Chirality in Drell-Yan (plus $L \leftrightarrow R$) production of lepton pairs .*

5.4 Transverse Spin in QCD

The simple structure of (5.6) and (5.7) shows that transverse spin effects and longitudinal spin effects are on a completely equivalent footing in perturbative QCD. On the other hand, h_1 was unknown in the early days of QCD when only deep inelastic lepton scattering was studied in detail.

Not knowing about h_1, many authors, beginning with Feynman[44], have attempted to interpret g_T as the natural transverse spin distribution function. Since g_T is twist-three and interaction dependent, this attempt led to the erroneous impression that transverse spin effects were inextricably associated with off-shellness, transverse momentum and/or quark-gluon interactions The resolution contained in the present analysis is summarized in Table 5 where the symmetry between transverse and longitudinal spin effects is apparent. Only ignorance of h_1 and h_L prevented the appreciation of this symmetry at a much earlier date.

Table 5. *The transverse and longitudinal spin distribution functions through twist-three.*

	Longitudinal Spin	Transverse Spin
Twist-2	$g_1(x,Q^2)$	$h_1(x,Q^2)$
Twist-3	$h_L(x,Q^2)$	$g_T(x,Q^2)$

Since experiments to measure h_1 are being planned, now is the time for theorists to make predictions. At this time, however, not much is known about either the general behavior of h_1 or its form in models. Here is a summary, presented in parallel with g_1 for the purpose of comparison.

- Inequalities:

$$\begin{aligned} |g_1(x,Q^2)| &\leq f_1(x,Q^2) \\ |h_1(x,Q^2)| &\leq f_1(x,Q^2) \end{aligned} \tag{5.8}$$

for each flavor of quark and antiquark. These follow from the positivity of parton probability distributions (see (5.6) and (5.7)). Another inequality, proposed by Soffer[46] has attracted attention recently,

$$f_1^a(x,Q^2) + g_1^a(x,Q^2) \geq 2|h_1^a(x,Q^2)|. \tag{5.9}$$

valid for each flavor (a) of quark and antiquark. Soffer's inequality is invalidated by QCD radiative corrections,[12] in much the same way as the Callan-Gross relation, $F_2 = 2xF_2$. Despite this problem, the inequality may prove to be a useful qualitative guide to the magnitude of h_1. A recent discussion of QCD radiative corrections to Soffer's inequality may be found in [47].

- Physical interpretation: The structure function $h_1(x, Q^2)$ measures transversity. It is chirally odd and related to a bilocal generalization of the tensor operator, $\bar{q}\sigma_{\mu\nu}i\gamma_5 q$. On the other hand $g_1(x, Q^2)$ measures helicity. It is chirally even and related to a bilocal generalization of the axial charge operator, $\bar{q}\gamma_\mu\gamma_5 q$. Although h_1 is spin-dependent, it is not directly related to the quark or nucleon spin. It would be very useful to have a better idea of the dynamical and relativistic effects which generate differences between g_1 and h_1.
- Sum rules: If we define a "tensor charge" in analogy to the axial vector charge measured in β–decay,

$$2S^i\delta q^a(Q^2) \equiv \langle PS|\bar{q}\sigma^{0i}i\gamma_5\frac{\lambda^a}{2}q\Big|_{Q^2}|PS\rangle, \tag{5.10}$$

where λ^a is a flavor matrix and Q^2 is a renormalization scale, then $\delta q^a(Q^2)$ is related to an integral over $h_1^a(x, Q^2)$,

$$\delta q^a(Q^2) = \int_0^1 dx(h_1^a(x, Q^2) - h_1^{\bar{a}}(x, Q^2)) \tag{5.11}$$

where h_1^a and $h_1^{\bar{a}}$ receive contributions from quarks and antiquarks, respectively. The analogous expressions for $g_1(x, Q^2)$ involve axial charges,

$$2s^i\Delta q^a(Q^2) \equiv \langle PS|\bar{q}\gamma^i\gamma_5\frac{\lambda^a}{2}q\Big|_{Q^2}|PS\rangle \tag{5.12}$$

$$\Delta q^a(Q^2) = \int_0^1 dx(g_1^a(x, Q^2) + g_1^{\bar{a}}(x, Q^2)). \tag{5.13}$$

Note the contrast: $h_1(x, Q^2)$ is not normalized to a piece of the angular momentum tensor, so h_1, unlike g_1, cannot be interpreted as the fraction of the nucleons' spin found on the quarks' spin. Note the sign of the antiquark contributions: δq^a is charge-conjugation odd, whereas Δq^a is charge-conjugation even. δq^a gets no contribution from quark-antiquark pairs in the *sea*. All tensor charges δq^a have non-vanishing anomalous dimensions [48], but none mix with gluonic operators under renormalization because they are chirally odd and gluon operators are even. In contrast, the flavor non-singlet axial charges, $\Delta q^a, a \neq 0$, have vanishing anomalous dimensions, whereas the singlet axial charge Δq^0 has an anomalous dimension arising from the triangle anomaly.[49]
- Evolution: It is worth re-emphasizing that h_1 has unusual QCD evolution properties. All of the local operators associated with h_1 have non-vanishing leading order anomalous dimensions. On the other hand, no gluon operators contribute to h_1 in any order, because h_1 is chiral odd. So h_1 is a *non-singlet* structure function – it evolves homogeneously with Q^2, but none of its moments are Q^2 independent.
- Models: h_1 and g_1 are identical in non-relativistic quark models, but differ in relativistic models like the bag model - see Fig. 26.

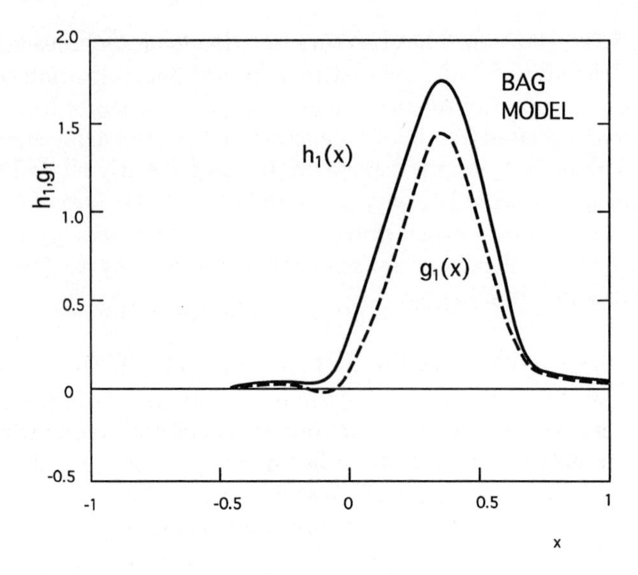

Fig. 26. *Bag model calculation for h_1 and g_1 from [8, 9].*

5.5 Twist-Three: Physics with $g_2(x, Q^2)$

There are several reasons to be particularly interested in the transverse spin dependent structure function, $g_2(x, Q^2)$

1. it can be measured in deep inelastic scattering;
2. it is unique among higher twist distributions in that it dominates the cross section in a specific kinematic domain — at 90° all twist-two effects decouple, see (2.22);
3. it is related to interesting quark gluon matrix elements;
4. it should obey an interesting sum rules.

For a review of the properties of g_2, see ref. [15]

Operator Product Expansion The spin structure functions g_1 and g_2 parameterize the antisymmetric part of the hadronic tensor $W_{\mu\nu}$ as shown in (2.15). Applying the methods of Sect. 2 we can relate the moments of the antisymmetric part of $W_{\mu\nu}$ to the matrix elements of quark operators accurate through twist-three. Consider the antisymmetric part of the forward Compton amplitude $T_{\mu\nu}$:

$$\begin{aligned} T_{[\mu\nu]} &= \frac{1}{2}\left[T_{\mu\nu} - T_{\nu\mu}\right], \\ &= \langle PS|\mathcal{T}_{[\mu\nu]}(q)|PS\rangle, \quad \text{where} \\ \mathcal{T}_{[\mu\nu]}(q) &= \int d^4\xi e^{iq\cdot\xi} T(J_{[\mu}(\xi)J_{\nu]}(0)). \end{aligned} \tag{5.14}$$

The leading twist (twists-two and -three) contributions come from the series of operators

$$\mathcal{T}_{[\mu\nu]} = i\epsilon_{\mu\nu\lambda\sigma}q^{\lambda} \sum_{n=0,2,4\ldots} q^{\mu_1}\ldots q^{\mu_n}\Theta^{\sigma}_{\{\mu_1\ldots\mu_n\}}\left(\frac{-2}{q^2}\right)^{n+1} C_n(q^2). \tag{5.15}$$

with

$$\Theta_{\sigma\{\mu_1\ldots\mu_n\}} \equiv i^n\bar{\psi}\gamma_\sigma\gamma_5 D_{\{\mu_1}\ldots D_{\mu_n\}}\psi - \text{traces},^{\star} \tag{5.16}$$

Here { , } denotes symmetrization and [,] denotes antisymmetrization of the enclosed indices. The string of symmetric indices originates in the expansion of the bilocal operator about $\xi = 0$, and contraction of each index with q^μ selects the totally symmetric part. On the other hand, the index on the γ–matrix is not symmetrized. $C_n(q^2)$ is the coefficient function and is unity to lowest order. Flavor factors such as squared charges are suppressed.

Equation (5.16) can be split into a totally symmetric part and a part of lower symmetry,

$$\Theta_{\sigma\{\mu_1\ldots\mu_n\}} \equiv \Theta_{\{\sigma\mu_1\ldots\mu_n\}} + \Theta_{[\sigma\{\mu_1]\mu_2\ldots\mu_n\}}, \tag{5.17}$$

where the first term has twist-two (dimension $n+3$, spin $n+1$):

$$\Theta_{\{\sigma\mu_1\ldots\mu_n\}} = \frac{1}{n+1}\left\{\Theta_{\sigma\{\mu_1\ldots\mu_n\}} + \Theta_{\mu_1\{\sigma\mu_2\ldots\mu_n\}} + \Theta_{\mu_2\{\mu_1\sigma\ldots\mu_n\}} + \ldots\right\}, \tag{5.18}$$

and the second term, with no totally symmetric part, has twist-three (dimension $n+3$, spin n):

$$\begin{aligned}\Theta_{[\sigma\{\mu_1]\mu_2\ldots\mu_n\}} = \frac{1}{n+1}\Big\{&\Theta_{\sigma\{\mu_1\ldots\mu_n\}} - \Theta_{\mu_1\{\sigma\mu_2\ldots\mu_n\}} \\ &+ \Theta_{\sigma\{\mu_1\ldots\mu_n\}} - \Theta_{\mu_2\{\mu_1\sigma\ldots\mu_n\}} + \ldots\Big\}.\end{aligned} \tag{5.19}$$

The proton matrix elements of these operators are

$$\begin{aligned}\langle PS|\Theta_{\{\sigma\mu_1\ldots\mu_n\}}|PS\rangle &= \{S_\sigma P_{\mu_1}\ldots P_{\mu_n} + S_{\mu_1}P_\sigma\ldots P_{\mu_n} + \ldots\}\frac{a_n}{n+1}, \\ \langle PS|\Theta_{[\sigma\{\mu_1]\mu_2\ldots\mu_n\}}|PS\rangle &= \{(S_\sigma P_{\mu_1} - S_{\mu_1}P_\sigma)P_{\mu_2}\ldots P_{\mu_n} \\ &\quad + (S_\sigma P_{\mu_2} - S_{\mu_2}P_\sigma)P_{\mu_1}\ldots P_{\mu_n} + \ldots\}\frac{d_n}{n+1}.\end{aligned} \tag{5.20}$$

---

* The terms denoted "traces" are whatever is necessary to subtract from the displayed term in order to render the resulting operator traceless and will be suppressed in the following.

Wandzura – Wilczek Decomposition of g_2 Extracting the relation between these matrix elements and the moments of g_1 and g_2 is an exercise in the methods of Sect. 2. Substituting these matrix elements into the definition of $T_{[\mu\nu]}$, writing a dispersion relation in terms of g_1 and g_2 and equating terms in the Taylor expansion in $1/x$ in the BJL–limit, we find

$$\int_0^1 dx x^n g_1(x,Q^2) = \frac{1}{4} a_n \quad n = 0,2,4,\ldots,$$

$$\int_0^1 dx x^n g_2(x,Q^2) = \frac{1}{4}\frac{n}{n+1}(d_n - a_n)), \quad n = 2,4\ldots, \tag{5.21}$$

Notice that the same operators which determine g_1 make an appearance in the moments of g_2. It follows that g_2 can be decomposed into two parts, one which is fixed by g_1, and another — the "true twist-three" part — associated with the operator of mixed symmetry,

$$g_2(xQ^2) = -g_1(x,Q^2) + \int_x^1 \frac{dy}{y} g_1(y,Q^2) + \bar{g}_2(x,Q^2). \tag{5.22}$$

Wandzura and Wilczek proposed this decomposition in 1977.[50] They went further and suggested that $\bar{g}_2$ might be zero. From another perspective, $\bar{g}_2$ is the *interesting* part of g_2. It's moments,

$$\int_0^1 dx x^n \bar{g}_2(x,Q^2) = \frac{n}{4(n+1)} d_n(Q^2), \tag{5.23}$$

are twist-three and measure quark-gluon correlations, as was pointed out by Shuryak and Vainshteyn[51]. They showed that the QCD equations of motion can be used to trade the antisymmetry of $\Theta_{[\sigma\{\mu_1]\mu_2\ldots\mu_n\}}$ for factors of the gluon field strength $G_{\mu\nu}$ and the QCD coupling constant g. The result is:

$$\Theta_{[\sigma\{\mu_1]\mu_2\ldots\mu_n\}} = \frac{g}{8}\mathcal{S}_n\left\{\sum_{l=0}^{n-2} i^{n-2}\bar{\psi}D_{\mu_1}..D_{\mu_l}\tilde{G}_{\sigma\mu_{l+1}}D_{\mu_{l+2}}..D_{\mu_{n-1}}\gamma_{\mu_n}\psi\right.$$
$$\left.+\frac{1}{2}\sum_{l=0}^{n-3} i^{n-3}\ \bar{\psi}D_{\mu_1}..D_{\mu_l}(D_{\mu_{l+1}}G_{\sigma\mu_{l+2}})D_{\mu_{l+3}}..D_{\mu_{n-1}}\gamma_{\mu_n}\gamma_5\psi\right\}, \tag{5.24}$$

where $\tilde{G}_{\alpha\beta} = \frac{1}{2}\epsilon_{\alpha\beta\lambda\sigma}G^{\lambda\sigma}$ and $\mathcal{S}_n$ symmetrizes the indices $\mu_1\ldots\mu_n$.

Equation (5.24) is quite formidable. A simpler example might help explain how manipulation of the equations of motion exposes the interaction dependence of higher twist operators. Consider the twist-three operator,

$$X_{\mu\nu} = \bar{\psi}D_\mu D_\nu\psi \tag{5.25}$$

as an example. X is clearly twist-three (dimension five, spin no greater than two). One might be tempted to make a "parton model" for the matrix element of X, by replacing $D_\mu \to \partial_\mu$ and evaluating X in a beam of collinear quarks. That, however, would be a mistake, since application of the identity $D_\mu = \frac{1}{2}\{\gamma_\mu, \not{D}\}$, and the QCD equations of motion, $\not{D}\psi = \bar{\psi}\not{D} = 0$ and $[D_\mu, D_\nu] = gG_{\mu\nu}$, yields

$$X_{\mu\nu} = \frac{g}{2}\bar{\psi}G_{\mu\lambda}\gamma^\lambda\gamma_\nu\psi. \tag{5.26}$$

So it is clear that X and its matrix elements are interaction dependent and measure a quark-gluon correlation in the target hadron. This is a special case of a general result that operators of twist ≥ 3 can always be written in a form in which they are manifestly interaction dependent. In particular one has:

$$\int_0^1 x^2\bar{g}_2(x, Q^2) \propto \langle PS|\frac{1}{8}gS_{\mu_1\mu_2}\bar{\psi}\tilde{G}_{\sigma\mu_1}\gamma_{\mu_2}\psi|PS\rangle. \tag{5.27}$$

Model builders or lattice enthusiasts who want to predict g_2 must confront such matrix elements.

The Burkhardt Cottingham Sum Rule A striking consequence of the light-cone analysis of g_2 is the apparent sum rule,[52]

$$\int_0^1 dx g_2(x, Q^2) = 0, \tag{5.28}$$

which follows directly from (5.21) with $n = 0$. If true, the sum rule requires g_2 to have a node (other than at 0 and 1).

The Burkhardt-Cottingham (BC) Sum Rule looks to be a consequence of rotation invariance. To see this, return to (5.4) and consider the $\gamma_\mu\gamma_5$ case in the laboratory frame (where $S^0 = 0$). First, set $\hat{S} = M\hat{e}_3$, contract the free Lorentz index with n^μ and integrate overall x, leaving

$$\int_{-1}^1 dx g_1(x, Q^2) = \langle P\hat{e}_3|\bar{q}(0)\gamma^3\gamma_5 q(0)|_{Q^2}|P\hat{e}_3\rangle. \tag{5.29}$$

Next repeat the process with $\hat{S} = M\hat{e}_1$ and $\mu = 1$, with the result,

$$\int_{-1}^1 dx g_T(x, Q^2) = \langle P\hat{e}_1|\bar{q}(0)\gamma^1\gamma_5 q(0)|_{Q^2}|P\hat{e}_1\rangle. \tag{5.30}$$

The right hand sides of these two equations are equal in the rest frame by rotation invariance, so

$$\int_{-1}^1 dx g_T(x, Q^2) = \int_{-1}^1 dx g_1(x, Q^2), \tag{5.31}$$

whence

$$\int_{-1}^{1} dx g_2(x, Q^2) = 0, \tag{5.32}$$

apparently a consequence of rotation invariance.

The subtlety in this derivation is that the integral in (5.32) goes from -1 to 1 including $x = 0$. As we have defined it, $g_2(x, Q^2)$ is the limit of a function of Q^2 and ν and therefore might contain a distribution (δ–functions, *etc.*) at $x = 0$. Suppose g_2 has a δ–function contribution at $x = 0$,

$$g_2(x, Q^2) = g_2^{\text{observable}}(x, Q^2) + c\delta(x). \tag{5.33}$$

Then since experimenters cannot reach $x = 0$, the BC sum rule reads

$$\int_{0}^{1} dx g_2^{\text{observable}}(x, Q^2) = -\frac{1}{2}c, \tag{5.34}$$

which is useless.

This pathology — a δ–function at $x = 0$ — is not as arbitrary as it looks. Instead it is an example of a disease known as a *"$J = 0$ fixed pole with non-polynomial residue"*. First studied in Regge theory,[53, 27] a $\delta(x)$ in $g_2(x, Q^2)$ corresponds to a *real constant term in a spin flip Compton amplitude which persists to high energy.* There is no fundamental reason to exclude such a constant. On the other hand the sum rule is known to be satisfied in QCD perturbation theory through order $O(g^2)$. The sum rule has been studied recently by several groups who find no evidence for a $\delta(x)$ in perturbative QCD.[54, 55, 56] So at least provisionally, we must regard this as a reliable sum rule. At least one other sum rule of interest experimentally, the Gerasimov, Drell, Hearn Sum Rule for spin dependent Compton scattering has the same potential pathology. For further discussion of the BC sum rule see ref. [15].

The Evolution of g_2 The following discussion can be regarded as a "theoretical interlude" and may be omitted by the casual reader.

The Q^2 evolution of quark distribution functions is an unavoidable complication in perturbative QCD. Data are inevitably taken at different Q^2, making it difficult to evaluate sum rules (which require data at some definite Q^2) without some information about Q^2–evolution. At leading twist the subject is well understood and we have ignored it. The evolution of g_2 is more complicated. Equation (5.4) provides as deceptively simple operator representation of g_T,

$$\begin{aligned} g_T(x, q^2) &= g_1(x, q^2) + g_2(x, q^2) \\ &= \int \frac{d\lambda}{4\pi} e^{i\lambda x} \langle PS_\perp | \bar{\psi}(0) \not{S}_\perp \gamma_5 \psi(\lambda n) |_{Q^2} | PS_\perp \rangle \end{aligned} \tag{5.35}$$

It looks as though g_T is determined by a single operator which evolves homogeneously. However, the operator in (5.35) is equivalent to the series of quark gluon operators given in (5.24). Worse still, all these operators mix under renormalization. The number of operators that mix grows linearly with n. Back in the 1970's, Ahmed and Ross calculated the anomalous dimension matrix element for the evolution of the transverse quark axial current operator in (5.35).[57] In 1982 Shuryak and Vainshteyn pointed out that this was but one element of an $n \times n$ anomalous dimension *matrix*[51]. Later Ratcliffe, Lipatov *et al.* and others calculated the full anomalous dimension matrix.[58, 59, 60, 61]

Underlying this matrix renormalization group evolution is a somewhat simpler parton picture. The fundamental objects for studying the evolution of g_2 are *two-variable* parton distribution functions defined by double light-cone Fourier transforms [7].

$$\begin{aligned}
&\int \frac{d\lambda}{2\pi}\frac{d\mu}{2\pi} e^{i\lambda x + i\mu(y-x)} \langle PS|\bar{q}(0) iD^{\alpha}(\mu n) \not{n} q(\lambda n)|_{Q^2}|PS\rangle = \\
&= 2i\varepsilon^{\alpha\beta\sigma\tau} n_{\beta} S_{\sigma} p_{\tau} G(x, y, Q^2) + \ldots \\
&\int \frac{d\lambda}{2\pi}\frac{d\mu}{2\pi} e^{i\lambda x + i\mu(y-x)} \langle PS|\bar{q}(0) iD^{\alpha}(\mu n) \not{n}\gamma_5 q(\lambda n)|_{Q^2}|PS\rangle = \\
&= 2i\varepsilon^{\alpha\beta\sigma\tau} n_{\beta} S_{\sigma} p_{\tau} \tilde{G}(x, y, Q^2) + \ldots
\end{aligned} \quad (5.36)$$

where ... denote other tensor structures of twist greater than three. The functions $G(x, y, Q^2)$ and $\tilde{G}(x, y, Q^2)$ are generalizations of parton distributions describing the amplitude to find quarks and gluons with momentum fractions x, y and $x - y$ in the target nucleon. They can be represented diagrammatically as in Fig. 27. These "higher twist distribution functions" share many properties

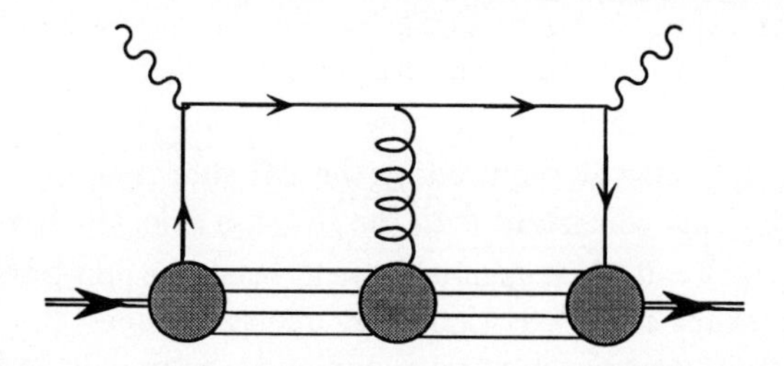

Fig. 27. *Quark/gluon diagram contributing to the two-variable parton distribution functions G and $\tilde{G}$.*

with leading twist. Only good light-cone components and collinear momenta appear – transverse momentum has been eliminated in favor of the interactions which generate it. The variables x, y, and $x - y$ take on only physical values, $-1 \leq (x, y, x - y) \leq 1$, corresponding to emission or absorption of quarks, antiquarks or gluons.[40]

At any value of Q^2, $g_T(x, Q^2)$ can be projected out by integrating out the variable y,

$$g_T(x,Q^2) = \frac{1}{2x}\int_{-1}^{1} dy\{G(x,y,Q^2) - G(y,x,Q^2) + \tilde{G}(x,y,Q^2) - \tilde{G}(y,x,Q^2)\} \tag{5.37}$$

The reason for introducing G and $\tilde{G}$ is that they obey natural generalizations of the Gribov-Lipatov-Altarelli-Parisi (GLAP) evolution equations. Schematically,

$$\frac{d}{d\ln Q^2}G(x,y,Q^2) = \frac{\alpha_{QCD}}{2\pi}\int_x^1\int_y^1 dx'dy'\mathcal{P}(x,x',y,y')G(x',y',Q^2) \tag{5.38}$$

for some "splitting function", $\mathcal{P}$. In order to evolve g_T from one Q^2 to another it is necessary to know G and $\tilde{G}$ as a function of both x and y. However, measurement of g_T does not supply that information. g_T is, in a sense, a "compressed" structure function. Information essential to evolution has been integrated out of it. The associated "open" distributions, G and $\tilde{G}$ evolve simply. g_T does not.

The situation is not entirely hopeless, however. It appears that the evolution of g_2 may simplify in the $N_c \to \infty$ limit, especially at large x.[62] Ali, Braun and Hiller showed that in this limit $\bar{g}_2$ obeys a standard GLAP equation, albeit with non-standard anomalous dimensions. As measurements of g_2 improve, theorists will be forced to expend more effort to clarify the nature of Q^2 evolution of g_T.

The Shape of g_2 With the first measurements of g_2 now available,[63] and more accurate measurements expected soon, it seems timely to review model predictions. First, here is an agenda of progressively more detailed questions for experimenters as they relate to theory

- Is g_2 zero?
  The answer is in on this one: the new SLAC data show $g_2 \neq 0$.
- Is $\int_0^1 dx g_2(x,Q^2)$ zero as required by the BC sum rule?
  The SLAC data are consistent with the BC sum rule, albeit with large errors.
- If $\int_0^1 dx g_2(x,Q^2) = 0$, then g_2 must have at least one non-trivial zero (besides $x=1$ and perhaps $x=0$). Is there just one such node?
  The SLAC data are consistent with one node, again within large errors.
- If there is only one non-trivial node, what is the sign of

$$M_2[g_2] \equiv \int_0^1 dx g_2(x,Q^2)? \tag{5.39}$$

  With only one non-trivial node, the sign of M_2 determines the gross structure of g_2. If M_2 is positive g_2 is negative at small x and positive at large x, and *vica versa* if M_2 is negative. The Wandzura-Wilczek (WW) contribution to g_2 gives $M_2^{WW} < 0$.
  The SLAC data favor $M_2 < 0$.

– Is there a signal for $\bar{g}_2$ or does the WW contribution account for all of g_2? The SLAC data agree rather well with the WW prediction, however the accuracy is such that a fairly significant $\bar{g}_2$ term would not yet be detectable. Some observers claim to be able to see a deviation from WW with $\overline{M}_2 > 0$.
– As more accurate data become available it will be possible to subtract away the WW contribution to reveal $\bar{g}_2$.

g_2 has been studied in quark models and using QCD Sum Rules. Perhaps the most thorough analysis, including QCD evolution, has been performed by Strattman.[64] He has taken several versions of the MIT bag model, calculated g_2 and then evolved the resulting distribution using the method of Ali, Braun and Hiller to experimentally interesting Q^2. His estimates for the proton are shown in comparison with the SLAC data in Fig. 28.

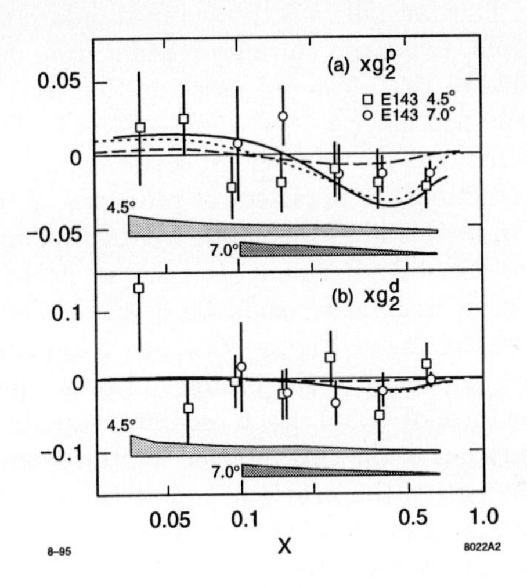

Fig. 28. *Data on g_2 from the SLAC E143 collaboration compared with the WW contribution and the bag model estimates of Strattmann.*

His estimates of $\bar{g}_2$ are small compared to g_2^{WW} and cannot be excluded by the existing data. The neutron g_2 is very small in quark and bag models for the same reason that the neutron's g_1 is so small: the correlations of charge and spin in the $SU(6)$ symmetric neutron tend to cancel for the neutron. The second moment of $\bar{g}_2$ has also been estimated using QCD sum rules.[65] Surprisingly ref. [65] finds $\overline{M}_2[g_2^{\text{proton}}]$ consistent with zero and $\overline{M}_2[g_2^{\text{neutron}}]$ negative and significantly different from zero. Existing data cannot rule out this behavior — we shall have to wait for the HERMES facility at HERA to provide more accurate data. $\bar{g}_2$ depends on quark gluon correlations within the nucleon, which are not likely to be perfectly described in such simple models, so these predictions should be more as a guide than a prediction of expected behavior.

6 The Drell-Yan Process

Although deep inelastic scattering has been the source of much insight into nucleon structure, it has many limitations: polarized targets can only be probed with electromagnetic currents (neutrino scattering from polarized targets being impractical); gluon distributions do not couple directly, but instead must be inferred from careful study of the evolution of quark distributions; chiral odd distributions like h_1 decouple. None of these limitations afflict deep inelastic processes with hadron initial states. These include not only the original Drell-Yan process, $pp \to \ell\bar{\ell}+X$ via one photon annihilation, but also generalizations to annihilation via $W^{\pm}$ and Z^0, and parton-parton scattering resulting in jet production or production of hadrons or photons at high transverse momentum. QCD predictions for all of these processes are obtained by combining quark/gluon distribution and fragmentation functions with hard scattering amplitudes calculated perturbatively. This formalism is treated in standard texts — our object is to explain how the spin, twist and chirality classification developed in previous sections can be applied to Drell-Yan processes. In this section we will treat only the "classic" Drell-Yan process, $pp \to \gamma^* + X \to \ell\bar{\ell} + X$. The generalizations of our spin/twist/chirality analysis to other processes is fairly straightforward and yields interesting predictions for a variety of processes. The original treatment of polarized Drell-Yan at this level was made by Ralston and Soper.[43]

When last we considered Drell-Yan in Sect. 2.4, we noted that the dominance of the leading light-cone singularity could be overwhelmed by rapid phase oscillations in the matrix element. Figure 20 shows a contribution to Drell-Yan which is obviously proportional to the product of quark distribution functions each of which has the form of (4.31). Each oscillates rapidly along the light-cone $\propto e^{i\alpha P\cdot\xi}$, where P is either of the two external hadron momenta. In this section we will use the formalism of the previous two sections to compute Fig. 20, to show that it generates large (scaling) contributions to Drell-Yan, and to classify them with respect to spin and twist.

6.1 Operator Analysis

We begin with the Drell-Yan tensor, $W_{\mu\nu}$, from (3.29), with momenta and spin more carefully labeled (and IN labels suppressed):

$$W_{\mu\nu} = \frac{s}{2}\int d^4\xi e^{iq\cdot\xi}\langle P_A S_A P_B S_B | J_\mu(0) J_\nu(\xi) | P_A S_A P_B S_B\rangle . \qquad (6.1)$$

The dominant contribution for the Drell-Yan process is shown in Fig. 29. From the diagram it seems that $W_{\mu\nu}$ reduces to a product of light-cone quark correlation functions, as indicated by the markings on the figure. The essential features are 1) that nothing propagates between the two currents, so there is no singularity as $\xi^\mu \to 0$; and 2) each current has a quark line landing on each hadron. The latter suggests that the diagram can be factored into products of quark-hadron amplitudes by making a Fierz transformation in order to couple the spin, color

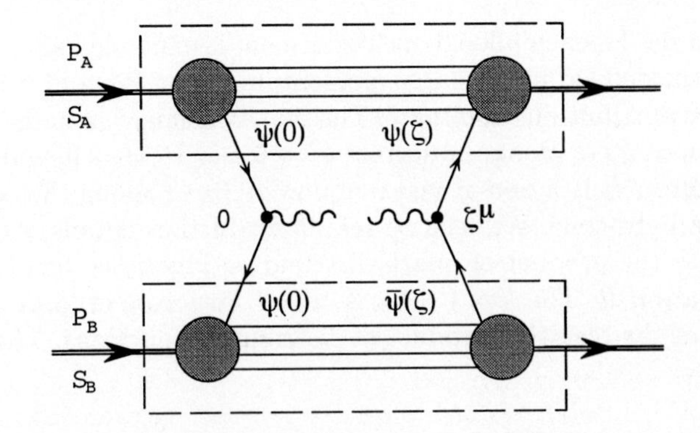

Fig. 29. *Dominant contribution to the Drell-Yan process.*

and flavor indices on the quarks in a more appropriate order. First we write out the currents in terms of quark fields, limiting ourselves to terms symmetric in $\mu \leftrightarrow \nu$ (which survive contraction with the lepton tensor, $L_{\mu\nu}$),

$$J^{\{\mu}(0)J^{\nu\}}(\xi) = -\bar{\psi}_k(0)\psi_l(\xi)\bar{\psi}_i(\xi)\psi_j(0)(\mathbf{1}\gamma^{\{\mu})_{kj}(\mathbf{1}\gamma^{\nu\}})_{il}. \tag{6.2}$$

Note the $-$ sign from anticommuting quark fields. We wish to recouple indices so quarks acting in the same hadron are coupled to one another — $(kj)(il) \to (kl)(ij)$. The color Fierz transformation is simple,

$$\mathbf{1}_{kj}\mathbf{1}_{il} = \frac{1}{3}(\mathbf{1}_{kl}\mathbf{1}_{ij} + 2\frac{\lambda^a_{kl}}{2}\frac{\lambda^a_{ij}}{2}), \tag{6.3}$$

where $\mathbf{1}$ is the 3×3 unit matrix in color space. [This result is easily derived by multiplying both sides by the matrices $\mathbf{1}_{kj}$ and λ_{kj}, in turn, and using multiplication properties of the λ's.] The second term in (6.3) vanishes when matrix elements are taken in color singlet hadron states. The Fierz transformation for the Dirac matrices is more complicated but better known,

$$\left(\gamma^{\{\mu}\right)_{kj}\left(\gamma^{\nu\}}\right)_{il} = \frac{1}{4}\Big[\left(\gamma^{\{\mu}\right)_{kl}\left(\gamma^{\nu\}}\right)_{ij} + \left(\gamma^{\{\mu}\gamma_5\right)_{kl}\left(\gamma^{\nu\}}\gamma_5\right)_{ij}$$
$$+ \left(\sigma^{\alpha\{\mu}\right)_{kl}\left(\sigma^{\nu\}}_{\ \ \alpha}\right)_{ij}\Big] + \frac{1}{4}g^{\mu\nu}[\cdots], \tag{6.4}$$

where the omitted terms are traces of the terms shown explicitly. One is left with bilocal light-cone correlation functions of the form: vector$\times$vector, axial vector$\times$axial vector and tensor$\times$tensor. The flavor Fierz transformation is straightforward and is left to the reader.

After reorganizing the product of currents to group fields that act in the same hadron together, the matrix element in (6.1) factors into the product of quark distribution functions. However, the coordinate interval ξ^μ is not constrained to be lightlike. In ref. [9] each bilocal matrix element is expanded about the tangent to the light-cone defined by the large momentum P_A or P_B, and it is shown that the light-cone contributions dominate. The matrix element in hadron A is a non-trivial function of $P_A \cdot \xi$ along the surface $\xi^2 = 0$ and $P_B \cdot \xi = 0$, while the matrix element in hadron B is a non-trivial function of $P_B \cdot \xi$ along the corresponding tangent to the light-cone. We refer to ref. [9] for further details. Accepting this, $W_{\mu\nu}$ reduces to the product of quark distribution functions, one for hadron A, another for hadron B. The $V \times V$, $A \times A$, $T \times T$ structure of (6.4) is reflected in the structure of the resulting product of distribution functions. The $V \times V$ part is described by:

$$W_V^{\mu\nu} = \frac{1}{3}(2\pi)^4 \delta^2(\mathbf{Q}_\perp) \sum_a e_a^2 f_1^a(x) f_1^{\bar{a}}(y) \left(p_A^\mu p_B^\nu + p_A^\nu p_B^\mu - g^{\mu\nu} p_A \cdot p_B\right), \quad (6.5)$$

where

$$P_A^\mu = p_A^\mu + \frac{M^2}{s} p_B^\mu \quad (6.6)$$

$$P_B^\mu = p_B^\mu + \frac{M^2}{s} p_A^\mu, \quad (6.7)$$

with $p_A^2 = p_b^2 = 0$ and $2p_A \cdot p_B = s$. This result is valid up to corrections of order $1/Q^2$ or $1/s$. There are no order $1/Q$ or $1/\sqrt{s}$ (twist-three) corrections — another example of the general result that only even twists appear in spin average deep inelastic phenomena. The $\delta(Q_\perp^2)$ in (6.5) reflects the fact that to leading twist and leading order in QCD radiative corrections, it is as though all partons move parallel to the parent hadron. Our calculation can only be used to study observables that integrate over $\mathbf{Q}_\perp$. Equation (6.5) is the original result of Drell and Yan: quarks that annihilate to form a virtual photon of squared-mass Q^2 and three-momentum Q^3 must have $xy = Q^2/s$ and $x - y = 2Q^3/\sqrt{s}$.

In the same spirit the $A \times A$ contribution is

$$\begin{aligned} W_A^{\mu\nu} = -\frac{1}{3}(2\pi)^4 \delta^2(\mathbf{Q}_\perp) \Bigg\{ & \sum_a e_a^2 g_1^a(x) g_1^{\bar{a}}(y) \frac{p_A \cdot S_B p_B \cdot S_A}{(p_A \cdot p_B)^2} \left(p_A^\mu p_B^\nu + p_A^\nu p_B^\mu \right. \\ & \left. - g^{\mu\nu} p_A \cdot p_B\right) + \sum_a e_a^2 g_1^a(x) g_T^{\bar{a}}(y) \frac{p_B \cdot S_A}{p_A \cdot p_B} \left(p_A^\mu S_{B\perp}^\nu + p_A^\nu S_{B\perp}^\mu\right) \\ & + \sum_a e_a^2 g_T^a(x) g_1^{\bar{a}}(y) \frac{p_A \cdot S_B}{p_A \cdot p_B} \left(p_B^\mu S_{A\perp}^\nu + p_B^\nu S_{A\perp}^\mu\right) \Bigg\}, \end{aligned} \quad (6.8)$$

The first line is twist-two and contributes only when both initial hadrons are longitudinally polarized. Not surprisingly, this contribution measures $g_1 \otimes g_1$. The latter two lines are twist-three, suppressed by the factor $S_\perp$, and contribute only when one hadron has longitudinal polarization and the other transverse. It is worthwhile relating the spin, twist and chirality structure of this result to the classification scheme developed in Sects. 3 and 4. Products of the form

$g_1 \otimes g_1$ and $g_1 \otimes g_T$ conserve quark chirality and contribute at orders of $1/Q^2$ which reflect the twist assignments we made in Sect. 4. Axial vector bilocals can only generate g_1 and g_T, so we can be confident that we have not missed other contributions. In fact, this result could have been written down *a priori*, up to coefficients of order unity simply by carefully considering the selections rules and twist assignments developed earlier in these notes. What is not at all obvious, however, is that other twist-three quark/gluon operators, of the form discussed in the previous section, and not directly related to g_T, do not arise when gluonic corrections to the diagram of Fig. 29 are computed. We take up this question below.

Finally, the $T \times T$ contribution takes the form,

$$\begin{aligned} W_T^{\mu\nu} = -\frac{1}{3}(2\pi)^4\delta^2(\mathbf{Q}_\perp)\Bigg\{ \Big[& S_{A\perp}\cdot S_{B\perp}\,(p_A^\mu p_B^\nu + p_A^\nu p_B^\mu - g^{\mu\nu} p_A\cdot p_B) \\ & + (p_A\cdot p_B)\,(S_{A\perp}^\mu S_{B\perp}^\nu + S_{A\perp}^\nu S_{B\perp}^\mu)\Big]\frac{1}{M^2}\sum_a e_a^2 h_1^a(x) h_1^{\bar{a}}(y) \\ & - \sum_a e_a^2 h_1^a(x) h_L^{\bar{a}}(y)\frac{p_A\cdot S_B}{p_B\cdot p_A}\,(p_A^\mu S_{A\perp}^\nu + p_A^\nu S_{A\perp}^\mu) \\ & - \sum_a e_a^2 h_L^a(x) h_1^{\bar{a}}(y)\frac{S_A\cdot p_B}{p_A\cdot p_B}\,(p_B^\mu S_{B\perp}^\nu + p_B^\nu S_{B\perp}^\mu)\Bigg\}. \end{aligned} \tag{6.9}$$

Here the first two lines are twist-two — they scale (modulo logarithms) in the deep inelastic limit. They contribute only when both initial hadrons are transversely polarized and provide a leading twist probe of transversity distributions. The second two lines are twist-three and contribute when one hadron is transversely polarized and the other longitudinally. Once again the classification scheme of Sects. 3 and 4 is illustrated. Note, for example, that transverse-longitudinal polarization receives contributions from both $g_1 \otimes g_T$ and $h_1 \otimes h_L$.

Unfortunately things are not quite as simple as they have been presented so far. Two questions of gauge invariance arise — one straightforward and the other rather subtle:

– The factorization of $W_{\mu\nu}$ into products of bilocal operators acting in different hadrons is not color gauge invariant. Evidence of this is the *absence* of the Wilson link,

$$\mathcal{P}\left(\exp i\int_0^\xi d\zeta^\mu A_\mu(\zeta)\right) \tag{6.10}$$

between the quark fields at 0 and ξ. The problem is that we have not included those diagrams that represent each quark propagating in the color field of the remnants of the other hadron. The same problem was discussed in Sect. 3 for the case of deep inelastic scattering, where it is easier to handle because the operator product expansion will preserve gauge invariance if treated with sufficient care. In this case we must find and analyze the diagrams which restore color gauge invariance through twist-three *and* be certain that they do not generate any contributions to $W_{\mu\nu}$ beyond rendering the identification of Drell-Yan matrix elements with quark distributions functions gauge invariant. An example of the type of diagram that does the trick is shown in Fig. 30. The interested reader is referred to ref. [9] for details.

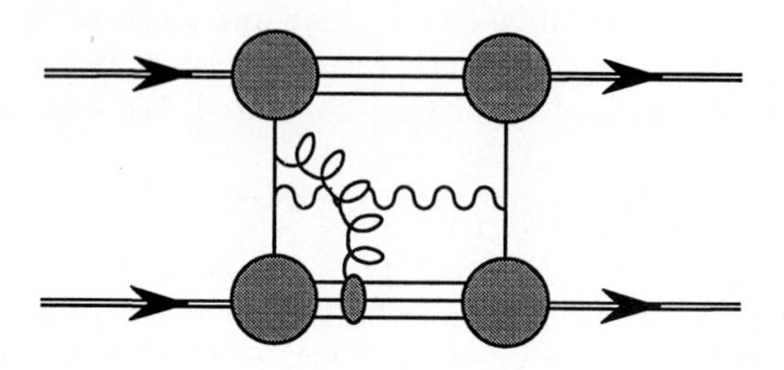

Fig. 30. *New class of diagrams, needed to restore color and electromagnetic gauge invariance.*

- The results we have just quoted ((6.9), (6.5), and (6.10)) violate electromagnetic gauge invariance at twist-three. It is easy to see that the twist-three terms do not satisfy $W_{\mu\nu}q^{\nu} = 0$. Once again the answer lies in diagrams like Fig. 30. In this case, when the Wilson link which appears in the bilocal operator is expanded to first order away from the light-cone, a set of contributions with explicit transverse gluon fields arise. Using the equations of motion these can be related back to the product of *good* and *bad* quark fields which define the twist-three distributions g_T and h_L. With sufficient care one finds exactly the terms necessary to restore electromagnetic gauge invariance. Once again a fuller discussion can be found in ref. [9].

6.2 Polarized Drell-Yan: a Brief Summary

Since the equations and the analysis in this section have become rather complicated, it is useful to extract the simple predictions for spin asymmetries to provide a summary and to reference the workers who originally derived these results.

In the polarized Drell-Yan process, three cases appear: longitudinal-longitudinal (LL), transverse-transverse (TT) and longitudinal-transverse (LT), as illustrated in Fig. 31. If instead of virtual photon production, we had considered Drell-Yan production of Z^0 or $W^{\pm}$, then a *longitudinal*, leading twist, parity violating single spin asymmetry proportional to $g_1 \otimes f_1$ appears. There is no analogous *transverse* asymmetry because the product $h_1 \otimes f_1$ cannot conserve quark chirality. The longitudinal asymmetry, which was first studied by Close and Sivers [66], is given by:

$$A_{LL} = \frac{\sum_a e_a^2 g_1^a(x) g_1^{\bar{a}}(y)}{\sum_a e_a^2 f_1^a(x) f_1^{\bar{a}}(y)}. \tag{6.11}$$

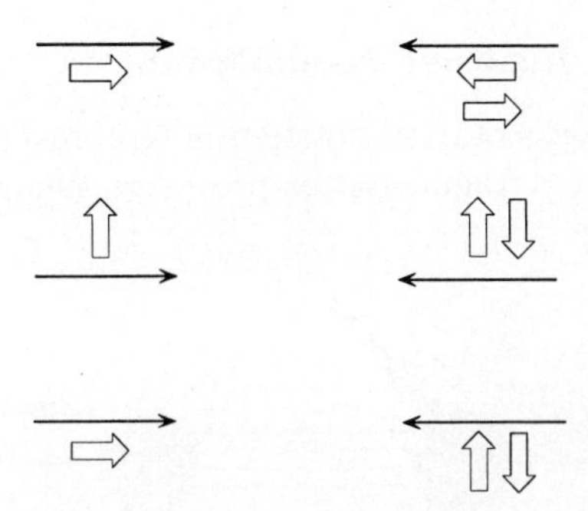

Fig. 31. *Various polarization configurations, studied in the Drell-Yan process. The small arrows denote the spin projections of the two nucleons relative to the direction of the nucleon beams (long arrows). For one of the nucleons the difference of the up and down projections is taken. One distinguishes longitudinal (top), transverse (middle) and longitudinal-transverse (bottom) scattering.*

Ralston and Soper first discovered transversity, defined a twist-two distribution and expressed the transverse asymmetry as:[43]

$$A_{TT} = \frac{\sin^2\theta\ \cos 2\phi}{1+\cos^2\theta} \frac{\sum_a e_a^2 h_1^a(x) h_1^{\bar{a}}(y)}{\sum_a e_a^2 f_1^a(x) f_1^{\bar{a}}(y)}. \tag{6.12}$$

The angles are defined in the lepton center of mass frame. The longitudinal-transverse asymmetry has been investigated by Jaffe and Ji [9] and can be written as:

$$A_{LT} = \frac{2\sin 2\theta \cos\phi}{1+\cos^2\theta} \frac{M}{\sqrt{Q^2}} \frac{\sum_a e_a^2 (g_1^a(x) y g_T^{\bar{a}}(y) - x h_L^a(x) h_1^{\bar{a}}(y))}{\sum_a e_a^2 f_1^a(x) f_1^{\bar{a}}(y)}. \tag{6.13}$$

Clearly, it is a twist-three observable, which in principle allows for a measurement of h_L.

7 Annihilation and Quark Fragmentation Functions

As a final application we give a brief introduction to the classification and uses of the spin dependent fragmentation functions which determine the distribution of final state hadrons in deep inelastic processes. There are strong reasons to want to develop a better understanding of hadron fragmentation processes. In Sect. 1 we mentioned the possibility of studying the spin structure of unstable hadrons like the Λ, ρ and D^*. Another reason is that parity violating processes like $W^\pm \to q\bar{q} \to$ hadrons provide probes of spin structure unavailable in deep inelastic scattering, where the analogous experiment would be neutrino scattering from a polarized target. A final reason is that the selection of a particular hadronic fragment can serve as a filter for an interesting quark or gluon distribution function. The material in this section is based primarily on refs. [10, 67] and [68] where more details can be found.

7.1 Single Particle Inclusive Annihilation

The simplest quark fragmentation function is represented diagrammatically in Fig. 32. More complicated fragmentation processes, such as coherent fragmenta-

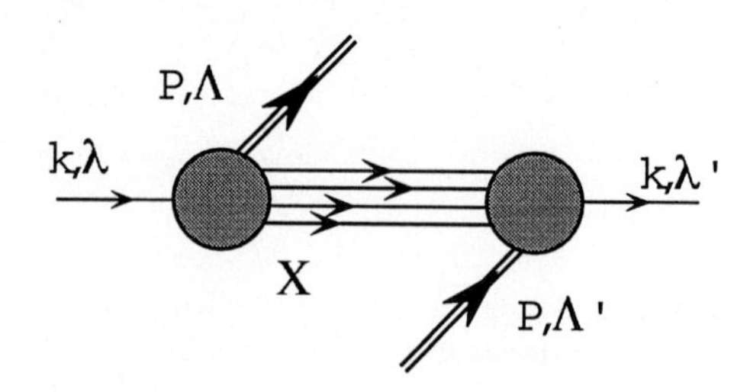

Fig. 32. *Quark fragmentation function in a helicity basis.*

tion of several quarks and gluons, do not contribute until order $1/Q^2$, beyond our interest. First we consider the helicity classification in analogy to Sects. 4.1–4.3. In the figure a quark of momentum k and helicity λ fragments into a hadron of momentum P and helicity Λ plus an unobserved final state X. The process then repeats in reverse as the unobserved system, X, plus the hadron of momentum P and helicity Λ' reconstitute the quark of momentum k and helicity λ'. The scattering $k + P \to k + P$ is forward, *i.e.* collinear. For definiteness, we take the momentum of the quark–hadron system to be aligned along the $\hat{e}_3$–axis. Then helicity is conserved as a consequence of angular momentum conservation about this axis: $\lambda - \Lambda = \lambda' - \Lambda'$. As in deep inelastic scattering, the initial and final hadron helicities Λ and Λ' need not be equal because the hadron need not have been in a helicity eigenstate; likewise for the quark. As in scattering, the quark lines may correspond either to good or bad light-cone components.

Some of the results of refs. [11] and [67] are as follows. Quark fragmentation functions of the form shown in Fig. 32 and the equivalent gluon fragmentation functions (without further active parton lines) are sufficient to characterize hadron production in hard processes, provided: i) one studies leading twist ($\mathcal{O}(1/Q^0)$) in any hard process, or ii) one studies an effect in deep inelastic lepton scattering at the lowest twist at which it arises, and one ignores QCD radiative corrections.[67] Each appearance of a bad component of the quark field costs one power of $\sqrt{Q^2}$ in the deep inelastic limit (*i.e.* it increases the twist by unity). As in scattering, for produced hadrons of spin-1/2, helicity differences are observed in longitudinal spin asymmetries; helicity flip is observed in transverse spin asymmetries. Since perturbative QCD cannot flip quark chirality (except through quark mass insertions which we assume to be negligible for light quarks), chirally–odd quark distribution and fragmentation functions must occur in pairs.

Fragmentation functions can be labeled uniquely by specifying the helicity of quarks and hadrons and the light-cone projection of the quarks in direct analogy to the classification of distribution functions. We denote fragmentation functions in a helicity basis by $\hat{A}$. Parity invariance of QCD requires: $\hat{A}_{\lambda\Lambda,\lambda'\Lambda'} = \hat{A}_{-\lambda-\Lambda,-\lambda'-\Lambda'}$. Time reversal invariance, which further reduces the number of independent quark distribution functions *does not* generate relationships among the $\{\hat{A}\}$ because it changes the *out*–state $(PX)_{\text{out}}$ in Fig. 32 to an *in*–state. As in the scattering case, we denote the appearance of bad light-cone components by an asterix on the appropriate helicity index.

As a simple example, consider production of a scalar meson like the pion. Through order $1/\sqrt{Q^2}$ there are three independent fragmentation functions: $\hat{A}_{\frac{1}{2}0,\frac{1}{2}0}$, $\hat{A}_{\frac{1}{2}0,\frac{1}{2}^*0}$, and $\hat{A}_{\frac{1}{2}^*0,\frac{1}{2}0}$. The first is twist-two and scales in the $Q^2 \to \infty$ limit, the latter two are twist-three and are suppressed by $1/\sqrt{Q^2}$ in the $Q^2 \to \infty$ limit. The first function, $\hat{A}_{\frac{1}{2}0,\frac{1}{2}0}$, is proportional to the traditional fragmentation function $D(z, Q^2)$. It has the same twist, light-cone, helicity and chirality structure as $f_1(x, Q^2)$, so to avoid an explosion of notation we denote it by $\hat{f}_1(z, Q^2)$ [We will follow the same convention for other fragmentation functions.]:

$$\hat{f}_1(z, Q^2) \propto \hat{A}_{\frac{1}{2}0,\frac{1}{2}0} \tag{7.1}$$

If we were studying quark *distribution* functions, the other two helicity amplitudes would be equal by time-reversal invariance. Here, there are two independent fragmentation functions.

$$\begin{aligned}
\hat{e}_1(z, Q^2) &\propto \hat{A}_{\frac{1}{2}0,\frac{1}{2}^*0} + \hat{A}_{\frac{1}{2}^*0,\frac{1}{2}0} \\
\hat{e}_{\bar{1}}(z, Q^2) &\propto \hat{A}_{\frac{1}{2}0,\frac{1}{2}^*0} - \hat{A}_{\frac{1}{2}^*0,\frac{1}{2}0}
\end{aligned} \tag{7.2}$$

We have found that the helicity classification of fragmentation functions is identical to that of distribution functions at leading twist. At twist-three, however, there are more fragmentation functions due to the absence of time reversal constraints.

The application to spin–1/2 is analogous to the classification of spin–1/2 distribution functions given in Table 2 of Sect. 4.2.2 except that each twist-three fragmentation function comes in two forms, one even and the other odd under time reversal. We suspect that the T-even functions are more important than the T-odd since the latter vanish in the absence of final state interactions. So fragmentation at twist-two requires $\hat{f}_1(z)$, $\hat{g}_1(z)$ and $\hat{h}_1(z)$ with helicity, transversity and chirality properties identical to the analogous distribution function. The only twist-three fragmentation function likely to be of much interest is the spin-averaged, chiral-odd, time reversal even function $\hat{e}_1(z)$.

In order to relate particular deep inelastic processes to quark distribution and fragmentation functions and to study them in models of non-perturbative QCD, it is necessary to have operator representations for them. This formalism is developed in refs. [11, 67]. Here we display the results for the three leading twist fragmentation functions for a spin-1/2 hadron, and for the twist-three spin average function, $\hat{e}_1$. Generalizations to spin-1 can be found in ref. [67]. The generic expression for a fragmentation function takes the form,

$$\hat{\Gamma}(z) = \mathrm{Tr}\left\{ \Gamma_{\alpha\beta} \sum_X \int \frac{d\lambda}{2\pi} e^{-i\lambda/z} \langle 0|\psi_\beta(0)|PX\rangle \langle PX|\bar{\psi}_\alpha(\lambda n)|0\rangle \right\}. \quad (7.3)$$

where $\Gamma_{\alpha\beta}$ stands for an arbitrary Dirac matrix. This result holds under the condition that the diagram of Fig. 32 dominates. To obtain $\hat{f}_1$, $\hat{g}_1$ and $\hat{h}_1$ one chooses $\Gamma = \not{n}, \not{n}\gamma_5$ and $\sigma^j_\nu n^\nu i\gamma_5$ respectively,

$$\hat{f}_1(z) = \frac{1}{2} \sum_X \int \frac{d\lambda}{2\pi} e^{-i\lambda/z} \langle 0|\not{n}\psi(0)|PX\rangle \langle PX|\bar{\psi}(\lambda n)|0\rangle ,$$

$$\hat{g}_1(z) = \frac{1}{2} \sum_X \int \frac{d\lambda}{2\pi} e^{-i\lambda/z} \langle 0|\not{n}\gamma_5\psi(0)|PSX\rangle \langle PSX|\bar{\psi}(\lambda n)|0\rangle ,$$

$$\frac{S_\perp{}^j}{M}\hat{h}_1(z) = \frac{1}{2} \sum_X \int \frac{d\lambda}{2\pi} e^{-i\lambda/z} \langle 0|\sigma^j_\nu n^\nu i\gamma_5\psi(0)|PS_\perp X\rangle \langle PS_\perp X|\bar{\psi}(\lambda n)|0\rangle \quad (7.4)$$

At twist-three the equations of motions can be used to express the structure functions in terms of independent degrees of freedom, quantized at $\xi^+ = 0$. For example one can obtain two expressions for the chiral odd, T-even spin independent fragmentation function $\hat{e}_1(z)$. One involving $\psi\bar{\psi}$ and another where bad light-cone components have been traded for transverse derivatives and gluon degrees of freedom,

$$M\hat{e}_1(z) = \frac{1}{2} \sum_X \int \frac{d\lambda}{2\pi} e^{-i\lambda/z} \langle 0|\psi(0)|PX\rangle\langle PX|\bar{\psi}(\lambda n)|0\rangle$$

$$\hat{e}_1(z) = \frac{z}{4M} \sum_X \int \frac{d\lambda}{2\pi} e^{-i\lambda/z} \{\langle 0|i\not{n}\not{D}_\perp(0)\psi_+(0)|PX\rangle\langle PX|\bar{\psi}_+(\lambda n)|0\rangle$$
$$- \langle 0|\psi_+(0)|PX\rangle\langle PX|\bar{\psi}_+(\lambda n) i\not{D}_\perp(\lambda n)\not{n}|0\rangle\}.$$

With these ingredients we are now prepared to explore a few applications of spin dependent fragmentation functions.

7.2 Polarized $q \to \Lambda$ Fragmentation Functions from $e^+e^- \to \Lambda + X$

This section is based on work done with M. Burkardt.[68] In the symmetric quark model, the Λ-baryon has a rather simple spin-flavor wavefunction. All its spin is carried by the s-quark, while the ud-pair is coupled to $S = 0$, $I = 0$: $\Delta u^\Lambda = \Delta d^\Lambda = 0$ and $\Delta s^\Lambda = 1$. While the quark model identifies the Λ-spin with the spin of the s-quark, the data on the quark spin structure of the nucleon suggests that the actual situation is more complex. If we take the latest SMC/SLAC data on the $\int_0^1 dx g_1^{ep}(x)$, combine it with β-decay data and assume exact $SU(3)_{flavor}$ symmetry for baryon axial charges, we find

$$\Delta u^\Lambda = \Delta d^\Lambda = \frac{1}{3}(\Sigma - D) = -0.23 \pm 0.06$$
$$\Delta s^\Lambda = \frac{1}{3}(\Sigma + 2D) = +0.58 \pm 0.07 \quad (7.5)$$

It would be exciting to test the $SU(3)_{flavor}$ assumption by observing deep inelastic scattering from a Λ target. Unfortunately we have to settle for observing the Λ as a fragment in annihilation processes. The parity violating, self analyzing decay of the final state Λ makes it particularly easy to study its polarization in fragmentation processes. Measurement of the helicity asymmetries for semi-inclusive production of Λ's in e^+e^- annihilation near the Z^0 resonance allows a complete determination of the spin-dependent fragmentation functions for the different quark flavors into the Λ. In the event that these could be measured it would be very interesting to compare the spin fractions measured in fragmentation, $\Delta\hat{s}^\Lambda$, $\Delta\hat{d}^\Lambda$, and $\Delta\hat{u}^\Lambda$, with the predictions, (7.5) in order to get a better understanding of the role of spin in the fragmentation process.

We are concerned here with twist-two helicity asymmetries, described by the fragmentation function $\hat{g}_1(z, Q^2)$. For simplicity we adopt a more conventional parton-model notation where we define $d_{q(L)}^{\Lambda(L)}$ to be the probability that a left handed Λ fragments into a left handed quark, *etc.* Then the unpolarized differential cross section for $e^-e^+ \to \Lambda + X$ is obtained by summing over the cross sections for $e^+e^- \to q\bar{q}$, weighted with the probability $d_q^\Lambda(z, Q^2)$ that a quark with momentum $\frac{1}{z}P$ fragments into a Λ with momentum P. As usual, we suppress the Q^2 dependence generated by radiative corrections in QCD,

$$\frac{d^2\sigma^\Lambda}{d\Omega\, dz} = \sum_q \frac{d\sigma^q}{d\Omega} d_q^\Lambda(z). \tag{7.6}$$

There is a single polarized fragmentation function for each flavor of quark or antiquark,

$$\begin{aligned} \Delta\hat{q}(z) &= d_{q(L)}^{\Lambda(L)}(z) - d_{q(L)}^{\Lambda(R)}(z) \\ &= d_{q(L)}^{\Lambda(L)}(z) - d_{q(R)}^{\Lambda(L)}(z) \\ \Delta\hat{\bar{q}}(z) &= d_{\bar{q}(L)}^{\Lambda(L)}(z) - d_{\bar{q}(L)}^{\Lambda(R)}(z) \\ &= d_{\bar{q}(L)}^{\Lambda(L)}(z) - d_{\bar{q}(R)}^{\Lambda(L)}(z), \end{aligned} \tag{7.7}$$

and furthermore isospin invariance requires that $\Delta\hat{u}(z) = \Delta\hat{d}(z)$ and $\Delta\hat{\bar{u}}(z) = \Delta\hat{\bar{d}}(z)$, so the number of independent fragmentation functions is reduces to four — *e.g.* $\Delta\hat{u}, \Delta\hat{\bar{u}}, \Delta\hat{s}$ and $\Delta\hat{\bar{s}}$.

As an example we quote the prediction for Λ production in e^-e^+ annihilation via photons. In this case one has to start from polarized e^- (or e^+) in order to fix the polarization of the quarks. One finds for the helicity asymmetric cross-section,

$$\begin{aligned} \frac{d^2\sigma^{e^-(L)e^+\to\Lambda(L)X}}{d\Omega\, dz} - \frac{d^2\sigma^{e^-(L)e^+\to\Lambda(R)X}}{d\Omega\, dz} = \frac{\alpha^2}{2s}\cos\theta \Big[&\frac{5}{9}\left(\Delta\hat{u}(z) + \Delta\hat{\bar{u}}(z)\right) \\ &+ \frac{1}{9}\left(\Delta\hat{s}(z) + \Delta\hat{\bar{s}}(z)\right)\Big] \end{aligned} \tag{7.8}$$

where L, R denotes the helicity of the e^- and the Λ. So annihilation into a single photon allows one combination of the four independent fragmentation functions to be measured. As described in ref. [68], annihilation at the Z^0 and just off the Z^0 peak where $\gamma - Z$ interference is largest will allow independent detection of all four quark and antiquark fragmentation functions. For further discussion of this process, backgrounds and experimental possibilities see ref. [68] and recent papers by the *Aleph* and *Delphi* collaborations.

7.3 Observing $h_1(x, Q^2)$ in Electroproduction

This section is based on work done with X. Ji.[10, 67] Chirally odd quark distributions are difficult to measure because they are suppressed in totally–inclusive deep inelastic scattering. So far, the only practical way to determine $h_1(x, Q^2)$ we have discussed has been Drell-Yan production with transversely polarized target and beam.

As an application of the fragmentation function formalism – one of many – we show how a chirally odd *fragmentation function* can be exploited to enable a measurement of $h_1(x, Q^2)$ to be obtained in polarized electroproduction of pions from a transversely polarized nucleon. This is an experiment which could be performed at several existing facilities. Related suggestions involving semi–inclusive production of Λ–hyperons and of two pions have been discussed previously.[69, 42] The proposal outlined here is simpler since it involves only one particle in the final state and does not require measurement of that particle's spin. The price we pay for this simplicity is suppression by a power of $\sqrt{Q^2}$.

We consider pion production in the current fragmentation region of deep-inelastic scattering with longitudinally polarized leptons on a polarized nucleon target. The simplest diagram for the process is shown in Fig. 33, where a quark struck by the virtual photon fragments into an observed pion plus other unobserved hadrons. The cross section is proportional to a trace and integral over the quark loop which contains the quark distribution function and fragmentation function. Due to chirality conservation at the hard (photon) vertex, the trace picks up only the products of the terms in which the distribution and fragmentation functions have the same chirality. When the nucleon is longitudinally polarized (with respect to the virtual-photon momentum), the twist-two, chirally even distribution $g_1(x)$ can couple with the twist-two chirally even fragmentation function $\hat{f}_1(z)$, producing a leading contribution $\mathcal{O}(1/Q^0)$ to the cross section. On the other hand, in the case of a transversely polarized nucleon, there is no leading-order contribution. At the next order, the nucleon's transversity distribution $h_1(x)$ can combine with the twist-three chirally odd fragmentation function $\hat{e}_1(z)$, and similarly $g_T(x)$ can combine with the chirally even transverse-spin distribution $\hat{f}_1(z)$. Both couplings produce $1/\sqrt{Q^2}$ contributions to the cross section. It is simple to see, however, that Fig. 33 alone does not produce an electromagnetically-gauge-invariant result. This is a typical example of the need to consider multi-quark/gluon processes beyond twist-two. In

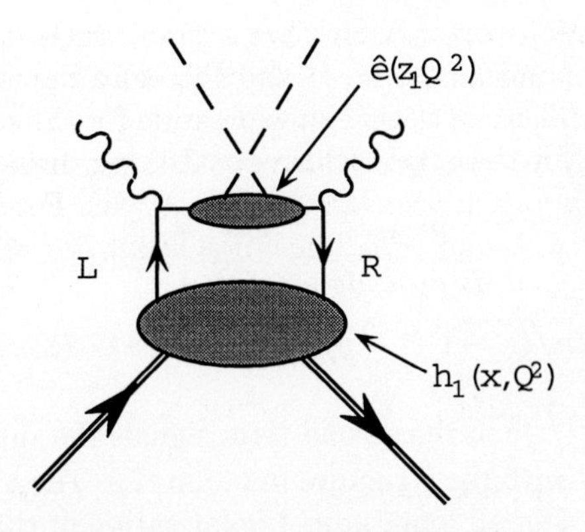

Fig. 33. *Single particle inclusive scattering ep → ehX. The labels L and R reflect the chiral odd nature of* h_1.

the present case (twist-three), however, the contributions from coherent scattering can be expressed, with novel use of QCD equations of motion, in terms of the distributions and fragmentation functions defined from quark bilinears. This is a specific example of another rule quoted in Sect. 6.2: "Parton diagrams (without further active parton lines) are sufficient to characterize hadron production in hard processes, provided: one studies an effect in deep inelastic lepton scattering at the lowest twist at which it arises, and one ignores QCD radiative corrections." The combined result is gauge invariant, as can be seen from the resulting nucleon tensor,

$$\hat{W}^{\mu\nu} = -i\epsilon^{\mu\nu\alpha\beta}\frac{q_\alpha}{\nu}[(S\cdot n)p_\beta \hat{G}_1(x,z) + S_{\perp\beta}\hat{G}_T(x,z)] \tag{7.9}$$

The two structure functions in $\hat{W}^{\mu\nu}$ are related to parton distributions and fragmentation functions,

$$\hat{G}_1(x,z) = \frac{1}{2}\sum_a e_a^2 g_1^a(x)\hat{f}_1^a(z)$$
$$\hat{G}_T(x,z) = \frac{1}{2}\sum_a e_a^2\left[g_T^a(x)\hat{f}_1^a(z) + \frac{h_1^a(x)}{x}\frac{\hat{e}^a(z)}{z}\right]$$

where the summation over a includes quarks and antiquarks of all flavors.

To isolate the spin-dependent part of the deep-inelastic cross section we take the difference of cross sections with left-handed and right-handed leptons, we use

$$\frac{d^2\Delta\sigma}{dE'd\Omega} = \frac{\alpha_{\rm em}^2}{Q^4}\frac{E'}{EM_N}\Delta\ell^{\mu\nu}\hat{W}_{\mu\nu} \tag{7.10}$$

It is convenient to express the cross section in terms of scaling variables in a frame where lepton beam defines the $\hat{e}_3$-axis and the $\hat{e}_1 - \hat{e}_3$ plane contains the nucleon polarization vector, which has a polar angle α. In this system, the scattered lepton has polar angles (θ, ϕ) and therefore the momentum transfer $\mathbf{q}$ has angles $(\theta, \pi - \phi)$. Then we obtain an expression for the semi-inclusive process quite similar to that for the total inclusive scattering defined by (2.22).

$$\frac{d^4 \Delta\sigma}{dx\, dy\, dz\, d\phi} = \frac{8\alpha_{\rm em}^2}{Q^2}\Big[\cos\alpha(1 - \frac{y}{2})G_1(x,z) + \cos\phi \sin\alpha\sqrt{(\kappa - 1)(1-y)}\, \Big(G_T(x,z) - G_1(x,z)(1 - \frac{y}{2})\Big)\Big] \tag{7.11}$$

where $\kappa = 1 + 4x^2M^2/Q^2$ in the second term signals the suppression by a factor of $1/\sqrt{Q^2}$ associated with the structure function G_T. The existence of G_1 in the same term is due to a small longitudinal polarization of the nucleon relative to $\mathbf{q}$ when its spin is perpendicular to the lepton beam.

Equation (7.11) is the main result of this section. As a check, we multiply by z, integrate over it and sum over all hadron species. Using the well-known momentum sum rule,

$$\sum_{\rm hadrons} \int dz z \hat{f}_1^a(z) = 1, \tag{7.12}$$

and the sum rule,

$$\sum_{\rm hadrons} \int dz \hat{e}_1^a(z) = 0 \tag{7.13}$$

which is related to the fact that the chiral condensate vanishes in the perturbative QCD vacuum, we get the result for *total* inclusive scattering, given in (2.22) The similarity between the inclusive and semi-inclusive cross sections suggests that they can be extracted conveniently from the same experiment.

The aim of this example was to show that an unfamiliar fragmentation function ($\hat{e}_1$) could be employed to obtain a measurement of an interesting, if unfamiliar, distribution function (h_1). It is apparent from (7.11) that we have been only partially successful: although the h_1^a distribution for each quark flavor appears in (7.11), the sum over flavors couples it to the unknown flavor dependence of $\hat{e}_1^a$. Perhaps flavor tagging can be used at large–z to identify the contributions of individual quark flavors. For x in the valence region (where one can ignore antiquarks in the nucleon), and $z \to 1$, the dominant fragmentation, $u \to \pi^+$, $d \to \pi^-$, $s \to K^-$, may allow one to trigger on the contributions of u, d and s quarks separately.[70] One might be concerned that the unknown fragmentation function, $\hat{e}_1$, might not respect the dominant fragmentation selection rules, which have only been tested for the spin-averaged, twist-two fragmentation function, $\hat{f}_1$. However, the coherent gluon interactions which distinguish the twist-three $\hat{e}_1$ from $\hat{f}_1$ are flavor independent and should not alter the selection rules. Although this may be a difficult path to measuring h_1, so are all the others. This one owes its existence to our improved understanding of the spin and twist dependence of quark fragmentation functions, including the spin-average twist-three fragmentation function $\hat{e}_1$.

References

[1] I. J. R. Aitchison, A. J. G. Hey, *Gauge Theories in Particle Physics, 2nd Edition* (Adam Hilger, Bristol, 1989).

[2] F. E. Close, *An Introduction to Quarks and Partons* (Academic Press, London, 1979).

[3] O. Nachtmann, *Elementary Particle Physics* (Springer-Verlag, Berlin, 1990)

[4] J. Collins, *Renormalization* (Cambridge University Press, Cambridge, 1984).

[5] T. Muta, *Foundations of Quantum Chromodynamics* (World Scientific, Singapore, 1987).

[6] R. G. Roberts, *The Structure of the Proton* (Cambridge University Press, Cambridge, 1990).

[7] R. L. Jaffe and X. Ji, *Phys. Rev.* **D43** (1991) 724.

[8] R. L. Jaffe and X. Ji, *Phys. Rev. Lett.* **67** (1991) 552.

[9] R. L. Jaffe and X. Ji, *Nucl. Phys.* **B375** (1992) 527.

[10] R. L. Jaffe and X. Ji, *Phys. Rev. Lett.* **71** (1993) 2547.

[11] K. Chen, G. R. Goldstein, R. L. Jaffe and X. Ji, *Nucl. Phys.* **B445** (1995) 380.

[12] G. R. Goldstein, R. L. Jaffe and X. Ji, *Phys. Rev.* **D52** (1995) 5006.

[13] H1 Collaboration, *Phys. Lett.* **B324** (1994) 241; ZEUS Collaboration, *Phys. Rev. Lett.* **75** (1995) 1006.

[14] C. G. Callan and D. J. Gross, *Phys. Rev. Lett.* **22** (1969) 156.

[15] R. L. Jaffe, *Comm. Nucl. Part. Phys.* **14** (1990) 239.

[16] S. G. Gorishny, A. L. Kataev, S. A. Larin *Phys. Lett.* **B259** (1991) 144.

[17] G. Bunce, J. Collins, S. Heppelmann, R. L. Jaffe, S. Y. Lee, Y. Makdisi, R. W. Robinett, J. Soffer, M. Tannenbaum, D. Underwood, A. Yokosawa, *Particle World* **3** (1992) 1.

[18] R. L. Jaffe, in *Relativistic Dynamics and Quark–Nuclear Physics*, M. B. Johnson and A. Picklesimer, eds. (Wiley, New York, 1986).

[19] R. K. Ellis, W. Furmanski, and R. Petronzio, *Nucl. Phys.* **B207** (1982) 1, *ibid.* **B212** (1983) 29.

[20] C. Itzykson and J. B. Zuber, *Quantum Field Theory* (McGraw–Hill, New York, 1980).

[21] S. Coleman, *Aspects of Symmetry* (Cambridge, 1985).

[22] B. L. Ioffe, it JETP Lett. **9** (1969) 163; **10** (1969) 143; *Phys. Lett.* **30B** (1969) 123.

[23] R. L. Jaffe, *Phys. Rev.* **D5** (1972) 2622; and in *Proceedings of the 1972 Erice School — Highlights in Particle Physics*, A. Zichichi, ed. (Edetrice Compositori, Bologna, 1973).

[24] J. D. Bjorken, *Phys. Rev. Lett.* **16** (1966) 408.

[25] K. Johnson and F. E. Low, *Prog. Theor. Phys. Suppl.* **37** (1967) 74.

[26] J. Ellis and R. L. Jaffe, *Scaling, Short Distances and the Light Cone,* Notes based on lectures presented at Univ. of Calif. Santa Cruz Summer School on Particle Physics, 1973, SLAC-PUB-1353, Dec 1973, unpublished.
[27] D. Broadhurst, J. Gunion and R. L. Jaffe, *Phys. Rev.* **D8** (1973) 566; *Ann. Phys.* **81** (1973) 88.
[28] J. D. Bjorken and S. D. Drell, *Relativistic Quantum Field Theory* (McGraw Hill, New York, 1964).
[29] T. P. Cheng and L. F. Li, *Gauge Theory of Elementary Particle Physics* (Clarendon Press, Oxford, 1984).
[30] J. C. Collins, D. E. Soper, *Nucl. Phys.* **B194** (1982) 445.
[31] N. N. Bogolyubov and D. V. Shirkov, *Introduction to the Theory of Quantized Fields* (Wiley Interscience, New York, 1959).
[32] R. K. Ellis, H. Georgi, M. Machacek, H. D. Politzer, G. G. Ross, *Phys. Lett.* **78B** (1978) 281; *Nucl. Phys.* **B152** (1979) 285.
[33] H. Fritzsch and M. Gell-Mann, in *Broken Scale Invariance and the Light-Cone* 1971 Coral Gables Conference on Fundamental Interactions at High Energies, M. Dal Cin, G. J. Iverson and A. Perlmutters, eds. (Gordon and Greach, New York, 1971).
[34] P. Hoodbhoy, A. V. Manohar and R. L. Jaffe, *Nucl. Phys.* **B312** (1989) 571.
[35] R. L. Jaffe and A. V. Manohar, *Nucl. Phys.* **B321** (1989) 343; E. Sather and C. Schmidt *Phys. Rev.* **D42** (1990) 1424.
[36] S. Weinberg, *Phys. Rev.* **150** (1966) 1313.
[37] L. Susskind, *Phys. Rev.* **165** (1968) 1535.
[38] J. Kogut and D. E. Soper, *Phys. Rev.* **D1** (1970) 2901.
[39] R. L. Jaffe and M. Soldate, in *Perturbative Quantum Chromodynamics* (Tallahassee, 1981), D. W. Duke and J. F. Owens, eds.; *Phys. Rev.* **D26** (1982) 49.
[40] R. L. Jaffe, *Phys. Lett.* **116B** (1982) 437; *Nucl. Phys.* **B229** (1983) 205.
[41] R. L. Jaffe and A. V. Manohar, *Phys. Lett.* **B223** (1989) 218.
[42] X. Artru and M. Mekhfi, *Z. Phys.* **C45** (1990) 669.
[43] J. Ralston and D. E. Soper, *Nucl. Phys.* **B152** (1979) 109.
[44] R. P. Feynman, *Photon-hadron Interactions*, (Benjamin, Reading, 1972).
[45] R. G. Roberts J. D. Jackson and G. G. Ross. *Phys. Lett.* **B226** (1989) 159.
[46] J. Soffer, *Phys. Rev. Lett.* **74** (1995) 1292.
[47] B. Kamal, A. P. Contogouris, and Z. Merebashvili, McGill University preprint, 1995.
[48] K. Sasaki J. Kodaira, S. Matusa and T. Uematsu, *Nucl. Phys.* **B159** (1979) 99.
[49] J. Kodaira, *Nucl. Phys.* **B165** (1979) 129.
[50] S. Wandzura and F. Wilczek, *Phys. Lett.* **72B** (1977) 195.
[51] E. V. Shuryak and A. I. Vainstein, *Nucl. Phys.* **B201** (1982) 141.
[52] H. Burkhardt and W. N. Cottingham, *Ann. Phys.* **56** (1970) 453.
[53] G. C. Fox and D. Z. Freedman, *Phys. Rev.* **182** (1969) 1628.

[54] I. Antoniadis and C. Kounnas, Ecôle Polytechnique preprint A399.0580 (1980), unpublished.
[55] M. Burkardt, University of Washington Preprint, hep-ph/9505226, May 1995.
[56] G. Altarelli, B. Lampe, P. Nason, and G. Ridolfi, *Phys. Lett.* **B334** (1994) 187.
[57] M. A. Ahmed and G. G. Ross, *Phys. Lett.* **B56** (1975) 385; *Nucl. Phys.* **B111** (1976) 441.
[58] P. G. Ratcliffe *Nucl. Phys.* **B264** (1986) 493.
[59] A. P. Bukhvostov, E. A. Kuraev and L. N. Lipatov, *JETP Letters* **37** (1983) 483.
[60] X. Ji and C. Chou, *Phys. Rev.* **D42** (1990) 3637.
[61] Y. Yasui, *Prog. Theor. Phys. Suppl.* **120** (1995) 239, and references therein.
[62] A. Ali, V. Braun and G. Hiller, *Phys. Lett.* **B226** (1991) 117.
[63] K. Abe, *et al.* The E143 Collaboration, SLAC-Pub-95-6982, August 1995.
[64] M. Strattman, *Z. Phys.* **C60** (1995) 369.
[65] I. I. Balitsky, V. M. Braun, and A. V. Kolesnichenko, *Phys. Lett.* **B242** (1990) 245; E. Stein, P. Gornicki, L. Mankiewicz, A. Schafer, and W. Greiner, *Phys. Lett.* **B343** (1995) 369.
[66] F. E. Close and D. Sivers, *Phys. Rev. Lett.* **39** (1977) 1116.
[67] X. Ji, *Phys. Rev.* **D49** (1994) 114.
[68] M. Burkardt and R. L. Jaffe, *Phys. Rev. Lett.***70** (1993) 2537.
[69] J. C. Collins, S. F. Heppelmann, G. A. Ladinsky, *Nucl. Phys.* **B420** (1994) 565.
[70] M. Veltri, M. Dueren, K. Rith, L. Mankiewicz, A. Schaefer, in *Proceedings of the Workshop on Physics at HERA* vol. 1, 447, W. Buchmuller and G. Ingelman, eds., 1992.

Quark–Gluon Structure of the Nucleon*

K. Rith[1]

Physikalisches Institut II, Universität Erlangen–Nürnberg, Erwin–Rommel Straße 1, 91058 Erlangen, Germany

Abstract. This review is an updated version of the lectures presented at the '92 Banz workshop. Originally it also contained a section on deep inelastic scattering at HERA. Since this subject is now extensively covered in the review by A. Levy, it has been skipped in these lecture notes. They therefore only summarise the actual status of unpolarised and polarised deep inelastic scattering of electrons and muons from fixed hydrogen and deuterium targets.

1 Unpolarised Deep Inelastic Scattering

The subject of unpolarised deep inelastic scattering is covered in several reviews [48, 101, 103, 177, 173], and nice pedagogical introductions can be found in many textbooks on particle physics [91, 130, 168, 128, 172, 169]. Therefore I will not repeat all the detailed arguments and derivations but will restrict myself to a short summary of the basic formulae and phenomena and will discuss the actual status of the field, especially the recent results from the NMC experiment at CERN.

1.1 Kinematics and Cross Sections

In deep inelastic scattering experiments one investigates the scattering of a point like lepton ℓ (electron, muon, neutrino) off a nucleon N (proton, neutron)

$$\ell + \mathrm{N} \to \ell' + \mathrm{X} \ ,$$

where the nucleon N is excited to a hadronic final state X with higher mass.

In lowest order pertubation theory the electroweak interaction occurs via the exchange of a virtual boson (γ, $\mathrm{Z^o}$, $\mathrm{W^{\pm}}$). In the following I will discuss mainly the electomagnetic interaction, where a virtual photon γ is exchanged between a charged lepton and the nucleon, and furthermore inclusive interactions, where only the scattered lepton is detected while the hadronic final state X remains unobserved.

We denote $\mathbf{k} = (E, \vec{k})$ and $\mathbf{k'} = (E', \vec{k'})$ as the four-momenta of incident and scattered lepton, $\mathbf{q} = (\nu, \vec{q})$ and $\mathbf{p} = (E_\mathrm{p}, \vec{p})$ as those of the exchanged virtual

---

* Lectures at the workshop "QCD and Hadron Structure" organised by the Graduiertenkolleg Erlangen-Regensburg, held on June 9th–11th, 1992 in Kloster Banz, Germany

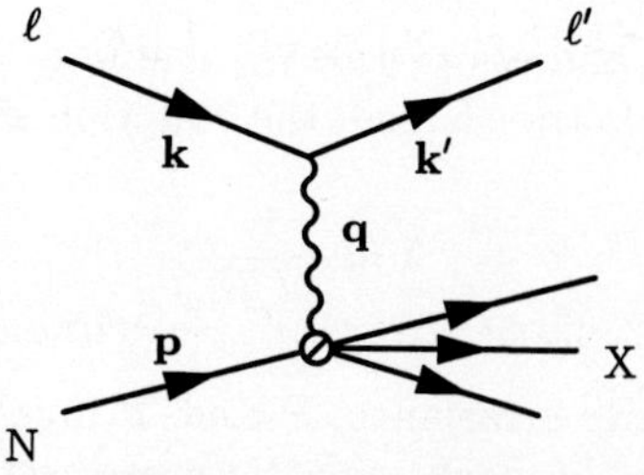

Fig. 1. *Deep inelastic charged lepton nucleon scattering in the approximation of one photon exchange*

photon and the target nucleon. Then the process, depicted in Fig. 1, can be characterised by the Lorentz-invariant quantities

$$q^2 = -Q^2 = (\mathbf{k} - \mathbf{k}')^2 \qquad \text{squared four momentum transfer} \tag{1}$$

$$s = (\mathbf{p} + \mathbf{k})^2 \qquad \text{squared center of mass energy} \tag{2}$$

$$W^2 = (\mathbf{p} + \mathbf{q})^2 \qquad \text{squared mass of the hadronic final state} \; . \tag{3}$$

We will only discuss fixed target experiments, $\mathbf{p} = (M, 0)$, where the lepton energies E, E' are much larger than the lepton masses m_ℓ, such that those can be neglected. In this case we obtain in the laboratory system

$$Q^2 \simeq 4EE' \sin^2 \frac{\theta}{2} = 2EE'(1 - \cos\theta) \tag{4}$$

$$s \simeq 2ME \tag{5}$$

$$W^2 = M^2 + 2M\nu - Q^2 \; , \tag{6}$$

where θ is the lepton scattering angle in the laboratory system, $\nu = E - E'$ the energy of the virtual photon transferred from the lepton to the nucleon and M the nucleon mass. We can also describe the process by two dimensionless scaling variables

$$y = \frac{\mathbf{p} \cdot \mathbf{q}}{\mathbf{p} \cdot \mathbf{k}} = \frac{\nu}{E} \; , \tag{7}$$

$$x = \frac{Q^2}{2\mathbf{p} \cdot \mathbf{q}} = \frac{Q^2}{2M\nu} \; . \tag{8}$$

Here y is the fractional energy-transfer from the lepton to the nucleon. x, the so-called Bjorken scaling variable, is a measure for the inelasticity of the process. For elastic scattering we have $W^2 = M^2$ and consequently $x = 1$, for inelastic processes $W^2 > M^2$ and $x < 1$. It should be noted that while for elastic scattering there is only one independent quantity Q^2 or ν, in the inelastic case ν and Q^2, respectively x and Q^2, can be varied independently.

Q^2 is a measure of the transverse size which can be resolved by a virtual photon with reduced wavelength $\lambda\!\!\!^-$. This quantity is not Lorentz-invariant but depends on the reference frame. In the laboratory frame it is ($\hbar = c = 1$):

$$\lambda\!\!\!^- = \frac{1}{|\vec{q}|} = \frac{1}{\sqrt{\nu^2 + Q^2}} \simeq \frac{1}{\nu} = \frac{2Mx}{Q^2} \; . \tag{9}$$

For example, if $x = 0.1$ and $Q^2 = 4\,\mathrm{GeV}^2$, one finds $\lambda\!\!\!^{-} \simeq 10^{-2}\,\mathrm{fm}$. In the Breit frame, where no energy is transferred, the equation simplifies to

$$\lambda\!\!\!^{-} = \frac{1}{\sqrt{Q^2}} \ . \tag{10}$$

If one requires Lorentz invariance, **P** and **T** invariance and conservation of the lepton current then the deep inelastic cross section in the energy E' and angle θ of the scattered electron averaged and summed over the spins can be written in the form

$$\frac{\mathrm{d}^2\sigma}{\mathrm{d}E'\,\mathrm{d}\Omega} = \frac{\alpha^2}{Q^4}\frac{E'}{E} L^{(s)\mu\nu} W^{(s)}_{\mu\nu} \ , \tag{11}$$

where

$$L^{(s)\mu\nu} = 2\left[k^\mu k'^\nu + k^\nu k'^\mu - (\mathbf{k}\cdot\mathbf{k}' - m_\ell^2)g^{\mu\nu}\right] \tag{12}$$

is the lepton tensor and

$$\begin{aligned} W^{(s)}_{\mu\nu} &= W_1(\nu,Q^2)\left(-g^{\mu\nu} + \frac{q^\mu q^\nu}{q^2}\right) \\ &+ W_2(\nu,Q^2)\frac{1}{M^2}\left(p^\mu - \frac{\mathbf{pq}}{\mathbf{q}^2}q^\mu\right)\cdot\left(p^\nu - \frac{\mathbf{pq}}{\mathbf{q}^2}q^\nu\right) \end{aligned} \tag{13}$$

represents the symmetric hadronic tensor which parametrises our ignorance of the form of the hadronic current (for a detailed derivation see [130] Chap. 6 and 8). In the laboratory system this leads to

$$\frac{\mathrm{d}^2\sigma}{\mathrm{d}E'\,\mathrm{d}\Omega} = \left(\frac{\mathrm{d}\sigma}{\mathrm{d}\Omega}\right)_{\mathrm{Mott}}\left\{W_2(\nu,Q^2) + 2W_1(\nu,Q^2)\tan^2\frac{\theta}{2}\right\} \tag{14}$$

with

$$\left(\frac{\mathrm{d}\sigma}{\mathrm{d}\Omega}\right)_{\mathrm{Mott}} = \frac{4\alpha^2 E'^2}{Q^4}\cos^2\frac{\theta}{2} \ . \tag{15}$$

The two structure functions $W_1(\nu,Q^2)$ and $W_2(\nu,Q^2)$ are the inelastic analogues to the electric and magnetic formfactors in elastic scattering from e.g. nucleons or nuclei, which in the Breit frame, where no energy but only momentum is transferred, can be interpreted as the Fourier-transforms of the spatial charge and magnetic distributions. From the measurement of the Q^2 dependence of the structure functions one can (via Fourier transformation) extract information about the spatial distribution of the scattering object. For the determination of both $W_{1,2}(\nu,Q^2)$ it is necessary to perform for each value of ν, Q^2 several measurements at different scattering angles θ and different incident lepton energies E.

For scattering on a spin-0 particle we have $W_1(\nu,Q^2) \equiv 0$. For elastic scattering on a pointlike spin-$\frac{1}{2}$ particle of mass m^* and charge $Q_f = e_f \cdot |e|$, where the charge distribution is a δ-function and consequently the Fourier transform is a constant, i. e. independent of Q^2, one obtains

$$2W_1^f(\nu, Q^2) = \frac{Q^2 e_f^2}{2m^{*2}}\,\delta\left(\nu - \frac{Q^2}{2m^*}\right) \tag{16}$$

$$W_2^f(\nu, Q^2) = e_f^2\,\delta\left(\nu - \frac{Q^2}{2m^*}\right) . \tag{17}$$

If one introduces the dimensionless structure functions

$$F_1^f(\nu, Q^2) = 2MW_1^f(\nu, Q^2) = \frac{Q^2 e_f^2 M}{2m^{*2}\nu}\,\delta\left(1 - \frac{Q^2}{2m^*\nu}\right) \tag{18}$$

$$F_2^f(\nu, Q^2) = \nu W_2^f(\nu, Q^2) = e_f^2\,\delta\left(1 - \frac{Q^2}{2m^*\nu}\right) , \tag{19}$$

then these point structure functions display the important property that they are only functions of the ratio $Q^2/(2m^*\nu)$ and not of Q^2 and ν independently.

1.2 Early Data, Interpretation, Consequences

Q^2 Independence of Structure Functions

The first deep inelastic scattering experiments were carried out in the late sixties at SLAC, using a linear electron accelerator with a maximum energy of 25 GeV. Above the resonance region ($W > 2$ GeV) one observed a cross section which had a much weaker Q^2 dependence than for elastic electron nucleon scattering or electro-excitation of the Δ resonance [80]. Fig. 2 shows the ratio $\frac{\mathrm{d}^2\sigma}{\mathrm{d}E'\mathrm{d}\Omega}/\left(\frac{\mathrm{d}\sigma}{\mathrm{d}\Omega}\right)_{\mathrm{Mott}}$ as a function of Q^2 at different values of W. It can be seen that this ratio depends only weakly on Q^2 for $W > 2$ GeV in clear contrast to the rapid drop for elastic scattering. Hence in deep inelastic scattering, the structure functions W_1 and W_2 are nearly independent of Q^2 for fixed values of the invariant mass W. This experimental observation was the basis of the quark-parton model (QPM) of Feynman and Bjorken [73, 74, 107, 108] where the deep inelastic scattering process is interpreted as the incoherent sum of elastic scattering from pointlike charged constituents of the nucleon. These were identified as the quarks introduced in the mid of the 60's [120, 188] to explain symmetries in hadron spectroscopy. The pedagogical argument goes as follows: Let the parton mass m^* be some fraction of the nucleon mass M, $m^* = xM$, and $q_f(x)\,\mathrm{d}x$ the probability that the parton mass between xM and $(x + \mathrm{d}x)M$ exists in the nucleon, then with (18) and (19) one obtains the Q^2 independent

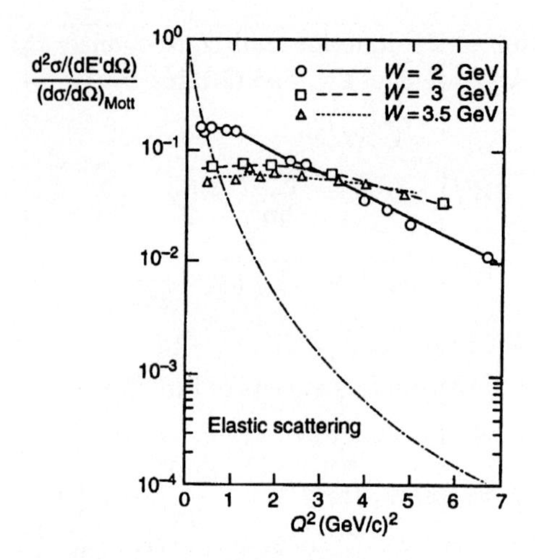

Fig. 2. *Electron proton scattering: measured cross sections normalised to the Mott cross section as a function of Q^2 at different values of the invariant mass W*

structure functions

$$\begin{aligned} F_2\left(\nu, Q^2\right) &= \sum_f \int_0^1 q_f(x) \cdot F_2^f\left(\nu, Q^2\right) \,\mathrm{d}x \\ &= \sum_f \int_0^1 e_f^2 x q_f(x) \delta\left(x - \frac{Q^2}{2M\nu}\right) \mathrm{d}x \\ &= \sum_f e_f^2 x q_f(x) = F_2(x) \ , \end{aligned} \tag{20}$$

and for spin-$\frac{1}{2}$ constituents:

$$F_1(\nu, Q^2) = \frac{1}{2} \sum_f e_f^2 q_f(x) = F_1(x) \ . \tag{21}$$

Such for scattering on pointlike spin-$\frac{1}{2}$ constituents one obtains the Callan-Gross-relation [84]

$$2xF_1(x) = F_2(x) \ . \tag{22}$$

In Fig. 3 the structure function $F_2(x, Q^2)$ from the early SLAC data [48] is displayed as a function of x for data covering a range of Q^2 between $2\,\mathrm{GeV}^2/\mathrm{c}^2$ and $18\,\mathrm{GeV}^2/\mathrm{c}^2$. The data exhibit scaling: For a fixed ratio $x = Q^2/2M\nu$, $F_2(x, Q^2)$ is essentially independent of Q^2. The tiny Q^2 dependence will be discussed in Chap. 1.3. The ratio $2xF_1/F_2$ is shown in Fig. 4 as a function of x. It can be seen that the ratio is, within the experimental error, consistent with unity as expected for scattering from pointlike spin-$\frac{1}{2}$ particles.

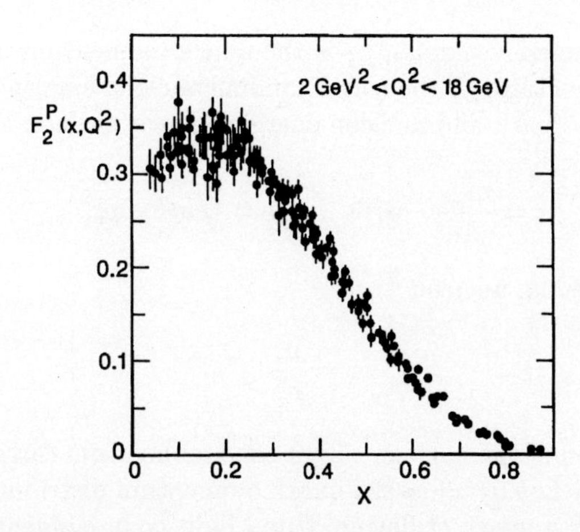

Fig. 3. *The structure function F_2 of the proton versus x, for Q^2 between 2 GeV^2 and 18 GeV^2*

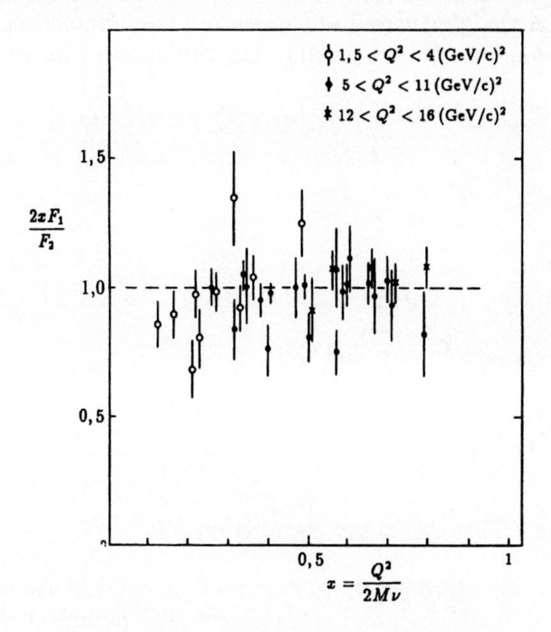

Fig. 4. *SLAC data for $2xF_1/F_2$ versus x*

In the derivation of (20), (21) we have used the assumption that the quark mass is a fraction x of the nucleon mass. Generally, on the light cone, x is defined as

$$x = \frac{p_0^{\mathrm{q}} + p_z^{\mathrm{q}}}{p_0^{\mathrm{N}} + p_z^{\mathrm{N}}} = \frac{\sqrt{m_{\mathrm{q}}^2 + p_z^{\mathrm{q}2} + p_{\mathrm{t}}^{\mathrm{q}2}} + p_z^{\mathrm{q}}}{\sqrt{M^2 + p_z^{\mathrm{N}2}} + p_z^{\mathrm{N}}} \tag{23}$$

where $m_{\rm q}$ is the quark rest mass, $p_z^{\rm q}$ is the quark momentum in z direction, $p_{\rm t}^{\rm q}$ its transverse momentum, $p_z^{\rm N}$ the corresponding nucleon longitudinal momentum and $p_0^{\rm q}$, $p_0^{\rm N}$ are the quark and nucleon energies. For a nucleon at rest it follows

$$x = \frac{m_{\rm eff}^{\rm q}}{M} \quad \text{with} \quad m_{\rm eff}^{\rm q} = \sqrt{m_{\rm q}^2 + p_{\rm t}^{\rm q2}} \ , \tag{24}$$

while for a fast moving nucleon

$$x = \frac{p_z^{\rm q}}{p_z^{\rm N}} \ , \tag{25}$$

i.e. x is the fraction of the nucleon's light cone momentum carried by the quark. Consequently we define $q_f(x)$ as the quark momentum distribution or the quark number density for a quark of flavour f and light cone momentum fraction x.

In our present understanding the nucleon consists of the valence quarks responsible for its quantum numbers and a sea of quark-antiquark pairs continuously created as virtual particles from gluons and annihilated to gluons, the field quanta of the strong interaction. Taking this into account we write

$$F_2(x, Q^2) \ = \ \sum_f e_f^2 x \left(q_f(x, Q^2) + \bar{q}_f(x, Q^2)\right) \ . \tag{26}$$

Using the Jacobian

$$\frac{{\rm d}^2\sigma}{{\rm d}Q^2\,{\rm d}x} = \frac{\nu}{x}\frac{{\rm d}^2\sigma}{{\rm d}Q^2{\rm d}\nu} = \frac{1}{2MEx}\frac{{\rm d}^2\sigma}{{\rm d}x{\rm d}y} = \frac{\nu}{x}\frac{\pi}{EE'}\frac{{\rm d}^2\sigma}{{\rm d}E'{\rm d}\Omega} \tag{27}$$

we obtain the cross section in x and Q^2:

$$\frac{{\rm d}^2\sigma}{{\rm d}Q^2\,{\rm d}x} = \frac{4\pi\alpha^2}{Q^4}\left\{\left(1 - y - \frac{Mxy}{2E}\right)\frac{F_2(x,Q^2)}{x} + y^2 F_1\left(x, Q^2\right)\right\} \ . \tag{28}$$

Virtual Photon Cross Sections $\sigma_{\rm L}$ and $\sigma_{\rm T}$, R

The cross section can alternatively be described in terms of the absorption cross section for longitudinally ($\sigma_{\rm L}$) and transversely ($\sigma_{\rm T}$) polarised virtual photons

$$\frac{{\rm d}^2\sigma}{{\rm d}x\,{\rm d}Q^2} = \Gamma\left(\sigma_{\rm T} + \varepsilon\sigma_{\rm L}\right) \ . \tag{29}$$

Γ describes the flux of virtual photons and ε is the degree of transverse polarisation of the virtual photon:

$$\varepsilon = \left(1 + \frac{1}{2}\left(1 - \frac{2m_\ell^2}{Q^2}\right)\frac{y^2 + \frac{Q^2}{E^2}}{1 - y - \frac{Q^2}{4E^2}}\right)^{-1} \simeq \frac{1-y}{1 - y + \frac{y^2}{2}} \ . \tag{30}$$

Often the structure function $F_1(x, Q^2)$ is replaced by the ratio of these cross sections

$$R(x, Q^2) = \frac{\sigma_L(x, Q^2)}{\sigma_T(x, Q^2)} \ . \tag{31}$$

In the Bjorken scaling limit longitudinally polarised photons do not interact with spin-$\frac{1}{2}$ quarks and consequently $\sigma_L(x, Q^2) = 0$ and $R(x, Q^2) = 0$ in this limit.

$R(x, Q^2)$ can also be expressed in terms of F_1 and F_2:

$$R(x, Q^2) = \frac{F_2(x, Q^2)\left(1 + \frac{Q^2}{\nu^2}\right) - 2xF_1(x, Q^2)}{2xF_1(x, Q^2)} = \frac{F_L(x, Q^2)}{2xF_1(x, Q^2)} \ , \tag{32}$$

where $F_L(x, Q^2)$ is called longitudinal structure function. We then can write the cross section (28) also in the form

$$\begin{aligned} \frac{d^2\sigma}{dQ^2\,dx} = \frac{4\pi\alpha^2}{Q^4}\frac{F_2(x, Q^2)}{x} \cdot \\ \cdot \left(1 - y - \frac{Q^2}{4E^2} + \left(1 - \frac{2m_l^2}{Q^2}\right) \cdot \frac{y^2 + \frac{Q^2}{E^2}}{2(1 + R(x, Q^2))}\right) \ . \end{aligned} \tag{33}$$

Proton and Neutron Structure Functions

In the quark parton model the proton is composed of two valence up (u) quarks ($e_u = \frac{2}{3}$) one valence down (d) quark ($e_d = -\frac{1}{3}$) and a 'sea' of quark-antiquark pairs, quantum fluctuations of the strong interaction. At the time of the SLAC measurements only the up, down and the strange (s) quark ($e_s = -\frac{1}{3}$) were known and therefore we will use only these three quarks in the argumetation of this paragraph. At higher beam energies as used later in the muon experiments ($E_\mu \lesssim 550\,\text{GeV}$) also the c-quark has to be taken into account, while the b- and t-quarks are so heavy that at the presently available energies for fixed target experiments their contribution to the structure functions can be safely neglected. In the following we will use $u(x, Q^2)$ for the momentum distribution $q_u(x, Q^2)$ of the u-quarks and so on. $u_v(x, Q^2)$ is the valence part of this distribution, $u_s(x, Q^2)$ the corresponding sea part and $\bar{u}(x, Q^2)$ the u-antiquark distribution which only contributes to the sea. Since quarks and antiquarks are always created from gluons in pairs of the same flavour we have

$$u_s(x, Q^2) = \bar{u}(x, Q^2) \ , \qquad d_s(x, Q^2) = \bar{d}(x, Q^2) \ , \qquad s_s(x, Q^2) = \bar{s}(x, Q^2) \ . \tag{34}$$

For brevity we will omit the Q^2 dependence of these distributions in the subsequent formulae, we will discuss it explicitely in a later chapter.

The following sum rules hold for the proton

$$\int_0^1 \{u^p(x) - \bar{u}^p(x)\}\,dx = \int_0^1 u_v^p(x)\,dx = 2 \tag{35}$$

$$\int_0^1 \{\mathrm{d^p}(x) - \bar{\mathrm{d}}^\mathrm{p}(x)\}\, \mathrm{d}x = \int_0^1 \mathrm{d_v^p}(x)\, \mathrm{d}x = 1 \tag{36}$$

$$\begin{aligned} &\int_0^1 \{\mathrm{u_s^p}(x) - \bar{\mathrm{u}}^\mathrm{p}(x)\}\, \mathrm{d}x \\ &= \int_0^1 \{\mathrm{d_s^p}(x) - \bar{\mathrm{d}}^\mathrm{p}(x)\}\, \mathrm{d}x \\ &= \int_0^1 \{\mathrm{s_s}(x) - \bar{\mathrm{s}}(x)\}\, \mathrm{d}x = 0\ . \end{aligned} \tag{37}$$

From (20) we then obtain for the proton

$$\begin{aligned} \frac{1}{x}F_2^\mathrm{p}(x) &= \frac{4}{9}\left(\mathrm{u^p}(x) + \bar{\mathrm{u}}^\mathrm{p}(x)\right) + \frac{1}{9}\left(\mathrm{d^p}(x) + \bar{\mathrm{d}}^\mathrm{p}(x)\right) + \frac{1}{9}\left(\mathrm{s}(x) + \bar{\mathrm{s}}(x)\right) \\ &= \frac{4}{9}\mathrm{u_v^p}(x) + \frac{1}{9}\mathrm{d_v^p}(x) + 2\left(\frac{4}{9}\bar{\mathrm{u}}^\mathrm{p}(x) + \frac{1}{9}\bar{\mathrm{d}}^\mathrm{p}(x) + \frac{1}{9}\bar{\mathrm{s}}(x)\right)\ . \end{aligned} \tag{38}$$

The neutron counterpart is

$$\frac{1}{x}F_2^\mathrm{n}(x) = \frac{4}{9}\mathrm{u_v^n}(x) + \frac{1}{9}\mathrm{d_v^n}(x) + 2\left(\frac{4}{9}\bar{\mathrm{u}}^\mathrm{n}(x) + \frac{1}{9}\bar{\mathrm{d}}^\mathrm{n}(x) + \frac{1}{9}\bar{\mathrm{s}}(x)\right)\ . \tag{39}$$

As neutron and proton are members of an isospin-doublett their quark content is related and we have

$$\mathrm{u^p}(x) = \mathrm{d^n}(x) \equiv \mathrm{u}(x), \qquad \mathrm{d^p}(x) = \mathrm{u^n}(x) \equiv \mathrm{d}(x)\ , \tag{40}$$

which leads to

$$\frac{1}{x}F_2^\mathrm{n}(x) = \frac{4}{9}\mathrm{d_v}(x) + \frac{1}{9}\mathrm{u_v}(x) + 2\left(\frac{4}{9}\bar{\mathrm{d}}(x) + \frac{1}{9}\bar{\mathrm{u}}(x) + \frac{1}{9}\bar{\mathrm{s}}(x)\right)\ . \tag{41}$$

$F_2^\mathrm{p}(x) - F_2^\mathrm{n}(x)$, Valence Quarks

Substracting (38) and (41) one obtains

$$\frac{1}{x}F_2^\mathrm{p}(x) - \frac{1}{x}F_2^\mathrm{n}(x) = \frac{1}{3}\left(\mathrm{u_v}(x) - \mathrm{d_v}(x)\right) + \frac{2}{3}\left(\bar{\mathrm{u}}(x) - \bar{\mathrm{d}}(x)\right)\ . \tag{42}$$

Thus if, and only if the sea is flavour symmetric, i. e. $\bar{\mathrm{u}} \equiv \bar{\mathrm{d}}$, then the contribution from sea quarks drops out and (42) contains only valence quark distributions.

In Fig. 5 the early SLAC data for $F_2^\mathrm{p}(x) - F_2^\mathrm{n}(x)$ are displayed as a function of x. The distribution falls to zero for $x \to 0$ and $x \to 1$ and has a maximum around $x = \frac{1}{3}$. This behaviour has often been interpreted as resulting from 3 valence quarks, each carrying in mean $\frac{1}{3}$ of the nucleon momentum and the sharply defined momentum at $x = \frac{1}{3}$ then washed out by the Fermi motion of the quarks inside the nucleon. This interpretation is wrong. As we will see later the distributions $\mathrm{u_v}$ and $\mathrm{d_v}$ each peak at x values around 0.17 and the peak at $x = \frac{1}{3}$ accidentally arises from the different x dependence of these two distributions.

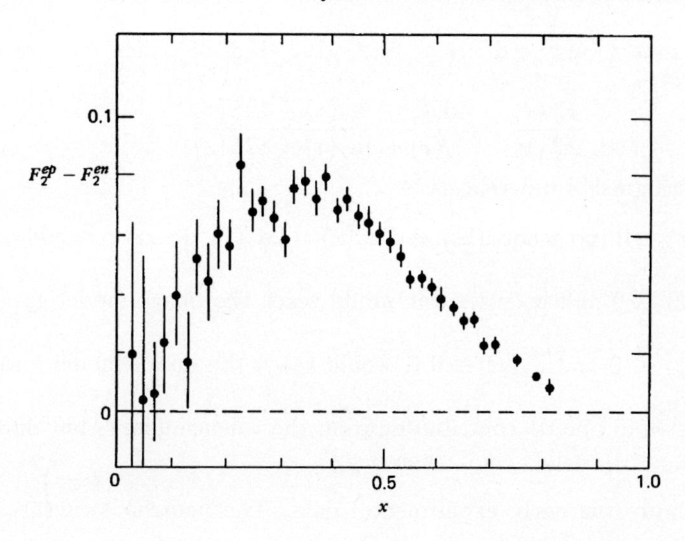

Fig. 5. *SLAC data for $F_2^{\mathrm{p}} - F_2^{\mathrm{n}}$ versus x*

Gottfried Sum Rule, Quark Charges

By integrating (41) over x and using (35) and (36) we obtain

$$\begin{aligned} S_{\mathrm{G}} = \int_0^1 \frac{1}{x}\left(F_2^{\mathrm{p}}(x) - F_2^{\mathrm{n}}(x)\right) \mathrm{d}x &= \frac{1}{3}(2-1) + \frac{2}{3}\int_0^1 \left(\bar{\mathrm{u}}(x) - \bar{\mathrm{d}}(x)\right) \mathrm{d}x \\ &= \frac{1}{3} \qquad \text{for } \bar{\mathrm{u}}(x) = \bar{\mathrm{d}}(x) \ . \end{aligned} \tag{43}$$

This is the Gottfried Sum Rule [126]. An early determination of this integral from the data shown in Fig. 5 resulted in $S_{\mathrm{G}} = 0.28 \pm ?$ [76], a result very near to the expected value of $\frac{1}{3}$. The '?' takes into account the uncertainty of the extrapolation of the data for $x \to 0$, which has a large weight due to the factor $\frac{1}{x}$ in the integral. Apart from this uncertainty the result agrees reasonably well with the expectation $\frac{1}{3}$ and one could conclude that the charge numbers $+\frac{2}{3}$ and $-\frac{1}{3}$ have been correctly attributed to the u- and d-quarks. We will discuss in Chap. 1.3 that the Gottfried integral S_{G} determined from recent high precision data is indeed substantially smaller than $\frac{1}{3}$. This does not change the conclusion about the quark charges but is being interpreted as an indication that the light quark-antiquark-sea is not flavour symmetric and that $\bar{\mathrm{u}} < \bar{\mathrm{d}}$.

$F_2^{\mathrm{n}}(x)\,/\,F_2^{\mathrm{p}}(x)$, Sea Quarks

If one makes the simple assumption that the sea is completely flavour symmetric $\bar{\mathrm{u}}(x) = \bar{\mathrm{d}}(x) = \bar{\mathrm{s}}(x) = \bar{S}(x)$, which in fact is not true for the light sea, as mentioned above, and wrong for the strange sea, since we know from neutrino scattering that $\frac{1}{2}\left(\bar{\mathrm{u}}(x)+\bar{\mathrm{d}}(x)\right) \simeq 2\bar{\mathrm{s}}(x)$ [103, 173, 65], then one obtains from (41) and (38)

$$\frac{F_2^{\mathrm{n}}(x)}{F_2^{\mathrm{p}}(x)} = \frac{4\mathrm{d_v}(x)+\mathrm{u_v}(x)+12S(x)}{\mathrm{d_v}(x)+4\mathrm{u_v}(x)+12S(x)} = r\ . \tag{44}$$

We can distinguish 4 different cases:

- for $S(x) \equiv 0$ (no seaquarks) and $\mathrm{u_v}(x) = 2\mathrm{d_v}(x)$ the ratio would be 2/3 for all x
- for $S(x) \equiv 0$ and $\mathrm{u_v}(x) \equiv 0$ it would reach the quark model upper bound $r = 4$
- for $S(x) \equiv 0$ and $\mathrm{d_v}(x) \equiv 0$ it would reach the quark model lower bound $r = 1/4$
- if there would be no contribution from the valence quarks but only the sea quarks, $\mathrm{u_v}(x) = \mathrm{d_v}(x) = 0$, then $r = 1$.

Figure 6 shows the early experimental data. The neutron structure function

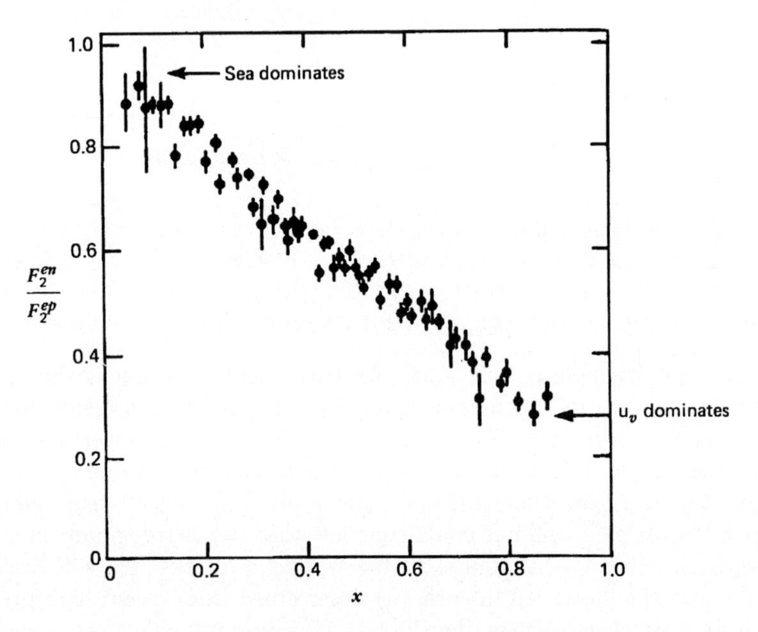

Fig. 6. *Early SLAC data for $F_2^{\mathrm{en}}/F_2^{\mathrm{ep}}$ versus x*

$F_2^{\mathrm{n}}(x)$ falls off with x more steeply than $F_2^{\mathrm{p}}(x)$ and is approximately given by [77] $F_2^{\mathrm{n}}(x) \simeq (1-0.75\,x)F_2^{\mathrm{p}}(x)$. The ratio $F_2^{\mathrm{n}}(x)/F_2^{\mathrm{p}}(x)$ approaches 1 for $x \to 0$, i. e. at very low momenta one observes dominantly sea quarks, and for $x \to 1$ it approaches the quark model lower bound of 1/4, i. e., if a quark carries a very large momentum fraction then it is the u-quark in the proton und the d-quark in the neutron, the quark which carries the quantum number of the corresponding nucleon. The d-quark distribution falls off faster with x than the u-quark distribution. Sea quarks dominate at low x.

Missing Momentum, Gluons

Finally, after having discussed the difference and ratio of neutron and proton structure functions, we can get another important information about the internal structure of the nucleon by looking at the integral over their sum. If there are no nuclear binding effects then the structure function F_2^{N} per nucleon for the deuteron is given by the sum of the structure functions for a free proton and a free neutron divided by two:

$$F_2^{\mathrm{N}}(x) = F_2^{\mathrm{d}}(x) = \frac{1}{2}\left(F_2^{\mathrm{p}}(x) + F_2^{\mathrm{n}}(x)\right) \tag{45}$$

We then obtain

$$\begin{aligned} F_2^{\mathrm{N}}(x) &= \frac{5}{18}x\left(\mathrm{u}(x) + \bar{\mathrm{u}}(x) + \mathrm{d}(x) + \bar{\mathrm{d}}(x) + \mathrm{s}(x) + \bar{\mathrm{s}}(x)\right) \\ &\quad -\frac{1}{6}\left(\mathrm{s}(x) + \bar{\mathrm{s}}(x)\right) \\ &\simeq \frac{5}{18}x\sum_f \left(q_f(x) + \bar{q}_f(x)\right) \ . \end{aligned} \tag{46}$$

The integral of $F_2^{\mathrm{N}}(x)$ is taken over all quark momenta weighted by their distribution functions and squared quark charges; hence, if the whole momentum of the nucleon would be carried by its charged constituents, the quarks, the integral would (apart from the small correction due to the s-quarks) yield the mean squared quark charge

$$\int_0^1 F_2^{\mathrm{N}}(x)\,\mathrm{d}x \simeq \frac{5}{18}\int_0^1 x\sum_f \left(q_f(x) + \bar{q}_f(x)\right)\mathrm{d}x = \frac{5}{18} \ . \tag{47}$$

Experimentally one finds only about 50% of this value. (The same result one obtains in neutrino nucleon interactions, see below). Thus roughly only half of the nucleon's momentum is carried by the charged quarks. The other half is carried by particles interacting neither electromagnetically nor weakly. They are identified with the gluons, the field quanta of the strong interaction [113]. This observation was the starting point of Quantum Chromodynamics (QCD), the field theory of the strong interaction. Consequently the momentum sum rule (47) has to be modified:

$$\int_0^1 \left\{\frac{18}{5}\sum_f \left(q_f(x) + \bar{q}_f(x)\right) + g(x)\right\} x\,\mathrm{d}x = 1 \ . \tag{48}$$

Here $g(x)$ is the momentum distribution of gluons inside the nucleon.

Neutrino Scattering, Separation of Valence and Sea Quarks

Until now we have only discussed charged lepton scattering. Valence- and sea quark distributions can not be separated in a direct way by these experiments alone: the virtual photon just couples to the charge of the quarks independent of their flavour. This is different in neutrino and antineutrino scattering. The charged current weak interaction takes place between members of the weak isospin dubletts which are lefthanded for particles and righthanded for antiparticles. Only selected flavour changes are allowed because of charge conservation (e. g. $\nu \mathrm{d} \to \mu^- \mathrm{u}$, $\nu \bar{\mathrm{c}} \to \mu^- \mathrm{s}$, $\bar{\nu} \mathrm{d} \to \mu^+ \mathrm{u}$ etc.)

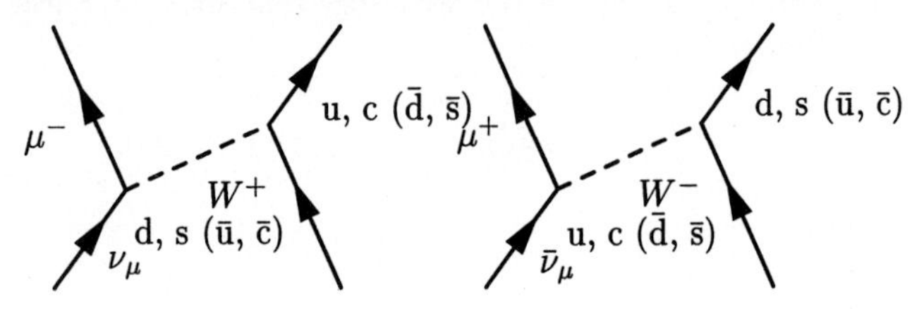

Interactions between particles of same helicity ($\nu \mathrm{q}$, $\bar{\nu} \bar{\mathrm{q}}$) can be distinguished from those between particles of opposite helicity ($\nu \bar{\mathrm{q}}$, $\bar{\nu} \mathrm{q}$) due to their different energy dependence. In the latter case backward scattering in the center-of-mass frame is forbidden by angular momentum conservation and the cross section is proportional to $(1-y)^2$, where $y = (E_\nu - E_\mu)/E_\nu$.

$S_3=0$ ν Quark Before the reaction ν Antiquark $S_3=-1$

$S_3=0$ μ⁻ Quark Scattering through 180° μ⁻ Antiquark $S_3=-1$

In terms of the individual neutrino quark cross sections the double differential cross section per nucleon for scattering on an isoscalar target can be written as

$$\frac{\mathrm{d}^2\sigma^\nu}{\mathrm{d}x\,\mathrm{d}y} = \frac{G_\mathrm{F}^2 ME}{\pi} P^2 \cdot 2x \left((\mathrm{u}+\mathrm{d}+2\mathrm{s}) + (1-y)^2(\bar{\mathrm{u}}+\bar{\mathrm{d}}+2\bar{\mathrm{c}}) \right) \tag{49}$$

$$\frac{\mathrm{d}^2\sigma^{\bar{\nu}}}{\mathrm{d}x\,\mathrm{d}y} = \frac{G_\mathrm{F}^2 ME}{\pi} P^2 \cdot 2x \left((\bar{\mathrm{u}}+\bar{\mathrm{d}}+2\bar{\mathrm{s}}) + (1-y)^2(\mathrm{u}+\mathrm{d}+2\mathrm{c}) \right) \ . \tag{50}$$

Here G_F is the Fermi weak interaction coupling constant and P^2 is the W propagator term

$$P^2 = \left(\frac{M_\mathrm{W}^2}{M_\mathrm{W}^2 + Q^2} \right)^2 \ . \tag{51}$$

Typical y-distributions for $\nu, \bar{\nu}$ interactions are shown in Fig. 7. For neutrino scattering the distribution is flat in y with a small $(1-y)^2$ admixture arising from the seaquarks, for antineutrino scattering the distribution is dominated by the $(1-y)^2$ term from the interaction with the valence quarks, a small y

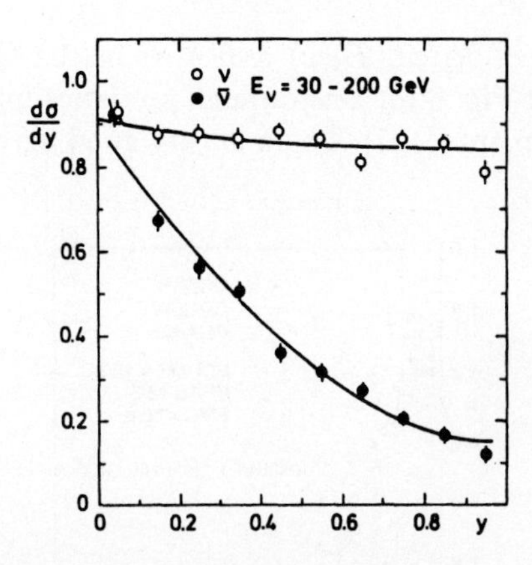

Fig. 7. *Differential cross sections $\mathrm{d}\sigma/\mathrm{d}y$ for neutrino and antineutrino scattering off nucleons as a function of y*

independent part arises from the seaquarks. Valence and sea quark distributions can directly be obtained by extrapolations to $y = 0$ and $y = 1$. These cross sections can also be written in terms of three structure functions $F_i(x, Q^2)$:

$$\begin{aligned}\frac{\mathrm{d}^2\sigma^{\nu,\bar{\nu}}}{\mathrm{d}x\,\mathrm{d}y} = \frac{G_\mathrm{F}^2 ME}{\pi} P^2 \Big\{ \left(1 - y - \frac{Mxy}{2E}\right) F_2^{\nu,\bar{\nu}}(x, Q^2) + \\ + \frac{y^2}{2} 2xF_1^{\nu,\bar{\nu}}\left(x, Q^2\right) \pm y\left(1 - \frac{y}{2}\right) xF_3^{\nu,\bar{\nu}}\left(x, Q^2\right) \Big\} \ .\end{aligned} \tag{52}$$

The structure function F_3 is a consequence of the V-A structure of the weak charged current. The term with F_3 has positive sign for neutrino scattering and negative sign for antineutrino scattering. Assuming $2xF_1 = F_2$, one obtains by comparison of equations (49) and (50) and (52) the expression for the structure functions per nucleon for an isoscalar target in terms of quark distributions

$$xF_3^\mathrm{N}(x) = \frac{1}{2}\left(xF_3^{\nu\mathrm{N}}(x) + xF_3^{\bar{\nu}\mathrm{N}}(x)\right) = x(\mathrm{u} + \mathrm{d} - \bar{\mathrm{u}} - \bar{\mathrm{d}}) = x(\mathrm{u_v} + \mathrm{d_v}) \tag{53}$$

$$F_2^{\nu\mathrm{N}}(x) = F_2^{\bar{\nu}\mathrm{N}}(x) = x(\mathrm{u} + \bar{\mathrm{u}} + \mathrm{d} + \bar{\mathrm{d}} + \mathrm{c} + \bar{\mathrm{c}} + \mathrm{s} + \bar{\mathrm{s}}) \simeq \frac{18}{5} F_2^{\mu\mathrm{N}}(x) \tag{54}$$

$$\bar{q}^{\bar{\nu}\mathrm{N}} = x(\bar{\mathrm{u}} + \bar{\mathrm{d}} + 2\bar{\mathrm{s}}) \ . \tag{55}$$

The quark, antiquark and valence quark distributions are separately measurable by appropriate combinations of neutrino and antineutrino cross sections: The valence quark distribution from

$$\frac{\mathrm{d}\sigma^\nu}{\mathrm{d}x} - \frac{\mathrm{d}\sigma^{\bar{\nu}}}{\mathrm{d}x} \ ,$$

the seaquark distribution from

$$3\frac{\mathrm{d}\sigma^{\bar{\nu}}}{\mathrm{d}x} - \frac{\mathrm{d}\sigma^\nu}{\mathrm{d}x} \ .$$

The x dependence of $F_2(x)$, $F_3(x)$ and $\bar{q}(x)$ in the Q^2 range $10 < Q^2 < 30\,(\mathrm{GeV}^2)$ is plotted in Fig. 8 for several high statistics muon and neutrino nucleon scattering experiments [167]. There is very good agreement between these

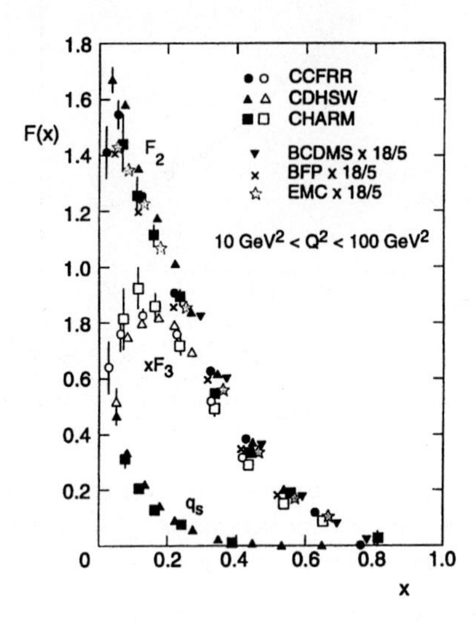

Fig. 8. *The structure functions F_2, xF_3 and the seaquark distribution q_s versus x from high energy neutrino and muon experiments.*

experiments. The main features of the data to be noticed are:

- $F_2^{\nu N}(x) \simeq 18/5 F_2^{\mu N}(x)$, as expected from the quark parton model,
- the seaquark distribution $\bar{q}^{\nu N}$ falls off steeply with x and is negligible for $x > 0.35$–0.4,
- at larger x only valence quarks contribute to F_2, they dominate for $x \gtrsim 0.1$.

The most precise measurement for the seaquark distribution $\bar{q}(x)$ has been obtained by the CCFR experiment at Fermilab [165]. The data is shown in Fig. 9 as a function of x at two values of Q^2. From detailed studies of di-muon events in neutrino and antineutrino scattering one can in addition obtain specific information about the distribution of strange quarks $\mathrm{s}(x)$, which only appear in the quark-antiquark sea. The most precise results were obtained by the CCFR collaboration [65]. The strange quark distribution is very similar in shape to the light antiquark distributions but suppressed by about a factor of two.

$$\frac{2\int x\bar{\mathrm{s}}(x)\,\mathrm{d}x}{\int x\left(\bar{\mathrm{u}}(x)+\bar{\mathrm{d}}(x)\right)\,\mathrm{d}x} = 0.477^{+0.051+0.017}_{-0.050-0.036}\,. \tag{56}$$

By integrating the cross sections (49) and (50) for neutrino and antineutrino scattering on an isoscalar target over x and y one can separate the contributions

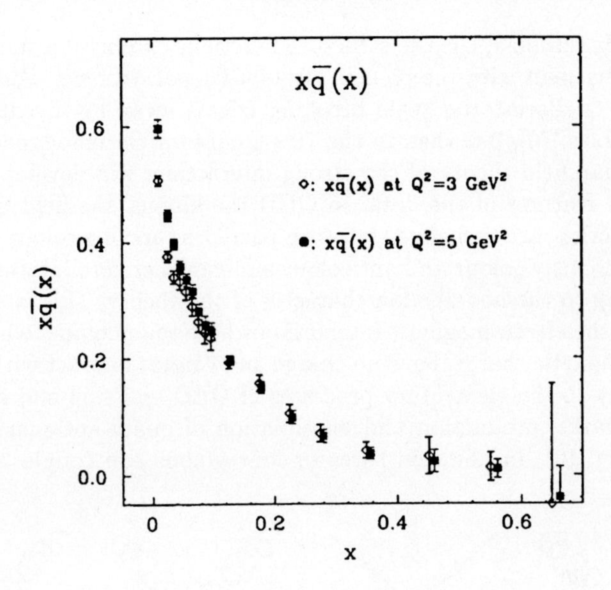

Fig. 9. *The antiquark distribution* $\bar{\mathrm{q}}(x)$ *versus* x *from the CCFR neutrino experiment*

of valence and sea quarks to the nucleon momentum [103], since the ratio of the total cross sections can be written as

$$\frac{\sigma(\bar{\nu})}{\sigma(\nu)} = \frac{\bar{Q} + \frac{1}{3}(V + Q_{\mathrm{s}})}{(V + Q_{\mathrm{s}}) + \frac{1}{3}\bar{Q}} = \frac{1 + \frac{4\bar{Q}}{V}}{3 + \frac{4\bar{Q}}{V}} \ , \tag{57}$$

where $Q_{\mathrm{s}} = \bar{Q} = \sum_f \int_0^1 \bar{q}_f(x)\,\mathrm{d}x$, $V = \int_0^1 (\mathrm{u_v}(x) + \mathrm{d_v}(x))\,\mathrm{d}x$, and the factor $\frac{1}{3}$ arises from the y integration of $(1-y)^2$.

From the experimental results

$$\begin{aligned} &\sigma(\nu)/E_\nu = 0.67 \times 10^{-38}\,\mathrm{cm}^2/\mathrm{GeV} \\ &\sigma(\bar{\nu})/E_\nu = 0.34 \times 10^{-38}\,\mathrm{cm}^2/\mathrm{GeV} \\ &\int_0^1 F_2^{\nu\mathrm{N}}\,\mathrm{d}x \simeq 0.48 \end{aligned} \tag{58}$$

one can derive

$$V = 0.31, \qquad \frac{\bar{Q}}{V} = 0.265, \qquad S = Q_s + \bar{Q}_s = 0.17 \ , \tag{59}$$

i. e. valence quarks carry about 31% of the nucleon's momentum, sea quarks about 17% and gluons about 52% at $Q^2 \simeq 10\,\mathrm{GeV}^2$.

1.3 Deep Inelastic Lepton Nucleon Scattering and Pertubative QCD

Already the very early experiments at SLAC showed that the charged quarks carry only about half of the nucleon's momentum and that scaling was only approximately fullfilled, i. e. the structure functions exhibited a small Q^2 dependence not consistent with the picture of pointlike constituents. For some period people tried to discuss the scale breaking effects away by inventing modified scaling variables [76], but then in the 70's Quantum Chromodynamics (QCD), the non-abelian field theory of the strong interaction, was developed which explains all the features of the data. In QCD the gluons, the field quanta of the strong interaction, are exchanged between particles carrying colour charge. They carry simultanously colour and anticolour and can therefore interact with other gluons leading to the non-abelian character of the theory. This is an important difference to the electromagnetic interaction where the photons, which couple to the electromagnetic charge, have no charge and cannot interact with each other.

In analogy to the elementary processes of QED emission and absorption of gluons by quarks, production and annihilation of quark-antiquarks pairs take place (see Fig. 10). In addition three or four gluons can couple to each other

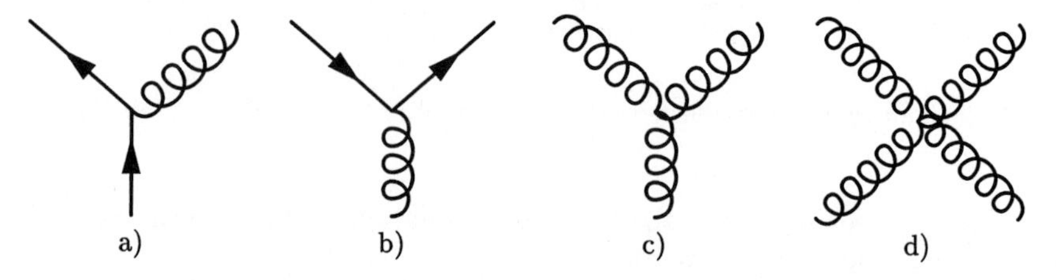

Fig. 10. *The fundamental interaction diagrams of the strong interaction: a) Emission of a gluon by a quark, b) splitting of a gluon into a quark-antiquark pair, c,d) "self-coupling" of gluons*

with a coupling strength which in first order pertubative calculation in QCD is given as

$$\alpha_s^{(1)}(Q^2) = \frac{4\pi}{\beta_0 \ln(Q^2/\Lambda_{(1)}^2)} \tag{60}$$

and in next to leading order as

$$\alpha_s^{(2)}(Q^2) = \alpha_s^{(1)}(Q^2)\left(1 - \frac{\beta_1}{\beta_0^2}\frac{\ln\ln(Q^2/\Lambda_{(2)}^2)}{\ln(Q^2/\Lambda_{(2)}^2)}\right) \tag{61}$$

$$\text{with} \qquad \beta_0 = 11 - \frac{2}{3}n_f\,,\ \beta_1 = 102 - \frac{38}{3}n_f\ . \tag{62}$$

Here n_f denotes the number of quark types involved at a given Q^2. Since a heavy virtual quark-antiquark pair has a very short lifetime and range it can be resolved only at very high Q^2. Hence n_f depends on Q^2, with n_f = 3–6. The parameter $\Lambda_{(1)}$ ($\Lambda_{(2)}$) is the QCD scale parameter which gives the limit where pertubative QCD can be applied. Since typical hadronic radii are in the order of $\sim$ 1 fm one expects Λ to be in the order of 200 MeV. The Q^2 dependence of the coupling strength corresponds to a dependence on spatial separation. For very small distances and corresponding high values of Q^2, the interquark coupling decreases logarithmically, vanishing asymptotically. This is the reason why at high Q^2 the DIS process can be considered as the incoherent sum of elastic scattering from free quarks. Due to the process depicted in Fig. 10 and the running coupling constant the quark and gluon momentum distributions and hence the structure functions become Q^2 dependent. Figure 11 is an attempt to visualise the underlying physics. At a certain value of $Q^2 = Q_0^2$ one measures a

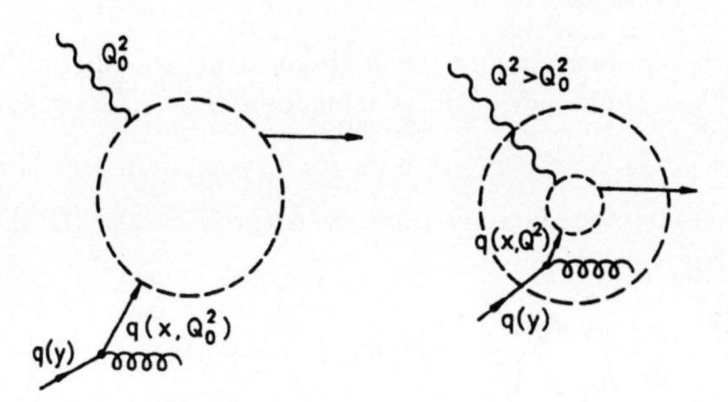

Fig. 11. *Q^2 dependence of quark distributions due to increased resolving power of the virtual photon with increasing Q^2*

quark distribution $q(x, Q^2)$. If one increases the space-time resolution by going to higher $Q^2 > Q_0^2$, one is able to see that the quark momentum has changed by gluon radiation. A quark with momentum fraction x can originate from a parent quark with a larger momentum fraction y. The probability that this happens is proportional to $\alpha_s(Q^2)P_{qq}(\frac{x}{y})$, where P_{qq} is a so-called splitting function. But a quark with momentum x can also arise from a gluon with higher momentum y, the probability for this process is proportional to another splitting function $P_{qg}(\frac{x}{y})$. Similary the gluon momentum is modified by contributions from quarks or other gluons of higher momentum radiating gluons.

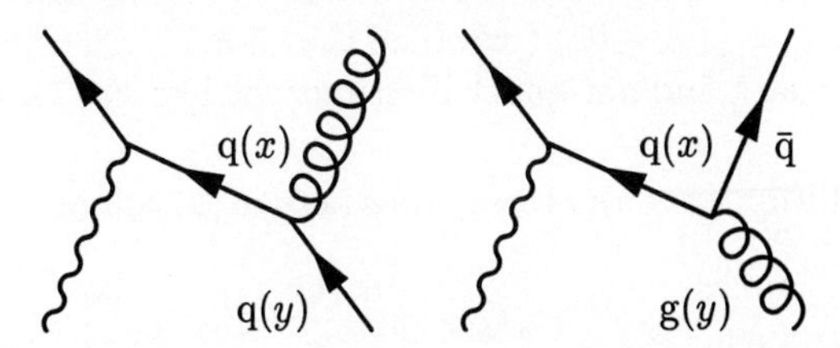

The net effect of this processes is that the structure functions are depleted with increasing Q^2 at high x and enhanced at low x as it is shown in Fig. 12 as a function of x for different Q^2 and as a function of Q^2 for fixed values of x. The coupling of $g(x)$ and $q(x)$ and their logarithmic Q^2 dependence is expressed

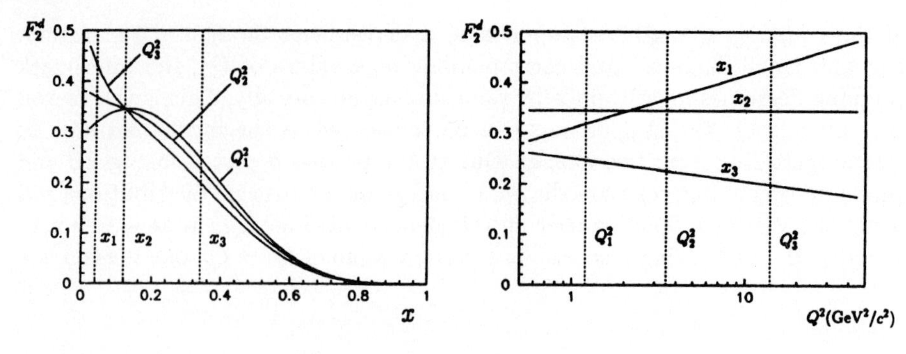

Fig. 12. *Schematic representation of the deuteron structure function F_2^{d} as a function of x at various values of Q^2* (left), *and as a function of Q^2 at constant x* (right)

by the coupled Dokshitzer-Gribov-Lipatov-Altarelli-Parisi (DGLAP) equations [100, 127, 152, 26]

$$\frac{\mathrm{d}q(x,Q^2)}{\mathrm{d}\ln Q^2} = \frac{\alpha_{\mathrm{s}}}{2\pi}\int_x^1 \frac{\mathrm{d}y}{y}\left\{q(y,Q^2)P_{\mathrm{qq}}\left(\frac{x}{y}\right) + g(y,Q^2)P_{\mathrm{qg}}\left(\frac{x}{y}\right)\right\} \tag{63}$$

$$\frac{\mathrm{d}g(x,Q^2)}{\mathrm{d}\ln Q^2} = \frac{\alpha_{\mathrm{s}}}{2\pi}\int_x^1 \frac{\mathrm{d}y}{y}\left\{\sum_f q_f(y,Q^2)P_{\mathrm{gq}}\left(\frac{x}{y}\right) + g(y,Q^2)P_{\mathrm{gg}}\left(\frac{x}{y}\right)\right\}\ . \tag{64}$$

The splitting functions can be calculated in QCD (for details see e. g. Chap. 5.2 of [172]).

If one knows the shape of the distributions $q(x,Q_0^2)$, $g(x,Q_0^2)$ at one certain value of $Q^2 = Q_0^2$ then from (63,64) one can calculate $q(x,Q^2)$ and $g(x,Q^2)$ at any other value of Q^2. Note that perturbative QCD can only predict the Q^2-dependence of the distributions but not their shape in x.

It is convenient to split the structure functions into contributions from non-singlet (NS) and singlet (S) quark distributions. Valence quark distributions are NS distributions. They do not depend on the gluon distribution and their Q^2 evolution is in leading order just given by $\alpha_{\mathrm{s}}(Q^2)$ and the splitting function P_{qq}. Examples are the structure functions $xF_3^{\nu\mathrm{N}} = x(u_{\mathrm{v}}+d_{\mathrm{v}})$ obtained from $\nu,\bar{\nu}$ scattering and $F_2^{\mathrm{p}} - F_2^{\mathrm{n}} = \frac{1}{3}(u_{\mathrm{v}} - d_{\mathrm{v}})$ (provided that $\bar{u} = \bar{d}$). The singlet distribution is the sum of all quark and antiquark distributions like $F_2^{\nu\mathrm{N}}$ and approximately F_2^{eN}.

Using the notation $(a\otimes b)(x) = \int_x^1 \mathrm{d}y\, a\,(x/y)\, b\,(y)\,/y$ one obtains in leading order:

$$\frac{\mathrm{d}}{\mathrm{d}\ln Q^2}\, q^{\mathrm{NS}}(x,Q^2) = \frac{\alpha_{\mathrm{s}}(Q^2)}{2\pi} P_{\mathrm{qq}}\otimes q^{\mathrm{NS}} \tag{65}$$

$$\frac{\mathrm{d}}{\mathrm{d}\ln Q^2}\begin{pmatrix} q^{\mathrm{S}}(x,Q^2) \\ g(x,Q^2)\end{pmatrix} = \frac{\alpha_{\mathrm{s}}(Q^2)}{2\pi}\begin{pmatrix} P_{\mathrm{qq}} & 2n_f P_{\mathrm{qg}} \\ P_{\mathrm{gq}} & P_{\mathrm{gg}}\end{pmatrix}\otimes\begin{pmatrix} q^{\mathrm{S}} \\ g\end{pmatrix}\ . \tag{66}$$

In next to leading order one obtains similar expressions with more and different splitting functions (there are e. g. also contributions from splitting functions

$P_{q_i\bar{q}_j}$) which depend on the renormalisation scheme used. (For details see e.g. Chap. 5.4 of [172] and [133, 98, 109].) The analysis of the pattern of scale breaking of the structure functions in terms of these equations in principle allows the determination of the gluon distribution $g(x)$ and its Q^2 dependence and of the QCD scale parameter Λ or of the coupling $\alpha_s(Q^2)$, respectively. Note, however, that both $\alpha_s(Q^2)$ and $g(x, Q^2)$ enter in the Q^2 dependence of q^S and are therefore coupled. Thus one usually determines $\alpha_s(Q^2)$ from the logarithmic slope of q^{NS} and then $g(x, Q^2)$ from the Q^2 dependence of q^S.

1.4 Second and Third Generation Experiments

The experiments at SLAC were continued through the 70's and plenty of high statistics deep inelastic data were collected covering a x-range $0.07 \leq x < 0.9$ and a Q^2 range $1 < Q^2 < 18\,\mathrm{GeV}^2$. End of the 70's the Tevatron at FNAL and the SPS at CERN came into operation and a series of high energy neutrino experiments (CDHS, CHARM, CCFRR) and muon experiments (BCDMS, EMC, E665) using lepton beam energies up to 550 GeV have been performed to study the quark gluon structure in detail up to high values of $Q^2 \simeq 200\,\mathrm{GeV}^2$, to compare the data with expectations from QCD and especially to precisely determine the strong coupling constant $\alpha_s(Q^2)$ and the gluon distribution $g(x, Q^2)$. The results of these experiments have been summarised in several excellent reviews [103, 177, 173]. The data were in reasonable agreement but obviously had larger systematic errors than anticipated, since although in general they showed the same pattern of scale breaking they differed in details. Especially the two CERN muon experiments EMC and BCDMS showed large discrepancies in their x dependence as can be seen from Fig. 13, where the Q^2 averaged ratio of the structure function F_2 from EMC and BCDMS is plotted as a function of x. Obviously at small values of x the EMC data were substantially below and at large x above those from BCDMS.

It also turned out that despite all the efforts the neutron structure function F_2^n and its Q^2 depedence was rather badly known. The main reason for this limited data quality was that for the determination of F_2^n a subtraction of deuteron and proton data is required.

$$F_2^n = 2F_2^d - F_2^p \; . \tag{67}$$

Normally these data were taken under rather different apparative conditions (e.g. in different years with different detector performances and efficiencies) and since two numbers of similar size needed to be subtracted the data were harmed a lot by systematic errors. Therefore mid of the 80's the New Muon Collaboration (NMC) at CERN started a new round of experiments using the upgraded EMC forward spectrometer [22].

The NMC Experiment

The main aim of the experiment, which was performed during the years 1986–1989 was to obtain high precision data over a large range of x and Q^2 with much

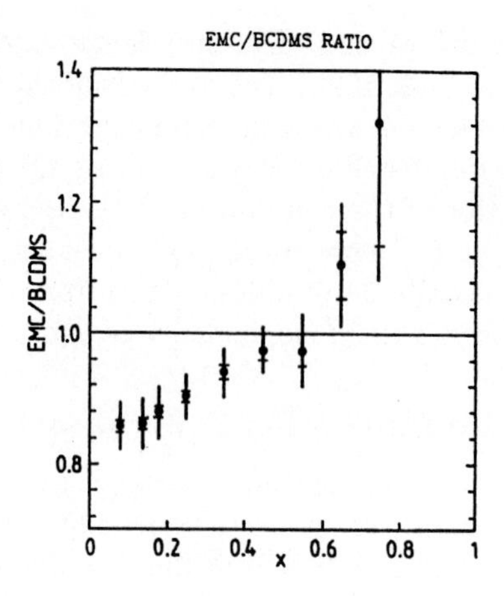

Fig. 13. *Ratio of the structure functions F_2 measured by EMC and BCDMS*

reduced systematic errors compared to the previous experiments. To achieve these goals additional tracking detectors were installed, the angular acceptance was increased down to very low scattering angles of about 3 mrad, the muon energy was calibrated with an additional spectrometer, a separate normalisation trigger was installed and a special 'complementary target' setup was used which will be discussed in some more detail below. Hydrogen, Deuterium, Lithium, Beryllium, Carbon, Aluminium, Calcium, Iron, Tin and Lead were used as target material. From the nuclear targets details of the EMC effect – the modification of quark and gluon distributions due to the nuclear environment – were studied. A discussion of the nuclear data is beyond the scope of this review and can be found e. g. in [38]. The experimental setup is shown in Fig. 14.

The proportional chambers P4A–P5C were additionally installed to improve the track reconstruction at small angles. It turned out that these detectors were crucial for the data quality and the understanding of the discrepancy between the EMC and the BCDMS data. Detailed studies of reconstruction efficiencies showed that the efficiency of the large drift chambers W4–W5 in front of the calorimeter/hadron absorber were affected substantially by correlated low energy background originating from backward particles from hadron showers in the calorimeter/hadron absorber. This background depended on the hadron multiplicities and therefore on ν. Despite enormous efforts and extensive Monte Carlo studies it was not possible to safely correct the data for this effect. Therefore only the central part of the detector covered by the chambers P4–P5 was used for the determination of the NMC structure functions. Fortunately data were taken at four different muon beam energies (90 GeV, 120 GeV, 200 GeV, 280 GeV) and therefore a complete coverage of the kinematic range could be achieved even with this geometrical restriction.

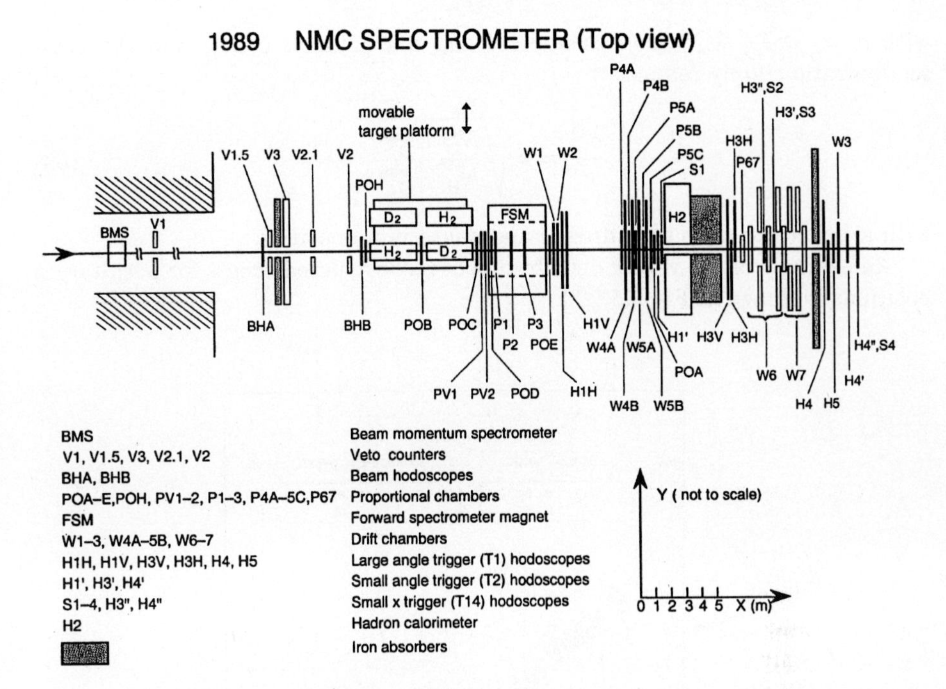

Fig. 14. *The NMC muon spectrometer*

The Complementary Target Setup

The complementary target setup of NMC was invented to reduce the systematic errors in the comparison of data from different target materials substantially. The principle of the setup is shown in Fig. 14 and the top part of Fig. 15 for the example of hydrogen (H) and deuterium (D).

Both a deuterium target and a hydrogen target are in the beam at the same time. In position 1 the D target is upstream and the H target is downstream. They see the same muon flux I_1 but their geometrical acceptance is different $A_{\mathrm{H}}^{\mathrm{up}} \neq A_{\mathrm{D}}^{\mathrm{down}}$ due to their different distance from the spectrometer magnet. Within short time intervalls this row of targets is interchanged with a second one where H and D targets have changed places. In this arrangement both targets again see the same muon flux I_2, and $A_{\mathrm{D}}^{\mathrm{up}} \neq A_{\mathrm{H}}^{\mathrm{down}}$. If this procedure is repeated frequently enough one can safely assume that (at least for subsequent runs) the performance of the spectrometer has not changed and $A_{\mathrm{H}}^{\mathrm{up}} = A_{\mathrm{D}}^{\mathrm{up}}$, $A_{\mathrm{H}}^{\mathrm{down}} = A_{\mathrm{D}}^{\mathrm{down}}$.

Calculating the cross section ratio $\sigma_{\mathrm{D}}/\sigma_{\mathrm{H}}$ from counting rates, acceptances, fluxes and target areal densities T one obtaines

$$\frac{\sigma_{\mathrm{D}}^{\mathrm{up}}}{\sigma_{\mathrm{H}}^{\mathrm{down}}} \cdot \frac{\sigma_{\mathrm{D}}^{\mathrm{down}}}{\sigma_{\mathrm{H}}^{\mathrm{up}}} = \frac{N_{\mathrm{D}}^{\mathrm{up}} A^{\mathrm{down}} I_1 T_{\mathrm{H}}^{\mathrm{down}}}{A^{\mathrm{up}} I_1 T_{\mathrm{D}}^{\mathrm{up}} N_{\mathrm{H}}^{\mathrm{down}}} \cdot \frac{N_{\mathrm{D}}^{\mathrm{down}} A^{\mathrm{up}} I_2 T_{\mathrm{H}}^{\mathrm{up}}}{A^{\mathrm{down}} I_2 T_{\mathrm{D}}^{\mathrm{down}} N_{\mathrm{H}}^{\mathrm{up}}} \tag{68}$$

with $\sigma = \frac{d^2\sigma}{dQ^2\,dx}\Delta Q^2 \Delta x$. Acceptances and muon fluxes cancel and the cross section ratio simply reduces to

$$\frac{\sigma^{\mathrm{D}}}{\sigma^{\mathrm{H}}} = \kappa \sqrt{\frac{N_{\mathrm{D}}^{\mathrm{up}} N_{\mathrm{D}}^{\mathrm{down}}}{N_{\mathrm{H}}^{\mathrm{up}} N_{\mathrm{H}}^{\mathrm{down}}}} \tag{69}$$

with κ being the ratio of hydrogen and deuterium densities.

An example of the vertex distribution of reconstructed events for both target positions is shown in Fig. 15 [23].

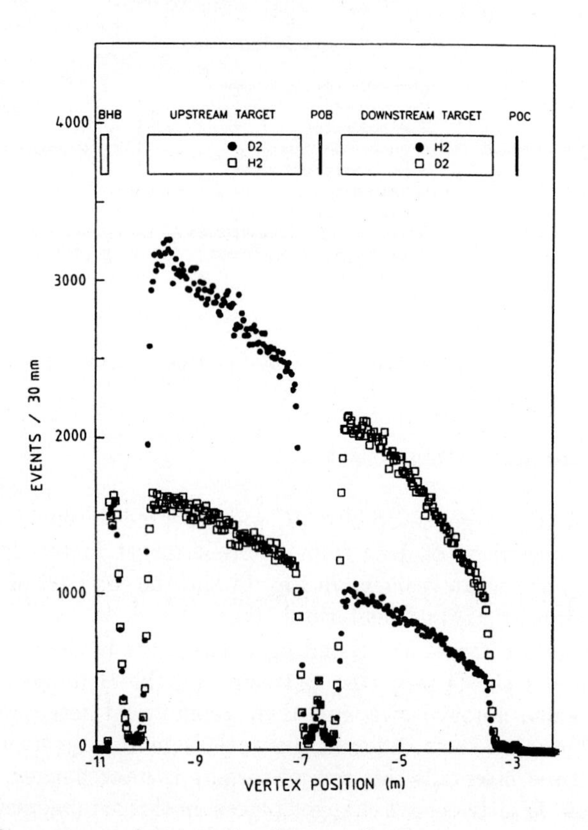

Fig. 15. *Reconstructed vertex distributions for the two target arrangements*

The Proton and Deuteron Structure Functions F_2^{p} and F_2^{d}

From the cross sections measured at beam energies of 90, 120, 200 and 280 GeV NMC has determined the structure functions F_2^{p} and F_2^{d}, first using the large angle trigger of the experiment [40] and recently also for the data taken with a small angle trigger at 200 and 280 GeV in 1989 [41]. The extraction of $F_2(x, Q^2)$ from (33) requires (apart from standard radiative corrections and corrections for geometrical acceptance, detector efficiency, reconstrution efficiency etc.) the knowledge of $R(x, Q^2)$. $R(x, Q^2)$ can be determined if for the same x and Q^2 data are available for different beam energies (see below), in the other cases assumptions about the value of R and its x, Q^2 dependence must be made. For the data shown below the SLAC parametrisation of R has been used [185]. In Fig. 16 and Fig. 17 the measured $F_2^{\mathrm{p}}(x, Q^2)$ and $F_2^{\mathrm{d}}(x, Q^2)$ averaged over the four beam energies are shown as a function of Q^2 for fixed values of x The error bars represent the statistical errors, the bands, apart from an overall 2.5% normalisation error, the systematic uncertainties.

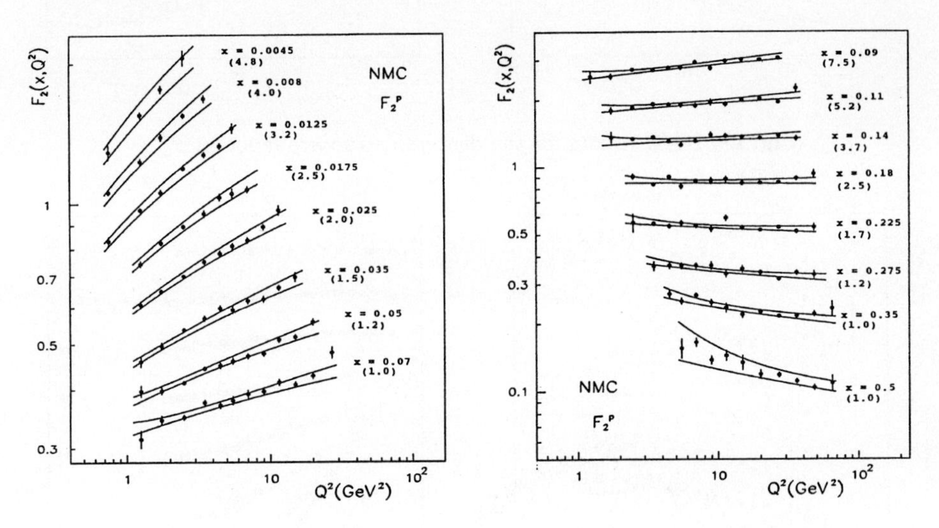

Fig. 16. *NMC results for the proton structure function* F_2^{p}

The data at each x bin have been scaled by the factor indicated in brackets. The data cover a range of $0.0045 < x < 0.5$, $0.7\,\mathrm{GeV}^2 < Q^2 < 65\,\mathrm{GeV}^2$. They exhibit the pattern of scale breaking expected from QCD. At low values of x F_2 rises with Q^2, at large values of x it decreases with Q^2. The data are in good agreement with those of SLAC [186] and BCDMS [67, 68] as shown in Fig. 18 for the deuteron.

Data points for $x < 0.07$ are exclusively from NMC. Recently, final results extending down to $x = 0.0008$ have become available from the fixed target muon scattering experiment E665 at Fermilab [142] which also agree well with the NMC data in the region of overlap. In Fig. 19 a comparison of the NMC data is shown with the results by the H1 [19] and ZEUS [99] collaborations at HERA from their 1994 data. Again, especially at low x, the agreement is very good and the high Q^2 data from HERA and low Q^2 data from NMC extrapolate nicely to each other.

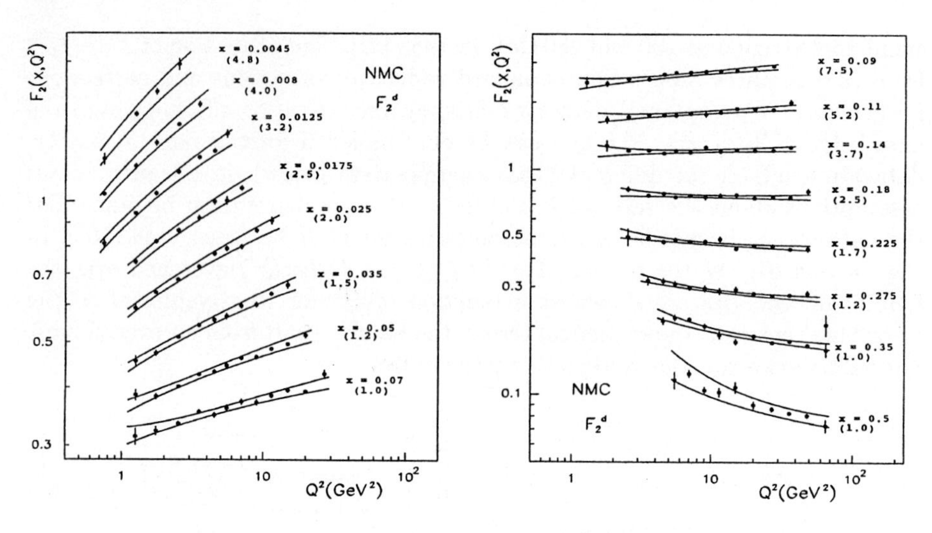

Fig. 17. *NMC results for the deuteron structure function* F_2^{d}

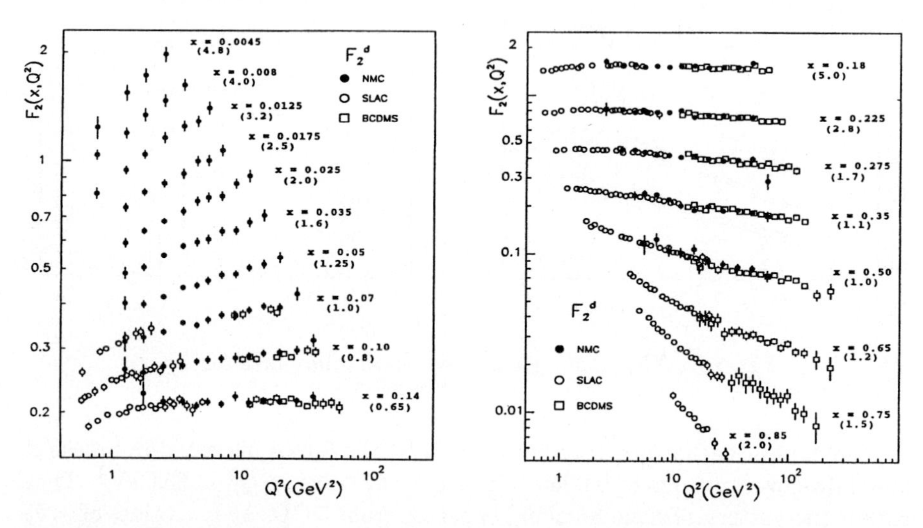

Fig. 18. *NMC results for the deuteron structure function* F_2^{d} *compared with results from BCDMS and SLAC*

Parametrisation of F_2^{p} and F_2^{d}

The NMC data [40] without the recent additional low angle data [43] have been used together with the results from SLAC and BCDMS to obtain parametrisations of the structure functions F_2^{p} and F_2^{d} and their uncertainities using the

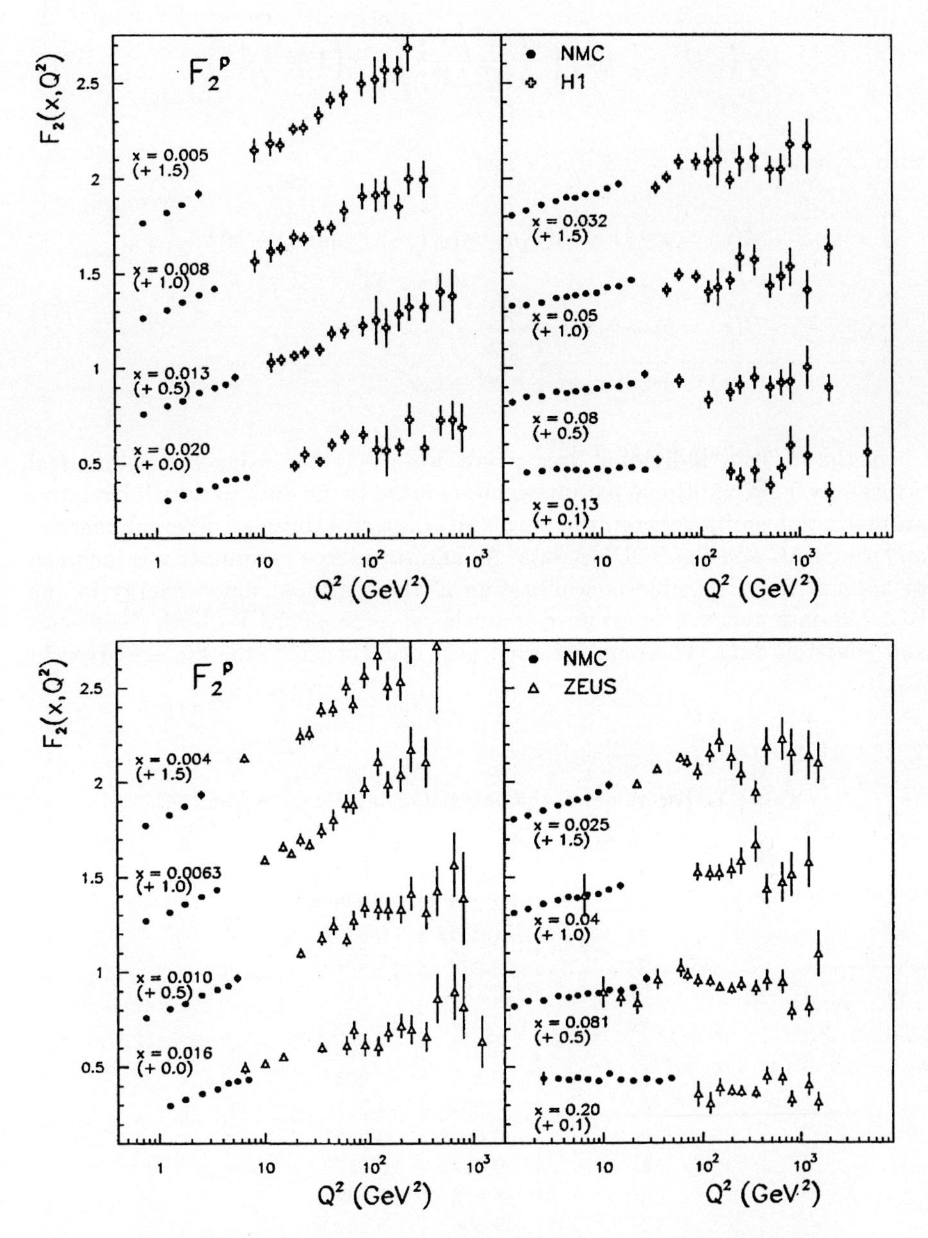

Fig. 19. *NMC data for F_2^p compared with data from H1 and ZEUS*

15-parameter function [164]

$$F_2\left(x, Q^2\right) = A(x) \left(\frac{\ln\left(Q^2/\Lambda^2\right)}{\ln\left(Q_0^2/\Lambda^2\right)} \right)^{B(x)} \cdot \left(1 + \frac{C(x)}{Q^2}\right) , \tag{70}$$

with $Q_0^2 = 20\,\mathrm{GeV}^2$, $\Lambda = 0.250\,\mathrm{GeV}$ and

$$\begin{aligned} A(x) &= x^{a_1}(1-x)^{a_2}\{a_3 + a_4(1-x) + a_5(1-x)^2 \\ &\qquad + a_6(1-x)^3 + a_7(1-x)^4\} \\ B(x) &= b_1 + b_2 x + \frac{b_3}{x + b_4} \\ C(x) &= c_1 x + c_2 x^2 + c_3 x^3 + c_4 x^4 \end{aligned} \tag{71}$$

In the fits, the individual data points were weighted using their statistical errors only. Five additional parameters were fitted to the data to describe relative normalisation shifts between the four NMC data sets taken at different energies and the SLAC and the BCDMS data. An additional free parameter was included to account for a possible miscalibration of the scattered muon energy in the BCDMS data and was determined from the fit to be +0.2% for both the proton and deuteron data. The parameters of (71) resulting from the fits are given in Table 1.

Table 1. *The values of the parameters of (71) for F_2^{p} and F_2^{d}*

Parameter	F_2^{p}	F_2^{d}
a_1	−0.02778	−0.04858
a_2	2.926	2.863
a_3	1.0362	0.8367
a_4	−1.840	−2.532
a_5	8.123	9.145
a_6	−13.074	−12.504
a_7	6.215	5.473
b_1	0.285	−0.008
b_2	−2.694	−2.227
b_3	0.0188	0.0551
b_4	0.0274	0.0570
c_1	−1.413	−1.509
c_2	9.366	8.553
c_3	−37.79	−31.20
c_4	47.10	39.98

QCD Analysis of xF_3, $F_2^{\rm p}$ and $F_2^{\rm d}$

The Q^2 evolution of a nonsinglet distribution does not depend on the gluon distribution and allows a direct determination of $\alpha_s(Q^2)$, using the nonsinglet DGLAP evolution equation (65). The presently most precise determination of a nonsinglet structure function is the measurement of the structure function $xF_3(x, Q^2)$ of the CCFR collaboration [170] from neutrino iron scattering. The data are shown in Fig. 20.

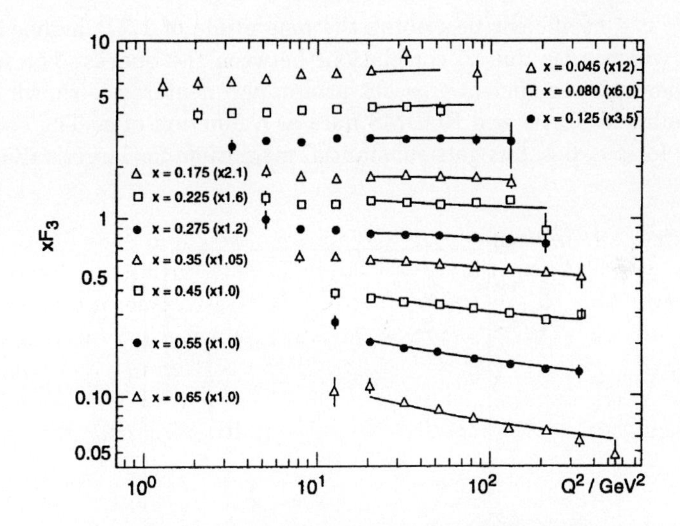

Fig. 20. *The structure function xF_3 from CCFR.*

The analysis has been performed in NLO in the $\overline{\rm MS}$ scheme including target mass corrections. Only data above $Q^2 = 15\,{\rm GeV}^2$ and $W^2 > 10\,{\rm GeV}^2$ have been used for the analysis to exclude nonpertubative effects. They obtained for 4 quark flavours ($n_{\rm f} = 4$) the following value of the QCD parameter Λ:

$$\Lambda^{(4)}_{\overline{\rm MS}} = (210 \pm 28({\rm stat}) \pm 41({\rm syst}))\ {\rm MeV}\ . \tag{72}$$

With this value of $\Lambda^{(4)}_{\overline{\rm MS}}$ one obtains

$$\alpha_s(M_Z^2) = 0.111 \pm 0.002({\rm stat}) \pm 0.003({\rm syst}) \pm 0.004({\rm scale})\ . \tag{73}$$

It should be noted, however, that a recent reanalysis [174] of the CCFR data results in a substantially higher value of the strong coupling constant, namely $\alpha_s(M_Z^2)_{\rm new} = 0.119 \pm 0.002({\rm exp}) \pm 0.004({\rm theor})$.

A detailed singlet analysis for $F_2^{\rm p}$ and $F_2^{\rm d}$ has been performed for the SLAC and BCDMS data [181] and somewhat later also for the NMC data [37] using the same computer code which performs a vectorised full numerical integration of the evolution equations in NLO in the $\overline{\mathrm{MS}}$ renormalisation and factorisation scheme [98, 115, 116]. The fit was performed simultaneously on the measured values of F_2^{p}, which has both flavour singlet and nonsinglet components and of $F_2^{\mathrm{d}} = \left(F_2^{\mathrm{p}} + F_2^{\mathrm{n}}\right)/2$ which is nearly a pure singlet structure function. Since the data extend to rather low Q^2 targetmass corrections [166, 122] and higher twist contributions have been taken into account

$$F_2 = F_2^{\mathrm{NLO+TMC}} \left(1 + \frac{C_{\mathrm{HT}}(x)}{Q^2}\right) , \tag{74}$$

where C_{HT} is a coefficient describing the magnitude of $1/Q^2$ higher twist contributions which arise due to correlations between the quarks. The magnitude of the higher twist coefficient C_{HT} for proton and deuteron is shown in Fig. 21 for the combined SLAC and BCDMS data as a function of x. The coefficient is very small for $x \leq 0.4$, but gets substantial magnitude for larger values of x.

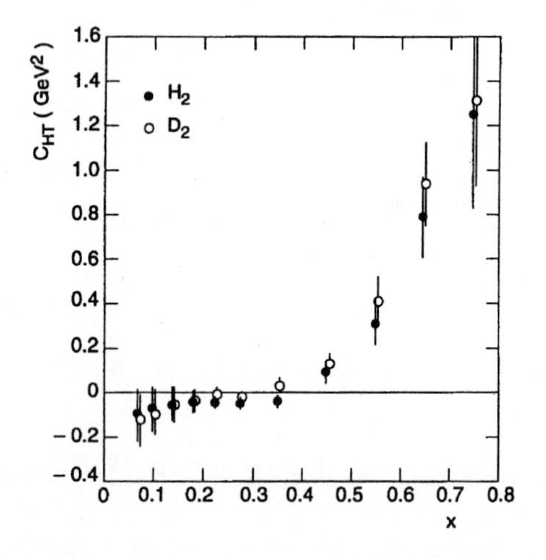

Fig. 21. *Higher twist coefficient C_{HT} for proton and deuteron versus x from the combined SLAC and BCDMS data.*

Thus the Q^2 dependence at large values of x and small values of Q^2 is substantially influenced by higher twist effects. The value of $\Lambda_{\overline{\mathrm{MS}}}^{(4)}$ for this analysis was

$$\Lambda_{\overline{\mathrm{MS}}}^{(4)} = (263 \pm 42)\ \mathrm{MeV} , \tag{75}$$

which results in

$$\alpha_{\mathrm{s}}(M_{\mathrm{Z}}^2) = 0.113 \pm 0.003(\mathrm{exp}) \pm 0.004(\mathrm{scale}) . \tag{76}$$

The NMC data extend to much lower values of x and therefore are much more sensitive to shape and size of the gluon distribution $g(x)$. The analysis [37] has been performed in two steps. In the first step the value of $\alpha_s(M_Z^2) = 0.113$ from the above analysis, which corresponds to $\alpha_s(Q_0^2 = 7\,\text{GeV}^2) = 0.240$ has been fixed to obtain the quark and gluon distributions with best precision. The value of Q_0^2 was chosen to $7\,\text{GeV}^2$ and the parametrisation used in the fit was

$$xq^{\text{NS}}\left(x, Q_0^2\right) = Ax^{\alpha}\left(1-x\right)^{\beta}, \tag{77}$$

$$xq^{\text{d}}\left(x, Q_0^2\right) = Bx^{\gamma}\left(1-x\right)^{\delta}\left(1+b_1 v+b_2 v^2\right), \tag{78}$$

with $v = 0.1 - x$,

$$xg\left(x, Q_0^2\right) = C\left(1-x\right)^{\eta}\left(1+c_1 w+c_2 w^2+c_3 w^3\right), \tag{79}$$

with $w = 0.1\ln(1+e^{10-100x})$. The results for $xq^{\text{NS}}(x, Q_0^2), xq^{\text{d}}(x, Q_0^2)$ are shown in Fig. 22 for $Q_0^2 = 7\,\text{GeV}^2$.

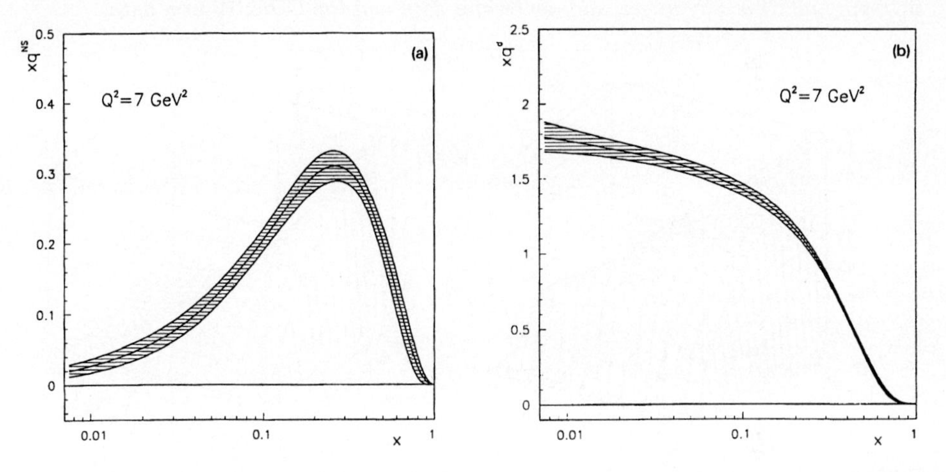

Fig. 22. *a) The non-singlet quark distribution, b) The deuteron quark distribution from the NMC analysis*

In Fig. 23 the gluon distribution obtained from this analysis is compared to previous determinations from deep inelastic scattering data from BCDMS, SLAC and from CDHSW [70].

The NMC data substantially improve the knowledge of the gluon distribution especially at low x in the range $0.01 < x < 0.1$. Measurements at HERA allow to determine the gluon distribution down to x values of around 10^{-4} (see the discussion in the review of Levy in these proceedings [151]). We have mentioned above that, once the shape of the parton distributions is known at one value of Q_0^2, it is possible to predict their value at any other value of Q^2 from the DGLAP evolution equations (65) and (66). Figure 24 is a three dimensional representation of the deuteron structure function F_2^{d}, here given as $xq(x) =$

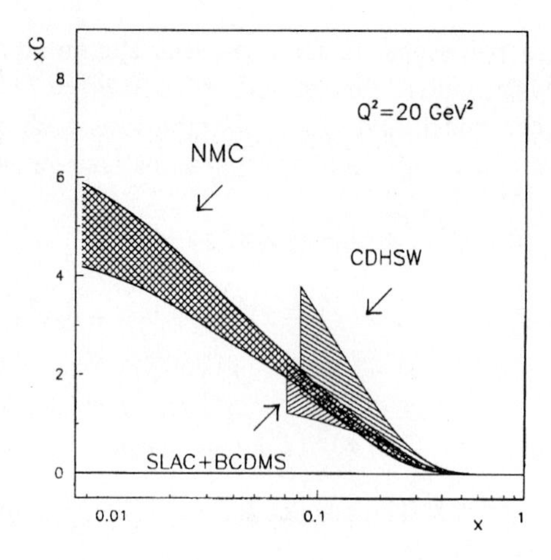

Fig. 23. *The gluon distribution from the NMC analysis compared to those from the BCDMS and SLAC hydrogen and deuterium data and the CDHSW iron data.*

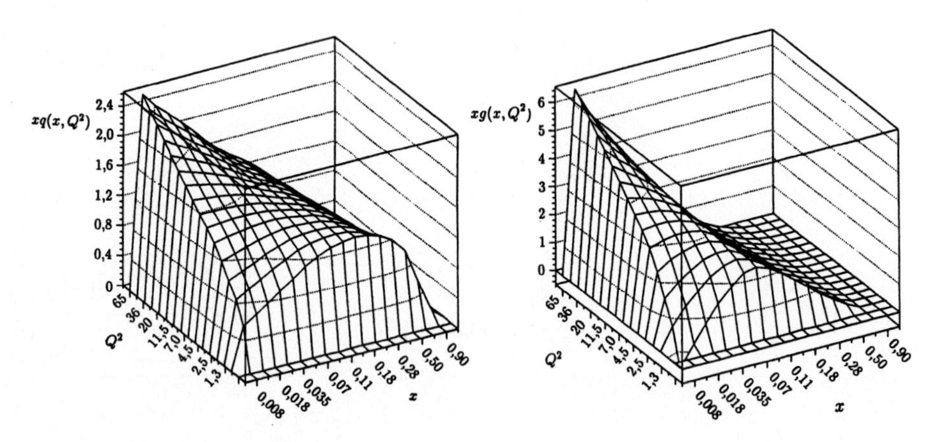

Fig. 24. *The Q^2 evolution of the deuteron structure function $xq(x)$ and the gluon structure function $xg(x, Q^2)$ as obtained from the NMC analysis.*

$\sum_f x\,(q_f(x) + \bar{q}_f(x))$ and of the gluon structure function $G(x) = xg(x)$ as they emerge from this analysis[71].

The drastic change of these distributions with Q^2 is clearly seen. At small values of x, $F_2(x)$ and $G(x)$ increase rapidly with Q^2. At small values of Q^2, the shape of $xq(x)$ approaches that of a valence quark distribution, since the sea quark distribution becomes less and less significant. It is interesting to note that at low values of Q^2 also the gluon distribution assumes a shape very similar to that of a valence quark distribution.

In a second part of the NMC analysis also $\alpha_s(Q_0^2)$ has been determined from the NMC data leaving it a free parameter in the fit. This resulted in a value for the strong coupling constant

$$\alpha_s(7\,\mathrm{GeV}^2) = 0.264 \pm 0.018(\mathrm{stat}) \pm 0.070(\mathrm{sys}) \pm 0.013(\mathrm{scale}) \quad (80)$$

which corresponds to

$$\alpha_s(M_Z^2) = 0.117^{+0.011}_{-0.016} \; . \quad (81)$$

The Q^2 dependences of the data agree over the whole measured x range with those predicted by QCD. The average logarithmic slopes $\mathrm{d}\ln F_2/\mathrm{d}\ln Q^2$ were determined in each bin of x separately, both from the data and the QCD fit. They are shown for the proton data in Fig. 25.

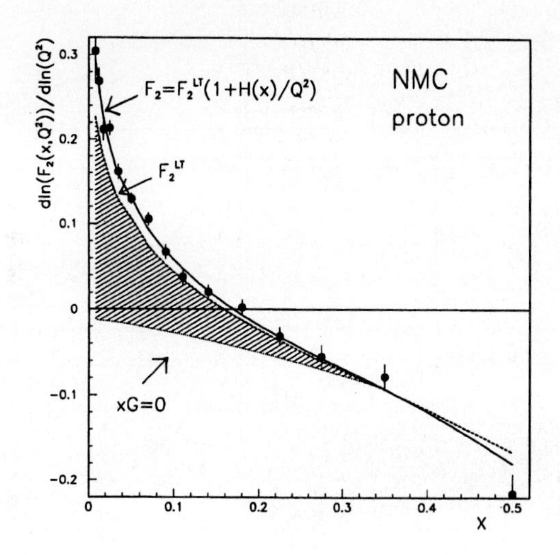

Fig. 25. *The logarithmic slopes* $\mathrm{d}\ln F_2(x,Q^2)/\mathrm{d}\ln Q^2$ *from the NMC proton data compared to those of the QCD fit*

Good agreement is observed over the entire x range. The dotted curves in the figure indicate the Q^2 evolution due to quarks only and the area between the dashed and dotted curves represent the contribution of gluons. It is clear that for most of the NMC data the Q^2 evolution is driven by the gluon distribution.

Results for $R(x, Q^2)$

While now there is excellent information available for the structure function $F(x,Q^2)$, $R(x,Q^2)$ is still rather poorly known. This is due to the fact that the influence of $R(x,Q^2)$ to the cross section (33) is rather small and even negligible in large regions of the kinematic domain. Very precise measurements at low values of Q^2 have been performed by the experiment E140 at SLAC [185]. The data are shown in Fig. 26 as a function of Q^2 for different values of x together with high Q^2 results from the high energy muon experiments EMC [50] and BCDMS [67, 68] and the neutrino experiment CDHSW [70].

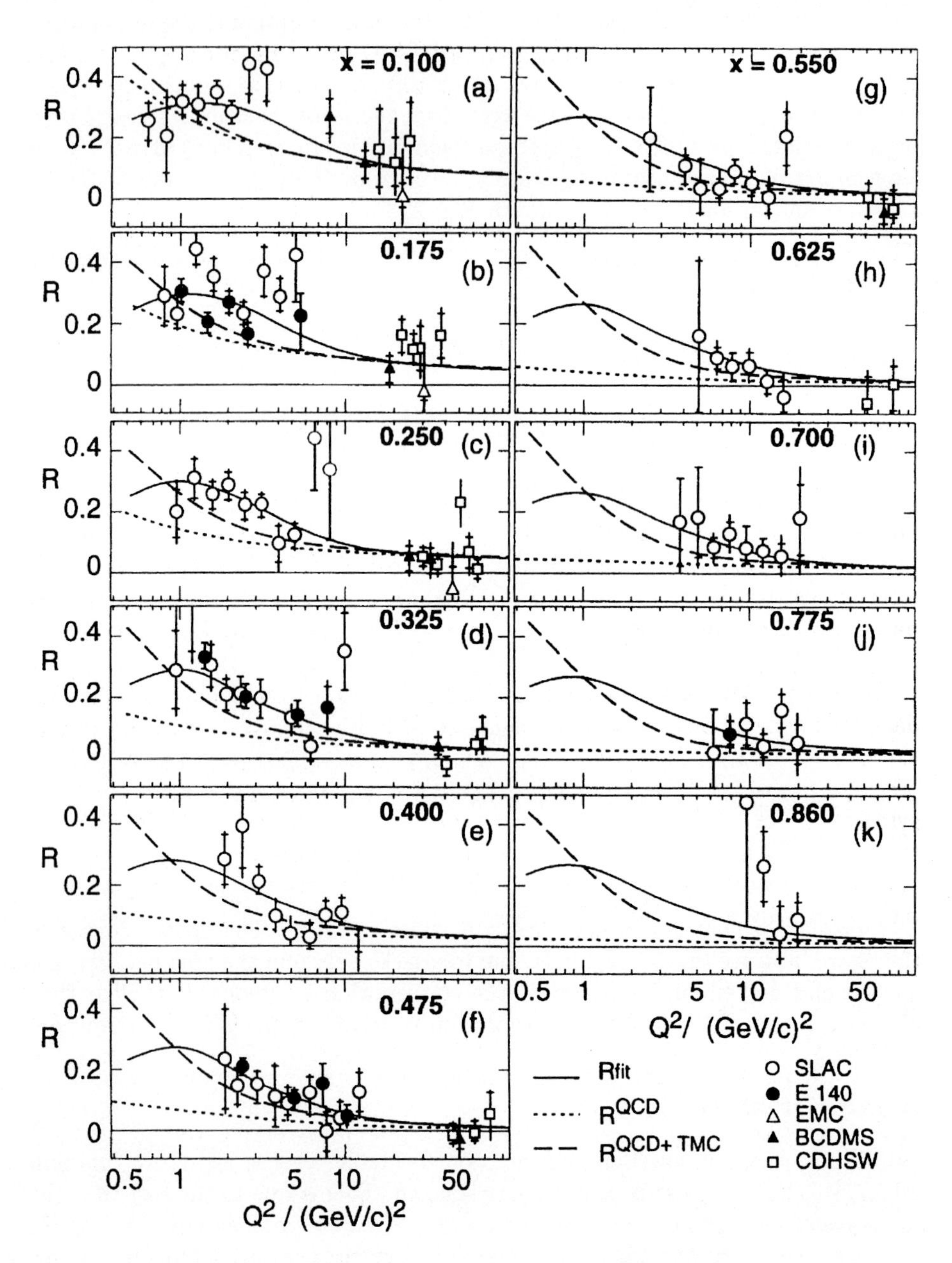

Fig. 26. *The ratio $R(x, Q^2)$ as a function of Q^2 for different x bins*

Note that these data only cover a range of $x > 0.1$. It is obvious that despite the enormeous statistics of E140 $R(x, Q^2)$ is still not known too well. The short dashed line is a QCD prediction, the long dashed line a QCD prediction including target mass corrections and the full line a parametrisation to the data. It is this parametrisation which is generally used for the determination of the unpolarised structure function $F_2(x, Q^2)$ and the polarised structure function $g_1(x, Q^2)$ (see below). $R(x, Q^2)$ is connected with the longitudinal structure function $F_\mathrm{L}(x, Q^2)$ by

$$R(x, Q^2) = \frac{F_\mathrm{L}(x, Q^2) + 4\left(\frac{Mx}{Q}\right) F_2(x, Q^2)}{F_2(x, Q^2) - F_\mathrm{L}(x, Q^2)} . \qquad (82)$$

The Q^2 dependence of $F_\mathrm{L}(x, Q^2)$ is in the next to leading order QCD given by the relation [27]

$$\begin{aligned} F_\mathrm{L}(x, Q^2) = {} & \frac{\alpha_\mathrm{s}(Q^2)}{2\pi} x^2 \cdot \\ & \cdot \int_x^1 \frac{\mathrm{d}y}{y^3} \left\{ \frac{8}{3} F_2^\mathrm{S}(y, Q^2) + \frac{40}{9}\left(1 - \frac{x}{y}\right) \cdot y \cdot g(y, Q^2) \right\} , \end{aligned} \qquad (83)$$

where F_2^S is the singlet structure function and $g(y, Q^2)$ the gluon distribution. So in principle a measurement of $R(x, Q^2)$ or $F_\mathrm{L}(x, Q^2)$ should allow a determination of the gluon momentum distribution $g(x, Q^2)$. The data shown in Fig. 26 make clear that at present such an attempt is hopeless. Instead one can use parametrisations of $F_2^\mathrm{S}(x, Q^2)$ and of $g(x, Q^2)$ to make a QCD prediction for $R(x, Q^2)$.

Recently NMC has presented results for $R(x, Q^2)$ from their combined full data set averaged over proton and deuteron [43]. Figure 27a shows these results for R as a function of x at an average Q^2 ranging from $\langle Q^2 \rangle = 1.4$ to $20.6\,\mathrm{GeV}^2$ for different x-bins. The error bars indicate the quadratic sum of statistical and systematic uncertainties. The systematic errors are 1.5 to 3 times larger than the statistical ones; they are dominated by the normalisation uncertainty and are largely correlated. Also shown is the QCD prediction (full curve) and the parametrisation from SLAC data (dashed curve). The agreement is reasonably good.

In Fig. 27b the NMC R measurements are compared to the results from BCDMS and CDHSW. Again good agreement is observed. The new NMC data improve the knowledge of R for $x < 0.1$ considerably.

For the results presented in Fig. 27a the proton and deuteron data have been averaged. This is justified since NMC has also demonstrated that the difference $\Delta R = R^\mathrm{d} - R^\mathrm{p}$ is small [31, 44]. This quantity can be obtained from the ratio $F_2^\mathrm{d}/F_2^\mathrm{p}$ which, using the complementary target method, can be measured with small statistical and systematic errors. We have from (33)

$$\frac{\sigma^\mathrm{d}}{\sigma^\mathrm{p}}(x, Q^2, E) = \frac{F_2^\mathrm{d}}{F_2^\mathrm{p}}(x, Q^2) \cdot \frac{1 + R^\mathrm{p}(x, Q^2)}{1 + R^\mathrm{d}(x, Q^2)} \cdot \frac{1 + \varepsilon R^\mathrm{d}(x, Q^2)}{1 + \varepsilon R^\mathrm{p}(x, Q^2)} . \qquad (84)$$

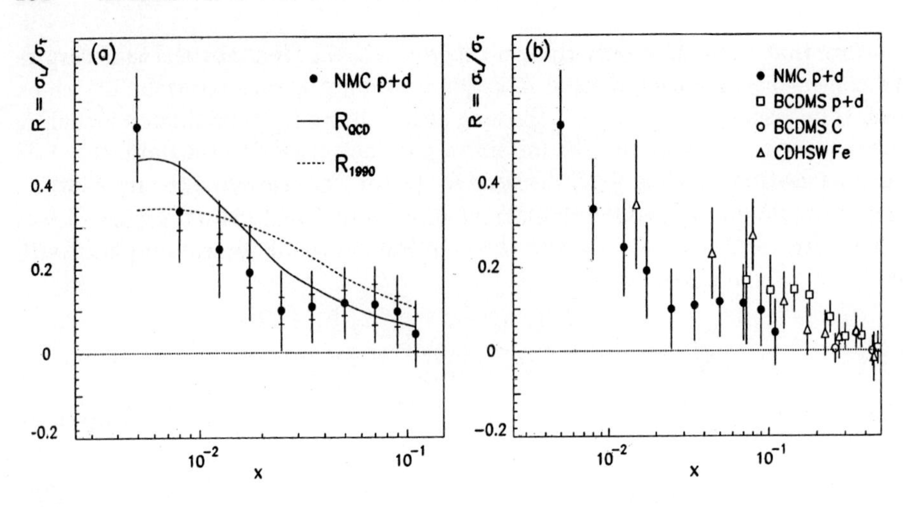

Fig. 27. *a) The ratio $R = \sigma_{\mathrm{L}}/\sigma_{\mathrm{T}}$ measured by NMC, averaged over proton and deuteron, b) Comparison of the NMC data with previous results*

The dependence of this ratio on the incident energy E appears only through ε, the polarisation of the virtual photon (see (30)). Expanding (84) one obtains to first order in ΔR:

$$\frac{\sigma^{\mathrm{d}}}{\sigma^{\mathrm{p}}}(x, Q^2, E) \simeq \frac{F_2^{\mathrm{d}}}{F_2^{\mathrm{p}}}\left(1 - \frac{1-\varepsilon}{(1+\bar{R})(1+\varepsilon\bar{R})} \cdot \Delta R\right) \quad , \tag{85}$$

where $\overline{R} = \frac{1}{2}(R^{\mathrm{d}} + R^{\mathrm{p}})$.

The results for ΔR [44] are shown in Fig. 28. They cover a range $0.003 < x < 0.35$.

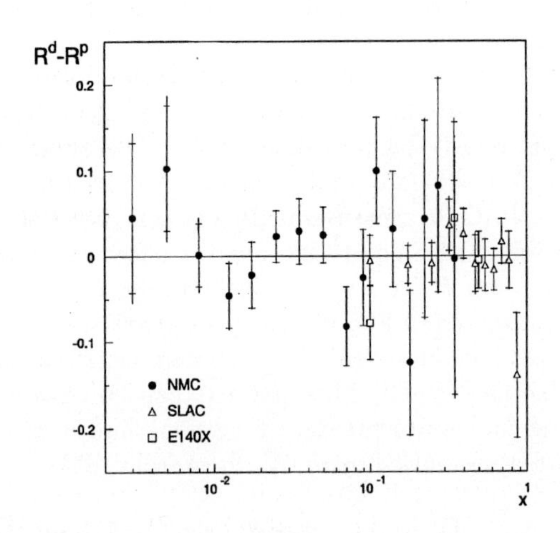

Fig. 28. *NMC results for $R^{\mathrm{d}} - R^{\mathrm{p}}$ compared with earlier data from SLAC*

The values of ΔR are small, no significant x dependence of ΔR is observed. Averaging the data over x one obtains

$$\Delta R = 0.004 \pm 0.012(\mathrm{stat}) \pm 0.011(\mathrm{syst}) \tag{86}$$

compatible with zero, at $\langle Q^2 \rangle = 5\,\mathrm{GeV}^2$. In Fig. 28 also the data from SLAC experiment E140 [180] and from a reanalysis of older SLAC data [185] are shown, which agree well with the present value. In Fig. 29 the data are compared with predictions from pertubative QCD.

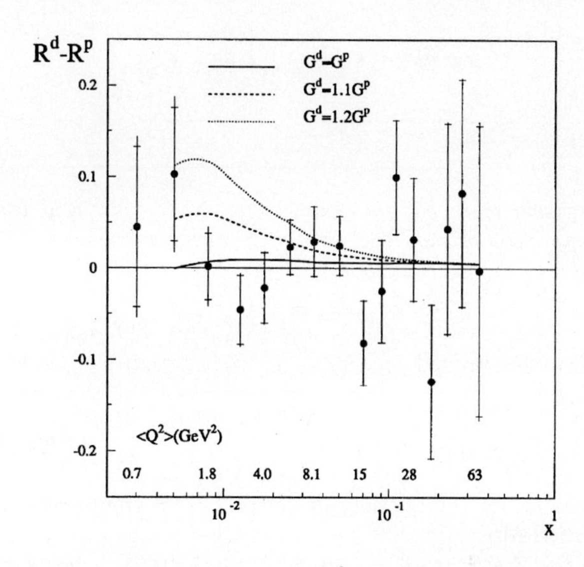

Fig. 29. *Comparison of the NMC data for $R^{\mathrm{d}} - R^{\mathrm{p}}$ to predictions from pertubative QCD*

The solid line was calculated using the same gluon distribution for the proton and the deuteron, the dashed and dotted ones assuming an increase of $xg(x)$ for the deuteron of 10% and 20% respectively. From these results it is justified to use the averaged proton and deuteron values for the determination of R shown in Fig. 27 and to assume that $g(x)$ for the proton and the deuteron do not differ by more than 10%.

x and Q^2 Dependence of $F_2^{\mathrm{n}}/F_2^{\mathrm{p}}$

Structure function ratios can be determined from the measured cross section ratios once ΔR and $\overline{R}$ are known. As ΔR is compatible with zero, NMC has taken the structure function ratio $F_2^{\mathrm{d}}/F_2^{\mathrm{p}}$ to be equal to the cross section ratio $\sigma^{\mathrm{d}}/\sigma^{\mathrm{p}}$. In most of the x bins the data cover nearly two decades in Q^2, with little dependence on Q^2. A comparison to results from SLAC and BCDMS for $F_2^{\mathrm{n}}/F_2^{\mathrm{p}}$ is shown in Fig. 30 for three x bins and demonstrates good agreement.

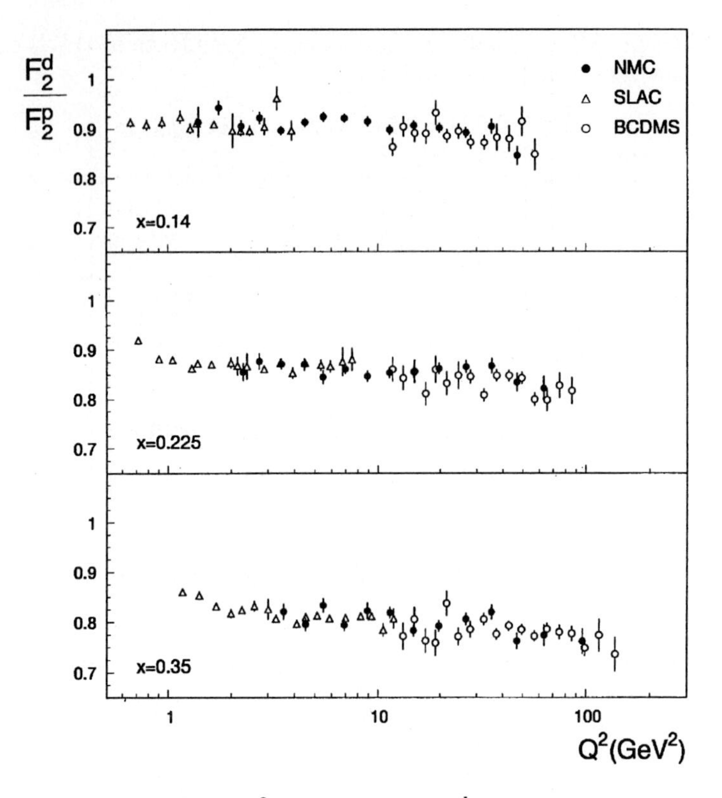

Fig. 30. *NMC results for the Q^2 dependence of $F_2^{\mathrm{d}}/F_2^{\mathrm{p}}$ compared with earlier data from SLAC and BCDMS*

To investigate possible Q^2 dependences, the data were fitted in each x bin with a linear function of $\ln Q^2$

$$F_2^{\mathrm{n}}/F_2^{\mathrm{p}} = b_1 + b_2 \ln Q^2 \ . \tag{87}$$

Figure 31 shows the fitted slope parameters b_2 as a function of x together with the two NLO QCD calculations, including target mass corrections from SLAC, BCDMS and NMC discussed above.

The measured slopes are consistent with these pertubative QCD calculations although there may be deviations at $x > 0.1$ due to higher twist effects being different in the proton and in the deuteron, possibly due to nuclear effects in the deuteron. Neglecting nuclear effects in the deuteron, the neutron structure function is given by

$$F_2^{\mathrm{n}} = 2F_2^{\mathrm{d}} - F_2^{\mathrm{p}} \ , \tag{88}$$

and the ratio of the neutron and proton structure functions by

$$\frac{F_2^{\mathrm{n}}}{F_2^{\mathrm{p}}} = 2\frac{F_2^{\mathrm{d}}}{F_2^{\mathrm{p}}} - 1 = 2\frac{\sigma^{\mathrm{d}}}{\sigma^{\mathrm{p}}} - 1 \ . \tag{89}$$

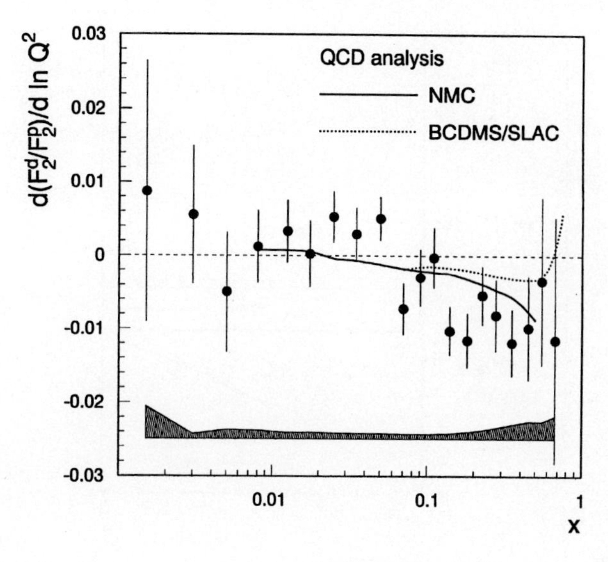

Fig. 31. *The slope parameter* $b_2 = \mathrm{d}(F_2^{\mathrm{d}}/F_2^{\mathrm{p}})/\mathrm{d}\ln Q^2$ *measured by NMC*

The results for $F_2^{\mathrm{d}}/F_2^{\mathrm{p}}$ averaged over Q^2 according to (89) are shown in Fig. 32, the data approach one for $x \to 0$ and fall down to about 0.45 for $x \sim 0.7$. In

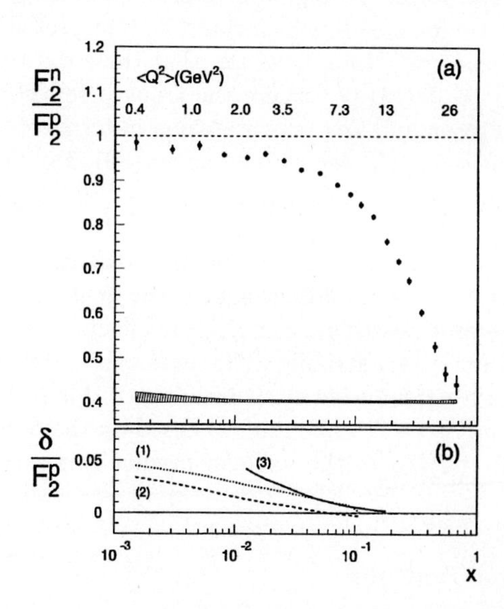

Fig. 32. *NMC results for the structure function ration* $F_2^{\mathrm{n}}/F_2^{\mathrm{p}}$ *as a function of* x

Fig. 33 results for the x dependence of $F_2^{\mathrm{n}}/F_2^{\mathrm{p}}$ from the Fermilab E665 collaboration [8] are compared to the NMC results. There is fair agreement between

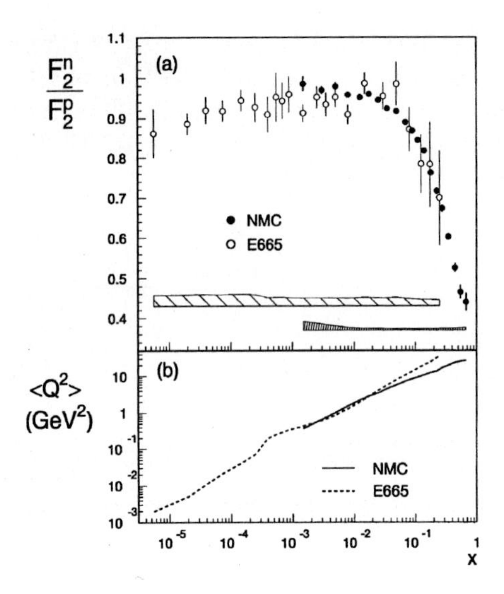

Fig. 33. *Comparison of the x dependence of the NMC data for F_2^n/F_2^p with the results of the E665 collaboration*

the two experiments while the accuracy of the NMC result is much higher. The average Q^2 is quite similar in the region of overlap. The E665 results indicate a sizable drop of the ratio for $x \to 0$, which could be interpreted as a nuclear shadowing effect in the deuteron. Note, however, that these data are at very low Q^2 down to values of $Q^2 < 0.01\,\mathrm{GeV}^2$ where the transverse resolution corresponds to the size of the deuteron and any interpretation in terms of the quark model is questionable. The ratio F_2^n/F_2^p determined using (89) may deviate significantly from the free nucleon ratio $(F_2^n/F_2^p)_{\mathrm{free}}$ due to nuclear effects in the deuteron (see e.g. [38]). At small x F_2^d may be reduced by shadowing effects which are also observed in the real photon cross section on the deuteron ([51, 52, 160, 60]). Near $x = 1$ the effect of the kinematic range for the deuteron extending to $x = 2$ and of Fermi motion and possibly other nuclear effects must become apparent. Note that for the region $x \to 1$ striking differences exist for predictions of F_2^n/F_2^p [85, 18, 111, 161]. In a pertubative QCD framework this ratio should approach the value $\frac{3}{7}$ in the limit $x \to 1$, while models based on the dominance of a scalar di-quark component $(ud)_{s=0}$ in the nucleon predict this ratio to approach the value $\frac{1}{4}$.

Higher Twist Contributions

We have mentioned before that the NLO QCD analysis of the combined SLAC and BCDMS data indicated a substantial contribution to the Q^2 dependence of structure functions at large x, low Q^2 due to higher twist effects. Such higher twist effects arise due to correlations between quarks which invalidate the assumption of the quark parton model that the deep inelastic cross section is the incoherent sum of cross sections for elastic scattering from free, uncorrelated quarks. The measurement of $F_2^{\rm n}/F_2^{\rm p}$ allows to investigate whether the higher twist effects are identical or different for proton and neutron. If we assume that they are different then we can write

$$\frac{F_2^{\rm n}}{F_2^{\rm p}} = \left(2\frac{F_2^{\rm d}}{F_2^{\rm p}} - 1\right)^{\rm LT+TMC} \left(1 + \frac{C^{\rm p} - C^{\rm n}}{Q^2}\right) . \tag{90}$$

For such an analysis the SLAC, BCDMS and the earlier NMC data [31] have been combined, and for each x bin the logarithmic slope $\mathrm{d}\left(F_2^{\rm n}/F_2^{\rm p}\right)/\mathrm{d}\ln Q^2$ has been determined. In Fig. 34 the measured slopes are shown together with the QCD prediction. There are substantial deviations from the QCD prediction. Figure 35 shows the result for the difference of the higher twist coefficients

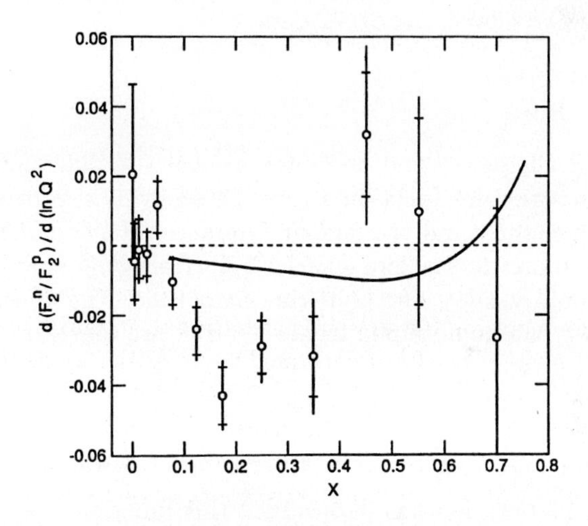

Fig. 34. *The difference of the higher twist coefficients of proton and neutron from a combined analysis of NMC, BCDMS and SLAC data*

$C^{\rm p} - C^{\rm n}$. In the region $x > 0.2$ the higher twist coefficient of the proton seems to be substantially smaller than that of the neutron. Note, however, that the data show in their x dependence some similarity to the nuclear EMC effect where also in the region $x \simeq 0.65$ the modification of quark distributions due to nuclear effects is largest. It is therefore not unreasonable to speculate that this difference is due to nuclear effects in the deuteron. Whether this is really the case can only be tested in future experiments where the spectator nucleon is being tagged and therefore one knows whether the scattering process occured on a proton or a neutron.

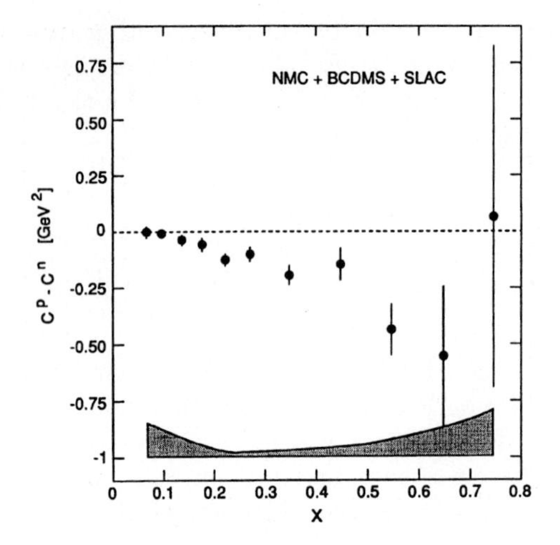

Fig. 35. *Difference $C^{\mathrm{p}} - C^{\mathrm{n}}$ of higher twist coefficients for proton and neutron from the combined NMC, BCDMS and SLAC data.*

The Gottfried Sum

We have seen in Chap. 1.2 that the integral over the difference of the proton and the neutron structure function F_2, weighted by $1/x$, provides information about the quark charges and the flavour symmetry of the quark-antiquark sea. For a flavour symmetric sea and quarks with fractional charges $\pm\frac{2}{3}$ and $\pm\frac{1}{3}$ this integral should yield $\frac{1}{3}$, the Gottfried Sum Rule. The nonsinglet structure function $F_2^{\mathrm{p}} - F_2^{\mathrm{n}}$ can be obtained from the structure function ratio $F_2^{\mathrm{n}}/F_2^{\mathrm{p}}$ as

$$F_2^{\mathrm{p}} - F_2^{\mathrm{n}} = F_2^{\mathrm{d}} \frac{1 - F_2^{\mathrm{n}}/F_2^{\mathrm{p}}}{1 + F_2^{\mathrm{n}}/F_2^{\mathrm{p}}} \ , \tag{91}$$

with $F_2^{\mathrm{n}}/F_2^{\mathrm{p}} = 2F_2^{\mathrm{d}}/F_2^{\mathrm{p}} - 1$.

The NMC collaboration has determined this function from their measured ratio $F_2^{\mathrm{d}}/F_2^{\mathrm{p}}$ and a parametrisation to their structure function F_2^{d} [39]. The results presented here were evaluated at $Q^2 = 4\,\mathrm{GeV}^2$. No corrections were applied for target mass, higher twist or nuclear effects. The difference $F_2^{\mathrm{p}} - F_2^{\mathrm{n}}$ (full symbols and scale to the right) and the Gottfried integral $S_{\mathrm{G}} = \int_x^1 \mathrm{d}x'(F_2^{\mathrm{p}} - F_2^{\mathrm{n}})/x'$ (open symbols and scale to the left) are shown in Fig. 36. The value of the Gottfried sum at $Q^2 = 4\,\mathrm{GeV}^2$ is found to be

$$S_{\mathrm{G}}(0.004 - 0.8) = 0.221 \pm 0.008(\mathrm{stat}) \pm 0.019(\mathrm{syst}) \ . \tag{92}$$

With the most recent data included the result is [42]

$$S_{\mathrm{G}}^{\mathrm{new}}(0.004 - 0.8) = 0.2281 \pm 0.0065(\mathrm{stat}) \ , \tag{93}$$

agreeing very well with the above value. To evaluate the contributions to S_{G} from the unmeasured regions at high and low x, extrapolations of $F_2^{\mathrm{p}} - F_2^{\mathrm{n}}$ to

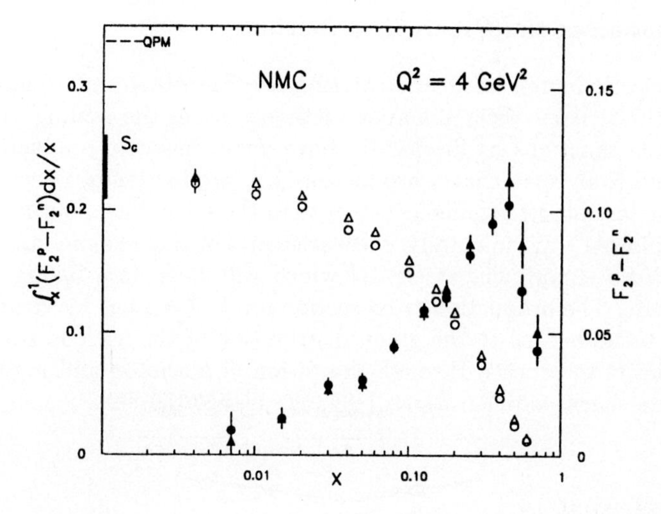

Fig. 36. *The difference $F_2^p - F_2^n$ and the Gottfried sum from the NMC data*

$x = 1$ and $x = 0$ were made which amounted to 0.001 ± 0.001 for the region $x > 0.8$ and 0.013 ± 0.005(stat) for $x < 0.004$, where a Regge-like behaviour was assumed. Summing all contributions one obtains for the Gottfried sum

$$S_G = 0.235 \pm 0.026 \ . \tag{94}$$

The error is the result of combining the statistical and systematic errors in quadrature, and including the effects on the systematic uncertainties of the extrapolations. Obviously this result deviates by about four standard deviations from the quark model expectation. Taking shadowing effects in the deuteron into account, the value of S_G would be further reduced to about 0.223 ± 0.033 while taking higher twist effects into account would increase S_G by about 10%. In the frame of the quark parton model one can, taking isospin to be a good symmetry, explain this result by the assumptions that the light quark-antiquark sea is not flavour symmetric. One obtains:

$$\int (\bar{d} - \bar{u}) \ dx = 0.165 \pm 0.059 \ . \tag{95}$$

It has been pointed out [176] that the nonpertubative processes of nucleon dissociation into π–N and π–Δ can lead to such a flavour asymmetric sea. Here the process $\mathrm{p} \to \mathrm{n} + \pi^+$ is favored over $\mathrm{p} \to \Delta^{++} + p^-$, which in quark terms corresponds to favoring $\mathrm{u} \to \mathrm{d} + \mathrm{u}\bar{\mathrm{d}}$ over $\mathrm{d} \to \mathrm{u} + \mathrm{d}\bar{\mathrm{u}}$. In another approach the result could be due to a small admixture of vector diquarks, without requiring a flavour asymmetric sea. A more detailed investigation of a number of the above effects and a review of the literature is given in [144, 110].

Determination of the Gluon Distribution

In deep inelastic lepton nucleon scattering one can obtain informations on the gluon distribution by three different methods: From the scaling violations of the structure functions as discussed above, from inelastic production of J/Ψ particles and from open charm production, i. e. production of D mesons. Theoretically the last process is much cleaner than the second one, but has, however, not yet exploited experimentally. Measurements of this channel are forseen by the COMPASS experiment at CERN which will start data taking at the end of this decade. The production cross section for J/Ψ mesons by virtual photons is thought to be related to the gluon distribution of the nucleon since the process is believed to proceed through the fusion of a photon and a gluon via an intermediate charm-anticharm pair [184, 57, 49, 53, 156, 69].

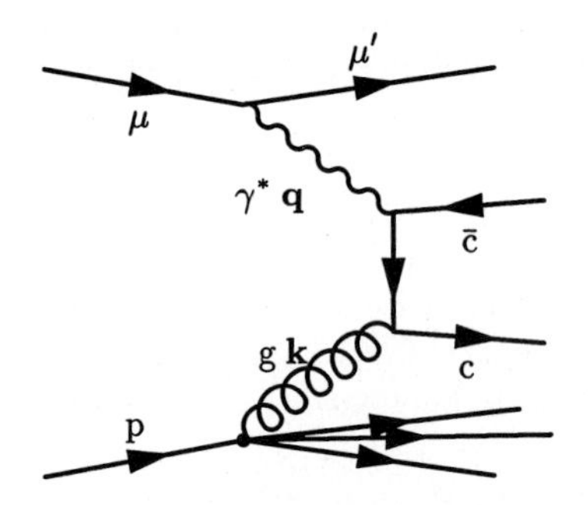

Fig. 37. *The photon-gluon fusion diagram*

The J/Ψ is identified via its decay into $\mu^+\mu^-$ pairs which has a branching ratio of 6%. To distinguish this QCD process from the fluctuation of a photon into a vector meson one demands that the J/Ψ obtaines some transverse momentum $p_\perp$ by gluon radiation and carries only a fraction of the energy of the virtual photon: $z = E_{J/\Psi}/\nu \lesssim 0.9$. Under certain assumptions then the cross section for these inelastic J/Ψ's is proportional to the gluon distribution $g(x)$. In Fig. 38 the results from EMC [47] and NMC [24] for $xg(x)$ obtained by this method is shown. Another method to obtain information about the gluon distribution is the production of direct photons with large transverse momentum in pp or $p\bar{p}$ scattering which proceeds via a gluon from one nucleon interacting with a quark from the other nucleon.

Parametrisation of Parton Densities

An important by-product of the QCD studies determining $\alpha_s(Q^2)$ is the extraction of parton densities at a fixed reference value of Q_0^2. These then can be evolved in Q^2 and used as input for phenomenological studies e. g. in hadron-hadron collisions. For the determination of these parton distributions the results of different

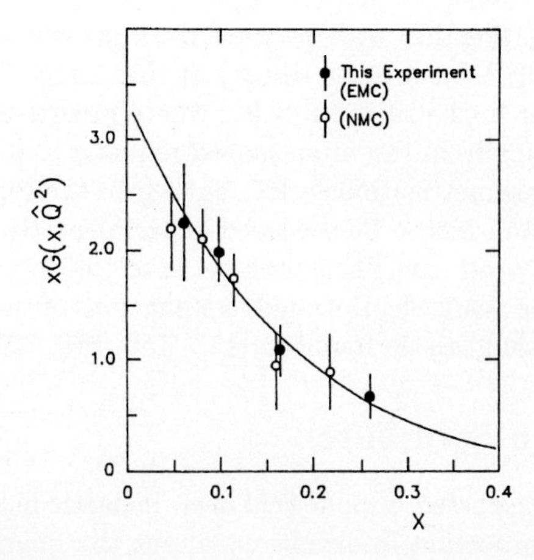

Fig. 38. *NMC and EMC results for the gluon structure function $xg(x)$ deduced from inelastic J/Ψ-production*

types of experiments like deep inelastic scattering, lepton pair production in pp and p$\bar{\text{p}}$ collisions (Drell-Yan-Process), direct photons with large transverse momenta or charge asymmetries of the decay leptons in $W^{\pm}$ production have been used as input. One should note however, that these parametrisations have very little predictive power for extrapolations in kinematically not yet accessed regions, since, as we pointed out above, QCD can not predict the x-dependence of parton distributions but only their Q^2 dependence once the distributions are known at a fixed value of Q^2. As an example we show in Fig. 39 a comparison of

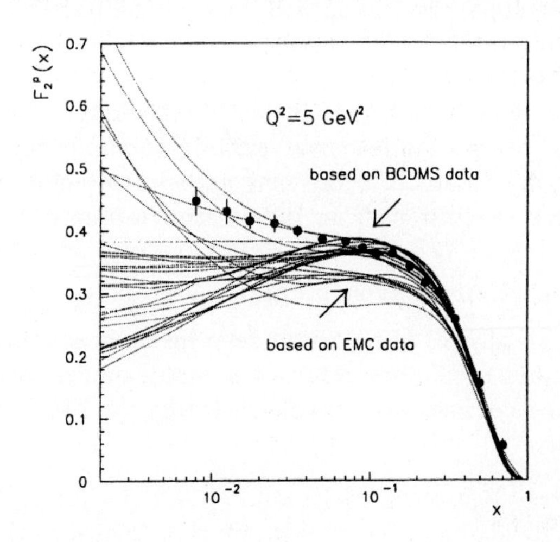

Fig. 39. *Comparison of NMC structure function data with a collection of structure function parametrisations.*

the NMC data for F_2^{p} together with parametrisations which where contained in a package from the CERN computer library at that time. The parametrisations were all rather similar in the region $x > 0.1$ where plenty of data existed before but differed substantically in the unmeasured region $x < 0.1$. Clearly the NMC data constrain the parametrisations a lot, but again the region at even lower x cannot be predicted and has to be explored experimentally by the HERA data. Of course the groups, producing their parametrisations, improve them as soon as new data have become available. Detailed explanations of these parametrisations and the procedures used can be found in [158, 159, 146, 123, 124].

2 Nucleon Spin Structure

So far we have only discussed unpolarised deep inelastic lepton nucleon scattering which provides important informations about the quark-gluon structure of the nucleon and the strong interaction. Additional informations can be obtained from polarised deep inelastic scattering. These experiments allow to extract details about the internal spin structure of the nucleon (for an early review see [134]) which are very important for the understanding of the strong force since this has a substantial contribution arising from the chromomagnetic spin part of the wavefunction. They are, however, by far more difficult than unpolarised deep inelastic scattering experiments since they require both a longitudinally polarised electron or muon beam and a polarised target. Therefore the experimental information is still rather limited. The pioneering experiments [21, 64] where performed mid/end of the 70's at SLAC, followed by the EMC muon experiment [45, 46] at CERN mid of the 80's. As will be discussed below the results of the muon experiment were in disagreement with 'naïve' expectations from the quark model and were the origin of the so called 'spin puzzle'. A huge number of theoretical papers dealing with this subject have been published since then, the related physics has been discussed at several dedicated workshops and summarised in excellent reviews as e. g. [171, 33, 154]. On the experimental side a new generation of polarised deep inelastic scattering experiments has been started/performed during the last five years. In the following I will summarise the physics, describe details of these new experiments and experimental data and will draw some conclusions from the present results.

2.1 Spin in the Nonrelativistic Quark Model

Before I discuss the physics of polarised deep inelastic scattering let me briefly recall the predictions of the non-relativistic quark model concerning magnetic moments and masses of hadrons to make clear why the EMC/SLAC result came as a surprise.

Magnetic Moments

The non-relativistic quark model is rather successful when its predictions for baryonic magnetic moments are compared with experimental results.

In Dirac theory the magnetic moment μ of a pointlike particle with mass M and spin-$\frac{1}{2}$ is

$$\mu_{\mathrm{Dirac}} = \frac{e\hbar}{2M} \ . \tag{96}$$

This relationship (and small corrections to it due to higher order corrections from QED and QCD) has been experimentally confirmed with high precision for both the electron and the muon.

For the proton and the neutron magnetic moments one obtains experimentally

$$\mu_{\mathrm{p}} = 2.79\,\mu_{\mathrm{N}} \ , \tag{97}$$

$$\mu_{\mathrm{n}} = -1.91\,\mu_{\mathrm{N}} \ , \tag{98}$$

with

$$\mu_{\mathrm{N}} = \frac{e\hbar}{2M_{\mathrm{p}}} \tag{99}$$

being the nuclear magneton. From this measurement it is evident that proton and neutron cannot be pointlike Dirac particles but must have an internal substructure.

In the nonrelativistic quark model the total symmetric SU(6) wavefunction of a proton with spin component $m_{\mathrm{s}} = +\frac{1}{2}$ relative to the quantisation axis is given by [169, 128]

$$\begin{aligned} |\mathrm{p}\uparrow\rangle = \frac{1}{\sqrt{18}}\{\, & 2\,|\mathrm{u}\uparrow\mathrm{u}\uparrow\mathrm{d}\downarrow\rangle + 2\,|\mathrm{u}\uparrow\mathrm{d}\downarrow\mathrm{u}\uparrow\rangle + 2\,|\mathrm{d}\downarrow\mathrm{u}\uparrow\mathrm{u}\uparrow\rangle \\ & - |\mathrm{u}\uparrow\mathrm{u}\downarrow\mathrm{d}\uparrow\rangle - |\mathrm{u}\uparrow\mathrm{d}\uparrow\mathrm{u}\downarrow\rangle - |\mathrm{d}\uparrow\mathrm{u}\uparrow\mathrm{u}\downarrow\rangle \\ & - |\mathrm{u}\downarrow\mathrm{u}\uparrow\mathrm{d}\uparrow\rangle - |\mathrm{u}\downarrow\mathrm{d}\uparrow\mathrm{u}\uparrow\rangle - |\mathrm{d}\uparrow\mathrm{u}\downarrow\mathrm{u}\uparrow\rangle\} \ . \end{aligned} \tag{100}$$

The neutron wave function is trivially found by exchanging the u- and d-quarks.

$$\begin{aligned} |\mathrm{n}\uparrow\rangle = \frac{1}{\sqrt{18}}\{\, & 2\,|\mathrm{d}\uparrow\mathrm{d}\uparrow\mathrm{u}\downarrow\rangle + 2\,|\mathrm{d}\uparrow\mathrm{u}\downarrow\mathrm{d}\uparrow\rangle + 2\,|\mathrm{u}\downarrow\mathrm{d}\uparrow\mathrm{d}\uparrow\rangle \\ & - |\mathrm{d}\uparrow\mathrm{d}\downarrow\mathrm{u}\uparrow\rangle - |\mathrm{d}\uparrow\mathrm{u}\uparrow\mathrm{d}\downarrow\rangle - |\mathrm{u}\uparrow\mathrm{d}\uparrow\mathrm{d}\downarrow\rangle \\ & - |\mathrm{d}\downarrow\mathrm{d}\uparrow\mathrm{u}\uparrow\rangle - |\mathrm{d}\downarrow\mathrm{u}\uparrow\mathrm{d}\uparrow\rangle - |\mathrm{u}\uparrow\mathrm{d}\downarrow\mathrm{d}\uparrow\rangle\} \ . \end{aligned} \tag{101}$$

The proton magnetic moment in the ground state, with $l = 0$, is a simple vectorial sum of the magnetic moments of the three quarks

$$\vec{\mu}_{\mathrm{p}} = \vec{\mu}_{\mathrm{u}} + \vec{\mu}_{\mathrm{u}} + \vec{\mu}_{\mathrm{d}} \tag{102}$$

and has the expectation value

$$\mu_{\mathrm{p}} = \langle \vec{\mu}_{\mathrm{p}} \rangle = \langle \chi_{\mathrm{p}} |\, \vec{\mu}_{\mathrm{p}} \,| \chi_{\mathrm{p}} \rangle \ , \tag{103}$$

where χ_{p} is the spin part of the wavefunction. From (103) we deduce:

$$\mu_{\mathrm{p}} = \frac{2}{3}\left(\mu_{\mathrm{u}} + \mu_{\mathrm{u}} - \mu_{\mathrm{d}}\right) + \frac{1}{3}\mu_{\mathrm{d}} = \frac{4}{3}\mu_{\mathrm{u}} - \frac{1}{3}\mu_{\mathrm{d}} \ , \tag{104}$$

where $\mu_{\rm u,d}$ are the quark magnetons

$$\mu_{\rm u,d} = \frac{e_{\rm u,d} \cdot |e|\,\hbar}{2m_{\rm u,d}} \tag{105}$$

and $m_{\rm u,d}$ the constituent quark masses. Similarly we obtain for the neutron

$$\mu_{\rm n} = \frac{4}{3}\mu_{\rm d} - \frac{1}{3}\mu_{\rm u} \tag{106}$$

and analogously for the Σ (uus)

$$\mu_{\Sigma^+} = \frac{4}{3}\mu_{\rm u} - \frac{1}{3}\mu_{\rm s} , \tag{107}$$

while for the Λ, where the spins of the u- and d-quark couple to zero, the spin and magnetic moment are determined solely by the s-quark

$$\mu_\Lambda = \mu_{\rm s} . \tag{108}$$

If we assume the masses of the light quarks to be equal to each other, then we have $\mu_{\rm u} = -2\,\mu_{\rm d}$ and we obtain

$$\mu_{\rm p} = \frac{3}{2}\mu_{\rm u}, \qquad \mu_{\rm n} = -\mu_{\rm u} . \tag{109}$$

We thus obtain the following prediction for their ratio

$$\frac{\mu_{\rm n}}{\mu_{\rm p}} = -\frac{2}{3} , \tag{110}$$

which is in good agreement with the experimental result of -0.685. From the measured magnetic moments one can deduce the constituent quark mass

$$m_{\rm u,d} = \frac{Mp}{2.79} = 336\,\mathrm{MeV} .$$

This result was one of the cornerstones of the development of the quark model and was taken as an important proof that the model is correct.

Experimentally one can determine baryon magnetic moments from magnetic resonance or, for the strange baryons, from the precession of the spin in a magnetic field and the decay distribution [81, 132, 167].

Table 2 summarises the experimental results and the theoretical predictions from the quark model. The agreement is impressive, but one should note that there are deviations on the 10–15 percent level, which can not be explained consistently with one theoretical ansatz [132].

For both proton and Σ^+ the contribution of the u-quarks alone is already $2.48\,\mu_{\rm N}$. The experimental result for $\mu_{\Sigma^+} = 2.458\,\mu_{\rm N}$ is smaller than this value and since the charge of the s-quark is $-\frac{1}{3}|e|$, a meaningless negative mass would be required for the strange quark from equation (109) to explain the experimental value. Obviously further effects, such as relativistic ones, pion or gluon exchange and those due to quark orbital angular momenta, must be taken into account. Thus the quark model works reasonably well, but maybe this is just an accident and the situation is much more complicated than it looks at the first glance.

Table 2. *Experimental and theoretical values of the baryon magnetic moments.*

Baryon	μ/μ_N (Experiment)		Quark model:	μ/μ_N
p	+2.792847386	±0.000000063	$(4\mu_u - \mu_d)/3$	—
n	−1.91304275	±0.00000045	$(4\mu_d - \mu_u)/3$	—
Λ^0	−0.613	±0.004	μ_s	—
Σ^+	+2.458	±0.010	$(4\mu_u - \mu_s)/3$	+2.67
Σ^0			$(2\mu_u + 2\mu_d - \mu_s)/3$	+0.79
$\Sigma^0 \to \Lambda^0$	−1.61	±0.08	$(\mu_d - \mu_u)/\sqrt{3}$	−1.63
Σ^-	−1.160	±0.025	$(4\mu_d - \mu_s)/3$	−1.09
Ξ^0	−1.250	±0.014	$(4\mu_s - \mu_u)/3$	−1.43
Ξ^-	−0.6507	±0.0025	$(4\mu_s - \mu_d)/3$	−0.49
Ω^-	−1.94	±0.22	$3\mu_s$	−1.84

Hadron Masses

That spin plays an important role in QCD can be seen from the striking difference of the masses of light mesons, where the $J = 1$ states have much larger masses than their $J = 0$ partners. The pion π (spin $= 0\hbar$, $m_{\pi^\pm} = 140\,\text{MeV}$) and the rho ρ (spin $= 1\hbar$, $m_{\rho^\pm} = 767\,\text{MeV}$) are composed by the same two quarks and differ only by their spin orientation. The gap between the singlet state $\pi^\pm$ and the triplet state $\rho^\pm$ is about 4.5 times the rest energy of the pion, to be compared with the small hyperfine splitting in the electromagnetic case where for instance the energy difference between the singlet and triplet 1s state of positronium is about $8 \times 10^{-4}\,\text{eV}$ or about 1.6×10^{-10} times the total energy of the ground state.

Similarly the Δ^+ particle (spin $= \frac{3}{2}\hbar$, $M_\Delta = 1239\,\text{MeV}$) and the proton (spin $= \frac{1}{2}\hbar$, $M_p = 939\,\text{MeV}$) are composed of the same three quarks (uud) and differ only by the spin orientation of the latter, which leads to a ratio of their masses of about 1.3.

The absolute masses of the light mesons can be described by the phenomenological formula

$$M_{q\bar{q}} = m_q + m_{\bar{q}} + \Delta M_{ss}, \tag{111}$$

where the term ΔM_{ss} arises from the strong chromomagnetic spin-spin interaction potential

$$V_{ss}(q\bar{q}) = \frac{8\pi}{9}\,\alpha_s\,\frac{\vec{\sigma}_q\vec{\sigma}_{\bar{q}}}{m_q m_{\bar{q}}}\,\delta(x), \tag{112}$$

with the expectation value of $\vec{\sigma}_q \cdot \vec{\sigma}_{\bar{q}}$

$$\vec{\sigma}_q\vec{\sigma}_{\bar{q}} = \begin{cases} -3 & \text{for } s = 0 \\ +1 & \text{for } s = 1 \end{cases} . \tag{113}$$

For details see for instance [128, 169].

From fits to the masses of the low lying meson states one can extract from (111) and (112) the light constituent quark masses.

Similarly the mass difference of the total spin $S = \frac{3}{2}$ and $S = \frac{1}{2}$ baryons can be traced back to the spin-spin interaction

$$V_{\mathrm{ss}}(\mathrm{q}_i\bar{\mathrm{q}}_j) = \frac{4\pi}{9}\,\alpha_{\mathrm{s}}\,\frac{\vec{\sigma}_i\vec{\sigma}_j}{m_i m_j}\,\delta(\vec{x}), \tag{114}$$

where one has to sum over all quark pairs to obtain the mass term ΔM_{ss}. The expectation value for the sums over $\vec{\sigma}_i \cdot \vec{\sigma}_j$ is in this case

$$\sum_{\substack{i,j=1\\ i<j}}^{3} \vec{\sigma}_i\vec{\sigma}_j = \begin{cases} -3 & \text{for } S = \frac{1}{2} \\ +3 & \text{for } S = \frac{3}{2} \end{cases} . \tag{115}$$

Again a fit to the masses of the baryons yield the constituent quark masses m_{u}, m_{d} and m_{s}. The fitted baryon masses are within 1% of their true value [117].

These considerations can, however, be only a crude approximation to reality. From the previous discussions we know that the nucleon does not just consist of the three valence quarks, but we also have to take the sea of quark-antiquark pairs and the gluons into account, which dominantly generate the masses of the constituent quarks. (The bare quark masses are only 5–10 MeV for the u- and d-quarks and about 150 MeV for the s-quark [167].) We also have neglected completely the possible orbital momentum contributions of both gluons and quarks to the hadron spins.

2.2 The Polarised Cross Section and Asymmetries

Information about the internal spin structure of the nucleon can be extracted from polarised deep inelastic scattering experiments using a polarised nucleon target and a longitudinally polarised charged lepton beam (for transverse beam polarisation all effects are suppressed by m_ℓ/E_{beam}). In this case one also has to take into account the orientation of the polarisation vector $\vec{p}$ relative to the direction of the incoming lepton $\vec{k}$, denoted by α, and the angle ϕ between the polarisation plane (formed by $\vec{k}$ and $\vec{p}$) and the scattering plane (formed by $\vec{k}$ and $\vec{k}'$), the momentum vector of the scattered lepton. The quantities are illustrated in Fig. 40. The threefold differential cross section then becomes

$$\frac{\mathrm{d}^3\sigma}{\mathrm{d}\cos\theta\,\mathrm{d}\phi\,\mathrm{d}E'} = \frac{\alpha^2 E'}{Q^4 E} L^{\mu\nu} W_{\mu\nu} , \tag{116}$$

where $L^{\mu\nu}$, $W_{\mu\nu}$ are the leptonic and hadronic tensors which in addition to the symmetric parts (12, 13) contain an extra antisymmetric piece:

$$\begin{aligned} L^{\mu\nu} &= L^{(\mathrm{S})\mu\nu} + \mathrm{i}L^{(\mathrm{A})\mu\nu} , \\ L^{(\mathrm{A})\mu\nu} &= 2m_\ell \varepsilon^{\mu\nu\alpha\beta} s^\ell_\alpha q_\beta \end{aligned} \tag{117}$$

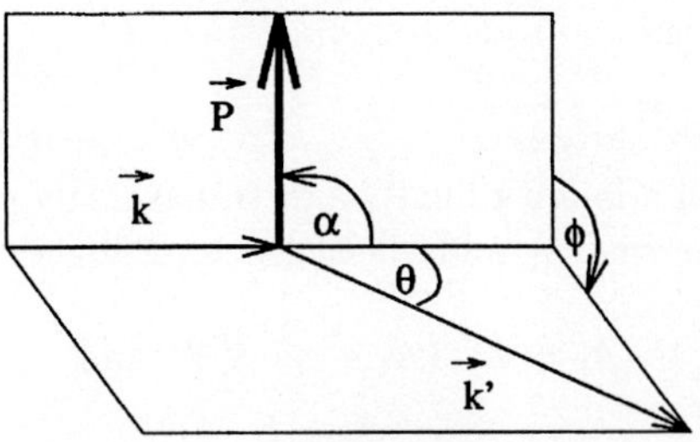

Fig. 40. *Definition of angles α and ϕ.*

$$
\begin{aligned}
W_{\mu\nu} &= W_{\mu\nu}^{(\mathrm{S})} + \mathrm{i}W_{\mu\nu}^{(\mathrm{A})}, \\
W_{\mu\nu}^{(\mathrm{A})} &= \frac{1}{\nu} g_1(\nu, Q^2)\varepsilon_{\mu\nu\lambda\sigma} q^{\lambda} s_{\mathrm{n}}^{\sigma} \\
&\quad + \frac{1}{\nu^2} g_2(\nu, Q^2)\varepsilon_{\mu\nu\lambda\sigma} q^{\lambda}(\mathbf{p}\cdot\mathbf{q} s_{\mathrm{n}}^{\sigma} - \mathbf{s}_{\mathrm{n}}\cdot\mathbf{q} p^{\sigma}).
\end{aligned}
\tag{118}
$$

Here s_ℓ^μ is the spin four-vector of the incoming lepton defined as

$$
2m_\ell s_\ell^\mu = \frac{1}{2}\bar{u}(\mathbf{k}, \mathbf{s}_\ell)\gamma^\mu\gamma_5 u(\mathbf{k}, \mathbf{s}_\ell) \tag{119}
$$

and $\mathbf{s}_n$ the polarisation vector of a spin-$\frac{1}{2}$ target (neglecting the lepton mass).

The spin-dependent part of the polarised deep inelastic cross section is given by the following formula [135]

$$
\begin{aligned}
\frac{\mathrm{d}^3(\sigma(\alpha) - \sigma(\pi + \alpha))}{\mathrm{d}x\,\mathrm{d}y\,\mathrm{d}\phi} = \frac{e^4}{4\pi^2 Q^2}\Bigg\{ & \cos\alpha\left[a_1 g_1 - \frac{y}{2}\gamma^2 g_2\right] - \\
& - \sin\alpha\cos\phi\sqrt{\gamma^2 a_2}\left[\frac{y}{2} g_1 + g_2\right]\Bigg\}.
\end{aligned}
\tag{120}
$$

γ is defined as $\gamma = \sqrt{Q^2}/\nu$, α is the polar angle of the target polarisation with respect to the beam direction, $a_1 = 1 - y/2 - y^2\gamma^2/4$ and $a_2 = 1 - y - y^2\gamma^2/4$. The difference $\sigma(\alpha) - \sigma(\pi + \alpha)$ describes the cross section difference which is achieved by reversing the target polarisation.

Spin Asymmetries

Experimentally, the spin structure functions $g_1(x)$ and $g_2(x)$ can be determined by a combined asymmetry measurement off a longitudinally ($\alpha = 0°, 180°$) and a transversely ($\alpha = 90°, 270°$) polarised target. Because of the mixing of g_1 and g_2 a precise determination of g_1 from a longitudinally polarised target alone is not possible. The experimentally measured cross section asymmetries $A_{\parallel}$ and $A_{\perp}$

$$A_{\parallel} = \frac{\sigma^{\uparrow\downarrow} - \sigma^{\uparrow\uparrow}}{\sigma^{\uparrow\downarrow} + \sigma^{\uparrow\uparrow}}; \qquad A_{\perp} = \frac{\sigma^{\uparrow\rightarrow} - \sigma^{\uparrow\leftarrow}}{\sigma^{\uparrow\rightarrow} + \sigma^{\uparrow\leftarrow}} \tag{121}$$

obtained by spin flip of a longitudinally and transversely polarised target can be related to the asymmetries A_1 and A_2 of the exchanged virtual photon by the relations

$$A_{\parallel} = D \cdot (A_1 + \eta \cdot A_2) \quad \text{and} \tag{122}$$

$$A_{\perp} = d \cdot (A_2 - \xi \cdot A_1) \ . \tag{123}$$

The kinematic factors D, d and η, ξ are defined by

$$D = \frac{y(2-y)\left(1+\gamma^2 y/2\right)}{y^2(1+\gamma^2)\left(1-2m_\ell^2/Q^2\right) + 2a_2(1+R)} \ , \tag{124}$$

$$\eta = \frac{\gamma a_2}{(1-y/2)\left(1+\gamma^2 y/2\right)} \ , \tag{125}$$

$$d = D \frac{\sqrt{a_2}}{1-y/2} \ , \tag{126}$$

$$\xi = \frac{\gamma\,(1-y/2)}{1+\gamma^2 y/2} \ . \tag{127}$$

D and d can be regarded as depolarisation factors of the virtual photon. $R(x, Q^2)$ is the ratio of cross sections for longitudinally and transversely polarised virtual photons.

Photon Absorption Cross Section

From the measured asymmetries $A_{\parallel}$ and $A_{\perp}$, the virtual photon asymmetries A_1 and A_2 can be calculated which are independent of the kinematics of the lepton and are directly related to the photon-nucleon absorption cross sections for a given x and Q^2:

$$A_1 = \frac{\sigma_{\frac{1}{2}} - \sigma_{\frac{3}{2}}}{\sigma_{\frac{1}{2}} + \sigma_{\frac{3}{2}}} = \frac{g_1 - \gamma^2 g_2}{F_1} \ , \tag{128}$$

$$A_2 = \frac{\sigma_{\mathrm{TL}}}{\sigma_{\mathrm{T}}} = \frac{\gamma(g_1 + g_2)}{F_1} \ . \tag{129}$$

Here $\sigma_{\frac{1}{2}}$ and $\sigma_{\frac{3}{2}}$ are the virtual photo-absorption cross sections when the projection of the total angular momentum of the photon-nucleon system along the incident photon direction is $\frac{1}{2}$ or $\frac{3}{2}$ respectively. $\sigma_{\mathrm{T}} = (\sigma_{\frac{1}{2}} + \sigma_{\frac{3}{2}})/2$ is the total transverse photo-absorption cross section and σ_{TL} is a term arising from the interference between transverse and longitudinal amplitudes. It follows that $\sigma_{\mathrm{TL}} \leq \sqrt{\sigma_{\mathrm{L}}\sigma_{\mathrm{T}}}$ and therefore there is a positivity limit on the value of A_2:

$$A_2 = \frac{\sigma_{\mathrm{TL}}}{\sigma_{\mathrm{T}}} \leq \sqrt{\frac{\sigma_{\mathrm{L}}\sigma_{\mathrm{T}}}{\sigma_{\mathrm{T}}^2}} = \sqrt{R(x, Q^2)} \ . \tag{130}$$

When the term proportional to A_2, which has an explicit $1/\sqrt{Q^2}$ dependence, is neglected in (122) and (125) then we obtain the relation between the polarised structure function g_1 and the asymmetry $A_\parallel$

$$g_1 \simeq \frac{F_1}{1+\gamma^2}\frac{A_\parallel}{D} , \tag{131}$$

with

$$F_1 = \frac{1+\gamma^2}{2x(1+R)}F_2 . \tag{132}$$

2.3 Polarised Structure Functions in the Quark Proton Model

In the quark parton model the polarised structure function g_1 has a transparent probabilistic interpretation. We denote $q_f^{+(-)}(x)$ as the quark number density for quarks of flavour f in the nucleon with light cone momentum fraction x and parallel (antiparallel) orientation of the parton spin with respect to the nucleon spin. A photon with positive helicity can, due to angular momentum conservation, only be absorbed by a quark with a spin orientation antiparallel to the photon spin, since the final state, a quark, has spin $\frac{1}{2}$ and hence cannot have spin projection $\frac{3}{2}$. In case that the spin orientation of the parent nucleon is antiparallel to the photon spin (cross section $\sigma_{1/2}$) one consequently probes the distribution $q^+(x)$, while in the case that photon and nucleon spin have the same orientation (cross section $\sigma_{3/2}$) one probes the distribution $q^-(x)$ and consequently

$$g_1(x) \propto \sigma_{1/2} - \sigma_{3/2} . \tag{133}$$

So we can (omitting the Q^2 dependence) identify

$$\begin{aligned} g_1(x) &= \frac{1}{2}\sum_f e_f^2\left[\left\{q_f^+(x)+\bar{q}_f^+(x)\right\}-\left\{q_f^-(x)+\bar{q}_f^-(x)\right\}\right] \\ &= \frac{1}{2}\sum_f e_f^2\left[\delta q_f(x)+\delta\bar{q}_f(x)\right] , \end{aligned} \tag{134}$$

with $\delta q_f(x) = q_f^+(x) - q_f^-(x)$.

In the unpolarised case we sum over both cases and recover (20)

$$\begin{aligned} F_1(x) &= \frac{1}{2}\sum_f e_f^2\left[\left\{q_f^+(x)+\bar{q}_f^+(x)\right\}+\left\{q_f^-(x)+\bar{q}_f^-(x)\right\}\right] \\ &= \frac{1}{2}\sum_f e_f^2\left[q_f(x)+\bar{q}_f(x)\right] , \end{aligned} \tag{135}$$

$$\text{with } q_f(x) = q_f^+(x) + q_f^-(x) \tag{136}$$

$$\text{and } F_1(x) \propto \sigma_{1/2} + \sigma_{3/2} . \tag{137}$$

The second spin dependent structure function $g_2(x,Q^2)$ does not have an equally transparent and probabilistic interpretation. Its knowledge is required for an unambiguous determination of $g_1(x,Q^2)$ from the cross section (120). In addition it contains further important information. From the operator product expansion it is given by [183, 175]

$$g_2(x,Q^2) = -g_1(x,Q^2) + \int_z^1 \frac{\mathrm{d}z}{z} g_1(z,Q^2) + \bar{g}_2(x,Q^2) \tag{138}$$

$$= g_2^{\mathrm{WW}}(x,Q^2) + \bar{g}_2(x,Q^2) \ , \tag{139}$$

where the Wandzura-Wilczek term $g_2^{\mathrm{WW}}(x,Q^2)$ corresponds to the twist-2 contribution and $\bar{g}_2(x,Q^2)$ is a twist-3 contribution which arises only for massive quarks and is sensitive to quark-gluon correlations.

Sum Rules

Important information about the spin structure of the nucleon can be extracted from the integrals of the structure functions as in the unpolarised case, where from $\int_0^1 (F_2^{\mathrm{p}}(x) + F_2^{\mathrm{n}}(x))\,\mathrm{d}x$ we learned that quarks carry only about 50% of the nucleon's momentum and from $\int_0^1 (F_2^{\mathrm{p}}(x) - F_2^{\mathrm{n}}(x))\mathrm{d}x/x$ we learned that quarks indeed carry the assigned fractional charges and that the light quark sea is not flavour symmetric, $\bar{\mathrm{u}} < \bar{\mathrm{d}}$.

We define as the first moment of g_1

$$I_1 = \int_0^1 g_1(x)\,\mathrm{d}x \ , \tag{140}$$

introduce the notation

$$\Delta q_f = \int_0^1 (\delta q_f(x) + \delta \bar{q}_f(x))\ \mathrm{d}x \tag{141}$$

and get from (134)

$$I_1 = \frac{1}{2} \sum_f e_f^2 \Delta q_f \ . \tag{142}$$

For the proton we then can write

$$\begin{aligned} I_1^{\mathrm{p}} &= \frac{1}{2}\left(\frac{4}{9}\Delta u + \frac{1}{9}\Delta d + \frac{1}{9}\Delta s \ldots\right) \\ &= \frac{1}{12}(\Delta u - \Delta d) + \frac{1}{36}(\Delta u + \Delta d - 2\Delta s) + \frac{1}{9}(\Delta u + \Delta d + \Delta s) + \ldots \\ &= \frac{1}{12}a_3 + \frac{1}{36}a_8 + \frac{1}{9}a_0 \ . \end{aligned} \tag{143}$$

Here a_0, a_3, a_8 are the proton expectation values of the axial vector current

$$A_j^\mu = \bar{\psi}\gamma^\mu\gamma_5(\frac{\lambda_j}{2})\psi, \qquad j = 0,\ldots,8 \ , \tag{144}$$

where the matrices λ_j are the generators of flavour SU(3) in the Gell-Mann standard notation.

From isospin symmetry one can obtain the corresponding formula for the neutron:

$$I_1^{\mathrm{n}} = \frac{1}{12}\left(-a_3 + \frac{1}{3}a_8\right) + \frac{1}{9}a_0 \ . \tag{145}$$

The matrix elements a_3 and a_8 can be also obtained from the weak decays of the hyperons in the spin-$\frac{1}{2}$ octet. Assuming SU(3) flavour symmetry all axial currents in this octet are given by two decay constants, F and D. These were determined experimentally. For details see [33] and references therein. The relations are

$$a_3 = F + D, \qquad a_8 = 3F - D \ . \tag{146}$$

The matrix element a_3 is, from isospin symmetry, equal to the decay constant $g_a = g_{\mathrm{A}}/g_{\mathrm{V}} = 1.2573 \pm 0.0028$ [167], the ratio of axialvector and vector coupling constants which can be obtained from the Gamov-Teller β decay of the neutron.

The other numerical values are [33]

$$F/D = 0.575 \pm 0.016, \qquad F = 0.459 \pm 0.008, \qquad D = 0.798 \pm 0.008 \ . \tag{147}$$

The axial singlet matrix element a_0 can, however, not be fixed from hyperon decays.

From (141) and (144) one obtains the Bjorken sum rule [72, 75], which contains only the triplet matrix element a_3

$$I_1^{\mathrm{p}} - I_1^{\mathrm{n}} = \frac{1}{6}a_3 = \frac{1}{6}g_a = 0.209 \text{ (Bjorken sum rule)} \ . \tag{148}$$

Separate sum rules for the proton and the neutron can only be derived when assumptions about a_0 are made. Ellis and Jaffe [104] assumed that the strange quarks are not polarised, $\Delta s + \Delta\bar{s} \equiv 0$, leading to $a_0 = a_8$ and thus

$$I_1^{\mathrm{p,n}} = \frac{1}{12}g_a\left\{\pm 1 + \frac{5}{3}\frac{3\frac{F}{D} - 1}{\frac{F}{D} + 1}\right\} \text{ (Ellis-Jaffe sum rules)} \ , \tag{149}$$

$$I_1^{\mathrm{p}} = 0.185, \qquad I_1^{\mathrm{n}} = 0.024 \ .$$

In the quark parton model the singlet axial charge a_0 is related to the contributions of all quarks to the nucleon's spin (in units of $\frac{1}{2}\hbar$)

$$a_0 = \Delta u + \Delta d + \Delta s = \Delta\Sigma \ . \tag{150}$$

Naïvely, from the discussions in Chap. 2.1, we expect that the nucleon spin is just the vectorial sum of the quark spins and thus $\Delta\Sigma = 1$. The value obtained from a_3 is $\Delta\Sigma = 0.579 \pm 0.026$ [93].

Such a reduction can be expected from relativistic effects, which also lead to a reduction of g_a from its quark model value $\frac{5}{3}$ to $\sim \frac{5}{4}$, and from the observation that quarks carry only $\sim 50\%$ of the nucleon's momentum.

2.4 QCD Evolution of Polarised Structure Functions

We have discussed in Chap. 1.3 that due to the Q^2 dependence of the strong coupling constant $\alpha_s(Q^2)$ and the QCD processes shown in Fig. 10 quark and gluon distributions become Q^2 dependent. The same is true in the polarised case, where one obtains, apart from the splitting functions, identical evolution equations for the polarised quark distributions $\delta q(x)$ and the polarised gluon distribution $\delta g(x)$. If we again split the polarised quark distributions into flavour-non-singlett (NS) and singlett (S) parts [55]

$$\delta q^{\mathrm{NS}}(x,Q^2) = \sum_{f=1}^{n_f} \left(\frac{e_f^2}{\langle e^2 \rangle} - 1 \right) \left(\delta q_f(x,Q^2) + \delta \bar{q}_f(x,Q^2) \right) \,, \tag{151}$$

$$\delta \Sigma(x,Q^2) = \sum_{f=1}^{n_f} \left(\delta q_f(x,Q^2) + \delta \bar{q}_f(x,Q^2) \right) \,, \tag{152}$$

with $\langle e^2 \rangle = \sum e_f^2 / n_f$, we obtain the analogues to equations (65), (66) [28]

$$\begin{aligned} \frac{\mathrm{d}}{\mathrm{d}\ln Q^2} \delta q^{\mathrm{NS}} &= \frac{\alpha_s}{2\pi} \Delta P_{\mathrm{qq}}^{\mathrm{NS}} \otimes \delta q^{\mathrm{NS}} \\ \frac{\mathrm{d}}{\mathrm{d}\ln Q^2} \begin{pmatrix} \delta\Sigma \\ \delta g \end{pmatrix} &= \frac{\alpha_s}{2\pi} \begin{pmatrix} \Delta P_{\mathrm{qq}}^{\mathrm{S}} & 2n_f \Delta P_{\mathrm{qg}}^{\mathrm{S}} \\ \Delta P_{\mathrm{gq}}^{\mathrm{S}} & \Delta P_{\mathrm{gg}}^{\mathrm{S}} \end{pmatrix} \otimes \begin{pmatrix} \delta\Sigma \\ \delta g \end{pmatrix} \,. \end{aligned} \tag{153}$$

The structure function g_1 is given by a convolution of the singlet and non-singlet coefficient functions, C_{S}, C_{NS}, C_{g} with the polarised parton distribution functions

$$g_1(x) = \frac{1}{2} \langle e^2 \rangle \left\{ C_{\mathrm{NS}} \otimes \delta q^{\mathrm{NS}} + C_{\mathrm{S}} \otimes \delta\Sigma + 2n_f C_{\mathrm{g}} \otimes \delta g \right\} \,. \tag{154}$$

The splitting and coefficient functions depend on x and $\alpha_s(Q^2)$ and can be expanded in power series in α_s

$$\begin{aligned} C(x,\alpha_s) &= C^{(0)}(x) + \frac{\alpha_s}{2\pi} C^{(1)} + \mathcal{O}(\alpha_s^2) \,, \\ P(x,\alpha_s) &= C^{(0)}(x) + \frac{\alpha_s}{2\pi} C^{(1)} + \mathcal{O}(\alpha_s^2) \,. \end{aligned} \tag{155}$$

At leading order $C_{\mathrm{S}}^{(0)}\left(\frac{x}{y}\right) = \delta\left(1 - \frac{x}{y}\right)$ and $C_g^{(0)}\left(\frac{x}{y}\right) = 0$, such that g_1 decouples from δg. The next to leading order coefficient functions $C^{(1)}$ were first calculated by Kodaira [139, 140]. Full next to leading order calculations became only recently available [187, 162, 182]. As in the unpolarised case the separation between the quark and gluon distributions become factorisation scheme dependent in next-to-leading order.

Similarities and differences of the spin-dependent and spin-averaged splitting functions are discussed e. g. in [147, 119]. Two independent facts need to be stated: Due to helicity conservation at the quark gluon vertex the process $\mathrm{g} \to \mathrm{q}\bar{\mathrm{q}}$

leads to a q, $\bar{q}$ pair with opposite helicity, i. e. zero polarisation. Consequently one cannot expect a rise of $g_1(x)$ at low x due to this process in contrast to the unpolarised case, where this process contributes substantially to the rise of $F_2(x, Q^2)$ and $F_1(x, Q^2)$ at low x. On the other hand helicity conservation at the quark gluon vertex for the process q → qg requires that the spin of the gluon must be compensated by orbital angular momentum L.

QCD Corrections to Sum Rules

Taking QCD effects into account the Bjorken and Ellis-Jaffe integrals become also Q^2 dependent, we obtain

$$I_1^{\mathrm{p}} - I_1^{\mathrm{n}} = \frac{1}{6} g_a C^{\mathrm{NS}}(Q^2) \tag{156}$$

$$I_1^{\mathrm{p,n}} = \frac{1}{12}\left(\pm a_3 + \frac{1}{3} a_8\right) C^{\mathrm{NS}}(Q^2) + \frac{1}{9} a_0 C^{\mathrm{S}}(Q^2) \ . \tag{157}$$

The nonsinglet and singlet coefficient functions can be expanded into power series in $\alpha_s(Q^2)/\pi$

$$C(\alpha_s) = 1 - C_1\left(\frac{\alpha_s}{\pi}\right) - C_2\left(\frac{\alpha_s}{\pi}\right)^2 - C_3\left(\frac{\alpha_s}{\pi}\right)^3 \dots \tag{158}$$

The coefficients C_1 were calculated in the $\overline{\mathrm{MS}}$ scheme by Kodaira [139, 140, 141]. Recently the coefficients were calculated up to third order for the non-singlet case [148] and up to the second order for the singlet case [149], and estimates exist for the constants C_4^{NS} and C_3^{S} [138]. They depend on the number of flavours n_f. For $n_f = 3$ they are:

$$\begin{aligned} &C_1^{\mathrm{NS}} = 1, \quad C_2^{\mathrm{NS}} = 3.5833, \quad C_3^{\mathrm{NS}} = 20.2153, \quad C_4^{\mathrm{NS}} = 130 \\ &C_1^{\mathrm{S}} = 1, \quad C_2^{\mathrm{S}} = 1.0959, \quad C_3^{\mathrm{S}} = 3.7 \ . \end{aligned} \tag{159}$$

Note that the coefficients increase substantially with increasing order. Nevertheless at not too small Q^2 the convergence of (158) is still reasonably fast. For $n_f = 3$ one gets from (158), (159)

$$\begin{aligned} &\alpha_s(Q^2 = 2.5\,\mathrm{GeV}^2) = 0.272 \ , \\ &\alpha_s(Q^2 = 10\,\mathrm{GeV}^2) = 0.205 \ , \\ &C^{\mathrm{NS}}(Q^2 = 2.5\,\mathrm{GeV}^2) = (1 - 0.087 - 0.027 - 0.013 - 0.007) = 0.866 \ , \\ &C^{\mathrm{NS}}(Q^2 = 10\,\mathrm{GeV}^2) = (1 - 0.065 - 0.015 - 0.0006 - 0.0002) = 0.9192 \, . \end{aligned} \tag{160}$$

Higher Twist Effects

The spin-dependent structure functions and their moments are, as the unpolarised structure functions, subject to higher twist contributions (see Chap. 1.4) proportional to $1/Q^{2n}$. These describe long-distance, nonperturbative effects like correlations between quarks and between quarks and gluons and involve details of the wavefunction of the quarks in the nucleon. Such effects could be quite large at the low Q^2 values of the present data. Several authors have estimated the higher twist effects to the Ellis Jaffe and Bjorken sum rules [54, 136, 163, 155, 106]. The calculations of such terms in different models yield very different results, but there is some consensus in the literature that the effects are small. For a recent review see [79].

The U(1) Anomaly

In the simple QPM approach the singlet axial charge a_0 and the contributions to it from the different flavors, a_u, a_d and a_s, are identified with the fraction of the nucleon's spin carried by the quarks: $\Delta\Sigma$, $\Delta\mathrm{u}$, $\Delta\mathrm{d}$ and $\Delta\mathrm{s}$. In QCD the situation is more complicated, as in contrast to the conserved vector current the axial currents are not conserved even for massless quarks due to an anomalous gluon contribution [29, 102, 86] owing to the Adler-Bell-Jackiw anomaly [15, 66] caused by the triangle graph

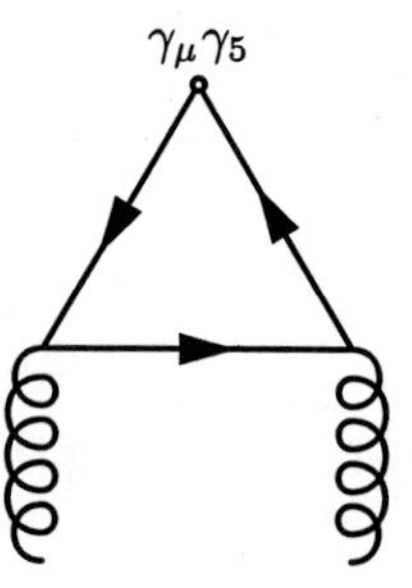

This affects only the singlet axial current while it cancels between flavors in the conserved nonsinglet current. One finds a non-vanishing divergence

$$\partial_\mu \bar{\Psi}\gamma^\mu\gamma_5\Psi = \frac{\alpha_s n_f}{2\pi}\partial_\mu \mathbf{k}^\mu = \frac{\alpha_s}{2\pi} n_f \mathrm{Tr}(\mathbf{G}_{\mu\nu}\tilde{\mathbf{G}}^{\mu\nu}) \tag{161}$$

where $\mathbf{k}^\mu$ is the axial gluon current, $\mathbf{G}_{\mu\nu}$ is the gluon field tensor and n_f is the number of active flavors. For a review in context of polarised scattering see [33, 87, 88]. The singlet axial charge $a_0(Q^2)$ contains then an extra piece proportional to $\alpha_s/2\pi\Delta g$, where Δg is the first moment of the polarised gluon distribution. The axial gluon current $\mathbf{k}^\mu$ is not gauge invariant and therefore the separation of a_0 into $\Delta\Sigma$ and Δg becomes factorisation scheme dependent. In the Adler-Bardeen [16] factorisation scheme

$$a_0(Q^2) = \Delta\Sigma - n_f\frac{\alpha_s(Q^2)}{2\pi}\Delta g(Q^2) \tag{162}$$

and $\Delta\Sigma$ is independent of Q^2. In other schemes $a_0(Q^2)$ is equal to $\Delta\Sigma$ but then it depends on Q^2.

2.5 Early Data, the EMC-SLAC 'Spin Puzzle'

The first deep inelastic polarised scattering experiments from a polarised proton target were performed end of the 70's at SLAC (target material butanol C_4H_9OH) [21, 64] and mid of the 80's by the EMC muon experiment at CERN (target material ammonia NH_3) [45, 46]. The EMC polarised target was subdivided into two halfes, which were longitudinally polarised in opposite direction to allow for simultanous data taking with two relative orientations of beam and target polarisation required to measure the asymmetry $A_{\parallel}$. The target polarisation was reversed about every two weeks.

In Fig. 41 the results for $g_1^{\mathrm{p}}(x)$ from these experiments is shown as a function of x at a fixed $Q^2 = 10.7\,\mathrm{GeV}^2$, the mean Q^2 of the data. It falls off quickly for $x \gtrsim 0.3$ and seems to approach a constant value for small values of x. Several observations should be noted:

- at large x the agreement between the SLAC and the EMC data is very good,
- the EMC data, taken at about ten times higher beam energy extend the x range from $x = 0.1$ down to $x = 0.015$, thus giving better constraints for the extrapolation towards $x = 0$, which is necessary for the determination of I_1^{p},
- the error bars of the low x data are quite large, although EMC required about 110 days of data taking to achieve this accuracy.

The reason for the large error bars is mainly of kinematical nature, since due to

$$\delta g_1 \propto \delta A_1 \cdot F_1 \propto \frac{\delta A_1 \cdot F_2}{x} \tag{163}$$

it blows up like $1/x$.

The error for the asymmetry A_1 is given by

$$\delta A_1 = \frac{1}{\sqrt{N^{\uparrow\downarrow} + N^{\uparrow\uparrow}}} (f \cdot p^{\mathrm{B}} \cdot p^{\mathrm{T}} \cdot D)^{-1} \tag{164}$$

where $N^{\uparrow\downarrow}$, $N^{\uparrow\uparrow}$ are the count rates per bin for the two target spin orientations, p^{B} is the beam polarisation, p^{T} the target polarisation, D the depolarisation factor (124) and f a dilution factor describing the fraction of deep inelastic events orginating from polarised protons in the target. For butanol this factor is about 10/74, for ammonia about 3/17. For typical values of these quantities ($f = 0.15$, $p^{\mathrm{B}} = p^{\mathrm{T}} = 0.8$, $D = 0.5$) δA_1 is more than 20 times bigger than the expectation from the count rates alone. It is evident that the best way to decrease the statistical error of g_1 substantially is to use targets with pure atomic species like polarised atomic hydrogen gas ($f = 1$ instead of 0.15).

The data in Fig. 41 tend to be constant for $x < 0.2$ as predicted from Regge theory [131, 105]. The contributions to the integral from the unmeasured regions

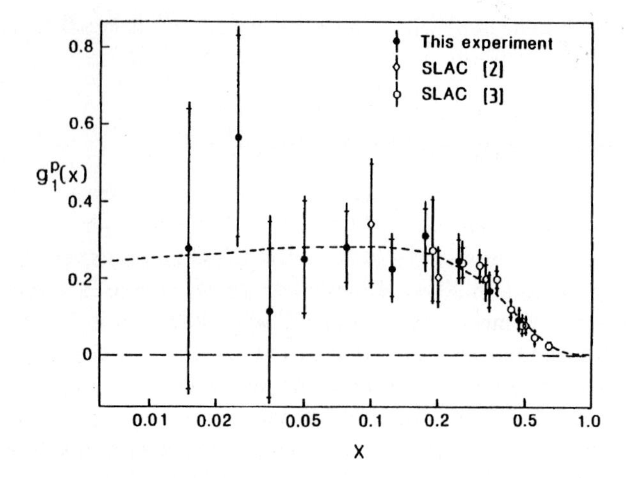

Fig. 41. *Early SLAC and EMC data for the polarised proton structure function $g_1^p(x)$ versus x.*

were estimated to be small:

$$\int_{0.7}^{1} g_1^p(x)\,dx = 0.001 \text{ and } \int_{0}^{0.01} g_1^p(x)dx = 0.002 \ . \tag{165}$$

From the EMC data alone a value for the Ellis-Jaffe integral

$$I_1^p = \int_0^1 g_1^p(x)\,dx = 0.123 \pm 0.013(\text{stat}) \pm 0.019(\text{syst}) \tag{166}$$

was obtained and for the combined EMC-SLAC data a slightly larger value, but with smaller error bars:

$$I_1^p = \int_0^1 g_1^p(x, Q^2)\,dx = 0.126 \pm 0.010(\text{stat}) \pm 0.015(\text{syst}) \ . \tag{167}$$

This value was much smaller than the value from the Ellis-Jaffe sum (149) which, using only first order QCD corrections in (157) with $\alpha_s(Q^2 = 10.7\,\text{GeV}^2) = 0.27$ and $F/D = 0.631 \pm 0.018$, was expected to be 0.189 ± 0.005.

Assuming the validity of the Bjorken sum rule and using slightly different F and D values than quoted above one obtained with this result for I_1^p three equations

$$\begin{aligned} a_3 &= \Delta u - \Delta d = 1.254 \pm 0.006 \\ a_8 &= \Delta u + \Delta d - 2\Delta s = 0.397 \pm 0.020 \\ a_0 &= \Delta u + \Delta d + \Delta s = 0.098 \pm 0.076 \pm 0.113 \end{aligned} \tag{168}$$

and could solve these equations for the mean z component of the spin carried by each of the three quark flavours in a proton with spin projection (in units of

$\hbar$) $s_z^{\mathrm{p}} = \frac{1}{2}$.

$$\begin{aligned}\langle s_z \rangle_{\mathrm{u}} &= \tfrac{1}{2}\Delta u = 0.391 \pm 0.016 \pm 0.023 \\ \langle s_z \rangle_{\mathrm{d}} &= \tfrac{1}{2}\Delta d = -0.239 \pm \ldots \\ \langle s_z \rangle_{\mathrm{s}} &= \tfrac{1}{2}\Delta s = -0.095 \pm \ldots\end{aligned} \tag{169}$$

and

$$\langle s_z \rangle_{\mathrm{quarks}} = \tfrac{1}{2}\Delta\Sigma = +0.060 \pm 0.047 \pm 0.069 \ . \tag{170}$$

From this very surprising result one had to conclude that the fraction of the nucleon spin originating from quark spins is very small, in contrast to the expectations of the Quark-Parton-Model discussed in Chap. 2.1 and Chap. 2.3. Furthermore the strange quarks, which were assumed in the derivation of the Ellis-Jaffe sum rule to be unpolarised not only seem to be polarised, but even more they seem to be polarised opposite to the polarisation of the parent nucleon.

The above result caused a lot of excitement and sometimes has been called 'proton spin crisis' or 'proton spin puzzle'. It has led to an extensive discussion about the internal spin structure of the nucleon and how quark and gluon spins and orbital angular momenta contribute to the nucleon spin, which most generally is given by helicity sum rule

$$\langle s_z \rangle_{\mathrm{Proton}} = \frac{1}{2}\Delta\Sigma + L_z^{\mathrm{quarks}} + \Delta g + L_z^{\mathrm{gluons}} = \frac{1}{2} \ , \tag{171}$$

where L_z^{quarks} and L_z^{gluons} denote the components of the orbital angular momentum of quarks and gluons along the quantisation axis.

It should be noted that recently leading order evolution equations for the quark and gluon orbital angular momenta have been derived [137] which couple them with each other and with the quark and gluon spins. For $Q^2 \to \infty$ they obtain the following solution for the asymptotic spin fraction carried by quarks and gluons

$$\begin{aligned}\Delta\Sigma + 2L_z^{\mathrm{quarks}} &= \frac{3n_f}{16 + 3n_f} = 2J_q = 0.43 \ , \\ 2\Delta g + 2L_z^{\mathrm{gluons}} &= \frac{16}{16 + 3n_f} = 2J_g = 0.57 \ .\end{aligned} \tag{172}$$

These are exactly the same values as those predicted for the asymptotic momentum fractions.

2.6 New Experiments

The unexpected EMC result triggered an enormous theoretical activity resulting in a plethora of publications which gave a lot of new insight into the problem. At the same time a new generation of experiments (SMC at CERN; E142, E143, E154 at SLAC and HERMES at HERA-DESY) has been performed/started using both polarised proton and neutron targets to clarify the 'spin-puzzle' experimentally, to disentangle the fractional contributions of quarks, gluons and orbital angular momenta to the helicity sum rule (171), and to answer a couple of questions like for example:

- what is the x-dependence and the first moment of the polarised neutron structure function $g_1^n(x)$?
- is the Bjorken sum rule violated?
- how large is the singlet axial change a_0, i. e. the contribution of quark spins to the nucleon spin?
- is the polarisation of seaquarks ΔS large and negative?
- does ΔS just cancel the contribution from valence quarks ΔV?
- is the gluon polarisation $\Delta g(x)$ large but compensated by large orbital angular momentum contributions L_z?
- is the nucleon spin mainly due to orbital angular momenta?

It can not be emphasised enough that these new experiments were only possible due to extraordinary technological developments to achieve both high beam and target polarisation. Each of these experiments is based on different technologies and represents a major effort with respect to technological ingeniuity, manpower, investment and large scale organisation.

The SMC Experiment

The measurements of the SMC collaboration were performed in the years 1992–1996 with the upgraded muon spectrometer [13, 6, 7, 9, 10, 12, 11] previously used by the EMC and NMC experiments (Fig. 42). The beam energies and targets used in these experiments are summarised in Table 3. The high energy

Table 3. *Data taking conditions of the SMC, SLAC and HERMES experiments*

Experim.	Year	E^B GeV	p^B %	I^B	N	Target Material	p^T %	f	T (N/cm^2)
SMC	1992	100	78	0.5 pA	d	butanol	22–38 $\parallel$	0.28	4.8×10^{25}
	1993	190	78	0.5 pA	p		86 $\parallel$	0.13	4.2×10^{25}
		100	78	0.5 pA	p		80 $\perp$		4.2×10^{25}
	1994	100	78	0.5 pA	d	butanol	49 $\parallel$	0.28	4.8×10^{25}
	1995	100	78	0.5 pA	d	butanol	50 $\parallel$	0.28	4.8×10^{25}
		190	78	0.5 pA	d	butanol	44 $\perp$	0.28	4.8×10^{25}
	1996	190	78	0.5 pA		NH_3	89 $\parallel$	0.15	4.2×10^{25}
E142	1992	25.51	39	4 μA	n	^3He	35 $\parallel, \perp$	0.11	3.6×10^{22}
E143	1993	29.1	84	0.6 nA	p	$^{15}NH_3$	72 $\parallel, \perp$	0.15	1×10^{24}
		29.1	84	0.6 nA	d	$^{15}ND_3$	25 $\parallel, \perp$	0.23	1.2×10^{24}
E154	1995	48.3	82	1 μA	n	^3He	38 $\parallel, \perp$	0.16	3.6×10^{22}
HERMES	1995	27.5	50	10–30 mA	n	^3He	50 $\parallel$	0.33	1×10^{15}
	1996	27.5		10–30 mA	p	H	90 $\parallel$	1	7×10^{13}

muon beam at CERN is produced by bombardement of a beryllium target with 540 GeV protons from the SPS and subsequent decay of momentum selected secondary pions and kaons into muons and neutrinos. Typical beam intensities

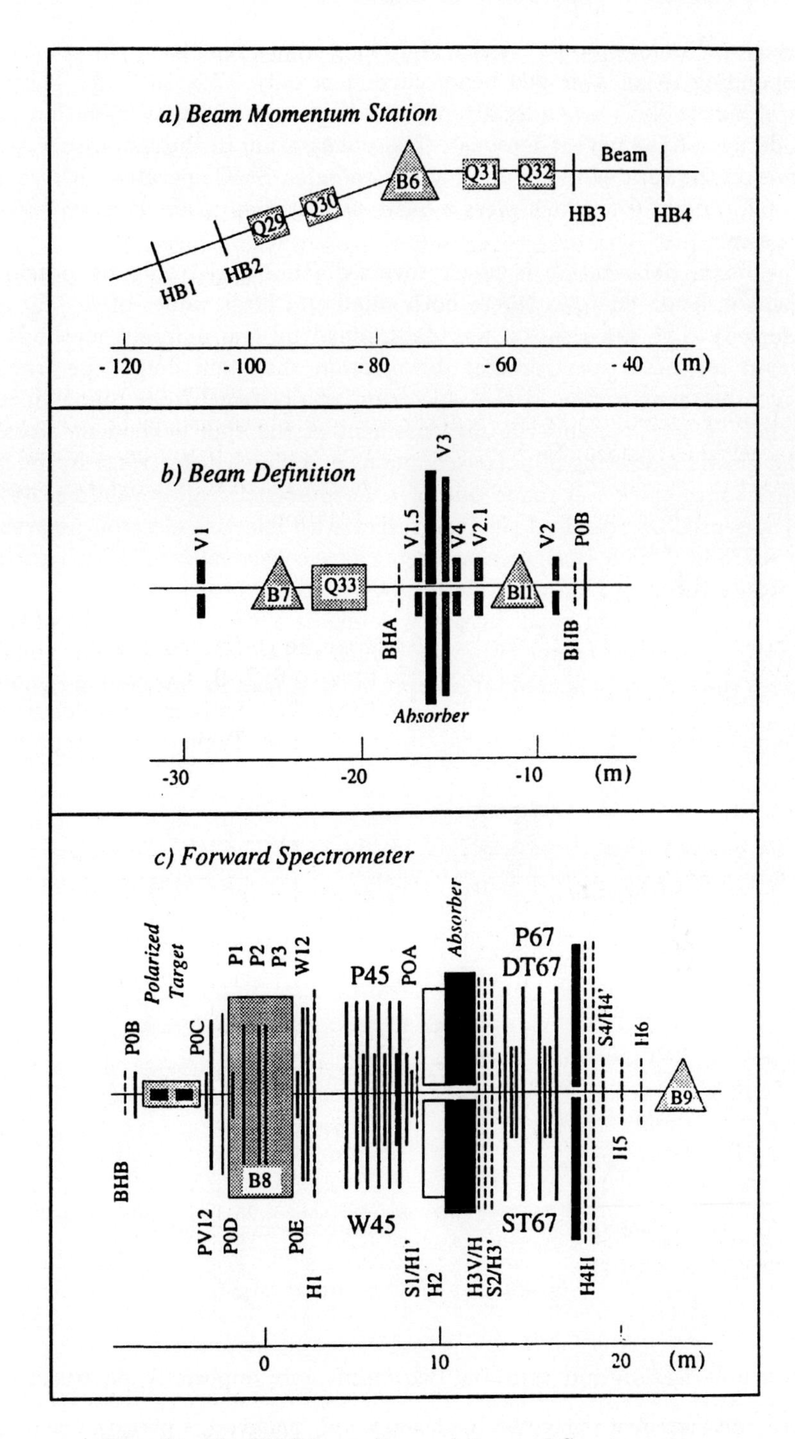

Fig. 42. *The SMC beam line and Forward Spectrometer.*

were 4.5×10^7 muons per SPS pulse (2.4 s long with a repetition period of 14.4 s) corresponding to an averaged beam current of only 0.5×10^{-12} A. The muon beam is 'naturally' longitudinally polarised because of parity violation in the weak decays of the parent mesons*. The polarisation in the laboratory system depends on the ratio of muon and hadron energies. SMC operates with a typical ratio of $E_\mu/E\pi \simeq 0.9$, which gives a beam polarisation of about $p_\mu \simeq -80\%$ for a μ^+ beam.

The beam polarisation is never reversed (this would only be possible at small muon beam energies where both small and large values of E_μ/E_π could be selected). The polarisation was determined by two different methods in a dedicated magnetic spectrometer downstream the main muon spectrometer: The measurement of the Michel spectrum of positrons from muon decay in flight, $\mu^+ \to e^+\bar{\nu}_\mu\nu_e$, and the measurement of the spin dependent cross section for elastic scattering of polarised muons on polarised electrons. In the latter case a 2.7 mm thick foil made out of a ferromagnetic alloy (49%Fe, 49%Co, 2%V) was used as polarised electron target with effective electron polarisation $|p_e| = 0.0756 \pm 0.0008$. Both measurements gave consistent results with a relative statistical error of $\sim 3.5\%$ and relative systematic error of $\sim 2.5\%$.

The very low muon beam intensity demands a very thick polarised target. The SMC solid state target (Fig. 43) (which is similar to the EMC polarised target) is the largest polarised target ever built. It uses the method of dynamical

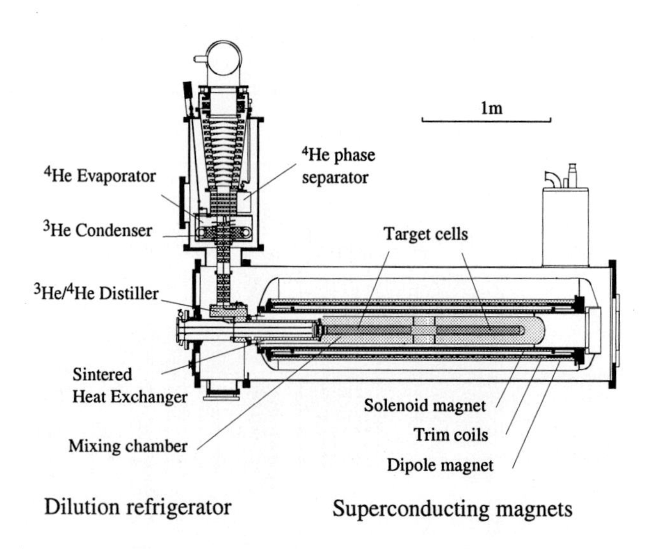

Fig. 43. *The SMC polarised target.*

nucleon polarisation and contains two 60 cm long oppositely polarised target

* In the rest system of the meson the muon is fully polarised, a positive muon having negative helicity.

cells, separated by a 30 cm gap, exposed to the same muon beam and thus allowing simultaneous data taking with the two relative orientations of beam and target polarisation required to measure the asymmetry $A_{\parallel}$. A superconducting magnet system provides a strong magnetic holding field of 2.5 T with a relative homogeneity of $\pm 3.5 \times 10^{-5}$ over the target volume. A perpendicular holding field was also used for reversal of the spin direction and for the measurement of $A_{\perp}$. The target material was cooled down by a ^{3}He-^{4}He dilution refrigerator to a temperature below 0.5 K and the target worked in the 'frozen spin' mode. The two spin directions were produced by irradiating the material with microwaves of slightly different frequencies just below or above the corresponding electron spin resonance frequency. Typical polarisations are 50% for the deuteron and 85% for the proton target. The polarisation was measured with nuclear magnetic resonance technique (NMR) with a typical relative accuracy of 3–5%. The spin direction was reversed every few hours by rotating the direction of the magnetic field.

In addition to the polarised material the target cells contained other materials, mostly the ^{3}He-^{4}He cooling liquid and the NMR coils for the polarisation measurement. The corresponding dilution factor f is shown for the proton target in Fig. 44 as a function of x. A typical value over the range $10^{-2} < x < 10^{-1}$

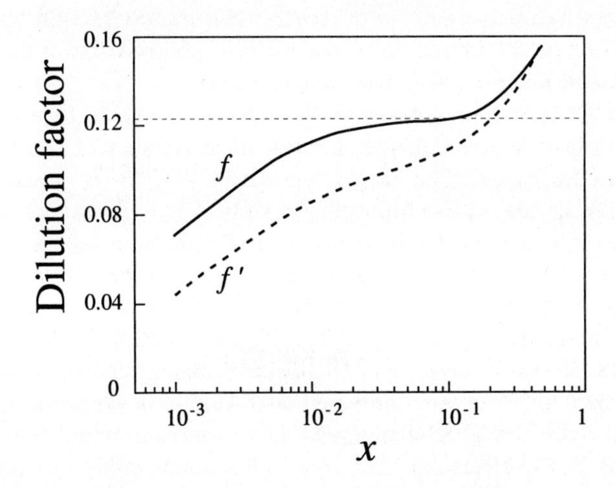

Fig. 44. *The dilution factor for the SMC proton target versus x without (f) and with (f') radiative corrections.*

is $f \simeq 0.12$, in the case of deuterated butanol it is about a factor of 2 larger. For proper calculation of the statistical errors the SMC collaboration recently also included the effect of radiative corrections (mainly from the unpolarised target material) in the dilution factor [12]. As can be seen from the dashed line in Fig. 44 the effective dilution factor f' is substantially smaller than f and becomes as low as $f' \sim 0.04$ at $x \simeq 10^{-3}$.

The SLAC Experiments E142, E143 and E154

The main parameters of the three SLAC experiments are summarised in Table 3. The first of these experiments, E142 [34, 35], was still performed with a 'standard' polarised electron source based on photoemission from an AlGaAs photo cathode illuminated by circularly polarised light from a dye laser [25]. This source delivered typically 2×10^{11} electrons per pulse at 120 Hz with an electron polarisation of about 36%. The electron helicity was changed randomly pulse by pulse by controlling the circular polarisation of the laser light. The beam polarisation was measured with a relative systematic accuracy of 3.1% by Møller scattering from thin magnetised Vacoflux foils. Atomic binding effects for the target electrons [150] were taken into account which modify the analysing power compared to free target electrons by up to 15% depending on the geometry of the polarimeter. A single arm polarimeter was used for E142, while E143 and E154 used in addition a double arm polarimeter with a relative precission of 2.5%.

E143 [1, 2, 3] and E154 [4] profited a lot from the development of electron sources with a strained-lattice GaAs cathode illuminated by a flash-lamp-pumped Ti-sapphire laser operated at 850 nm [157]. This technology was successfully pushed at SLAC with an enormeous effort both in manpower and capital investment for the measurements of electroweak parameters on the Z^0 peak at SLC. This new type of source delivered a beam polarisation of up to 86% with $(2\text{–}4) \times 10^9$ electrons per pulse at a rate of 120 Hz.

E142 and E154 used a polarised ^{3}He target, which is to a good approximation a polarised neutron target, for the measurement of the polarised neutron structure functions. The target, shown in Fig. 45, is based on a double chamber design. In the upper 'pumping chamber' the ^{3}He atoms were polarised through spin-exchange collisions with polarised rubidium atoms which were optically pumped by circularly polarised laser-light [89]. E142 used a system of five Ti-sapphire lasers pumped by argon-ion lasers producing 20 W cw optical power. E154 even increased the total laser power to 60 W by the additional use of three diode laser array systems. The electron beam passed through the lower 'target chamber', a 30 cm long glass cell with thin exit windows (E142: 110 μm, E154: 50 μm) operated at 8.6 bar and 0° C temperature resulting in a ^{3}He target density of 2.3×10^{20} atoms/cm^3. A small amount of N_2 was added to cause radiationless quenching of the Rb excited state atoms to increase the efficiency of optical pumping. The Rb atoms were contained almost entirely in the upper (heated) pumping cell. The extra material of the exit windows and the N_2 further reduced the number of polarisable nucleons in the target. The dilution factor of ^{3}He ($f^{^3\mathrm{He}} \simeq 1/3$) was further decreased by a slightly x-dependent factor (E142: $f \simeq 0.35 \cdot f^{^3\mathrm{He}}$; E154: $f \simeq 0.58 \cdot f^{^3\mathrm{He}}$).

The ^{3}He polarisation was measured with NMR technique, calibrated with signal from a water sample of the same geometrical size as the target cell. Typical values were $p_\mathrm{T} \sim 0.33$ (E142) and $p_\mathrm{T} \sim 0.38$ (E154) with a relative systematic accuracy of $\pm 7.1\%$, where the error is dominated by the calibration constant

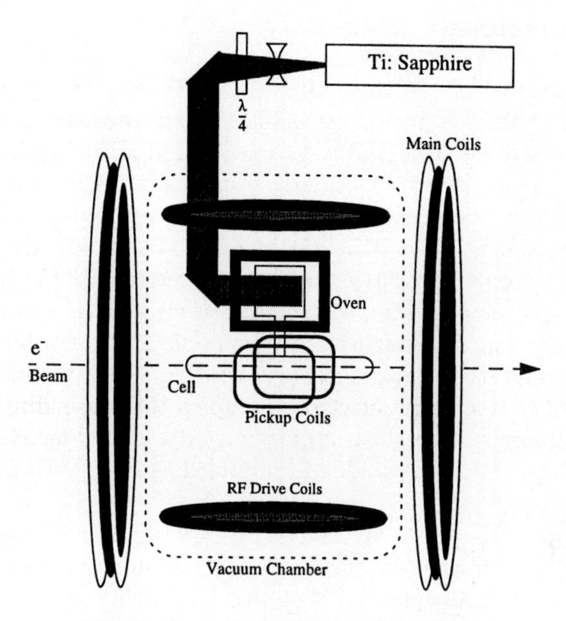

Fig. 45. *The E142 polarised* ^{3}He *target.*

derived from the water signal.

It should be stressed that the fabrication of useable glass cells is really an art. Very special treatments [35] during fabrication and filling with ^{3}He, Rb and N_2 was required to achieve good spin-relaxation properties of the glass walls and to avoid breaking of the cell. Relaxation times in the range of 50 to 65 hours were reached for the cells used in the experiments.

The E143 polarised target (Fig. 46) [97], which uses $^{15}NH_3$ and $^{15}ND_3$ as target material, is a cold solid target which is based on the same design principle as the one described above for SMC, although of much smaller size. It consists of only one cell of 3 cm length in the beam, the NH_3 and the ND_3 targets stacked on top of each other. The material was polarised by microwave irradiation at about 138 GHz using the mechanism of dynamical nuclear polarisation. The holding field of 4.8 T was provided by superconducting Helmholtz coils, the temperature was kept at 1 K by a ^{4}He evaporation refrigerator. Proton polarisations of 65–80% were achieved in 10–20 min which then slowly decreased to about 50–55% by radiation damage. The average deuteron polarisation was about 25% with a maximum greater than 40%. The dilution factor varied between 0.13 and 0.17 for the NH_3 target and between 0.22 and 0.25 for the ND_3 target. The electron beam was rastered over the 4.9 cm^2 front surface of the target to uniformly distribute heating and radiation damage. The radiation damage was repaired by a special annealing procedure, after typically 10 anneal cycles the target material was replaced.

The beam energies used by E142 were 19.4, 22.7 and 25.5 GeV. For E154 the

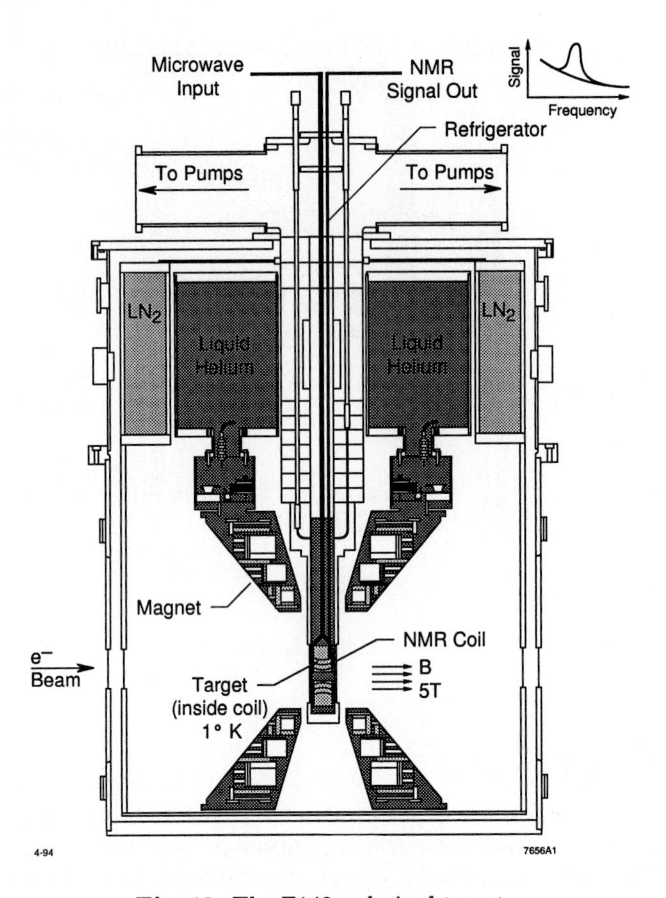

Fig. 46. *The E143 polarised target.*

beam energy was increased to 48.3 GeV. This allowed to extend the kinematic range to lower values of x and higher values of Q^2.

Scattered electrons were collected in two single-arm spectrometers (Fig. 47) at horizontal scattering angles Θ of 4.5° and 7° (E154: 2.75° and 5.5°) covering a momentum range from 7 to 20 GeV/c for both arms (E154: 10–35 GeV/c and 10–43 GeV/c), and solid angles of 0.097 msr and 0.435 msr. Each spectrometer was instrumented with a pair of threshold Čerenkov detectors, a segmented lead-glass calorimeter, six planes of segmented scintillation counters grouped into two hodoscopes and two pairs of lucite trigger counters. The main electron trigger consisted of a triple coincidence between the two Čerenkovs and the sum of the shower counter signals. Energy E' and scattering angle Θ were extracted from the summed pulse heights and the centroid position of the shower in the calorimeter. A large sample of deep inelastic scattering events was collected in the different experiments and used to determine the asymmetries and spin structure functions.

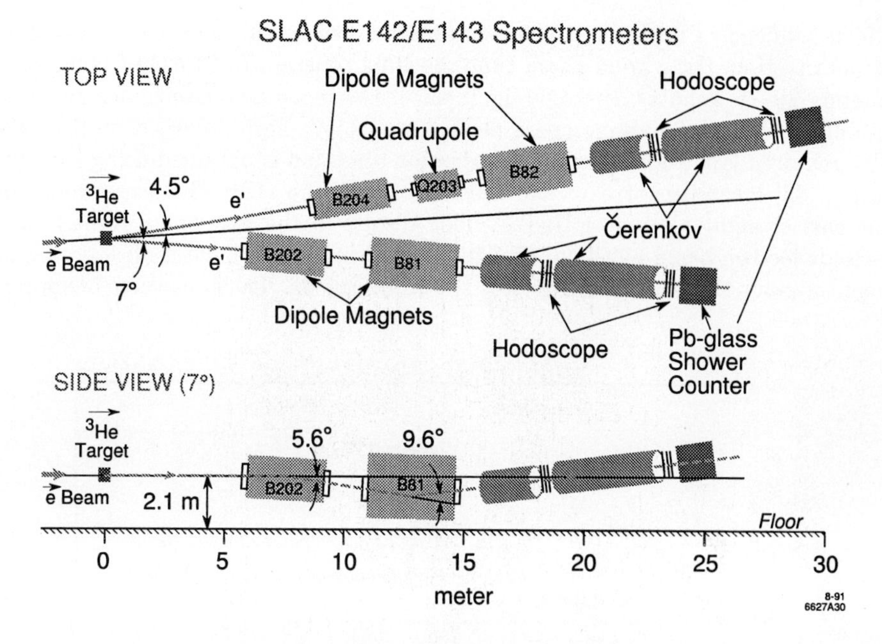

Fig. 47. *The E142/E143 spectrometers.*

The HERMES Experiment

The HERMES experiment at the positron/electron storage ring of the HERA accelerator system is designed to investigate not only inclusive but also semi-inclusive processes in polarised deep inelastic scattering using a large acceptance forward spectrometer. It is based on two novel technical achievements: longitudinal electron polarisation in a high energy storage ring, based on the asymmetry in the spin-flip probability for the emission of synchrotron radiation (Sokolov-Ternov effect [178], which leads to self-polarising of the beam, and the use of polarised internal gas targets of hydrogen, deuterium and ^{3}He consisting of a windowless thin-walled storage cell fed by high intensity sources of polarised atoms. This target technology has the advantage that the target atoms are pure atomic species and hence no dilution of the asymmetry occurs in the scattering from unpolarised target material resulting in the very favourable dilution factors of 1, 1 and 1/3 for the H, D and ^{3}He target. Furthermore the target spin can be reversed rapidly which is essential for minimising systematic errors. The HERA positron beam at present has an energy of 27.5 GeV and a peak current of 35–40 mA. As mentioned above the positron beam in the HERA storage ring becomes transversely polarised to a high level by the Sokolov-Ternov mechanism. For an ideal machine the maximum asymptotic degree of polarisation is 92.4%, the time constant τ for polarisation build up is proportional to ρ^3/E^5, where ρ is the bending radius in the magnetic field and E is the beam energy. In a realistic storage ring the polarisation build up is disturbed by depolarising effects influencing both the polarisation build up time and the achievable degree of polarisation. For a good beam tune the time constant at 27.5 GeV is approximately 20–25 minutes. Precise alignment of the machine quadrupoles and fine tuning of the orbit parameters is needed to achieve high polarisation [58, 60]. The required longitudinal beam polarisation direction is obtained using spin rotators [83] located upstream and downstream of the HERMES experiment in the east straight section of HERA. This results in the first longitudinally polarised electron beam in a high-energy storage ring [61]. The first observation of longitudinal polarisation at HERA is shown in Fig. 48. The transverse beam po-

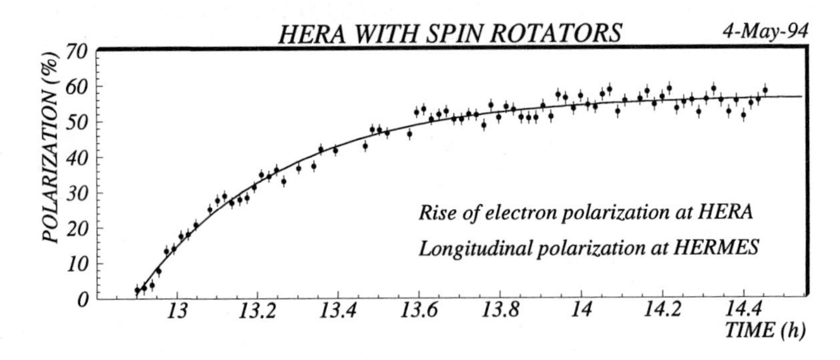

Fig. 48. *First observation of longitudinal electron polarisation at HERA.*

larisation is measured continuously using Compton back scattering of circularly polarised laser light. Values of the equilibrium polarisation in the range 40% to 65% are obtained under normal running conditions. For the '95 data taken with the ^{3}He target experimental data were analyzed only when the polarisation was above 40%. The average polarisation for the analyzed data was 55% with a fractional systematic error of 5.5% dominated by the uncertainty in the calibration of the beam polarimeter. In principle the direction of the beam polarisation can be reversed for every fill by moving the magnets of the spin rotators. This was, however, tried for the first time beginning of 1997. A single polarisation direction was used for the measurements in 1995/96.

In 1995 the polarised ^{3}He target was used. The ^{3}He atoms are polarised in a glass pumping cell by spin exchange collisions with a small sample ($\sim 10^{-6}$) of the ^{3}He atoms which are in the 2^3S_1 metastable state. The ^{3}S atoms are created by a weak RF discharge and polarised by optical pumping with 1083 nm laser light [95, 121]. Compared to the technique used by E142 no Rb and N_2 admixture is needed and the polarisation build up time is much faster. The polarised atoms diffuse from the glass pumping cell into a 400 mm long open ended thin-walled storage cell inside the positron ring, which is constructed of 128 μm thick ultra-pure aluminium and is cryogenically cooled to typically 25 K for the ^{3}He and to about 100 K for H, D. The atoms perform several hundred wall bounces before they leave the storage cell. This way the target thickness is increased by about two orders of magnitude compared to a free gas jet. This provides a target of pure atomic species with an areal density of approximately 3.3×10^{14} atoms/cm^2. The polarisation direction was defined by a 3.5 mT magnetic field parallel to the beam direction provided by a pair of Helmholtz-coils. The polarisation in the pumping cell is determined from the polarisation of emitted photons produced via atomic excitation by the RF discharge, the polarisation in the target cell was monitored by measuring the polarisation of photons emitted from atoms that were excited by the electron beam. The average value of the target polarisation was 46% with a fractional systematic uncertainty of 5%.

The HERMES atomic beam source (ABS) [179] for polarised hydrogen (H) and deuterium (D) is based on the Stern-Gerlach spin separation by multipole magnets (see Fig. 49). An intense atomic beam of thermal velocity is produced

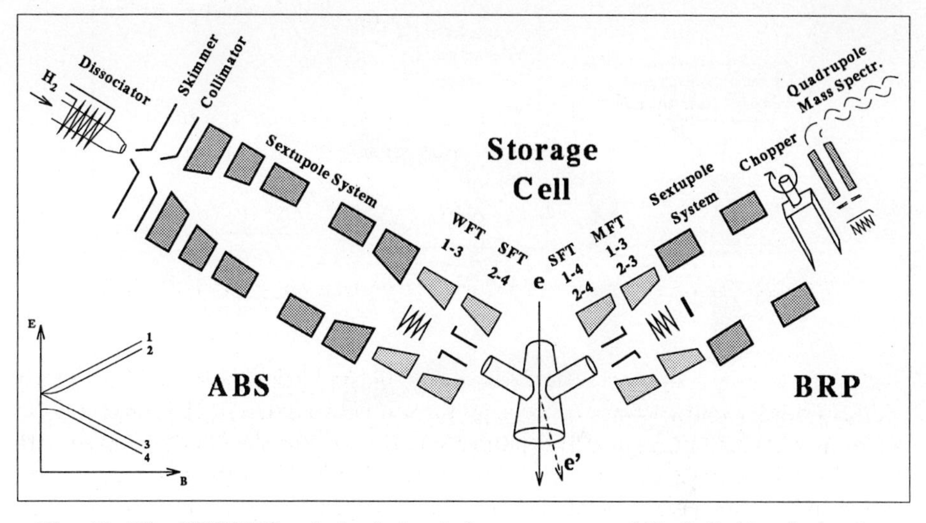

Fig. 49. *The HERMES polarised atomic beam source and Breit-Rabi polarimeter.*

by means of a RF-dissociator with a cooled nozzle and skimmer and a powerfull differential pumping system. The hyperfine-states with magnetic electron spin quantum number $m_j = +\frac{1}{2}$ are focused onto the entrance tube of the storage cell by a system of sextupole magnets, while those with $m_j = -\frac{1}{2}$ are defocused. High frequency transitions are used to populate the substates with the required magnetic nucleon spin quantum numbers and to provide high vector polarisation (P_z) or (for D) tensor polarisation (P_{zz}). They also provide rapid spin reversal. Typical polarisation values for H are above 90%, typical fluxes of polarised atoms are 6–8 $\times$ 10^{16} atoms/s which result in a target areal density of about 7×10^{13} nucleons/cm^2 for the chosen storage cell geometry.

A small fraction ($\sim$ 10%) of the polarised target atoms is extracted from the storage cell sideways as a sample beam and their polarisation is measured by a Breit-Rabi polarimeter [118]. This is an atomic beam resonance apparatus consisting again of sextupole magnets and high frequency transition units, a chopper system for background suppression and a quadrupole mass spectrometer. It allows to measure the relative population of the individual substates with 2% accuracy and to deduce electron and nuclear polarisation with an error of about 3%.

The high intensity electron/positron bunches of HERA produce a strong transient magnetic field transverse to the beam direction which can cause depolarisation of the H and D atoms in the storage cell. A substantial holding field of around 0.35 T for H and 0.15 T for D is needed to decouple the nuclear and electron spins and to avoid nuclear depolarisation. A simplified sketch of the HERMES target area is shown in Fig. 50. A differential pumping system

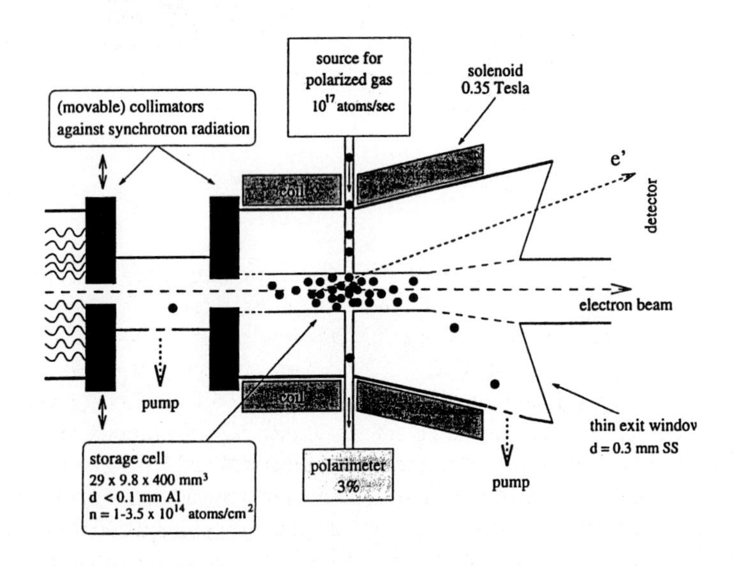

Fig. 50. *Simplified sketch of the HERMES target region.*

upstream and downstream of the target chamber assures that the vacuum in the ring is not affected by the injected gas. Meshes between the storage cell and the beam pipe avoid rapid changes of the geometry and prevent the electron bunches from inducing wake fields and resonant RF losses in the target chamber which could heat up and destroy the storage cell and other components. Two sets of collimators protect the storage cell against synchrotron radiation generated in the beam bending and focusing components and tails of the beam. With proper beam tuning the detector system was essentially free of electromagnetic background from such sources. Without these collimators about 100 W of synchrotron radiation would heat the storage cell. The first collimator is movable, it can be closed to an elliptical aperture of $\pm 2.6(\text{vert}) \times 6.5(\text{hor})\,\text{mm}^2$ which is by far the smallest aperture in the ring.

A schematic diagram of the HERMES spectrometer [129] is shown in Fig. 51. The spectrometer consists of a large dipole magnet surrounding the electron/

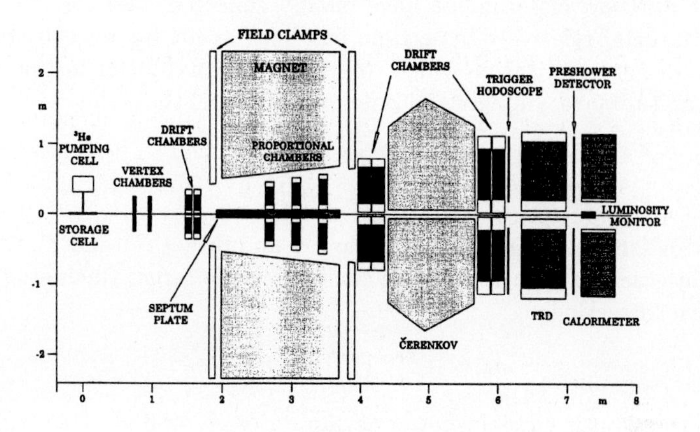

Fig. 51. *The HERMES spectrometer.*

positron and proton beam pipes of HERA. The beams are shielded from the magnetic field by a horizontal iron plate. The spectrometer is constructed as two identical halves, mounted above and below the beam pipes. The scattering angle acceptance is 40 mrad $< \Theta <$ 220 mrad. Tracking is done by tracking chambers (microstrip-gas-counters, drift chambers and multi-wire proportional chambers) before, inside and after the magnet. Particle identification is done using the combination of four detectors: a radiation-hard lead-glass calorimeter, a pre-shower section consisting of one radiation length of lead followed by a hodoscope, a transition radiation detector and a N_2 gas threshold Čerenkov counter. They serve to suppress the large background of hadrons, mostly pions from photoproduction, which in the worst part of the kinematic plane is about 400 times larger than the positron yield. The trigger is formed from a coincidence between the Pb-glass calorimeter and a pair of scintillator hodoscope planes and require an energy of greater then typically 3.5 GeV deposited in the calorimeter. This system provided positron identification with an average efficiency of 98% and hadron contamination < 1%. The Čerenkov counter serves to separate pions from other hadrons. In 1995 a conservative pion threshold of about 5.5 GeV was used to help in positron identification at low energies, in 1996/97 it was decreased to below 4 GeV. It is foreseen to convert the threshold Čerenkov counter to a dual radiator RICH counter in the winter shutdown 1997/98. It will have clear silica aerogel ($n = 1.03$, $\gamma_{\rm th} \simeq 4.2$) and C_4F_{10} ($n = 1.0005$, $\gamma_{\rm th} \simeq 32$) as radiators and will allow full π, K, p separation essentially for all momenta > 1 GeV/c. The luminosity of the experiment is measured by detecting Bhabha (Møller) scattered target electrons in coincidence with the scattered positrons (electrons) in a pair of very radiation-hard $NaBi(WO_4)_2$ electromagnetic calorimeters.

2.7 The Experimental Data

The Asymmetry A_1

So far the different experiments have mainly collected data for longitudinally polarised targets, relatively little time has been spent for measurements with transverse target polarisation which we will not discuss furter in this review.

The longitudinal asymmetry $A_{\parallel}$ is determined from the count rate asymmetry

$$A_{\parallel}^{\text{exp}} = \frac{N^{\uparrow\downarrow} - N^{\uparrow\uparrow}}{N^{\uparrow\downarrow} + N^{\uparrow\uparrow}} \tag{173}$$

by correcting for the polarisation of beam P^B and target P^T and f', the fraction of events originating from polarisable nucleons in the target (including the effect of radiative corrections)

$$A_{\parallel} = A_{\parallel}^{\text{exp}} \cdot (f' \cdot p^{\text{B}} \cdot p^{\text{T}})^{-1} \ . \tag{174}$$

$A_{\parallel}$ depends on both virtual photon asymmetries A_1 and A_2 (see (122)). Since the contribution of A_2 is kinematically suppressed, generally A_2 is assumed to vanish and the spin asymmetry A_1 is determined from (131). A_2 is only taken into account in the determination of the systematic error by using the upper limit from the experiments, shown in Fig. 52 for the proton. Radiative corrections to

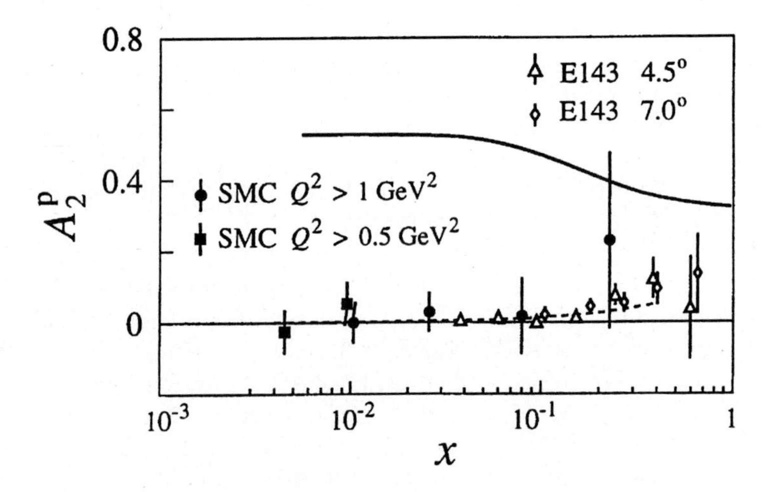

Fig. 52. *A_2^{p} versus x from SMC and E143.*

the asymmetry [143, 20] are much smaller than for the unpolarised measurements and are applied additively. In the case of solid state targets nuclear corrections to the structure functions of unpolarised nuclei of the target material and the corresponding radiative corrections have also to be applied.

The results for the virtual photon asymmetry A_1^{p} for $Q^2 > 1\,\mathrm{GeV}^2$ obtained by the EMC [45, 46], SMC [12], SLAC E89 [21], E130 [64] and E143 [1] experiments, as published in spring '97, are shown in Fig. 53 as a function of x. The

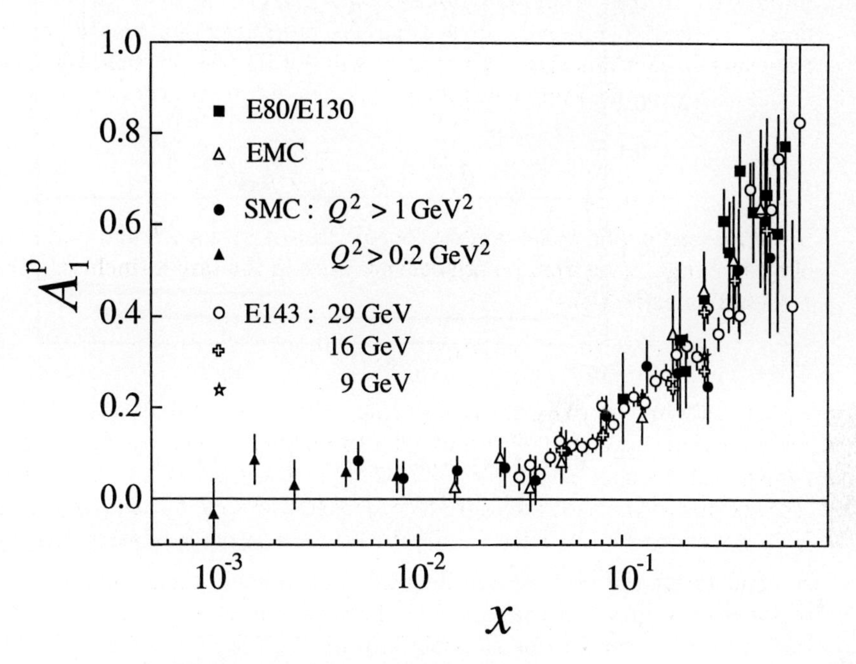

Fig. 53. *World data for A_1^{p} as a function of x.*

data cover a range $0.001 < x < 0.8$ and $0.2\,\mathrm{GeV}^2 < Q^2 < 72\,\mathrm{GeV}^2$. For each point in x the mean Q^2 value is different and typically the $\langle Q^2 \rangle$ for the data from the muon experiments is about an order of magnitude higher than for the corresponding ones from the electron experiments. Within the statistical accuracy of the data the agreement between the different experiments is excellent. For $x > 0.03$ the x dependence of A_1 is determined by the high statistics data from E143. The low-x region $x < 0.03$ can presently only be accessed by the muon experiments due to their higher beam energy. The SMC data for $x < 0.004$ are in the Q^2 range $0.2\,\mathrm{GeV}^2 < Q^2 < 1\,\mathrm{GeV}^2$ and are not used to evaluate g_1^{p} or its first moment since it is questionable whether QCD or the QPM can be applied at such low values of Q^2. At $x < 0.04$ the asymmetry seems to stay constant at a value of $A_1 \simeq 0.06$, then it rises smoothly to a value of $A_1 \simeq 0.8$ at the highest x point in agreement with the expectation $A_1 \to 1$ for $x \to 1$ (i.e. the quark carrying most of the nucleon's momentum also is carrying most of its spin) [82].

The results for A_1^{d} from SMC [11] and SLAC E143 [2] at the average Q^2 of each x bin are shown in Fig. 54.

The E143 data have much higher precision (although the data quality is not

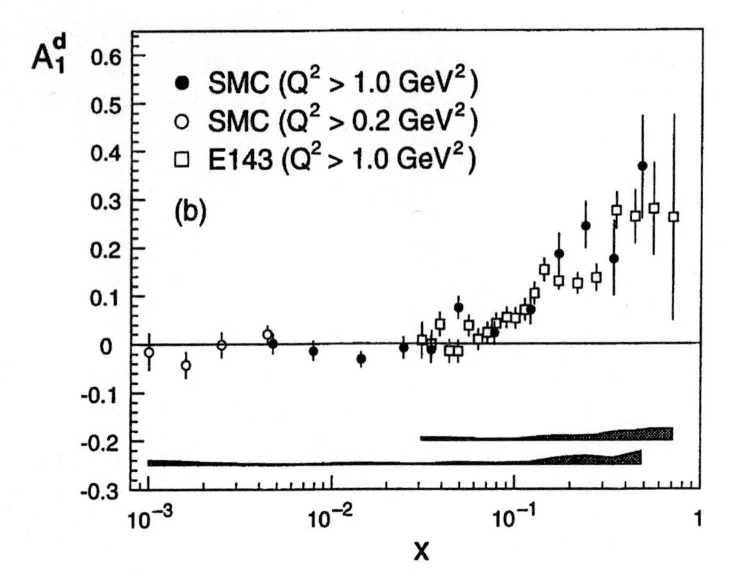

Fig. 54. A_1^{d} *versus* x *from SMC and E143.*

as good as for the proton) and determine completely the x dependence in the region of overlap, where the agreement between the two experiments is good. The rise at large x seems to be substantially weaker than for the proton. The very low x region $x < 0.03$ is again only covered by the SMC data. At $x < 0.05$ the asymmetry is very small, compatible with zero and possibly slightly negative. In Fig. 55 A_1^{d} is shown as a function of Q^2 [11] for fixed values of x in the region of overlap of the two experiments. No clear Q^2 dependence of A_1^{d} is visible within the accuracy of the data. The same conclusion can be drawn for the present data for A_1^{p} [3].

Finally the results for A_1^{n} obtained by the E142 [35], E154 [4] and HERMES [5] from ^{3}He targets is shown in Fig. 56. The wave function of ^{3}He is dominated by the configuration with the protons paired to zero spin. Therefore, most of the ^{3}He asymmetry is due to the neutron [114]. A correction for the non-zero polarisation of the protons (-0.028 ± 0.004) using the E143 results for A_1^{p} and the neutron polarisation (0.86 ± 0.02) [90] has to be applied. The agreement between all three experiments is excellent. The E154 data extend to lower values of x due to the higher beam energy and have much smaller error bars than those from HERMES and E142. The asymmetry is negative essentially over the whole x range covered by the data. The asymmetry at $x \simeq 0.5$ is much smaller than for the proton and slightly positive. This can be explained by assuming that the positive polarised $\mathrm{d_v^n}$-quark distribution is largely cancelled by the much smaller negative $\mathrm{u_v^n}$-quark distribution, which enters the cross section with a four times bigger charge weight.

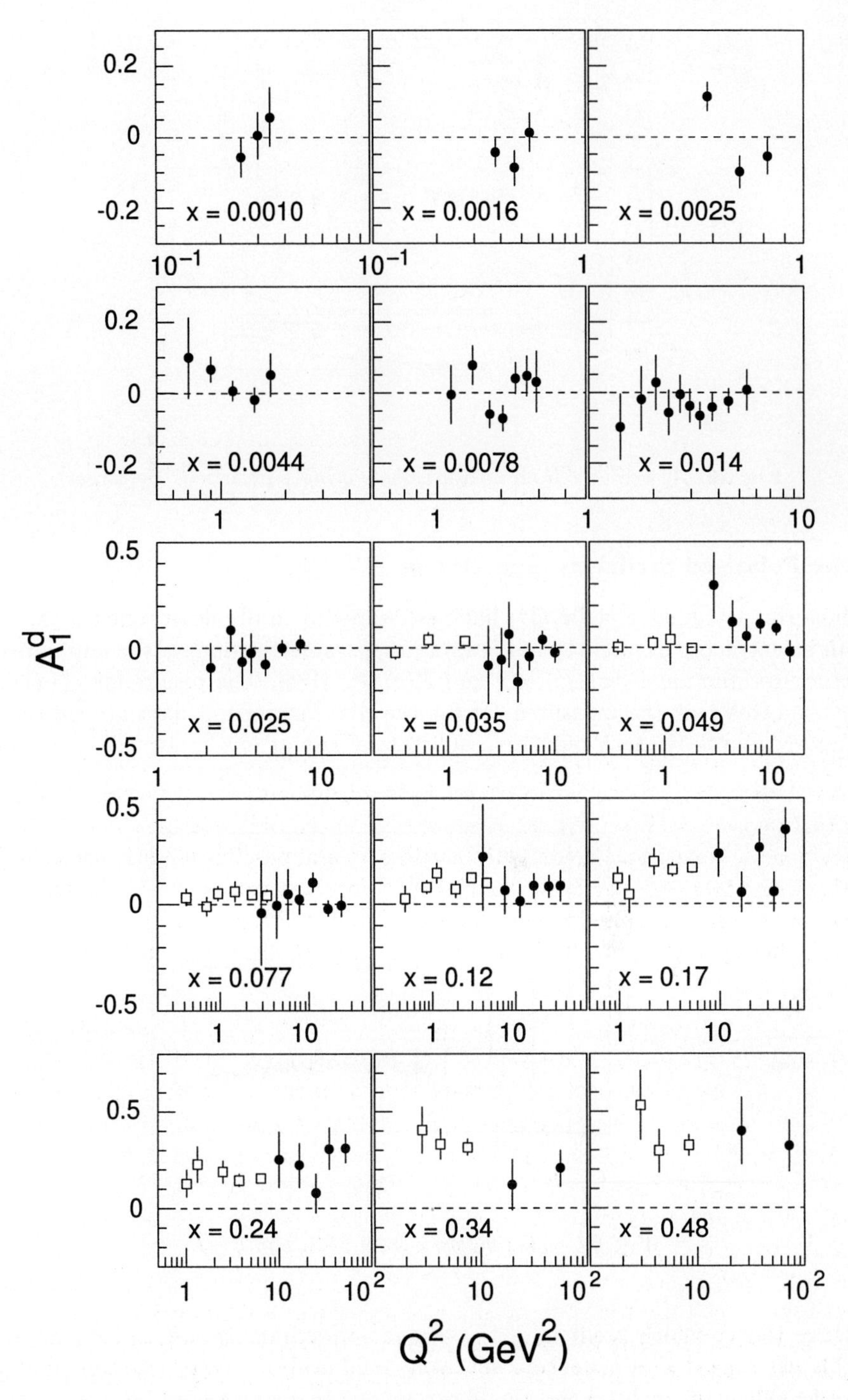

Fig. 55. *A_1^d from SMC and E143 as a function of Q^2.*

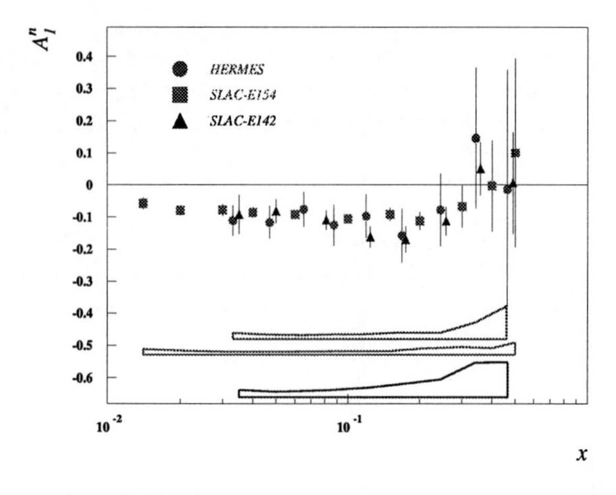

Fig. 56. *A_1^{n} versus x from measurements using a polarised ^{3}He target.*

The Polarised Structure Function g_1

Once the asymmetry $A_1(x, Q^2)$ has been experimentally determined $g_1(x, Q^2)$ can be extracted from (131) and (132) using parametrisations of the unpolarised structure function $F_2(x, Q^2)$ [40] and $R(x, Q^2)$ [185]. The results for $g_1^{\mathrm{p}}(x)$ for $Q^2 > 1\,\mathrm{GeV}^2$ at the measured Q^2 for the SMC and E143 data are shown in Fig. 57. The SMC data indicate a substantial rise at low x not expected from

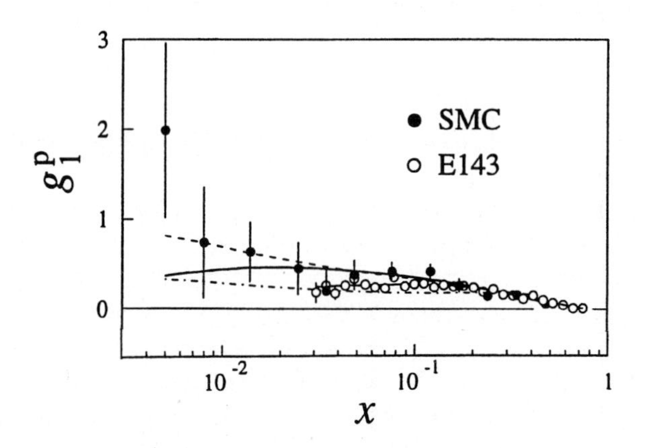

Fig. 57. *$g_1^{\mathrm{p}}(x)$ versus x from SMC and E143.*

Regge theory which predicts g_1 to become approximately constant for $x \to 0$. This rise caused a lot of speculations as it could lead to a very large contribution to the Ellis-Jaffe integral and would correspond to a counter-intuitive very large polarised sea. Recently, however, SMC has presented preliminary results from their high statistics '96 proton run and the combined data set from '93 and '96 [153]. Figure 58 shows the preliminary results for $g_1^{\rm p}$ together with the '93 data. It can be seen that the '96 data are above the old data essentially for all

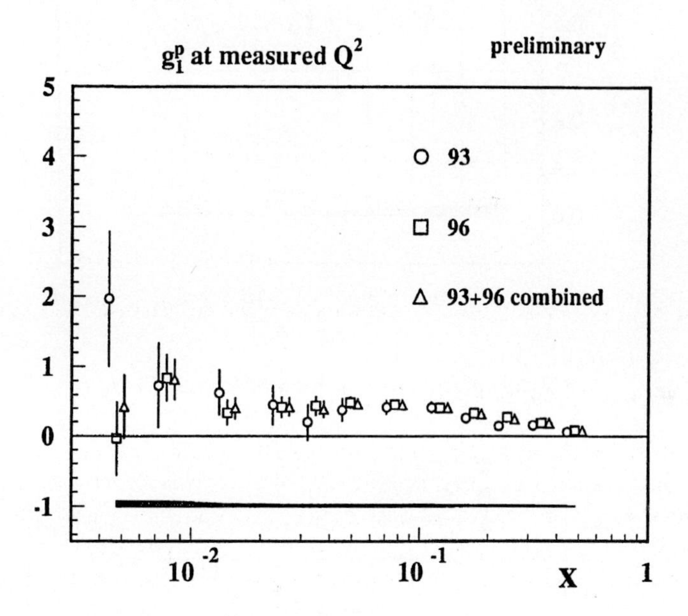

Fig. 58. *Preliminary SMC results for $g_1^{\rm p}$ from the '96 data compared to the results from the '93 data [Mag 97].*

higher x points. This leads to an increase of the Ellis-Jaffe integral by about 11 percent compared to the '93 value. On the other hand at low x the new data are substantially below the old ones. The combined data for $g_1^{\rm p}$ are (apart from one outlier at $x = 0.008$) flat for all $x < 0.1$, with a value of $g_1^{\rm p}(x) \simeq 0.4$.

Figure 59 shows the structure function $g_1^{\rm d}(x)$ from SMC and E143, this time not at the measured Q^2 but evolved to a fixed $Q_0^2 = 5\,{\rm GeV}^2$. (For the discussion of the evolution to a common Q^2 see below). Above $x \simeq 0.08$ the data behave quite smoothly while at lower values substantial fluctuations can be observed. This is mainly due to the fact that here the measured asymmetries are very small. Clearly these data do not yet determine the shape of $g_1^{\rm d}$ very well at low x and there is plenty of room for improvement. Figure 60 shows the structure function $g_1^{\rm n}$ extracted from the ^{3}He data. The agreement between the three experiments is excellent and a proof that systematic errors are well under control. The x dependence is completely determined by the high precision E154 data. $g_1^{\rm n}$ is negative over the whole x range covered by the experiments. It falls from zero at $x \simeq 0.5$ to a value of -0.5 at $x = 0.02$. The low x behaviour will be discussed below in some detail.

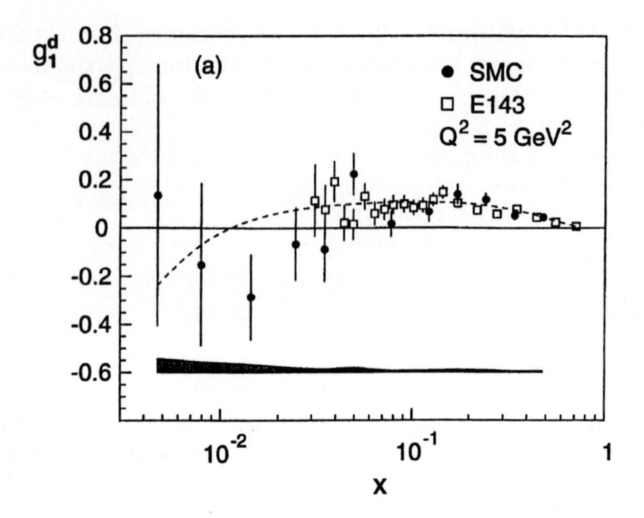

Fig. 59. $g_1^{\mathrm{d}}(x)$ *versus* x *from SMC and E143.*

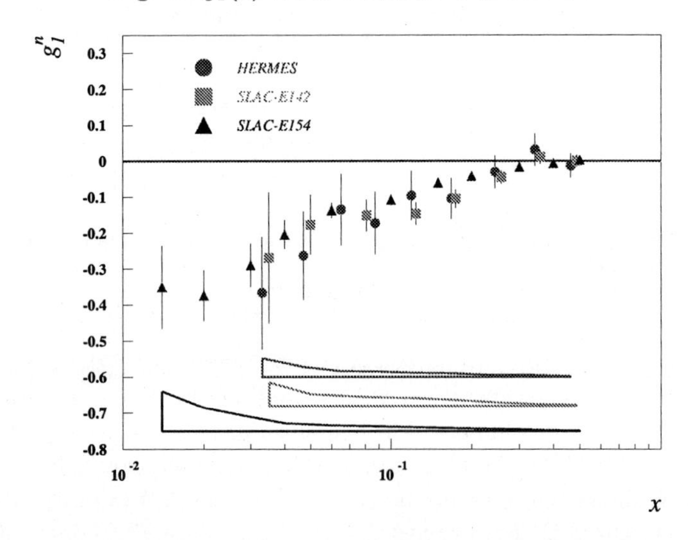

Fig. 60. $g_1^{\mathrm{n}}(x)$ *versus* x *from E142, E154 and HERMES.*

The First Moment of g_1

For a comparison of the data with the Ellis-Jaffe and Bjorken sum rules the integral

$$I_1(Q^2) = \int_0^1 g_1(x, Q^2)\,\mathrm{d}x \tag{175}$$

must be determined at fixed Q_0^2 and an extrapolation into the unmeasured regions at high and low x must be made. The mean Q^2 is different for each x value, therefore the data have to be evolved to a fixed Q_0^2. This requires an assumption about the Q^2 dependence of either A_1 or g_1. The existing data are consistent with the assumption that A_1 is independent of Q^2 over the range covered by the data. Therefore in early analyses the Q^2 evolution has been determined with (131) and (132) from the known Q^2 dependences of $F_2(x, Q^2)$ and $R(x, Q^2)$. However, perturbative QCD predicts the Q^2 dependences of g_1 and F_1 to differ substantially at low x. Therefore more recently the machinery of NLO QCD analysis of polarised structure functions [55, 125, 119] has been used [12, 35] to fit the available data and to determine from the fit the expected variation of $g_1(x, Q^2)$ with Q^2:

$$\delta g_1(x, Q^2, Q_0^2) = g_1^{\text{fit}}(x, Q_0^2) - g_1^{\text{fit}}(x, Q^2) \tag{176}$$

The measured $g_1(x, Q^2)$ are then evolved from Q^2 to Q_0^2 by adding this correction. As an example the results from different phenomenological analyses are shown for the '93 proton data in Fig. 61. Despite the different procedures the

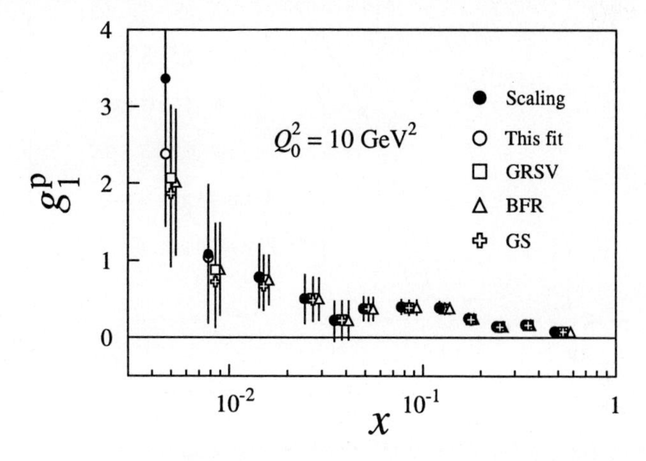

Fig. 61. *Q^2 evolution of $g_1(x, Q^2)$ from different procedures.*

differences in their results are small compared to the statistical errors of the data (apart from the lowest x point) and they are covered by the systematic error quoted for the evolution uncertainty. The extrapolation to $x \to 1$ is not critical, since F_1 is small and dropping very fast in the high x region. SMC assumed $A_1^{\text{p}} = 0.7 \pm 0.3$ and $A_1^{\text{d}} = 0.4 \pm 0.6$ in the region $0.7 < x < 1$, while E143 assumed $g_1^{\text{p,d}} \propto (1 - x)^3$ and HERMES assumed a linear rise in A_1^{n} from 0 at $x = 0.6$ to 1 at $x = 1$. However, the small x behaviour of $g_1(x)$ is theoretically not very well established and the evaluation of I_1 depends critically on the assumptions made for this extrapolation. From Regge theory [131, 105] it is expected that for $x \to 0$ g_1 behaves like $x^{-\alpha}$, where α is the intercept of the lowest contributing Regge trajectories, which is assumed to be in the range $[-0.5, 0]$. There are also several other predictions which assume a very different behaviour [17, 63, 94, 55]. In all analyses presented so far a simple Regge parametrisation has been used assuming $g_1(x) = \text{const.}$ and quoting a systematic error for α varying in the range $[-0.5, 0]$. While with the present data this procedure seems to be reasonable for $g_1^{\rm p}$ and $g_1^{\rm d}$ it turns out to be problematic for $g_1^{\rm n}$. Figure 62 shows the E154 data together with $g_1^{\rm n}$ extracted from the '93 proton and the deuteron data from SMC. Here $g_1^{\rm n}$ has been obtained assuming

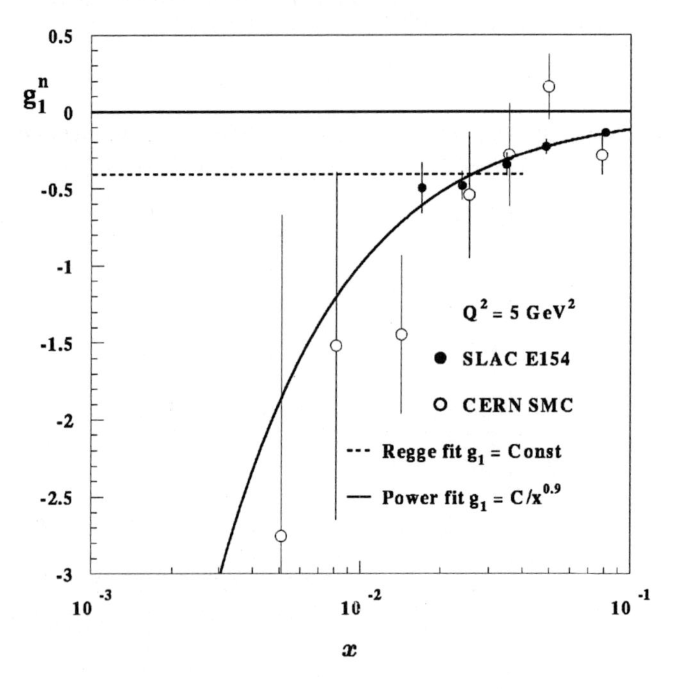

Fig. 62. $g_1^{\rm n}$ *versus* x *in the low* x *region from E154 and SMC.*

$$g_1^{\rm p} + g_1^{\rm n} = 2g_1^{\rm d}/(1 - 1.5w_{\rm D}) \tag{177}$$

where $w_{\rm D} = 0.05 \pm 0.01$ is the D-wave state probability in the deuteron. As well the E154 as the SMC data are falling over the whole measured x range. Therefore the value of the Regge type extrapolation depends critically on the number of low x points used to determine the constant and it might dramatically underestimate the contribution of the very low x region. An unconstrained power low fit $g_1^{\rm n} = C/x^a$ yielded $a = 0.9 \pm 0.2$ resulting in $I_1^{\rm n}$ as large as -0.2 and diverges for $a = 1$. Therefore E154 does not dare to quote a value for the Ellis-Jaffe integral. The situation might, however, not be as dramatic. Since the preliminary '96 proton data from SMC are much different from the published ones at low x also $g_1^{\rm n}$ is different. The preliminary data points for $g_1^{\rm n}$ as presented by [153] are approximately $0.5 \pm 1.2(x = 0.005)$, $-1.4 \pm 0.8(x = 0.008)$, $-1.05 \pm 0.4(x = 0.0125)$ and $-0.5 \pm 0.35(x = 0.025$ and $x = 0.035)$. These new data are very well consistent with g_1^{n} being a constant around -0.5 at low x. The only two points which seem to be substantially lower might just be affected by statistical fluctuations, e. g. a too high $g_1^{\mathrm{p}}(0.008)$ and a too low $g_1^{\mathrm{d}}(0.0125)$. Anyway due to their large error bars they do not really constrain the behaviour at low x. Clearly high quality data in the low x region are badly needed. At present the only realistic option to access this region experimentally would be polarised ep and ed collisions in HERA. This would require to accelerate and store polarised protons and deuterons in HERA [56]. This is presently under consideration [62].

Table 4. *Integrals over the polarised structure function g_1 from recent experiments.*

Exp.	Target	Q^2 GeV^2	I_1^{meas}	I_1^{high}	I_1^{low}	I_1	I^{EJ}
SMC	p('93)	10	0.130 (0.013)(0.009)	0.0015 (0.0007)	0.0042 (0.0016)	0.136 (0.013)(0.010)	0.170 (0.005)
SMC	p('93+'96)	10	0.145 (0.006)(0.009)	0.0015 (0.0007)	0.0029 (0.0007)	0.149 (0.006)(0.010)	0.170 (0.005)
SMC	d	10	0.0407 (0.0059)(0.0046)	0.0006 (0.0009)	0.0000 (0.0009)	0.0414 (0.0059)(0.0048)	0.071 (0.003)
	n('93p+d)					−0.046 (0.013)(0.010)	−0.017 (0.005)
E143	p	3	0.120 (0.004)(0.008)	0.001 (0.001)	0.006 (0.006)	0.127 (0.004)(0.010)	0.160 (0.006)
	d	3	0.040 (0.003)(0.004)	0.000 (0.001)	0.001 (0.001)	0.042 (0.003)(0.004)	0.069 (0.004)
	n	3				−0.037 (0.008)(0.011)	−0.011 (0.005)
E142	n(^{3}He)	3	−0.028 (0.006)(0.006)	0.003 (0.003)	−0.0053 (0.0053)	−0.031 (0.006)(0.009)	−0.011 (0.005)
E154	n(^{3}He)	5	−0.036 (0.004)(0.005)			−0.041 (0.004)(0.006)	
HERMES	n(^{3}He)	2.5	−0.036 (0.013)(0.005)	0.002 (0.002)	−0.005 (0.005)	−0.041 (0.013)(0.007)	−0.011 (0.005)

In Table 4 the actual results for I_1 from the different experiments are summarised where the abbreviations

$$
\begin{aligned}
I_1^{\text{meas}} &= \int_{x_{\text{min}}}^{x_{\text{max}}} g_1(x)\,\mathrm{d}x, & I_1^{\text{low}} &= \int_0^{x_{\text{min}}} g_1(x)\,\mathrm{d}x, \\
I_1^{\text{high}} &= \int_{x_{\text{max}}}^{1} g_1(x)\,\mathrm{d}x, & I_1 &= \int_0^1 g_1(x)\,\mathrm{d}x
\end{aligned}
\qquad (178)
$$

have been used. The numbers in brackets are the statistical and total systematic errors.

2.8 Comparison to Theoretical Predictions

Ellis-Jaffe Sum Rules

Keeping in mind that there might be substantial uncertainties in the values of the experimentally determined integrals due to the not well known behaviour at low x, we can compare these results to the expectations of the Ellis-Jaffe sum rules, which are given in the last column of Table 4 for the Q_0^2 value of the corresponding data. All experimental results differ by about 2–3 standard deviations from the Ellis-Jaffe value. As an example the Ellis-Jaffe prediction for the proton is shown in Fig. 63 as a function of Q^2 together with the experimental results. The SMC point does not yet include the preliminary '96 proton data.

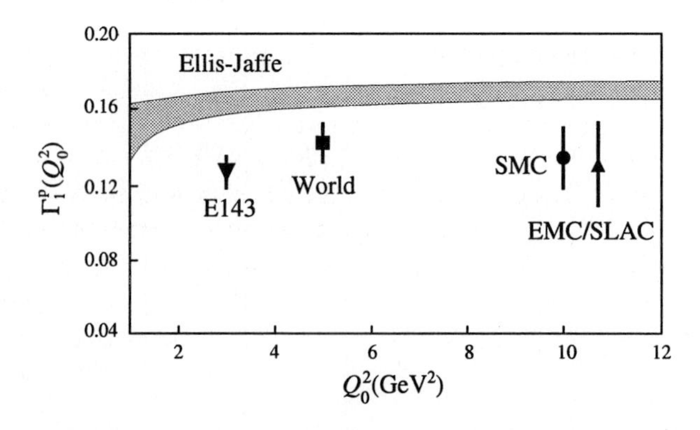

Fig. 63. *Experimental results and theoretical expectation for the proton Ellis-Jaffe integral versus* Q^2.

These move the point up, but at the same time the statistical error is reduced by about a factor of 2 and there remains a 2 standard deviations difference from the prediction. The dashed band is obtained from (157) using $g_a = 1.2601 \pm 0.0028$ and $F/D = 0.575 \pm 0.016$. One can see that the integral decreases with decreasing Q^2. This has to be taken into account when data from different experiments at different Q^2 are compared.

Bjorken Sum Rule

The relation between I_1^p, I_1^d and I_1^n and the sum rules is illustrated in Fig. 64. For

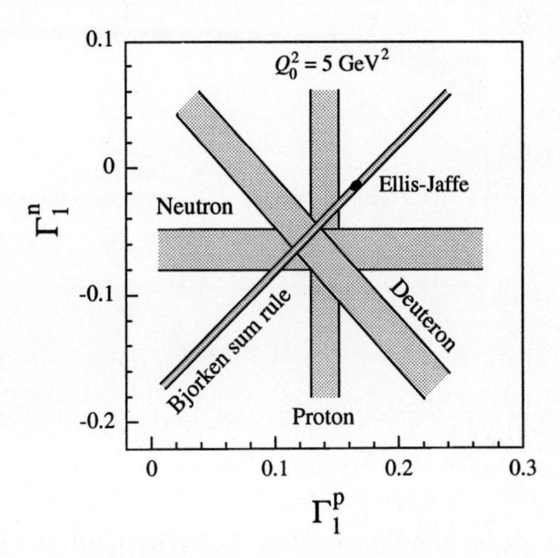

Fig. 64. *Comparison of the combined results for I_1^p, I_1^d and I_1^n with the predictions of the Bjorken and Ellis-Jaffe sum rules.*

this comparison performed by SMC [12], all data from the different experiments were evolved to a common $Q_0^2 = 5\,\mathrm{GeV}^2$ and then combined. The Ellis-Jaffe prediction is shown by the black ellipse in the narrow Bjorken sum rule band. Proton, deuteron and neutron results disagree with the Ellis-Jaffe sum rules but they confirm the Bjorken sum rule. This is shown in more detail in Fig 65 where the data for $I_1^p - I_1^n$ of different experiments are compared to the Bjorken sum rule [35] with third order QCD corrections (156,160) and no higher twist corrections. From this figure it can be seen that the QCD corrections reduce the integral substantially below its QPM value $g_a/6 = 0.21$. The agreement between theory and experiment is good.

2.9 Extracting Information About the Spin Content of the Proton

Axial Quark Charges

As discussed in Chap. 2.3 and 2.6 the first moments of the polarised structure functions can be used to extract the matrix elements of the axial vector currents a_0, a_3, a_8. This can be done either by using

$$I_1^p = \frac{C_1^{NS}(Q^2)}{12}\left(a_3 + \frac{1}{3}a_8\right) + \frac{C_1^S(Q^2)}{9}a_0(Q^2) \tag{179}$$

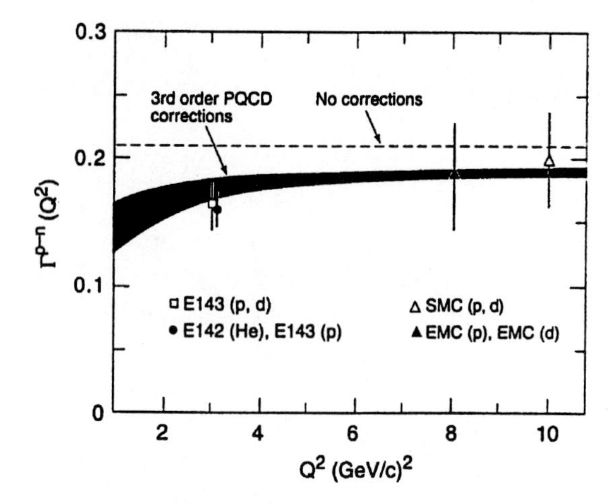

Fig. 65. *Comparison of experimental results and theoretical prediction for the Bjorken sum rule as a function of Q^2.*

together with $a_3 = g_A/g_V$ and $a_8 = 3F - D$, or without the informations from the hyperon decays from a combination of $\Gamma_1^{\rm p}$, $\Gamma_1^{\rm d}$ or $\Gamma_1^{\rm p}$, $\Gamma_1^{\rm n}$, respectively, together with g_A/g_V. The detailed results of such an analysis, which has been performed by several groups, depend somewhat on the selection and treatment of the data. But the main message essentially stays the same. Therefore I just cite a recent global analysis performed by SMC [12]. Below we give the results from the combined analysis of all proton data at $Q_0^2 = 5\,\mathrm{GeV}^2$ and in brackets those for an combined analysis of all proton, deuteron and neutron data. They obtain

$$a_0(Q_0^2 = 5\,\mathrm{GeV}^2) = 0.37 \pm 0.11(0.29 \pm 0.06) \tag{180}$$

(which changes to 0.41 ± 0.11 when the preliminary '96 SMC proton data are included in the fit). This value of a_0 is about twice as large as the value (168) derived from the EMC/SLAC result using first order QCD corrections only. The individual contributions from the different quark flavours can be evaluated from (146). They are

$$\begin{aligned} a_{\rm u} &= 0.85 \pm 0.04 \quad (0.82 \pm 0.02) \\ a_{\rm d} &= -0.41 \pm 0.04 \quad (-0.43 \pm 0.02) \\ a_{\rm s} &= -0.07 \pm 0.04 \quad (-0.10 \pm 0.02) \end{aligned} \tag{181}$$

Very similar results have been obtained by E142 [Ant 96] when using only their neutron data for such an analysis. In the QPM, $a_f = \Delta q_f$ and these results correspond to the values of $\Delta\Sigma$, Δu, Δd and Δs in (169) from the naïve analysis.

The Polarised Quark and Gluon Distributions

However, as discussed in Chap. 2.4 due to the U(1) anomaly of the singlet axial current the axial charges receive a gluon contribution. The quark spin contributions $\Delta\Sigma$, Δu, Δd and Δs to the proton spin (171) differ from the axial charges a_0, a_u, a_d, a_s in a scheme dependent way. In the Adler-Bardeen scheme [55], where Δq_f is independent of Q^2, they are related by

$$a_f = \Delta q_f - \frac{\alpha_s(Q^2)}{2\pi}\Delta g(Q^2) \tag{182}$$

and the determination of the quark spin contributions to the nucleon spin requires an input value for Δg. The allowed values for $Q_0^2 = 5\,\mathrm{GeV}^2$ are shown in Fig. 66 as a function of Δg [12]. A value of $\Delta\Sigma = 0.57$ and $\Delta s = 0$ assumed for

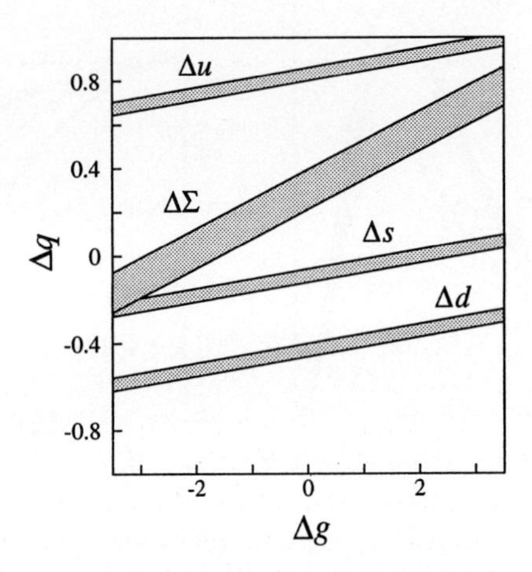

Fig. 66. *Quark spin contributions to the proton spin as a function of the gluon contribution at $Q_0^2 = 5\,\mathrm{GeV}^2$ in the Adler-Bardeen scheme.*

the derivation of the Ellis-Jaffe sum rules corresponds to $\Delta g(Q_0^2 = 5\,\mathrm{GeV}^2) = 2$. This consequently would require that also orbital angular momentum contributions L_z are large. For $\Delta g = 0$ quark spins contribute only by about 30 to 40 percent to the spin of the nucleon and strange quarks carry a small polarisation opposite to the polarisation of the parent nucleon. Therefore the new high quality data have not changed the conclusions from the original EMC/SLAC finding. Only the error bars have become much smaller.

Polarised Parton Distributions From QCD Fit

An attempt has been made by several authors to extract informations about the polarised parton distributions and their first moments by fitting the measured structure functions $g_1^{\mathrm{p,n,d}}$ using the QCD formalism described in Chap. 2.4. Several such polarised parton densities are collected and compared in [145]. It should be noted that presently such fits can only give hints about size and magnitude of polarised parton distributions as the quality of the data is much worse than in the unpolarised case where the Q^2 dependence of the structure functions has been measured with high precision for each value of x over a large range of Q^2 (see Chap. 1.4). In the polarised case we have only one or very few Q^2 points for each x value (and these from different experiments). Depending on the data sets used and also on the flexibility of the input parametrisations rather different results have been obtained. This is demonstrated in Fig. 67 [145], where a variety of possible solutions is shown. While $\delta u(x)$ seems to be rather well determined

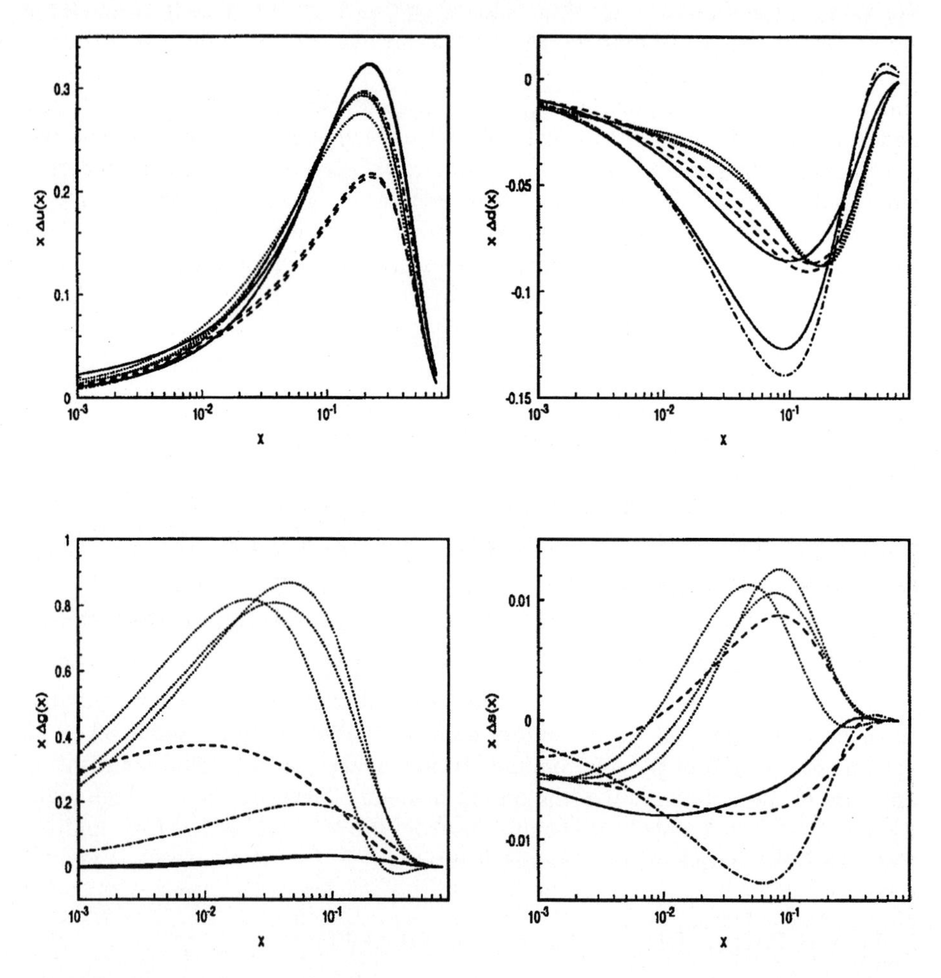

Fig. 67. *Polarised parton distributions at* $Q^2 = 15\,\mathrm{GeV}^2$ *from different QCD analyses.*

as well in shape as in magnitude, the fits give a much larger possible range for $\delta d(x)$. $\delta s(x)$ is essentially undetermined and could have as well positive as negative sign. Very different solutions are also possible for δg from such fits. From the most recent analysis [30] the following parameters have been obtained:

$$\begin{aligned} a_0(Q^2 = \infty) &= 0.10^{+0.17}_{-0.10} \ , \\ \Delta g(Q^2 = 1\,\mathrm{GeV}^2) &= 1.6 \pm 0.9 \ , \\ \alpha_s(M_z^2) &= 0.118^{+0.010}_{-0.024} \ . \end{aligned} \tag{183}$$

2.10 Polarised Parton Distributions From Semi-Inclusive Data

We have seen in Chap. 2.8 that the determination of $\Delta\Sigma$ and Δq_f from inclusive data requires assumption about the polarised gluon distribution which is very difficult to access experimentally. QCD fits hardly constrain the z dependence and magnitude of the polarised gluon and seaquark distributions. A detailed understanding of the spin structure of the nucleon therefore requires that the x-dependence and the moments of the individual polarised quark distributions, separated into contributions from valence and seaquarks, the polarised gluon distribution and possibly also the orbital angular momentum contributions are determined separately. This can be achieved when not only the scattered lepton is detected, but also the leading hadron which contains the struck quark [112, 92]. Information can be obtained for the different contributions to the nucleon spin in the following way:

- $\Delta V(\Delta \mathrm{u_v}, \Delta \mathrm{d_v})$: Spin dependent differences in production of π^+ and π^-
- $\Delta S(\Delta\bar{\mathrm{u}}, \Delta\bar{\mathrm{d}}, \Delta\bar{\mathrm{s}})$: Spin dependent π^- or K^- production
- Δg: Spin dependent J/Ψ or D, D^* production
- L_z: possibly from azimuthal distributions of hadrons for transversaly polarised targets or from deeply virtual Compton scattering (DVCS).

For example ΔV is determined as follows. The number of π^+ produced from a proton target with the beam and target spin antialigned, $N^{\pi^+}_{\uparrow\downarrow}$, is proportional to

$$\begin{aligned} N^{\pi^+}_{\uparrow\downarrow} \propto\ & \frac{4}{9}\mathrm{u}^+(x) D^{\pi^+}_{\mathrm{u}}(z) + \frac{1}{9}\mathrm{d}^+(x) D^{\pi^+}_{\mathrm{d}}(z) + \frac{1}{9}\mathrm{s}^+(x) D^{\pi^+}_{\mathrm{s}}(z) \\ & \frac{4}{9}\bar{\mathrm{u}}^+(x) D^{\pi^+}_{\bar{\mathrm{u}}}(z) + \frac{1}{9}\bar{\mathrm{d}}^+(x) D^{\pi^+}_{\bar{\mathrm{d}}}(z) + \frac{1}{9}\bar{\mathrm{s}}^+(x) D^{\pi^+}_{\bar{\mathrm{s}}}(z) \ . \end{aligned} \tag{184}$$

The fragmentation functions $D^{\pi^+}_{\mathrm{q}}$ are the probabilities that the struck quark will manifest itself as a π^+ with energy fraction $z = E^{\pi^+}/\nu$. Hadrons with high z have a higher probability of containing the struck quark. There are favoured fragmentation functions like $D^{\pi^+}_{\mathrm{u}}$, which I will denote as D_1, and unfavoured fragmentation functions like $D^{\pi^+}_{\mathrm{d}}$, denoted as D_2. Expressions similar to (184) can be written for the other spin orientation (where the q^+ are replaced by q^- but the fragmentation functions do not change) and for π^- production (where only the fragmentation functions change). If isospin and charge symmetry are assumed to reduce the number of fragmentation functions, then terms such as

$$\frac{4}{9}(\mathrm{u}^+ - \bar{\mathrm{u}}^+)(D_1 - D_2) = \frac{4}{9}{\mathrm{u_v}}^+(D_1 - D_2) \tag{185}$$

appear in the difference of π^+ and π^- production in a particular spin orientation ($N_{\uparrow\downarrow}^{\pi^+-\pi^-}$) and the strange quark contributions cancel. Furthermore, taking the difference for both spin orientations we obtain the combination between the polarised valence quark distributions and the fragmentation functions

$$N_{\uparrow\downarrow}^{\pi^+-\pi^-} - N_{\uparrow\uparrow}^{\pi^+-\pi^-} \propto \left(\frac{4}{9}\delta u_v - \frac{1}{9}\delta d_v\right)(D_1 - D_2) \ . \tag{186}$$

Dividing this difference by the sum of the yields for the different spin orientations, the fragmentation functions cancel and we get

$$A_p^{\pi^+-\pi^-} = \frac{4\delta u_v - \delta d_v}{4u_v - d_v} \ , \qquad A_d^{\pi^+-\pi^-} = \frac{\delta u_v + \delta d_v}{4u_v + d_v} \tag{187}$$

for proton and deuteron targets. The unpolarised valence quark distributions are very well known from the fits to the unpolarised data so it is possible to deduce the polarised valence quark distributions and their moments from pion production on two different targets. The HERMES experiment has specially been designed to explore this kind of physics and after the upgrade with a RICH counter in '97/98 it will have full particle identification for all momenta bigger than about 1 GeV. It should be able to determine the valence quark distributions to the nucleon spin to about 10 percent.

The SMC experiment has made a first study of semi-inclusive hadrons in polarised deep inelastic scattering on hydrogen and deuterium [14]. The SMC spectrometer does not have the capability to identify pions so the data include all hadrons. The asymmeties for positive and negative hadrons are then given by

$$A^\pm = \frac{\sum_f e_f^2 \delta q_f D_q^\pm}{\sum_f e_f^2 q_f D_q^\pm} \ , \tag{188}$$

where

$$D_q^\pm = \sum_{h^\pm} \int_{z_{\min}}^1 D_q^{h^\pm}(z)\, dz \ , \tag{189}$$

with $z_{\min} = 0.2$. The fragmentation functions are the same for the polarised and unpolarised case, provided the hadrons are spinless or it is summed over their spin orientations. They were measured by the EMC in exactly the same energy region for up and down quarks fragmenting into charged pions, kaons and protons [36]. By making assumptions about the polarised sea like $\delta\bar{u} = \delta\bar{d}$ SMC was able to extract the quark helicity distribution functions from a combination of the hadron asymmetries measured with the proton and the deuteron targets and the corresponding inclusive asymmetries. Figure 68 shows the measured virtual-photon asymmetries for the production of hadrons and Fig. 69 the extracted polarised valence-quark $\delta u_v(x)$, $\delta d_v(x)$ and antiquark $\delta\bar{q}(x)$ helicity distributions (Δu_v in the figure is equivalent to δu_v in our notation). The polarisations of the

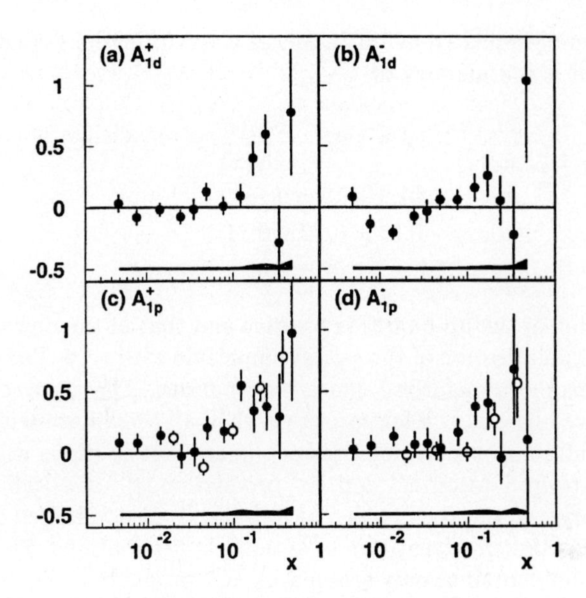

Fig. 68. *Virtual photon asymmetries for the production of positive and negative hadrons from proton and deuteron targets from SMC.*

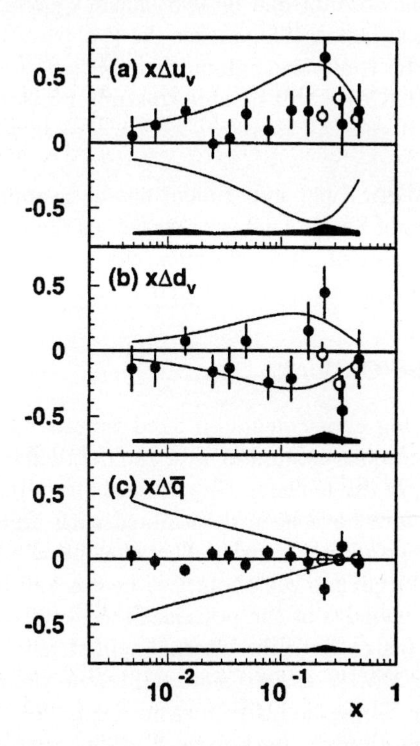

Fig. 69. *Polarised quark distributions extracted by SMC from hadron asymmetries.*

valence quarks obtained from this analysis are within the statistical accuracy independent of x and amount to

$$\begin{aligned} &\frac{\delta \mathrm{u_v}(x)}{\mathrm{u_v}(x)} = 0.5 \pm 0.1 \quad \frac{\delta \mathrm{d_v}(x)}{\mathrm{d_v}(x)} = -0.6 \pm 0.2 \\ &\Delta \mathrm{u_v} = 1.01 \pm 0.19 \pm 0.14 \\ &\Delta \mathrm{d_v} = -0.57 \pm 0.22 \pm 0.11 \\ &\Delta \bar{\mathrm{u}} = \Delta \bar{\mathrm{d}} = -0.02 \pm 0.09 \pm 0.03 \ . \end{aligned} \tag{190}$$

The polarisation of the up-quarks is positive and that of the down quarks negative, while the polarisation of the sea is compatible with zero. The solid lines are the limits from the unpolarised quark distributions. The accuracy of the data will be improved by about a factor of two when all SMC data will be included.

Detailed informations on polarised sea quark distributions will be obtained by HERMES from data on π^- and K^- production. π^- is a $(\bar{\mathrm{u}}\mathrm{d})$ combination. Since the charge of the $\bar{\mathrm{u}}$ is twice that of the d, it dominates the scattering process which is therefore sensitive to antiquark distributions. The K^- is a $(\bar{\mathrm{u}}\mathrm{s})$ combination and containes only sea-quarks. K^- production should therefore be very sensitive to sea quarks and the sensitivity to the strage quark distribution is enhanced.

J/Ψ and D mesons contain one or two charm quarks and result from the process of photon-gluon-fusion. This makes the (polarised) charmed meson production sensitive to the (polarised) gluon distribution. After the upgrade with the RICH detector HERMES will be able to study especially D production via the π-K-decay channel and will therefore have some sensitivity to the polarised gluon distribution.

Recently the COMPASS [96] experiment has been approved at CERN as an upgraded continuation of SMC. Among other things the experiment will concentrate on the measurement of $\Delta g/g$ via open charm production. The projected precision for the measurement in the range $0.07 < x < 0.4$ is $\delta(\Delta g/g) \simeq 0.10$.

2.11 Summary and Outlook

Deep inelastic scattering experiments on fixed targets have provided us in the last years with an enormeous amount of data and detailed informations about the quark-gluon structure of the nucleon. Unpolarised quark and gluon distributions and their Q^2 dependence have been determined with high accuracy, the information about $R(x, Q^2)$ and higher twist effects has been substantially improved. Recent inclusive measurements with polarised beams and targets have considerably improved our knowledge of the polarised structure function $g_1(x, Q^2)$ and give us already some insight into the internal spin structure of the nucleon.

The E155 and HERMES experiments will soon yield inclusive data with even higher precision. These inclusive measurements are, however, insufficient for a precise determination of the polarised gluon distribution or the flavour decomposition of the nucleon spin. Semi-inclusive data are required to obtain these informations. SMC has already produced first low statistics results, precision measurements on this subject will be performed by HERMES and later by COMPASS. A combination of these experiments, which cover a different kinematic range, will presumably yield precise informations about the polarised gluon distribution.

First measurements on the second spin structure function $g_2(x)$, which might contain informations about a twist-3 component arising from quark-gluon correlations are now just becoming available, much better data can be expected in the future from E155 and HERMES. HERMES has also the ability to measure additional structure functions like the higher multipole structure function $b_1^{\mathrm{d}}(x)$, which arises from binding effects in the deuteron, or $\Delta^{\mathrm{d}}(x)$, which is sensitive to gluon components in the wavefunction of the deuteron, or the chiral odd structure function $h_1(x)$.

Finally the polarisation of the HERA proton ring, which is presently under discussion, will allow measurements at high Q^2 and low x and such to study the spin structure of the nucleon in a new kinematic domain.

Acknowledgements

I am very grateful to W. Ratzka, who produced an early version of these lecture notes, S. Bernreuther and F. Schmidt, who helped typing parts of the present version and especially S. Meinlschmidt, who had the burden of typing most of the text, incorporating the numerous figures and preparing the final document.

References

[1] E143, K. Abe et al., Phys. Rev. Lett. **74** (1995) 346.
[2] E143, K. Abe et al., Phys. Rev. Lett. **75** (1995) 25.
[3] E143, K. Abe et al., Phys. Lett. **B364** (1995) 61.
[4] E154, K.Abe et al., preprint hep-ex/9705012 (May 1997).
[5] HERMES, K. Ackerstaff et al., Phys. Lett. **B404** (1997) 383.
[6] SMC, D. Adams et al., Phys. Lett. **B329** (1994) 399,erratum Phys. Lett. **B339** (1994) 332.
[7] SMC, D. Adams et al., Phys. Lett. **B336** (1994) 125.
[8] E665, M. R. Adams et al., Phys. Rev. Lett. **75** (1995) 1466.
[9] SMC, D. Adams et al., Phys. Lett. **B357** (1995) 248.
[10] SMC, D. Adams et al., Phys. Lett. **B369** (1996) 93.
[11] SMC, D. Adams et al., Phys. Lett. **B369** (1997) 338.
[12] SMC, D. Adams et al., preprint hep-ex/9702005, (Feb 97).
[13] SMC,B. Adeva et al., Phys. Lett. **B302** (1993) 533.
[14] SMC, B. Adeva et al., Phys. Lett. **B369** (1996) 93.
[15] S. I. Adler, Phys. Rev. **177** (1969) 2426.
[16] S. I. Adler and W. Bardeen, Phys. Rev. **182** (1969) 1517.
[17] M. A. Ahmed and G. G. Ross, Phys. Lett. **B56** (1975) 385.

[18] J. Ahrens et al, Nucl. Phys. **A446** (1985) 2096.
[19] H1, S. Aid et al, Nucl. Phys. **B470** (1996) 3.
[20] I. V. Akushevich and N. M. Shumeiko, J. Phys. **G20** (1984) 513.
[21] E-80, M. J. Alguard et al., Phys. Rev. Lett. **37** (1976) 1261.
[22] EMC, O. C. Allkofer et al., Nucl. Instr. and Meth. **179** (1981) 445.
[23] NMC, D. Allasia et al., Phys. Lett. **B249** (1990) 366.
[24] NMC, D. Allasia et al., Phys. Lett. **B258** (1991) 493.
[25] R. Alley et al., Nucl. Inst. Meth. Phys. Res. **A365** (1995) 1.
[26] G. Altarelli, G. Parisi, Nucl. Phys. **B126** (1977) 298.
[27] G. Altarelli and G. Martinelli, Phys. Lett. **B76** (1978) 89.
[28] G. Altarelli, Phys. Rep. **81** (1982) 1.
[29] G. Altarelli and G. G. Ross, Phys. Lett. **B212** (1988) 391.
[30] G. Altarelli et al., preprint hep-ph/9701289 (Jan. 97).
[31] NMC, P. Amaudruz et al., Phys. Lett. **B294** (1992) 120.
[32] NMC, P. Amaudruz et al., Nucl. Phys. **B371** (1992) 3.
[33] M. Anselmino, A. Efremov and E. Leader, Phys. Rep. **261** (1995) 1.
[34] E142, P. L. Anthony et al., Phys. Rev. Lett. **71** (1993) 959.
[35] E142, P. L. Anthony et al., Phys. Rev. **D54** (1996) 6620.
[36] EMC, M. Arneodo et al., Nucl. Phys. **B321** (1989) 541.
[37] NMC, M. Arneodo et al., Phys. Lett. **B309** (1993) 222.
[38] M. Arneodo, Phys. Rep. **240** (1994) 301.
[39] NMC, M. Arneodo et al., Phys. Rev. **D50** (1994) R1.
[40] NMC, M. Arneodo et al., Phys. Lett. **B364** (1995) 107.
[41] NMC, M. Arneodo et al., Nucl. Phys. **B 481** (1996) 3.
[42] NMC, M. Arneodo et al., Nucl. Phys. **B 481** (1996) 23.
[43] NMC, M. Arneodo et al., Nucl. Phys. **B483** (1997) 3.
[44] NMC, M. Arneodo et al., Nucl. Phys. **B487** (1997) 3.
[45] EMC, J. Ashman et al., Phys. Lett. **B206** (1988) 364.
[46] EMC, J. Ashman et al., Nucl. Phys. **B328** (1989) 1.
[47] EMC, J. Ashman et al., Z. Phys. **C56** (1992) 21.
[48] W. B. Atwood, *Lectures on Lepton Nucleon Scattering and Quantum Chromodynamics*, Progress in Physics, Vol. 4, Birkhäuser 1982.
[49] EMC, J. J. Aubert et al., Nucl. Phys. **B213** (1983) 1.
[50] EMC, J. J. Aubert et al., Nucl. Phys. **B259** (1985) 189.
[51] B. Badelek and J. Kwiecinski, Nucl. Phys. **B370** (1992) 278.
[52] B. Badelek and J. Kwiecinski, Phys. Rev. **D50** (1994) R4.
[53] R. Baier and R. Rückl, Nucl. Phys **B218** (1983) 289.
[54] I. I. Balitzky et al., Phys Lett. **B242** (1990) 245, erratum Phys. Lett. **B318** (1993) 648.
[55] R. D. Ball et al., Phys. Lett. **B378** (1996) 255.
[56] R. Ball et al., *Proceedings of the Workshop 1995/96 'Future Physics at HERA'*, G. Ingelmann, A. de Roeck and R. Klanner, eds, 777.
[57] V. Barger et al., Phys. Lett. **B91** (1980) 253.
[58] D. P. Barber et al., Nucl. Instr. Meth. **A329** (1993) 79.
[59] D. P. Barber et al., Nucl. Instr. Meth. **A338** (1994) 166.

[60] V. Barone et al., Phys. Lett. **B321** (1994) 137.
[61] D. P. Barber et al., Phys. Lett. **B343** (1995) 436.
[62] D. P. Barber et al., ibid, 1205.
[63] S. D. Bass and P. V. Landshoff, Phys. Lett. **B336** (1994) 537.
[64] E-130, G. Baum et al., Phys. Rev. Lett. **51** (1983) 1135.
[65] CCFR, A. Bazarko et al., Z. Phys. **C65** (1995) 189.
[66] J. S. Bell and R. Jackiw, Nuovo Cim. **51A** (1969) 47.
[67] BCDMS, A. C. Benvenuti et al., Phys. Lett. **B223** (1989) 485.
[68] BCDMS, A. C. Benvenuti et al., Phys. Lett. **B237** (1990) 592.
[69] E. L. Berger and D. Jones, Phys. Rev. **D23** (1981) 1521.
[70] CDHSW, P. Berge et al., Z. Phys. **C49** (1991) 2712.
[71] I. G. Bird, Ph. D. Thesis, Vrije Universiteit te Amsterdam (1992).
[72] D. Bjorken, Phys. Rev. **148** (1966) 1467.
[73] J. D. Bjorken, Phys. Rev. **179** (1969) 1547.
[74] J. D. Bjorken and E. A. Paschos, Phys. Rev. **185** (1969) 1975.
[75] D. Bjorken, Phys. Rev. **D1** (1970) 1376.
[76] E. D. Bloom, *Proceedings of the 6th Int. Symp. on Electron and Photon Interactions at High Energies*, Bonn 1973, H. Rollnik and W. Pfeil, eds, North Holland 1974, 227.
[77] A. Bodek et al., Phys. Rev. **D20** (1979) 1471.
[78] M. A. Braun and M. V. Tokarev, Phys. Lett. **B320** (1994) 381.
[79] V. M. Braun, preprint hep-ph/9505317 (May 1995).
[80] M. Breidenbach et al., Phys. Rev. Lett. **23** (1969) 935.
[81] L. Brekke and J. L. Rosner, Comm. Nucl. Part. Phys. **18** (1988) 83.
[82] S. J. Brodsky et al., Nucl. Phys. **B441** (1995) 197.
[83] J. Buon and K. Steffen, Nucl. Instr. Meth. **A245** (1986) 248.
[84] C. G. Callan and D. J. Gross, Phys. Rev. Lett. **22** (1969) 156.
[85] D. O. Caldwell et al., Phys. Rev. Lett. **42** (1979) 553.
[86] R. D. Carlitz et al., Phys Lett. **B214** (1988) 229.
[87] H. Y. Cheng, preprint hep-ph/9512267 (Dec. 1995).
[88] H. Y. Cheng, Int. J. Mod. Phys. **A11** (1996) 5109.
[89] T. E. Chupp et al., Phys. Rev. **C45** (1992) 915.
[90] C. Ciofi degli Atti et al., Phys. Rev. **C48** (1993) R 968.
[91] F. E. Close, *An Introduction to Quarks and Partons*, Academic Press, 1979.
[92] F. E. Close and R. G. Milner, Phys. Rev. **D44** (1991) 3691.
[93] F. Close and R. Roberts, Phys. Lett. **B336** (1993) 165.
[94] F. E. Close and R. G. Roberts, Phys. Lett. **B336** (1994) xxx.
[95] F. D. Colgrove et al., Phys. Rev. **132** (1963) 2561.
[96] COMPASS proposal, CERN/SPSLC 96-14, SPSC/P297 (Geneva, March 1996), CERN/SPSLC 96-30 (Geneva,May 1996).
[97] D. G. Crabb et al., Nucl. Inst. Meth. **A356** (1995) 20.
[98] G. Curci, W. Furmanski and R. Petronzio, Nucl. Phys. **B175** (1980) 27.
[99] ZEUS, M. Derrick et al., Zeit. Phys. **C72** (1996) 399.
[100] Y. L. Dokshitzer, Sov. Phys. JEPT **46** (1977) 461.
[101] J. Drees and M. Montgomery, Ann. Rev. Nucl. Part. Sci. **33** (1983) 383.

[102] A. V. Efremov and O. V. Teryaev, JINR preprint E2-88-287, Dubna (1988).
[103] F. Eisele, Rep. on Propr. in Phys. **49** (1986) 233.
[104] J. Ellis and R. L. Jaffe, Phys. Rev. **D9** (1974) 1444.
[105] J. Ellis and M. Karliner, Phys. Lett. **B213** (1988) 73.
[106] J. Ellis et al., Phys. Lett. **B366** (1996) 268.
[107] R. P. Feynman, Phys. Rev. Lett. **23** (1969) 1415.
[108] R. P. Feynman, *Photon Hadron Interactions*, Benjamin 1972.
[109] E. G. Floratos, C. Kounnas and R. Lacaze, Nucl. Phys. **B192** (1981) 417.
[110] S. Forte, Phys. Rev. **D47** (1993) 1842.
[111] L. L. Frankfurt and M. I. Strikman, Phys. Lett. **B76** (1978) 333.
[112] L. L. Frankfurt et al., Phys. Lett. **B230** (1989) 141.
[113] H. Fritzsch, M. Gell-Mann and H. Leutwyler, Phys. Lett. **B47** (1974) 365.
[114] J. L. Friar et al., Phys. Rev. **C42** (1990) 2310.
[115] W. Furmanski and R. Petronzio, Phys. Lett. **B97** (1980) 427.
[116] W. Furmanski and R. Petronzi, Z. Phys. **C11** (1982) 293.
[117] S. Gasiorowicz and J. L. Rosner, Am. J. Phys. **49** (1981) 954.
[118] H. G. Gaul and E. Steffens, Nucl. Inst. Meth. **A316** (1992) 297.
[119] Th. Gehrmann and W. J. Stirling, Phys. Rev. **B365** (1996) 347.
[120] M. Gell-Mann, Phys. Lett. **8** (1964) 214.
[121] T. R. Gentile and R. D. McKeown, Phys.Rev. **A47** (1993) 456.
[122] H. Georgi and D. Politzer, Phys. Rev. **D14** (1976) 1829.
[123] M. Glück et al., Nucl. Phys. **B422** (1994) 37.
[124] M. Glück et al., Z. Phys. **C67** (1995) 433.
[125] M. Glück et al., Phys. Rev. **D53** (1996) 4775.
[126] K. Gottfried, Phys. Rev. Lett. **18** (1967) 1174.
[127] V. N. Gribov, L. N. Lipatov, Sov. J. Nucl. Phys **15** (1972) 438.
[128] D. Griffiths, *Introduction to Elementary Particles*, John Wiley & Sons, 1987.
[129] HERMES Collaboration, Technical Design Report, DESY-PRC 93/06.
[130] F. Halzen, A. D. Martin, *Quarks and Leptons*, John Wiley & Sons, 1984.
[131] R. L. Heimann, Nucl. Phys. **B64** (1973) 429.
[132] K. Heller, *Proceedings of the Seventh Lake Louise Winter Institute 'Symmetry and Spin in the Standard Model'* B. A. Campbell, L. G. Greeniaus, A. N. Kamal, F. C. Khanna, eds, World Scientific 1992, 47.
[133] R. T. Herrod and S. Wada, Phys.Lett **B96** (1980) 195.
[134] V. W. Huges and J. Kuti, Ann. Rev. Nucl. Sci. **33** (1983) 611.
[135] R. L. Jaffe, Comm. Nucl. Part. Phys. **19** (1990) 239.
[136] X. Ji and P. Unrau, Phys. Lett. **B333** (1994) 228.
[137] X. Ji et al., Phys. Rev. Lett. **76** (1996) 740.
[138] A. L. Kataev and V. V. Starshenko, Mod. Phys. Lett. **A10** (1995) 235.
[139] J. Kodaira et al., Nucl. Phys. **B159** (1979) 99.
[140] J. Kodaira et al., Phys. Rev. **D20** (1979) 627.
[141] J. Kodaira, Nucl. Phys. **B165** (1980) 129.
[142] A. V. Kotwal, Ph. D. Thesis, Harvard University (1995) and E665, preprint Fermilab-Pub-95/396-E.

[143] T. V. Kukhto and N. M. Shumeiko, Nucl. Phys. **B219** (1983) 412.
[144] S. Kumano and J. T. Londergan, Phys. Rev. **D44** (1991) 717.
[145] G. Ladinsky, *Proceedings of the Workshop on the Prospects of SPIN PHYSICS at HERA*, J. Blümlein and W. D. Nowak, eds, 285.
[146] M. Lai et al., (CTEQ), hep-ph/9606399.
[147] B. Lampe, Fortschr. Phys. **43** (1995) 673.
[148] S. A. Larin and J. A. M. Vermaseren, Phys. Lett **B259** (1991) 345.
[149] S. A. Larin, Phys. Lett. **B334** (1994) 192.
[150] L. G. Levchuk, Nucl. Inst. Meth. Phys. Res. **A345** (1994) 496.
[151] A. Levy, *Low-x physics at HERA*, these proceedings.
[152] L. N. Lipatov, Sov. J. Nucl. Phys. **20** (1975) 94.
[153] SMC, A. Magnon, Contribution to DIS97 and private communication.
[154] A. Manohar, *Proceedings of the Seventh Lake Louise Winter Institute 'Symmetry and Spin in the Standard Model'* B. A. Campbell, L. G. Greeniaus, A. N. Kamal, F. C. Khanna, eds, World Scientific 1992, 1.
[155] L. Mankiewicz et al., preprint hep-ph/9510418 (Oct. 1995).
[156] A. D. Martin et al., Phys. Lett. **B191** (1987) 200.
[157] T. Marŭyama et al., Phys. Rev. **B46** (1992) 4261.
[158] A. D. Martin et al., Phys. Rev. **D50** (1994) 6734.
[159] A. D. Martin et al., hep-ph/9606345 (June 1996).
[160] W. Melnitchouk and A. Thomas, Phys. Rev. **D47** (1993) 3783.
[161] W. Melnitchouk et al., Phys. Lett. **B335** (1994) 11.
[162] R. Mertig and W. L. van Neerven, Z. Phys. **C70** (1996) 637.
[163] M. Meyer-Hermann et al., preprint hep-ph/9605229 (May 1995).
[164] A. Milsztajn et al., Z. Phys. **C49** (1991) 527.
[165] CCFR, S. R. Mishra et al., Phys. Rev. Lett. **68** (1992) 3499.
[166] O. Nachtmann, Nucl. Phys. **B63** (1973) 237.
[167] Particle Data Group, Rev. of Particle Properties, Phys. Rev. **D54** (1996) 1.
[168] A. Perkins, *Introduction to High Energy Physics*, Addison Wesley, 1986.
[169] B. Povh, K. Rith, Ch. Scholz and F. Zetsche, *Particles and Nuclei*, Springer 1995.
[170] CCFR, P. Z. Quintas et al., Phys. Rev. Lett. **71** (1993) 1307.
[171] E. Reya, Lecture Notes in Physics, Springer 1994, 175.
[172] R. G. Roberts, *The structure of the proton*, Cambridge University Press, 1990.
[173] F. Sciulli et al., Ann. Rev. Nucl. Part. Sci. **39** (1989) 2259.
[174] CCFR, W. G. Seligman et al., preprint hep-ex/9701017 (Jan 97).
[175] E. V. Shuriak and A. F. Vainshtein, Nucl. Phys. **B201** (1982) 142.
[176] A. Signal et al., Mod. Phys. Lett. **A6** (1991) 271.
[177] T. Sloan, G. Smadja and R. Voss, Phys.Rep. **162** (1988) 45.
[178] A. A. Sokolov and I. M. Ternov, Sov. Phys. Dokl. **8** (1964) 1203.
[179] F. Stock et al., Nucl. Inst. Meth. **A343** (1994) 334.
[180] SLAC E140x, L. H. Tao et al., Z. Phys. **C70** (1996) 83.
[181] M. Virchaux, A. Milsztajn, Phys. Lett. **B274** (1992) 221.

[182] W. Vogelsang, Phys. Rev. **D54** (1996) 2023.
[183] S. Wandzura, F. Wilczek, Phys. Lett. **B72** (1977) 195.
[184] T. Weiler, Phys. Rev. Lett. **44** (1980) 305.
[185] L. Whitlow et al., Phys. Lett. **B250** (1990) 193.
[186] SLAC, L. Whitlow et al., Phys. Lett. **B282** (1992) 475.
[187] E. B. Zijlstra and W. L. van Neerven, Nucl. Phys. **B417** (1994) 61.
[188] G. Zweig, CERN-TH 401, TH 412 (1964).

Low-x Physics at HERA*

A. Levy[1];
Notes in cooperation with M. Ferstl and A. Gute[2]

[1] Raymond and Beverly Sackler Faculty of Exact Sciences, School of Physics, Tel Aviv University, Tel Aviv, Israel
[2] Physikalisches Institut II, Universität Erlangen–Nürnberg, Erwin–Rommel Straße 1, 91058 Erlangen, Germany

1 General Introduction

The dynamical structure of the proton [1] evolved from the pioneering deep inelastic scattering (DIS) experiments at SLAC, through higher energy fixed target lepton–proton DIS measurements till the present results of the HERA ep collider. The proton has three valence quarks and a vast number of additional sea partons, each carrying a fraction x of the proton's momentum. This information is obtained by 'looking' at the proton with a probe of virtuality represented by its negative squared mass Q^2. The higher the Q^2 the smaller the objects 'inside' the proton that can be 'observed'. These objects carry a fraction x of the proton's momentum. The regions in the x–Q^2 plane studied before HERA started to operate were up to about $300\,\mathrm{GeV}^2$ in Q^2 and down to about 10^{-2} in x. The HERA collider has extended the plane in both directions by more than two orders of magnitude, as can be seen in Fig. 1.

What do we expect to learn in this new kinematic region? How does the proton 'look' when probed at very high Q^2? Can one detect substructure in the partons or in the electron? Are there 'exotic' particles such as leptoquarks? Can one detect supersymmetric particles? What is the x distribution of the partons within the proton when probed at different values of Q^2? How many partons are there as x becomes smaller and smaller?

The above is only a partial list of questions hoped to be answered by the HERA data. The topic of these lectures is 'Low–x physics at HERA'. Studying the low–x region actually means studying the high probe–proton center of mass energy W. We will discuss the new results obtained in this wider kinematic region. Since high energy phenomena have been well described within the Regge picture, there will be a chapter devoted to this subject. Next, the DIS kinematics will be introduced and the proton structure functions will be defined. This will lead to the chapter discussing the QCD factorization theorem, the definitions of the parton distributions and their evolution with Q^2, with a special emphasis

* Lectures at the workshop "Nonperturbative QCD" organised by the Graduiertenkolleg Erlangen-Regensburg, held on October 10th–12th, 1995 in Kloster Banz, Germany

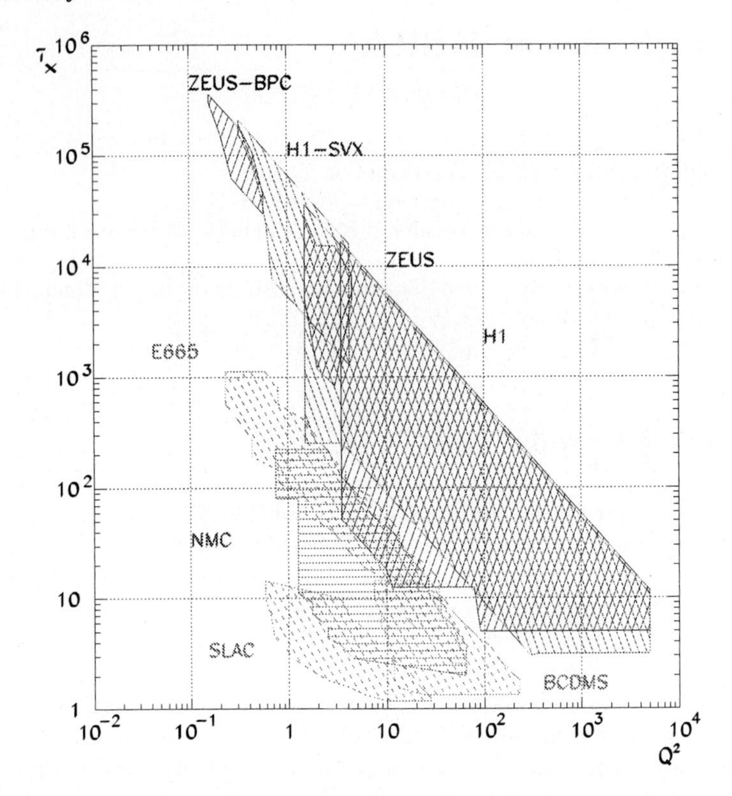

Fig. 1. *A contour of the x–Q^2 plane, indicating the regions covered by the fixed target experiments and those at the HERA collider.*

on their behaviour at low x. The DGLAP and the BFKL evolution equations will be reviewed, different parton density parameterizations will be compared and some methods to obtain the gluon density distribution in the proton will be described. The following chapter discusses the structure function of the photon, the parton density distributions in the photon and the picture of the photon, real and virtual, emerging from the HERA results. One of the unexpected results of HERA, diffraction in DIS, is preceded in the next chapter by a general introduction about diffraction in hadroproduction, the concept of the Pomeron is introduced and the large rapidity gap events in DIS are analysed from the point of view of a partonic structure of the Pomeron. The final chapter is an attempt to give an operational definition to 'hard' and 'soft' interactions and the interplay between the two. This will include a discussion about the energy behaviour of the total DIS cross section as well as that of exclusive production of vector meson in DIS.

Before indulging on this extensive program, I would like, as part of this general introductory chapter, to summarize the highlights of the HERA results so far. We start with a short description of the machine and the two detectors.

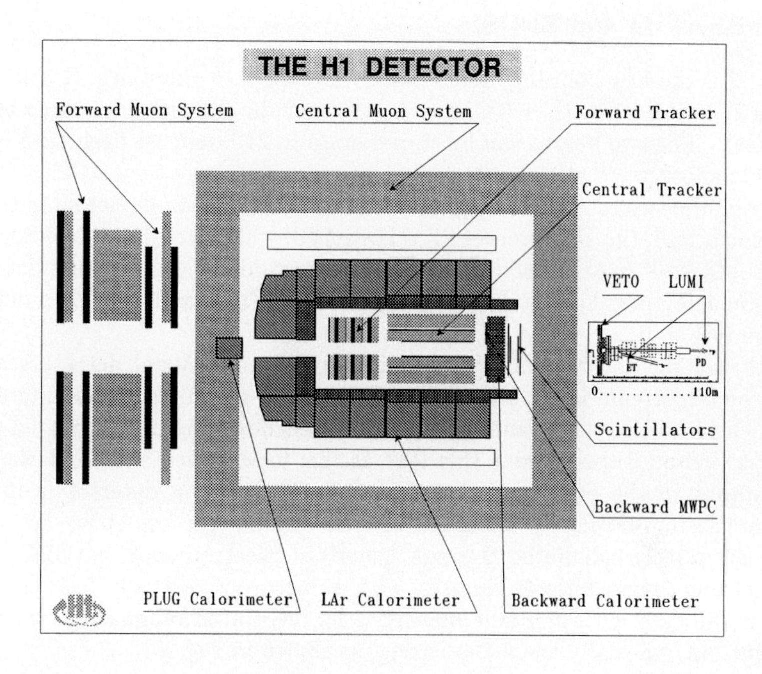

Fig. 2. *The H1 detector.*

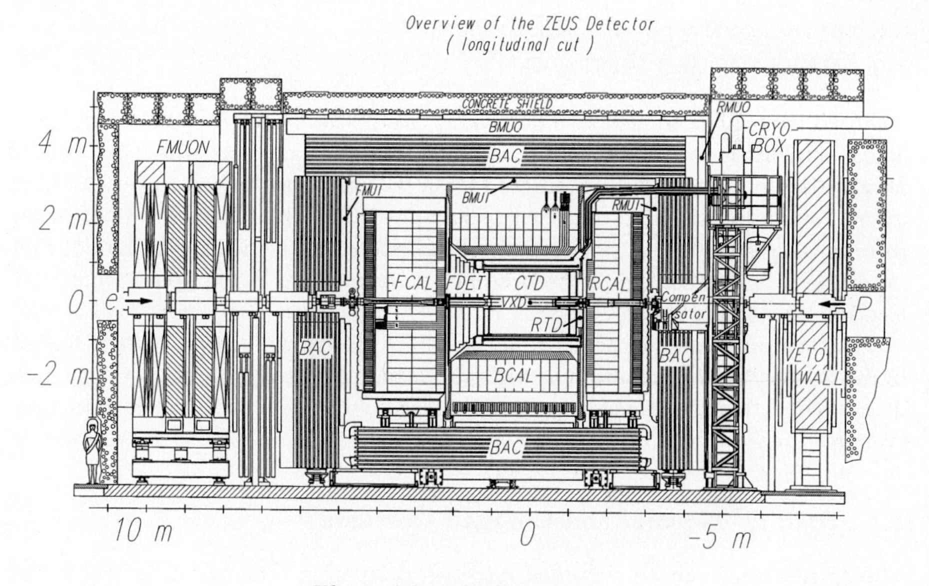

Fig. 3. *The ZEUS detector.*

1.1 HERA, H1 and ZEUS

HERA [2] is the $e^{\pm}p$ colliding beam facility at DESY in Hamburg. It collides at present 27.5 GeV $e^{\pm}$ with 820 GeV protons, providing a center–of–mass energy of 300 GeV. The two beams can be stored in up to 210 bunches each, and collide every 96 nanoseconds at two interaction points.

Each of the two interaction regions is surrounded by a 4π detector. In the so–called north hall, the H1 detector [3] is stationed, while the ZEUS detector [4] is in the south hall. Both detectors use a sampling calorimeter, tracking detectors and muon detectors. The H1 detector is depicted in Fig. 2 and the ZEUS detector is shown in Fig. 3.

Both the H1 and the ZEUS experiments have additional detectors downstream and upstream up to a distance of about 100 meters from the interaction point. Their purpose is to measure protons and neutrons in the proton beam direction (denoted throughout this text as the 'forward' region), and electrons and photons in the electron beam direction. The photon detector is used to measure the luminosity by using the bremsstrahlung process $ep \to ep\gamma$ which can be accurately calculated through Quantum Electrodynamics (QED). This measurement provides the luminosity with an accuracy of about 1–2 %.

As is the case with any new machine, the integrated luminosity starts at a low value but gradually keeps increasing, as shown in Fig. 4.

The present level of luminosities is very well suited for studying the low and intermediate Q^2 physics. It is planned to upgrade [5] HERA in about 2–3 years to increase the integrated luminosity by an order of magnitude by going to the low β mode.

1.2 HERMES and HERA–B

The two beams are in separate rings and can be used also for other purposes. Thus, in addition to the two collision points, the electron beam, which can have a polarization of about 60 %, interacts with a polarized stationary target. At that interaction point, situated in the east straight section, the HERMES detector [6] records the results of the interactions of the polarized particles. The aim of this experiment is to investigate the origin of proton and neutron spin in inelastic electron–nucleon collision.

The proton beam will be used by the HERA–B experiment [7] to be installed in the west area. This experiment will use the halo of the proton beam to produce B–mesons. The experiment objective is to study CP violation in the B–meson system.

1.3 High Q^2 Neutral and Charged Currents

The ep reactions can be classified into two categories. The events in which the outgoing lepton is an electron are called neutral current (NC) events. In this case the particle exchanged between the initial lepton and the proton is a neutral

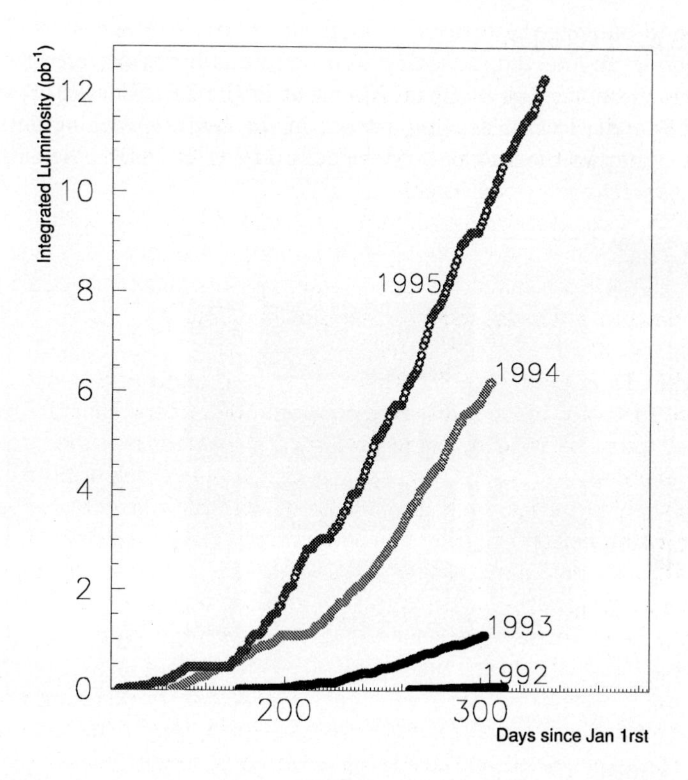

Fig. 4. *The delivered luminosity by HERA versus time for the years 1992–1995.*

particle, predominantly a γ at lower Q^2 values and when Q^2 becomes high enough the Z^0 starts to also contribute appreciably. The other class of events are those where the outgoing lepton is a neutrino. In this case the charged $W^{\pm}$ is exchanged and the events are called charged current (CC) events. These two classes of events can be represented by the two simple exchange diagrams below:

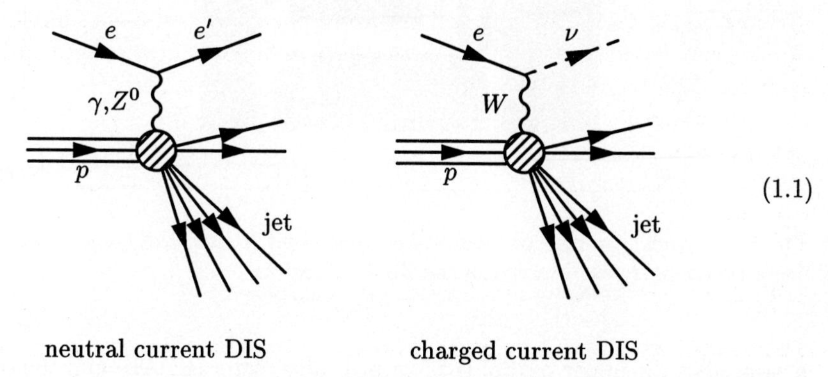

(1.1)

neutral current DIS charged current DIS

The NC events can be identified by observing the scattered electron which makes sure to balance the transverse momentum, p_T. In the case of CC events, the neutrino is undetected, resulting in a large missing transverse momentum. Figure 5 shows on the top a typical NC event in the ZEUS detector, where one sees the scattered electron leaving a track in the central tracking detector and depositing energy in the rear part of the calorimeter (RCAL). The current jet is

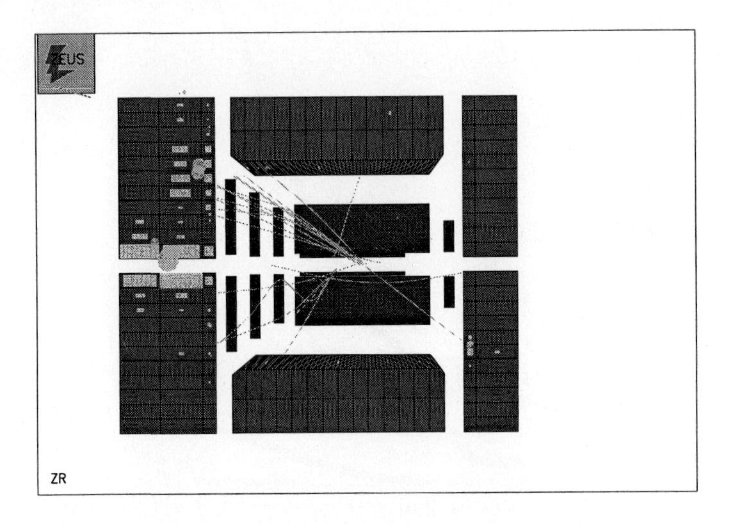

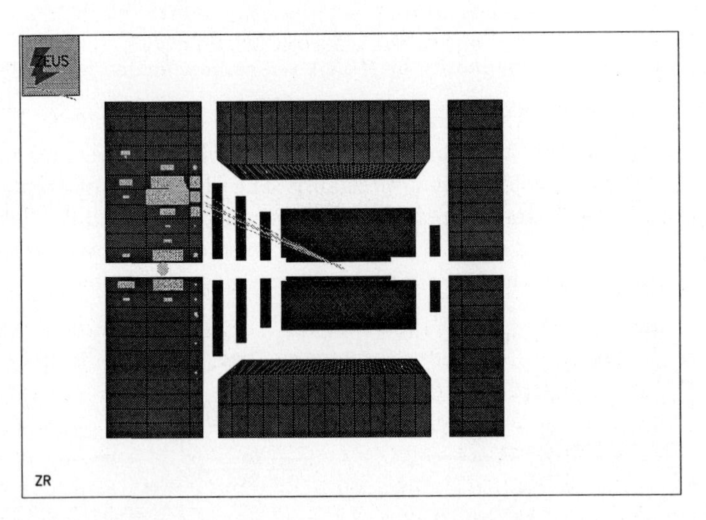

Fig. 5. *A typical example of a neutral current event (upper part) and charged current event (lower part) as observed in the ZEUS detector.*

produced in the upper part of the forward calorimeter (FCAL) and the remnants of the proton produce energy in FCAL around the beam pipe. In the lower part of the figure one sees an example of a CC event, identified by the large missing p_T. Also here one can see the current jet and the remnants depositing energy in the FCAL.

Using the transverse momentum information and the electron identification it is possible for both (e^+ and e^-) beams to isolate the NC and CC events. In the 1993 data sample 436 NC events and 23 CC events with $Q^2 > 400\,\mathrm{GeV}^2$ were found.

The cross–section for the unpolarized e^-p NC DIS can be expressed as [8]:

$$\frac{\mathrm{d}^2\sigma}{\mathrm{dx\,dQ}^2} = \frac{2\pi\alpha^2}{xQ^4}\left\{\left[1+(1-y)^2\right]\mathcal{F}_2 + \left[1-(1-y)^2\right]x\mathcal{F}_3\right\} \tag{1.2}$$

The NC structure function $\mathcal{F}_2$ can be written as:

$$\mathcal{F}_2^{NC} = \sum_f q_f^+ \times$$
$$\times\left[e_f^2 + 2v_e v_f e_f\left(\frac{Q^2}{Q^2+M_Z^2}\right) + \left(v_e^2+a_e^2\right)\left(v_f^2+a_f^2\right)\left(\frac{Q^2}{Q^2+M_Z^2}\right)^2\right] \tag{1.3}$$

where q_f^+ is the sum of the momentum density distributions of the quarks and antiquarks of flavour f, each having the electric charge e_f. The a's and the v's are axial and vector couplings of the electron or the quark to the Z^0 which has the mass M_Z.

For CC reactions, neglecting heavy quarks, the cross section is given by:

$$\frac{\mathrm{d}^2\sigma}{\mathrm{dxdQ}^2} = \frac{G_F^2}{2\pi}\left(\frac{M_W^2}{M_W^2+Q^2}\right)^2 2x\cos^2\theta_c\left[u+c+(1-y)^2\left(\bar{d}+\bar{s}\right)\right] \tag{1.4}$$

where G_F is the Fermi constant and M_W is the mass of the W boson.

While the CC events are produced purely by a weak interaction, the NC events will be dominanted by electromagnetic interactions at lower Q^2, while at higher Q^2 the weak interactions will start to play an important role. The expectations were thus that at some high enough value of Q^2 the cross sections of the NC and CC will be of comparable magnitude. These expectations have been borne out by the first HERA data, taken in 1993 where both collaborations, H1 [9] and ZEUS [10] have measured NC and CC events at high Q^2. The differential cross section of the NC and CC reactions as function of Q^2 are shown in Fig. 6.

One can see that in the region of Q^2 of the order of M_Z^2, the two cross sections are equal. In addition one observes that the shape of the CC cross section is sensitive to the W mass. By fitting the distribution to (1.4), one obtains the result $M_W = 76 \pm 16 \pm 13\,\mathrm{GeV}$.

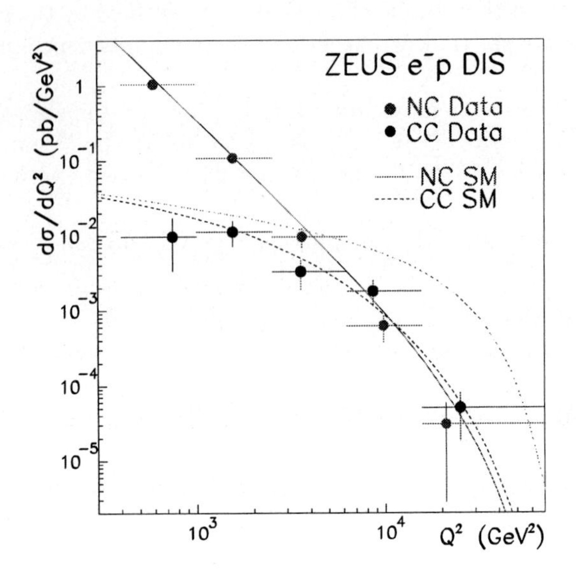

Fig. 6. *The differential cross section* $d\sigma/dQ^2$ *as function of* Q^2 *for NC and CC events as measured in the ZEUS detector using the 1993 data. The dashed line which does not describe the data assumes an infinite mass for the W.*

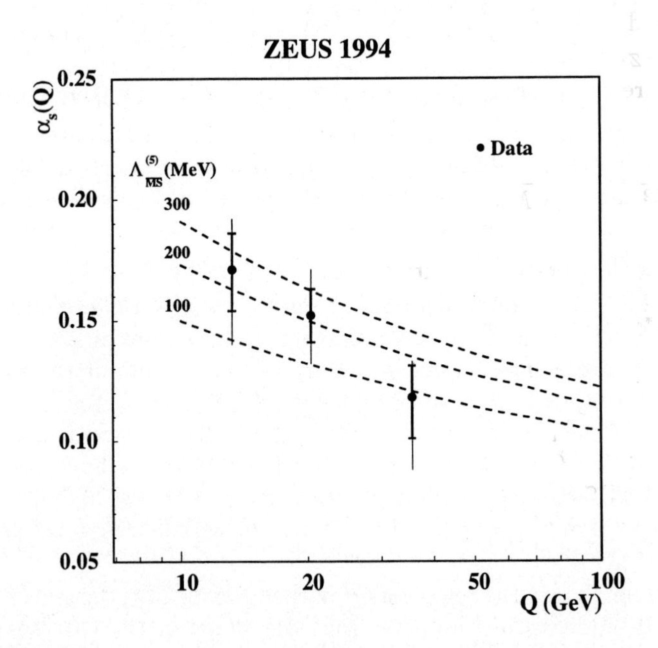

Fig. 7. *Measured value of* $\alpha_S(Q)$ *for three different* Q^2 *regions.*

1.4 Determination of the Strong Coupling α_S

Multi–jet production in NC DIS events can be used to determine the strong coupling constant α_S. In particular, the measured rate of $2+1$ jets, where the '+1' stands for the proton remnant jet, can be compared to theoretical calculations in which α_S is the only free parameter.

By using some kinematical cuts [11] which exclude the problematic region where higher order effects are important and jets are not well measured in the experiment, the value of α_S was determined [12, 13] in three Q^2 regions in the range $120 < Q^2 < 3600\,\mathrm{GeV}^2$, and was found to decrease with Q^2 (Fig. 7), consistent with the running of the strong coupling constant.

The value of α_S expressed at the Z^0 mass was determined to be:

$$\alpha_S(M_{Z^0}) = 0.117\ \pm 0.005(\mathrm{stat})\ {}^{+0.004}_{-0.005}(\mathrm{sys_{exp}})\ \pm 0.007(\mathrm{sys_{th}})$$

which is in good agreement with the results obtained from the e^+e^- event shape (0.121 ± 0.006) and Z^0 width (0.124 ± 0.007).

1.5 The Strong Rise of F_2 at Low x

From the measurements of the cross section of NC DIS events one can unfold the proton structure function [14] F_2 which is a function of x and of Q^2. The data from fixed target experiments indicated that for a given value of Q^2 the structure function F_2 rose slowly with decreasing x and seemed to level off with further decrease. The new data [15, 16] of HERA, which allowed to measure F_2 for higher Q^2 and lower x, showed that it rises sharply with decreasing x. This can be seen in Fig. 8 where the measurements of F_2 are shown for Q^2 values from $1.5\,\mathrm{GeV}^2$ up to $5000\,\mathrm{GeV}^2$ and down to x values close to 10^{-5}. The rise with decreasing x is associated to the increase of the gluon density as x gets smaller. One can see from the lines in Fig. 8 that QCD can accommodate this behaviour in the whole kinematic region where data exist.

Another way [17] to look at the behaviour of F_2 is through its connection to the total $\gamma^* p$ cross section:

$$\sigma_{tot}(\gamma^* p) = \frac{4\pi^2\alpha}{Q^2(1-x)} \frac{Q^2 + 4m_p^2 x^2}{Q^2} F_2(W^2, Q^2) \tag{1.5}$$

Figure 9 shows the behaviour of $\sigma_{tot}(\gamma^* p)$ as calculated from the data of $F_2(W^2, Q^2)$ through (1.5). Also included are measurements of the real γp cross sections as function of W^2. The curves are the expectations of the ALLM [18, 19] parameterization. While the real photoproduction cross section shows a mild rise with energy, one sees a steeper rise for the higher Q^2 data. There clearly is a transition. At which Q^2 value does it happen? Is it gradual or is it sharp? What does the transition mean? We will discuss these questions in later chapters.

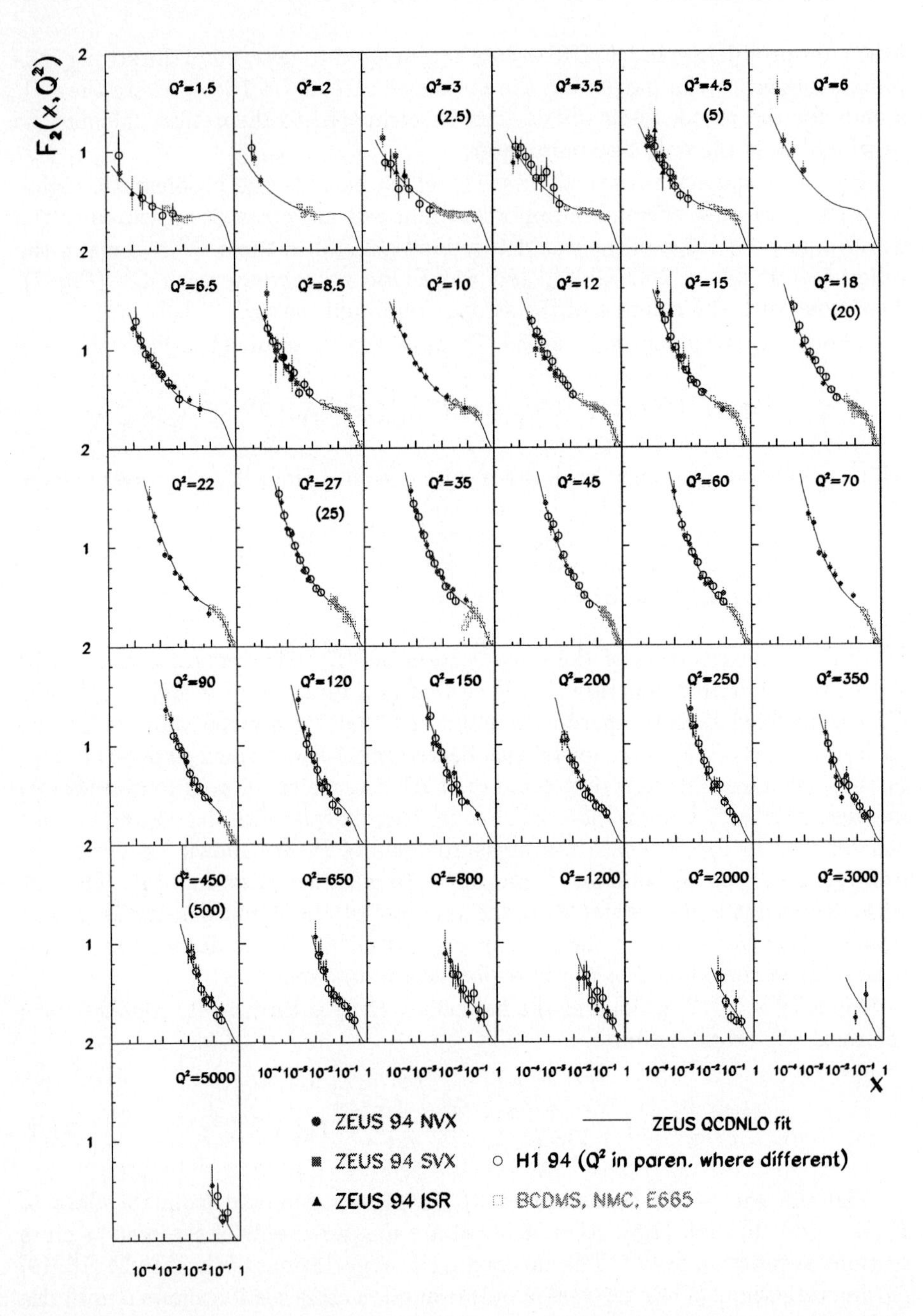

Fig. 8. *The structure function $F_2(x, Q^2)$ as function of Bjorken–x, for fixed Q^2 values.*

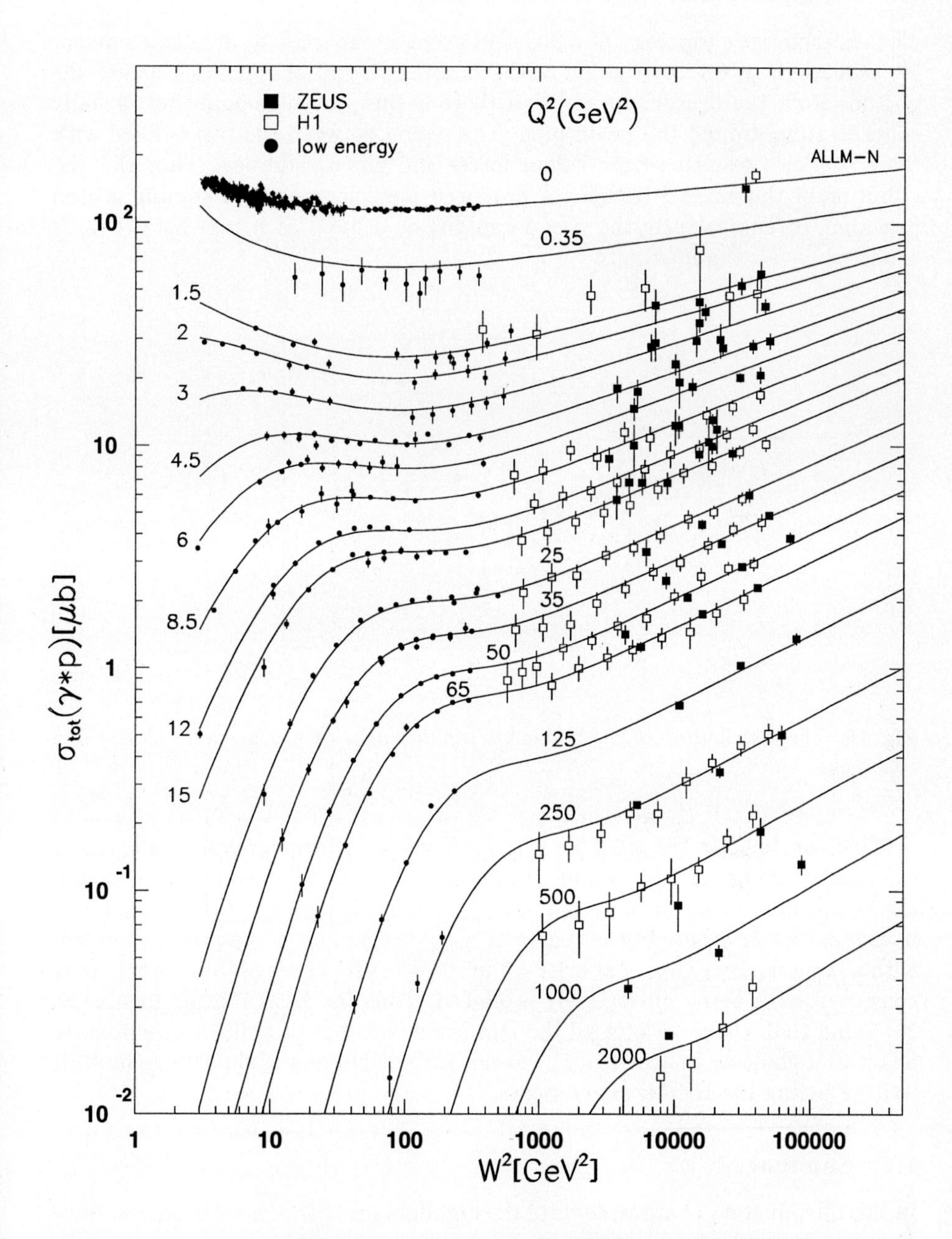

Fig. 9. *The total $\gamma^* p$ cross section as function of W^2, for different Q^2 values. The curves are the expectations of the ALLM parameterization.*

1.6 Large Rapidity Gap Events in DIS

The characteristic topology of a NC DIS event is expected to include a current jet as a result of the interaction of the probing virtual photon with one of the partons from the proton. In addition there is the proton remnant jet, usually concentrating around the beam pipe. The region between the two is filled with more particles resulting from colour forces and gluon radiation. Thus the distribution of the angle θ between a produced particle and the incoming proton direction, or equivalently the pseudorapidity η, defined as $\eta = -\ln\tan\theta/2$, is expected to fall exponentially.

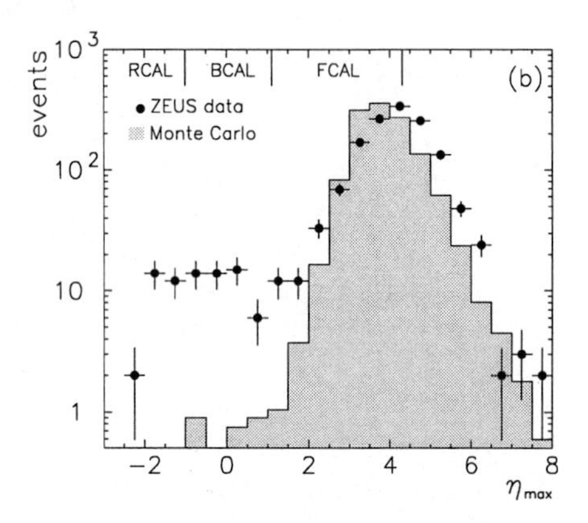

Fig. 10. *The distribution of the maximum rapidity $\eta_{\max}$ of a calorimeter cluster in a DIS event.*

One can look at the variable $\eta_{\max}$, defined as the maximum rapidity of a calorimeter cluster of DIS events, displayed in Fig. 10. While the shaded area is the behaviour which was expected of $\eta_{\max}$ as described above, the data [20] had a large excess of events in the region of $\eta_{\max} < 1.5$. This corresponds to events with a large rapidity gap, of at least 2.8 units. The properties of these events were consistent with being diffractively produced. Thus the HERA experiments [20, 21] found that about 10 % of all the DIS events are due to a diffractive process, a fact that came as a surprise and was not included in any of the DIS generators written before the HERA experiments.

1.7 Summary

In this introductory chapter, some of the highlights of HERA have been described in a very general way. The following have been mentioned:

- The first experimental evidence to the expectation of the electroweak theory that when Q^2 is close to M_Z^2, the electromagnetic and the weak force become of equal magnitude.

- From the jet rates of the NC DIS events one can determine α_S with high precision and can observe the running of α_S in one experiment.
- The proton structure function $F_2(x, Q^2)$ shows a dramatic increase as x decreases for a large span of Q^2 values.
- Large rapidity gap events were observed in NC DIS reactions, indicating the presence of diffractive mechanisms also at higher Q^2.

The first two items do not belong to the scope of these lectures and thus one is referred to the original publications for further details. The last two items will be expanded in the future chapters.

2 Introduction to Regge Theory

2.1 General Introduction

Our understanding of particle physics has evolved in two directions. The static properties of the hadronic spectrum profitted from the breakthrough of the SU(3) theory of Gell–Mann and Ne'eman [22], which relates particles of different internal quantum numbers but the same spin–parity (and mass, in perfect SU(3) symmetry). The hadron dynamics has been investigated by many theories. One of the successful ones which we will describe in more detail here is the Regge theory [23, 24].

The Regge theory investigates the dynamics of hadrons by studying the two–particle scattering $A + B \to C + D$. It relates the spin J and the mass M of particles with the same internal quantum numbers (strangeness, isospin, baryon number, etc.). When one plots all known particles in the Chew–Frautschi [25] plane (J vs M^2) they all seem to lie on a straight line called a Regge trajectory. As an example we show this plot for some of the meson particles (Fig. 11). A similar one exists for the baryons.

The Regge theory predicts some characteristics which can be tested experimentally for the behaviour of hadronic interactions at high energies. One such prediction is that the high energy behaviour of all processes where one exchanges the same quantum numbers should be similar.

2.2 One Pion Exchange (OPE)

The early description of two-body reactions was the picture of one pion exchange (OPE). For instance, the reaction $\pi p \to \rho p$ could be well described by the following diagram:

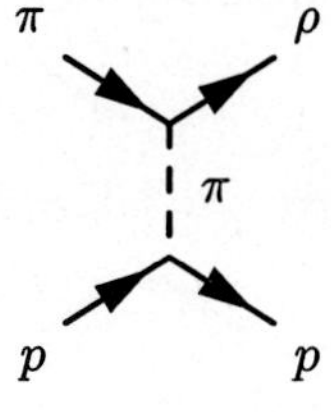

π ρ

π

p p (2.1)

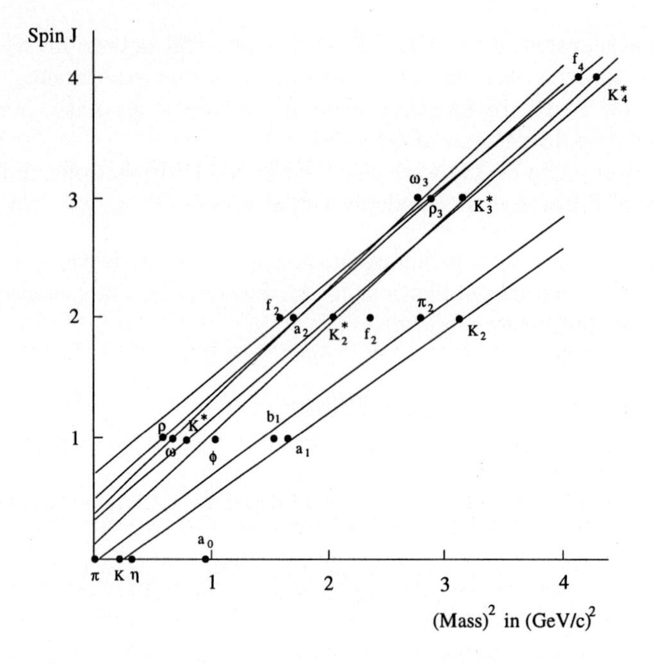

Fig. 11. *Chew–Frautschi plot: Spin J versus mass squared for different mesons. The lines are the corresponding Regge trajectories.*

It was realized soon that this picture is incomplete. There is no justification for ignoring the exchange of two or more pions or of other particles. In some reactions the exchange of a pion is even not possible, like in elastic πp scattering where, due to G–parity, one cannot exchange a pion but a ρ:

$$\pi \quad \pi \quad \rho \quad p \quad p \tag{2.2}$$

Such problems are avoided in the Regge theory. Here one exchanges one or more trajectories and instead of speaking about a particle that is exchanged one talks about a Reggeon exchange. The exchange of a Reggeon is equivalent to the exchange of many particles with different spins:

$$\mathbb{R} \;=\; J=0 \;+\; J=1 \;+\; J=2 \;+\; \ldots \tag{2.3}$$

2.3 s and t Channel

Before continuing with the development of the Regge theory, a short section to define two Lorentz invariant variables which are useful for further discussion. Let us denote by p_i the four–vector of particle i. The four momentum transfer squared t between A and C (or equivalently between B and D) is then defined as:

$$t = (p_C - p_A)^2 = (p_{\bar{A}} + p_C)^2 = (p_B - p_D)^2 = (p_B + p_{\bar{D}})^2 \tag{2.4}$$

Note that t can also be viewed as the center of mass squared of the crossed reaction $B + \bar{D} \to \bar{A} + C$. The center of mass squared s of the system $A + B$ (= $C + D$) is given by:

$$s = (p_A + p_B)^2 = (p_B - p_{\bar{A}})^2 = (p_C + p_D)^2 = (p_C - p_{\bar{D}})^2 \tag{2.5}$$

Thus s can also be interpreted as the four momentum transfer squared from B to $\bar{A}$ of the process $B + \bar{D} \to \bar{A} + C$.

One can therefore look at a given two–body reaction either in the s–channel or in the t–channel, as described in the following diagram:

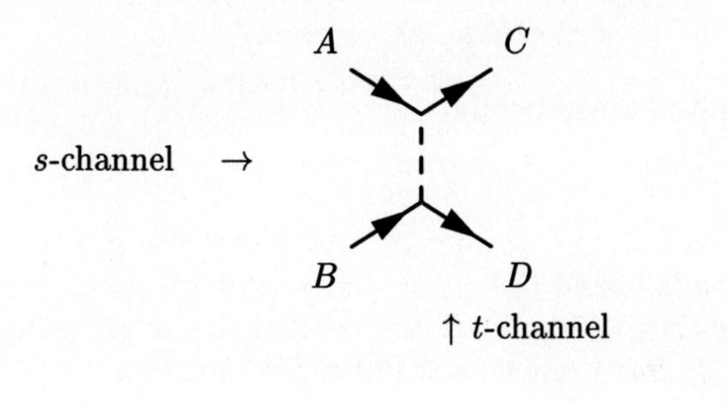

(2.6)

We can now resume the Regge theory discussion.

2.4 The Froissart Bound

A fundamental reason why single particle exchange as described above cannot be the correct description of the two–particle hadronic processes at high energies is the following. Assume an elastic collision between spinless particles, all of mass m, exchanging a meson of mass M and spin J:

m m

$\frac{1}{t-M^2}$

m m (2.7)

The transition amplitude can be written as:

$$T(s,t) \sim \frac{P_J(\cos\theta_t)}{t - M^2} \tag{2.8}$$

where P_J is the Legendre-function and θ_t the scattering angle in the center of mass system of the t–channel reaction $B + \bar{D} \to \bar{A} + C$. The angle can be expressed as:

$$\cos\theta_t = 1 + \frac{2s}{t - 4m^2} \tag{2.9}$$

At fixed t, as s grows, $\cos\theta_t \sim s$, and thus

$$T(s,t) \overset{s\to\infty}{\longrightarrow} s^J \tag{2.10}$$

For $J > 1$ this violates the Froissart [26] bound. What is the Froissart bound?

Froissart showed, that in the partial wave expansion of the scattering amplitude all partial waves with $l \geq l_{\max} = c\sqrt{s}\ln s$ (where c is some constant) are negligible. Summing the partial wave series and assuming maximum scattering in each partial wave he obtained:

$$\sigma \leq c\ln^2 s \quad \text{as} \quad s \to \infty. \tag{2.11}$$

The constant c is limited theoretically:

$$c \leq \frac{\pi}{m_\pi^2}. \tag{2.12}$$

Thus the cross section is bound by:

$$\sigma \leq \frac{\pi}{m_\pi^2}\ln^2 s \simeq (60\,\text{mb})\ln^2 \text{s}. \tag{2.13}$$

This bound is known as the Froissart bound, also referred to sometimes as the Martin [27]–Froissart bound.

2.5 Regge Trajectories

In this section we will describe the main steps leading to the Regge trajectory. Let us assume that there exists a bound state in the t–channel with angular momentum $l = L$ and mass $M_B < 2m$. The t–channel partial wave amplitude $f_L(t)$ has then a pole at $t = t_B \equiv M_B^2$. Similarly, if in the t–channel there is a resonance with mass M_R, width Γ and $l = L$, one gets a pole at a complex value $t = t_R \equiv M_R^2 - \mathrm{i}M_R\Gamma$.

The sequence $f_l(t)$, $\quad l = 0, 1, 2, ...$ can be generalized to a function $f(l,t)$ which should be equal to $f_l(t)$ for $l = 0, 1, 2, ...$ This function is defined also for complex l. The sequence of poles for $l = L_1$ at $t = t_1$, $l = L_2$ at $t = t_2, \ldots$ is interpreted as a single moving Regge pole at $l = \alpha(t)$. The function $\alpha(t)$ is a trajectory function such that $\alpha(t_1) = L_1, \ldots$.

When $\alpha(t)$ is equal to an integer value L at $t = t_L$, this corresponds to a bound state or resonance with $l = L$ and mass and width given by $t_l = M_L^2 - iM_L\Gamma$. The trajectory which gives a resonance with $l = L$ and complex t_L will also cause a pole at the real value $t = M_L^2$ when l has the complex value $L + \mathrm{Im}\,\alpha(t)$. Therefore the real values t_l where $\mathrm{Re}\,\alpha(t_l) = L$ give the masses $M_L^2 = t_l$ of the particles with spin L. This can be demonstrated in Fig. 12 where the ρ trajectory is determined from the charge exchange reaction $\pi^- p \to \pi^0 n$. When the trajectory gets to the mass of the ρ its value is equal to the spin of the ρ: $\alpha(M_\rho^2) = 1$. In the Chew–Frautschi plot shown earlier in Fig. 11 one sees more examples of trajectories having the same feature.

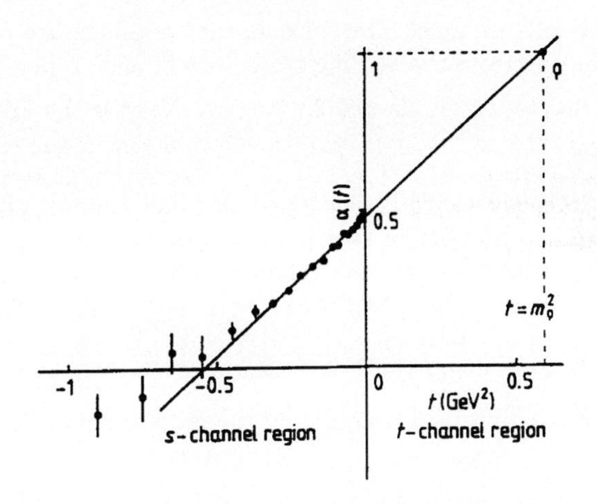

Fig. 12. *The ρ trajectory as determined from the charge exchange reaction $\pi^- p \to \pi^0 n$.*

Originally, Regge was interested in the behaviour of the t–channel scattering amplitude in the unphysical limit $\cos\theta_t \to -\infty$. The usual partial-wave series diverges when $|\cos\theta_t| = 1 + \epsilon$. So he used the function $f(l, t)$ to convert the sum into a contour integral in the complex l plane, which allowed $\cos\theta_t \to -\infty$. He obtained

$$T_t(t, s) \sim (\cos\theta_t)^{\alpha(t)} \tag{2.14}$$

where $\alpha(t)$ is the trajectory whose real part is largest for the given t.

It was Mandelstam who realised that the limit $\cos\theta_t \to -\infty$ in the t–channel reaction $B + \bar{D} \to \bar{A} + C$ corresponds to $s \to \infty$ in the s–channel reaction $A + B \to C + D$ (see Eqn. (2.9) with fixed (negative) t).

Since the transition amplitude fulfils $T_s(s, t) = T_t(t, s)$ one gets at fixed t:

$$T_s(s, t) \sim s^{\alpha(t)} \tag{2.15}$$

More precisely, at fixed t:

$$T_s(s, t) \overset{s\to\infty}{\longrightarrow} \gamma(t) \left(\frac{s}{s_0}\right)^{\alpha(t)} \tag{2.16}$$

where s_0 is a scale factor. The function $\gamma(t)$ is closely related to the residue of $f(l,t)$ at the pole $l = \alpha(t)$.

In the relativistic case one needs to consider the signature (+ for even J, - for odd J) and one gets:

$$T_s(s,t) = \gamma_\pm(t) \frac{1 \pm \mathrm{e}^{-i\pi\alpha(t)}}{\sin \pi\alpha(t)} \left(\frac{s}{s_0}\right)^{\alpha(t)} . \tag{2.17}$$

The functions $\gamma_\pm(t)$ factorize. This means that for $A + B \to C + D$ one can write:

$$\gamma(t) = \gamma_{\mathrm{AC}}(t)\gamma_{\mathrm{BD}}(t). \tag{2.18}$$

If several sets of t–channel internal quantum numbers are possible, one includes a contribution from the leading trajectory of each type.

2.6 Shrinkage

Before the Regge theory was fully developed, one had a simple phenomenological description of a large number of two–body reactions. The energy behaviour of the forward differential cross section of these reactions could be described by the form:

$$\frac{\mathrm{d}\sigma}{\mathrm{d}t}(s, t=0) = \frac{A}{s^2}\left(\frac{s}{s_0}\right)^{2\alpha_{\mathrm{eff}}} \tag{2.19}$$

The values of α_{eff} obtained from fitting the data are given in Table 1. Regge theory identifies α_{eff} with $\alpha(t=0)$ of the dominant Regge trajectory contributing to the reaction. The value $\alpha(t=0)$, also denoted sometimes by α_0, is called the intercept of the Regge trajectory.

Table 1. *Coefficients α_{eff} and the possible exchanged particles for various processes.*

Reaction	t-channel	$\approx \alpha_{\mathrm{eff}}$
$\pi^- p \to \pi^0 n$	ρ	0.5
$K^- p \to \bar{K}^0 n$	ρ, a_2	0.5
$\gamma p \to \pi^0 p$	ρ, ω	0.5
$\pi^- p \to K^0 \Lambda$	K^*, K_2^*	0.2
$K^- p \to \pi^0 \Lambda$	K^*, K_2^*	0.2
$\pi^- p \to p\pi^-$	Δ	0
$\pi^+ p \to p\pi^+$	N, Δ	0

The Chew–Frautschi plot shows that the trajectories are linear in t and can be expressed in a simple form:

$$\alpha(t) = \alpha_0 + \alpha' t \tag{2.20}$$

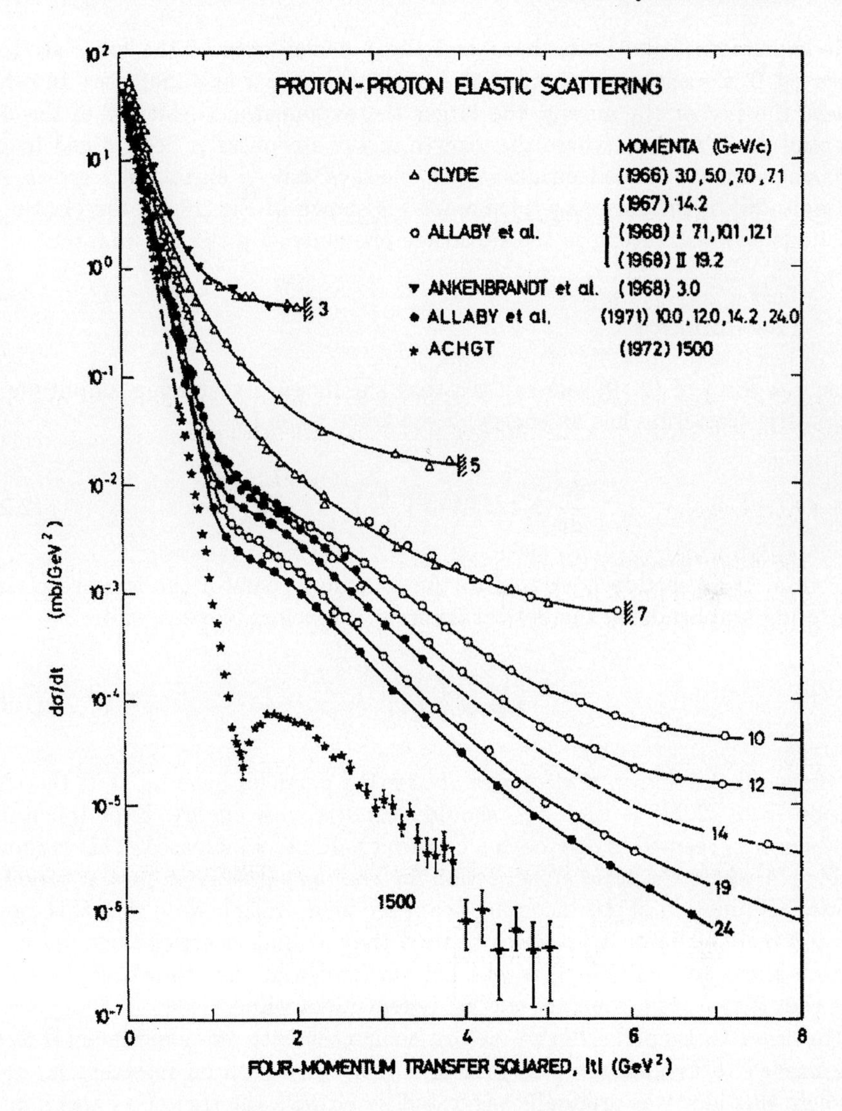

Fig. 13. $\frac{d\sigma}{dt}$ *for different s for pp reactions.*

The slope of the trajectory, α' is positive and has the value $\alpha' \approx 1\,\text{GeV}^{-2}$ for most trajectories.

If one pole dominates one can write the differential cross section in the form:

$$\frac{d\sigma}{dt} = \frac{\gamma(t)}{s^2}\left(\frac{s}{s_0}\right)^{2\alpha(t)} = \frac{\gamma(t)}{s^2}\left(\frac{s}{s_0}\right)^{2\alpha_0+2\alpha' t} \tag{2.21}$$

$$= \frac{\gamma(t)}{s_0^2}\left(\frac{s}{s_0}\right)^{2(\alpha_0-1)} e^{2\alpha' t \ln\left(\frac{s}{s_0}\right)} \tag{2.22}$$

This expression gives both the s and the t dependence of the cross section. Since $t < 0$, the exponential cutoff in t becomes sharper as s increases. In other words, the higher the energy, the larger the exponential coefficient of the differential cross section, where the growth of the steepness is determined by α'. This phenomena is called shrinkage and one says that $\frac{d\sigma}{dt}$ shrinks as s grows. An example [28] of the shrinkage phenomena is shown in Fig. 13 for the elastic pp reactions. We will return to the shrinkage phenomena in the last chapter.

2.7 The Pomeron

From the form of (2.19) one can see that the forward scattering amplitude of the elastic scattering has an energy dependence given by:

$$\frac{d\sigma}{dt}(s,t=0) \sim \left(\frac{s}{s_0}\right)^{2(\alpha_0-1)} . \tag{2.23}$$

The total cross section is related to the imaginary part of the forward elastic scattering amplitude by the optical theorem. Therefore one can write:

$$\sigma_{\mathrm{tot}} \sim \left(\frac{s}{s_0}\right)^{\alpha_0-1} . \tag{2.24}$$

Since all the known trajectories of existing particles have $\alpha_0 < 1$, the conclusion from (2.24) is that σ_{tot} should decrease with energy. Experimentally, however, σ_{tot} seemed to approach a constant value as s increased. This is shown in Fig. 14 where the total cross section for various particles on proton target is plotted as function of the incoming beam momentum [28]. Note that this figure already includes later data which showed that at high energies the total cross section starts to increase. This was not yet known at that time and the belief was that it reaches a constant energy–independent value.

In order to keep the Regge picture consistent with the experimental data, one needed to assume the existence of a trajectory with an intercept $\alpha_0 \simeq 1$. Though this idea was originally suggested by Gribov, the trajectory was named by Gell–Mann after Pomeranchuk, who derived in 1958 his theory [29] about the asymptotic behaviour of the differences of cross sections. The trajectory was first called the Pomeranchukon trajectory, and was later abbreviated to the Pomeron. This trajectory ($\mathbb{P}$) was assumed to have the form:

$$\alpha_{\mathbb{P}}(t) = \alpha_{\mathbb{P}}(0) + \alpha'_{\mathbb{P}}t \tag{2.25}$$

with an intercept of $\alpha_{\mathbb{P}}(0) \simeq 1$ and a slope [30] of $\alpha'_{\mathbb{P}} \simeq 0.25\,\mathrm{GeV}^{-2}$. These values are different from all those of the known trajectories, some of which are given in Table 1. So far no particle was found which corresponds to the Pomeron trajectory.

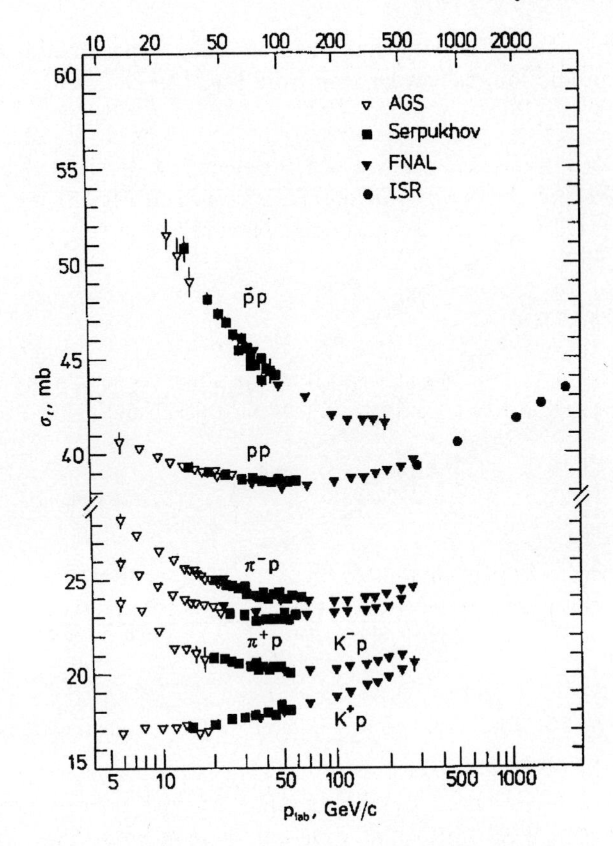

Fig. 14. *Total cross section measurements for various reactions.*

The Pomeranchuk Theorem Since we mentioned earlier the Pomeranchuk theorem, lets say a few words about it. Pomeranchuk studied [29] the high energy behaviour of the differences between the total cross sections, $\Delta\sigma$, defined as:

$$\Delta\sigma \equiv \sigma(\bar{A}B) - \sigma(AB) \tag{2.26}$$

for any particles A, $\bar{A}$ and B. The scattering should become purely 'diffractive' (in the optical sense) at high energies. The elastic scattering should be just the shadow of the inelastic reactions. In this picture, via the optical theorem, one expects the amplitudes to be almost purely imaginary. Under the assumption that

$$\frac{\mathrm{Re}\, T(s,0)}{\mathrm{Im}\, T(s,0)} \stackrel{s\to\infty}{\longrightarrow} 0 \tag{2.27}$$

and if $\sigma(\bar{A}B) \to C_1$ and $\sigma(AB) \to C_2$, Pomeranchuk proved that $C_1 = C_2$. This led to his theorem that at high energies the differences between particle and antiparticle cross sections should vanish:

$$\Delta\sigma \to 0 \tag{2.28}$$

This is known as the Pomeranchuk theorem. The experimental data seem to support this prediction, as can be seen from Fig. 15.

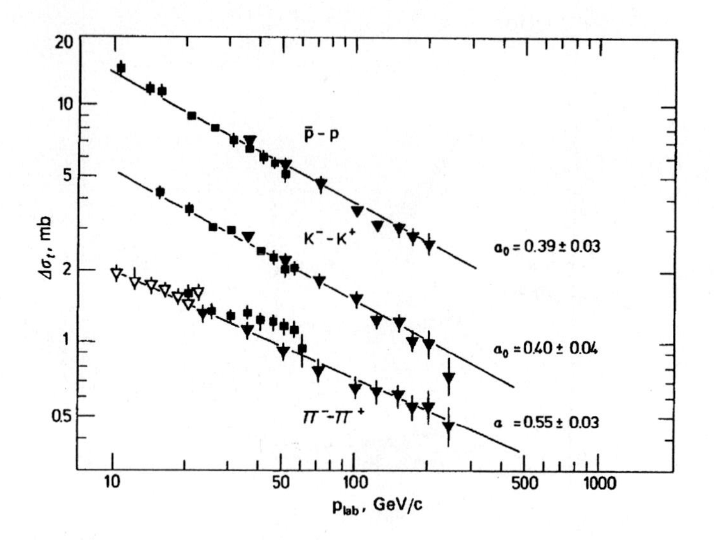

Fig. 15. *Total cross section differences for various reactions.*

2.8 High Energy Behaviour of $\sigma_{\rm tot}$

As we mentioned above, the total cross section actually starts rising with energy. What changes does one have to make to the Pomeron trajectory parameters? Clearly, in order to describe the rise, the intercept has to be bigger than 1. By how much? One way to find out is to fit the total cross section data to the form given in (2.24). However, we know that there are also other trajectories which can be exchanged in addition to the Pomeron.

Donnachie and Landshoff (DL) [31] attempted a global fit to all existing $\sigma_{\rm tot}$ data. They realized that all the Reggeon intercepts can be represented by one effective intercept having a value of $\alpha_{\mathbb{R}}(0) \sim 0.55$. Also, the elastic scattering data can be described by having two exchanges, a Reggeon and a Pomeron.

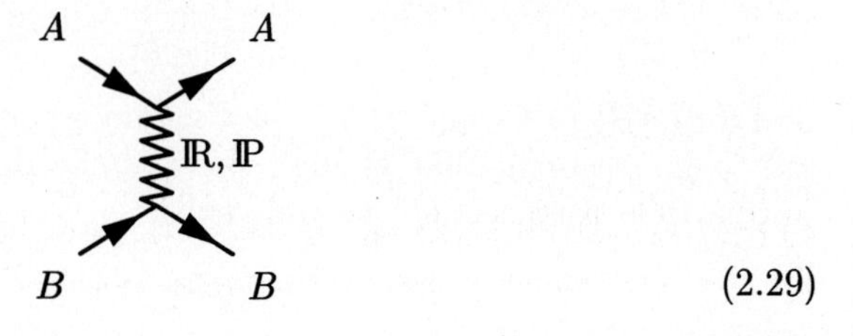

(2.29)

When a Reggeon is being exchanged, one usually exchanges quantum numbers, while in the case of the Pomeron exchange there is the exchange of vacuum quantum numbers. Using the optical theorem, one expects also the total cross section to be described by the sum of these two exchanges. Thus motivated, DL fitted the total cross section data to a sum of two terms:

$$\sigma_{\rm tot} = X s^{\alpha_{\mathbb{P}}(0)-1} + Y s^{\alpha_{\mathbb{R}}(0)-1} \tag{2.30}$$

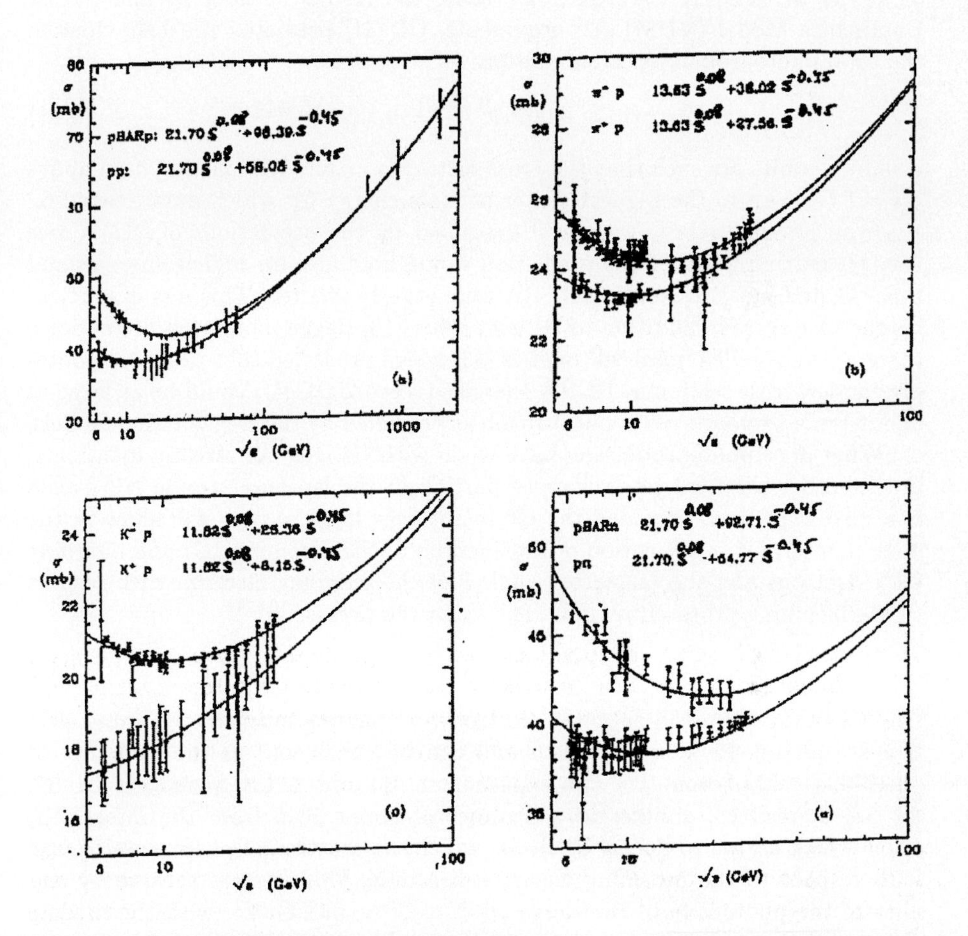

Fig. 16. *Fits of a simple Regge form containing two terms to total cross section measurements of pp, $\bar{p}p$, $\pi^{\pm}p$, $K^{\pm}p$, pn and $\bar{p}n$ reactions.*

In order to get the rising cross section at high energies they parametrized the Pomeron intercept as $\alpha_{\mathbb{P}}(0) = 1+\epsilon$. The value of the Reggeon intercept was fixed to 0.55 (actually 0.5475). In addition, DL used the Pomeranchuk theorem for the first term which describes the Pomeron exchange. Since at high energies only the

Pomeron term remains, they constrained the coefficients X to be the same for particle and for antiparticles. For example, they require $X(\pi^+) = X(\pi^-)$. The combined fit can be seen in Fig. 16 and gives a good description of the data. The resulting value for the Pomeron parameter was $\epsilon \simeq 0.08$ (actually 0.0808). The rising power of the total cross section is small enough and violates the Froissart bound only at energies of about 10^3 TeV.

$\sigma_{tot}(\gamma p)$ at HERA Energies By using the results of their fit and Vector Dominance Model (VDM) [32] arguments, DL [31] predicted the behaviour of the total photoproduction cross section:

$$\sigma_{tot}(\gamma p) = 0.0677 s^{0.0808} + 0.129 s^{-0.4525} \tag{2.31}$$

where the units are such that the cross section is in mb. All existing data above $\sqrt{s} = 6$ GeV up to the highest center of mass energy for which data existed on $\sigma_{tot}(\gamma p)$, about 20 GeV, were well described by the predictions of DL. It was very interesting whether the prediction would hold also for higher energies and thus the first measurements of HERA were eagerly awaited. This was in particular the case since in addition to DL and other [18] Regge motivated predictions, there were so–called 'mini-jet' models [33] which predicted that the total photoproduction cross section at HERA energies ($\sqrt{s} \sim 200$ GeV) could be as large as about 1 mb, compared to about 0.15 mb as predicted by the Regge based models.

What does photoproduction have to do with HERA? As already mentioned in the earlier chapter, the exchanged particle at the lepton vertex in NC events is a virtual photon, provided the Q^2 is not very large. As we will show in the next chapter, Q^2 is a function of the energies of the incoming (E) and outgoing (E') electrons and the scattering angle θ of the outgoing electron with respect to the incoming proton direction. The exact relation is:

$$Q^2 = 2EE'(1 + \cos\theta) \tag{2.32}$$

One sees that when the outgoing electron continues close to the incoming electron's direction ($\theta \approx \pi$), $Q^2 \approx 0$ and can be considered as the exchange of an almost–real photon. The two experiments, H1 and ZEUS, have each [34, 35, 36, 37] a small calorimeter at a distance of about 30 m from the interaction point which allows to detect electrons which are scattered by less than 5 mrad with respect to the incoming electron direction. This ensures that the virtuality of the photons is in the range $10^{-8} < Q^2 < 0.02\,\mathrm{GeV}^2$, with the median $Q^2 \sim 10^{-5}\,\mathrm{GeV}^2$. A diagramatic example of a photoproduction event is shown in Fig. 17. Since the cross section for photoproduction reactions is large compared to that of DIS events, even a low luminosity run can be sufficient for determining $\sigma_{tot}(\gamma p)$. Thus this was the first measurement [38, 39] done when the HERA collider was turned on in summer of 1992. Using the photoproduction events taken during a net running time of 7 minutes, the first measurement of $\sigma_{tot}(\gamma p)$ was obtained at $\sqrt{s} = 200$ GeV (Fig. 18). Although the measurement had a large uncertainty, it was enough to establish that there is no dramatic rise of the cross

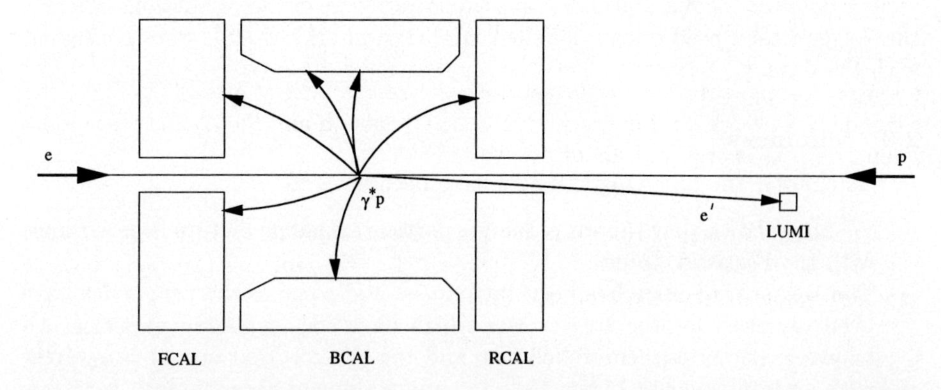

Fig. 17. *A diagramatic example of a photoproduction event in the ZEUS detector, where the scattered electron is detected in the small angle electron calorimeter LUMIE.*

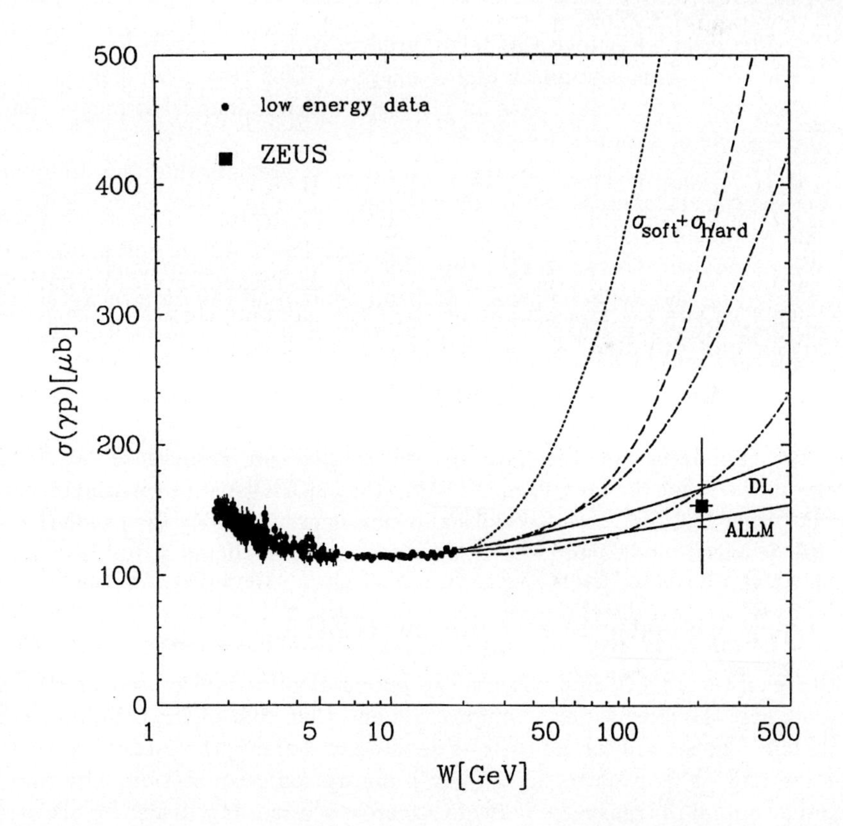

Fig. 18. *The HERA first measurement of $\sigma_{tot}(\gamma p)$, together with lower energy data. The curves are predictions of different models for the HERA energy range region.*

section between 20 and 200 GeV, thus excluding some of the predictions. Clearly the Regge based predictions, labelled DL [31] and ALLM [18], were consistent with the data.

2.9 Summary

In this chapter the following subjects have been covered:

- We have shown that the simple single particle exchange picture is in variance with the Froissart bound.
- The Regge trajectory has been introduced and some of the properties have been reviewed for the trajectories which have corresponding particles. All known trajectories seem to be linear and can be expressed as $\alpha(t) = \alpha_0 + \alpha' t$, with an intercept not bigger than 0.5 and a slope of about $1\,\text{GeV}^{-2}$.
- The shrinkage phenomena has been introduced for further discussion in the chapter on diffraction. As long as the slope of the Regge trajectory is non–zero, one expects to have a steeper fall of the differential cross section as one goes to higher energies.
- The Pomeron trajectory was introduced in order to explain the behaviour of the total cross section at higher energies. This trajectory has so far no corresponding particle. It has an intercept which is somewhat bigger than 1 and a slope of about $0.25\,\text{GeV}^{-2}$.
- The Pomeranchuk theorem was mentioned. It predicts that the differences between the total cross section of particles and antiparticles should approach zero at high energies.
- We introduced the two–term expression of Donnachie and Landshoff which is based on the Regge approach and can explain all the data on total cross sections. By using a fixed effective intercept of about 0.55 for the Reggeon trajectory, and by using the Pomeranchukon theorem for the Pomeron term, DL obtained an intercept of 1.08 for the Pomeron by performing a joint fit to all the data existing in 1992.
- After explaining how the total γp cross section can be measured at HERA, we showed that the measurement of $\sigma_{tot}(\gamma p)$ at HERA shows a mild increase, like other hadronic cross sections and thus consistent with the predictions of Regge based models and excluding those which predicted a dramatic rise.

3 Deep Inelastic Scattering at HERA

In this chapter we will first discuss the kinematical variables used in DIS, describing first the fixed target configuration and then that of the colliding beams at HERA. The second section will be devoted to defining the inelastic structure function and its relation to the total $\gamma^* p$ absorption cross section. The ratio of the longitudinal to transverse part of that cross section, R will also be discussed. In the third section we will mention shortly the important issue of radiative corrections and the chapter will finish with a section describing the experimental determination of the structure function.

3.1 Kinematics

We shall start with the most general case of a deep inelastic scattering of a lepton with mass m_l and four vector $k(E_l, \mathbf{k})$ on a proton with mass m_p and four vector $p(E_p, \mathbf{p})$, as depicted in Fig. 19. The outgoing lepton has a mass of m_l' with four vector $k'(E_l', \mathbf{k'})$ and the scattered hadrons emerge with a mass m_h and four vector $p_h(E_h, \mathbf{p_h})$. The exchanged particle can be a gauge boson γ, Z^0 or $W^{\pm}$, depending on the circumstances. The four vector of the exchanged boson is $q(q_0, \mathbf{q})$.

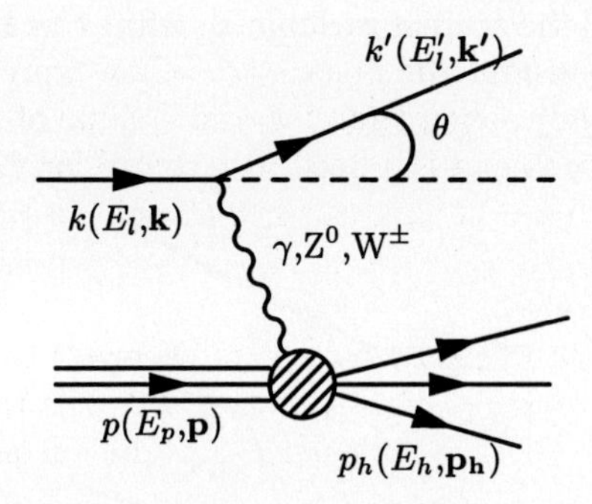

Fig. 19. *Deep inelastic lepton–proton–scattering.*

With these notations one can define the following variables:

$$q = k - k' = p_h - p \tag{3.1}$$

$$\nu \equiv \frac{p \cdot q}{m_p} \tag{3.2}$$

$$y \equiv \frac{p \cdot q}{p \cdot k} \tag{3.3}$$

$$W^2 = (p + q)^2 \tag{3.4}$$

$$s = (k + p)^2 \tag{3.5}$$

The meaning of the variables ν and y is most easily realized in the rest frame of the proton (see fixed target subsection). The variable W^2 is the center of mass squared of the gauge–boson proton system, and thus also the invariant mass squared of the hadronic final state. The variable s is the center of mass squared of the lepton proton system.

The four momentum transfer squared at the lepton vertex can be approximated as follows (for $m_l, m_l' \ll E, E'$):

$$q^2 = (k - k')^2 = m_l^2 + m_l'^2 - 2kk' \approx -2EE'(1 - \cos\theta) < 0 \tag{3.6}$$

Note that in this expression the scattering angle θ is defined with respect to the incoming lepton direction. The variable which is mostly used in DIS is the negative value of the four momentum transfer squared at the lepton vertex:

$$Q^2 \equiv -q^2 \tag{3.7}$$

One is now ready to define the other variable most frequent in DIS, namely the dimensionless scaling variable x:

$$x \equiv \frac{Q^2}{2p \cdot q} \tag{3.8}$$

The Physical Meaning of the Bjorken–x Variable In order to understand the physical meaning of the scaling variable x, defined by Bjorken, let us consider the diagram presented in Fig. 20.

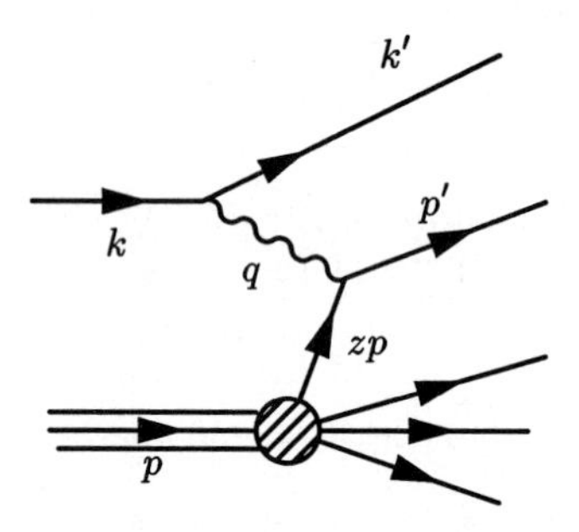

Fig. 20. *Explanation of the Bjorken x variable*

An exchanged boson with four momentum q interacts with a parton having a fraction z of the incoming proton four momentum, producing a parton with four momentum p'. Using the definitions in the previous section one can obtain the following:

$$(zp + q)^2 = p'^2 \tag{3.9}$$

$$z^2p^2 - Q^2 + 2zp \cdot q = m_{p'}^2 \tag{3.10}$$

which finally leads to:

$$z = \frac{Q^2 + m_{p'}^2 - z^2p^2}{2p \cdot q} = x\left(1 + \frac{m_{p'}^2 - z^2m_p^2}{2p \cdot q}\right) \tag{3.11}$$

If one can assume that the partons have zero mass, which is a good assumption in the infinite momentum frame, then one gets:

$$x = z \tag{3.12}$$

This means that x is the fraction of the proton momentum taken by the parton which is hit by the exchanged boson in the DIS interaction.

Fixed Target Kinematics ($\mathbf{p} = 0$) Before HERA started to operate, all DIS experiments [40, 41, 14] were on fixed targets ($\mathbf{p} = 0$). For this case, some of the variables defined earlier have a simple physical interpretation. The variable ν:

$$\nu \equiv \frac{p \cdot q}{m_p} = \frac{p \cdot (k - k')}{m_p} = \frac{m_p(E - E')}{m_p} = (E - E') = q_0 \tag{3.13}$$

Thus, for fixed target experiments, ν is the energy of the exchanged boson. Another expression for ν is:

$$\nu = \frac{p \cdot (p_h - p)}{m_p} = \frac{m_p(E_h - m_p)}{m_p} = E_h - m_p \tag{3.14}$$

which is the energy transfer at the hadronic vertex.

The second scaling variable defined by Bjorken is y. It has the following meaning in the case of a fixed target:

$$y \equiv \frac{p \cdot q}{p \cdot k} = \frac{m_p \nu}{p \cdot k} = \frac{\nu}{E} \tag{3.15}$$

which is the fraction of the incoming lepton energy carried by the exchanged boson, also called inelasticity. It can be calculated either at the lepton vertex or at the hadron vertex:

$$y = \begin{cases} \frac{E-E'}{E} & \text{leptons} \\ \frac{E_h - m_p}{E} & \text{hadrons} \end{cases} \tag{3.16}$$

Clearly one sees that the value of y is limited to:

$$0 \leq y \leq 1 \tag{3.17}$$

What are the limits on the value of the Bjorken–x? The Bjorken variable x can be written as

$$x \equiv \frac{Q^2}{2p \cdot q} = \frac{Q^2}{2m_p \nu} \tag{3.18}$$

On the other hand we can express W^2 as follows:

$$W^2 = (p+q)^2 = p^2 - Q^2 + 2p \cdot q = m_p^2 - 2p \cdot qx + 2p \cdot q = m_p^2 + 2p \cdot q(1-x) \tag{3.19}$$

Since the invariant mass squared of the hadronic system has to be equal or bigger than the proton mass squared, $W^2 \geq m_p^2$, one gets:

$$0 \leq x \leq 1 \tag{3.20}$$

which is consistent with the physical interpretation of x being the fraction of the proton's momentum carried by the struck parton.

The following relations are useful if one wants to calculate the kinematic limits which one can achieve on Q^2 and x in the fixed target experiments.

$$s = (p + k)^2 = p^2 + k^2 + 2p \cdot k = m_p^2 + \frac{2p \cdot q}{y} = m_p^2 + \frac{Q^2}{xy} \tag{3.21}$$

which leads to:

$$Q^2 = (s - m^2)xy \tag{3.22}$$

Thus the maximum value of Q^2, given a lepton beam of energy E, is:

$$Q^2_{max} = s - m_p^2 \approx 2m_p E \tag{3.23}$$

and the minimum value obtainable for x is:

$$x_{min} = \frac{Q^2}{s - m_p^2} = \frac{Q^2}{2m_p E} \tag{3.24}$$

The fixed target experiments used typically muon beams with energies of a few hundred GeV. Thus, for example, for $E = 200\,\text{GeV}$, and $Q^2 = 4\,\text{GeV}^2$, the minimum value of x is $x_{min} \approx 10^{-2}$.

HERA Kinematics At HERA, an electron (we will use electron to mean both electron or positron) beam of energy E collides with a proton beam of energy E_p. At present $E = 27.5\,\text{GeV}$ and $E_p = 820\,\text{GeV}$. The $+z$ axis is chosen in the proton beam direction and the scattering angle of the outgoing electron is measured with respect to the proton beam. The four vectors of the incoming electron (k), outgoing electron (k'), incoming proton (p) and outgoing hadrons (p_h) are defined as follows:

$$k = \begin{pmatrix} E \\ 0 \\ 0 \\ -E \end{pmatrix} \quad k' = \begin{pmatrix} E' \\ E'\sin\theta \\ 0 \\ E'\cos\theta \end{pmatrix} \quad p = \begin{pmatrix} E_p \\ 0 \\ 0 \\ E_p \end{pmatrix} \quad p_h = \begin{pmatrix} E_h \\ p_{xh} \\ p_{yh} \\ p_{zh} \end{pmatrix} \tag{3.25}$$

Since the detectors at HERA have an almost complete 4π coverage, one can determine the x and Q^2 DIS variables by more than one method [42]. One can use the outgoing electron, the outgoing hadrons or a combination of both. This allows therefore a consistency check on the determination by comparing the results from the different methods. Each method has its kinematical range in which it can determine the variables with a better precision than the others. In the following we will discuss the three different methods.

Electron variables: In this case, as used to be done in the fixed target experiments, only the four vector of the scattered electron is used:

$$Q^2 = -(k - k')^2 = 2EE'(1 + \cos\theta) \tag{3.26}$$

$$\begin{aligned} y \equiv \frac{p \cdot q}{p \cdot k} &= 1 - \frac{pk'}{pk} = 1 - \frac{pk'}{2EE_p} \\ &= 1 - \frac{E_pE' - E_pE'\cos\theta}{2EE_p} = 1 - \frac{E'}{2E}(1 - \cos\theta) \end{aligned} \tag{3.27}$$

Using Q^2 and y one can get x:

$$x = \frac{Q^2}{4EE_p y} \tag{3.28}$$

Some further useful relations in this case can be obtained in the following way:

$$\begin{aligned}
\frac{Q^2}{2EE'} &= 1 + \cos\theta \\
\frac{2E(1-y)}{E'} &= 1 - \cos\theta \\
\frac{Q^2 2E(1-y)}{2EE'^2} &= \sin^2\theta
\end{aligned} \tag{3.29}$$

Therefore one gets:

$$Q^2 = \frac{E'^2 \sin^2\theta}{1-y} = \frac{p_{Te}^2}{1-y} \tag{3.30}$$

Since in the NC case the p_T of the electron balances that of the hadronic final state, the last relation means also that $Q^2 = p_{Th}^2/(1-y)$, a relation to be used in the description of the hadronic method.

Hadronic variables: One can clearly determine the y, Q^2 and x variables if one measures all the outgoing hadrons. However, some of the hadrons escape through the uncovered region in the beam pipe, as shown in the sketch in Fig. 21. Nevertheless, as was shown by Jacquet and Blondel [43], one can still determine to a good approximation the above variables.

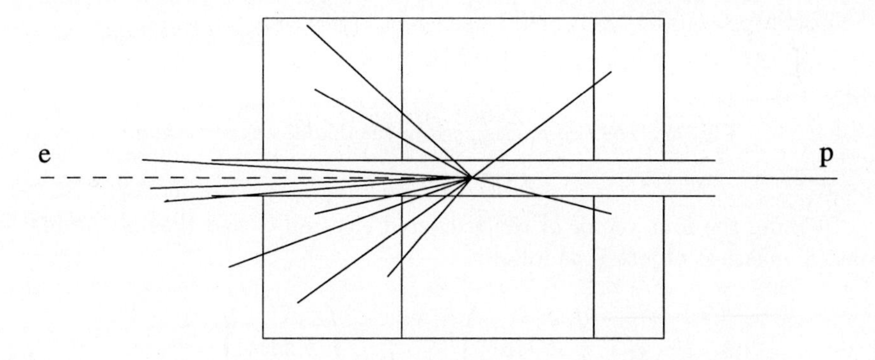

Fig. 21. *A sketch for understanding the Jacquet–Blondel method.*

The variable y can be calculated in the following way:

$$y = \frac{p \cdot (p_h - p)}{p \cdot k} = \frac{p \cdot p_h}{2EE_p} = \frac{E_p E_h - E_p p_{zh}}{2EE_p} = \frac{E_h - p_{zh}}{2E} \tag{3.31}$$

Since most of the missing hadrons which do not make it into the detector have a small p_T, their contribution to $E_h - p_{zh}$ is negligible and thus one gets a good estimate of y using this formula. When determined this way, it is usually denoted y_{JB}, namely y Jacquet–Blondel.

The other two variables can be now calculated in the following way:

$$Q^2 = \frac{p_{xh}^2 + p_{yh}^2}{1 - y_{JB}} \tag{3.32}$$

and

$$x = \frac{Q^2}{4EE_p y_{JB}} \tag{3.33}$$

The double angle method: When using a method based on a mixture of the electron and the hadron variables, one can choose different combinations. The one described here is called the double angle (DA) method and uses measurements of two angles. One is the scattering angle θ of the outgoing electron (see Fig. 22). The other angle is that of an object which has a simple meaning in the naive parton model: assuming that the struck parton is massless, it would scatter by an angle γ. In this interpretation the p_T of the proton remnant is zero. Note that these assumption are necessary only for the physical interpretation of the angle γ. The calculation is however exact.

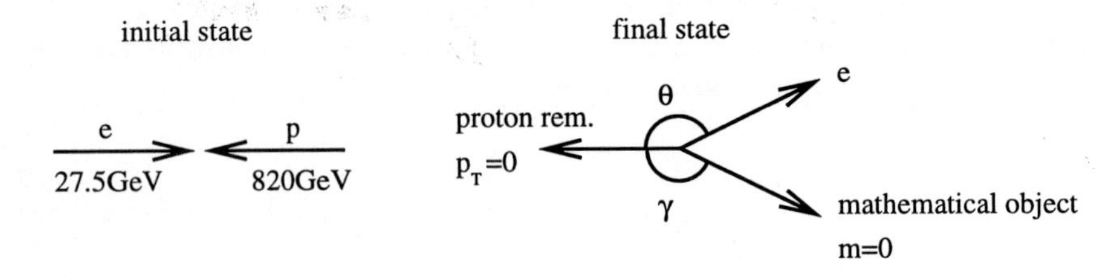

Fig. 22. *The two angles used in the double angle method.*

Defining the four vector of the scattered electron k' and that of the mathematical massless object Γ as follows:

$$k' = \begin{pmatrix} E' \\ E' \sin\theta \\ 0 \\ E' \cos\theta \end{pmatrix} \qquad \Gamma = \begin{pmatrix} \Gamma \\ \Gamma \sin\gamma \\ 0 \\ \Gamma \cos\gamma \end{pmatrix} \tag{3.34}$$

one gets from the scattered electron measurements:

$$Q^2 = \frac{E' \sin^2\theta}{1 - y} \tag{3.35}$$

and

$$y = 1 - \frac{E'}{2E}(1 - \cos\theta) \tag{3.36}$$

Using the measurements coming from the hadrons one obtains:

$$Q^2 = \frac{p_{Th}^2}{1-y} = \frac{\Gamma^2 \sin^2\gamma}{1-y} \tag{3.37}$$

and

$$y = \frac{E_h - p_{zH}}{2E} = \frac{\Gamma(1-\cos\gamma)}{2E} \tag{3.38}$$

which leads to the calculation of the angle γ:

$$\cos\gamma = \frac{p_{Th}^2 - (E_h - p_{zh})^2}{p_{Th}^2 + (E_h - p_{zh})^2} \tag{3.39}$$

We are now ready to define Q^2 and x using only the two angles (in addition to the given incoming energies of the beams). The variables calculated this way usually have the subscript DA to denote that they were obtained through the double angle method:

$$Q_{\mathrm{DA}}^2 = 4E^2 \left[\frac{\sin\gamma(1+\cos\theta)}{\sin\gamma + \sin\theta - \sin(\theta+\gamma)}\right] \tag{3.40}$$

and

$$x_{\mathrm{DA}} = \left(\frac{E}{E_p}\right)\left[\frac{\sin\gamma + \sin\theta + \sin(\theta+\gamma)}{\sin\gamma + \sin\theta - \sin(\theta+\gamma)}\right] \tag{3.41}$$

The advantage of the DA method is that one is less sensitive to a scale uncertainty in the energy measurement of the final state particles since the angle γ is obtained by ratios of energies.

Kinematical limits at HERA: What are the kinematical limits on x and Q^2 that one can reach at HERA? Since $s = 9 \times 10^4\,\mathrm{GeV}^2$, this is also the value of Q^2_{max}. With the present luminosities, the two experiments have enough data for measurements up to about $10^4\,\mathrm{GeV}^2$, as shown in Fig. 1 of the first chapter. Amazingly enough, the HERA experiments made big efforts to obtain data at very low Q^2, down to about $0.2\,\mathrm{GeV}^2$, for reasons to be mentioned later. Thus they also reached very low x values, close to 10^{-6}.

One of the aims of these measurements is to obtain the structure function of the proton, which is the subject of the next section.

3.2 Inelastic Structure Function

In analogy to the Rosenbluth formula [44] in the elastic case the deep inelastic cross–section can be written as [45]:

$$\frac{\mathrm{d}^2\sigma}{\mathrm{d}\Omega \mathrm{dE}'} = \frac{4\alpha^2 E'^2}{Q^4}\left[2W_1 \sin^2\frac{\theta}{2} + W_2 \cos^2\frac{\theta}{2}\right] \tag{3.42}$$

where $W_{1,2}(\nu, Q^2)$ are two structure functions which can be related to the absorption cross section of the virtual photon γ^*. In order to see the relation, note that the DIS cross section is obtained from a product of two tensors, a leptonic one and a hadronic tensor. The hadronic tensor $W_{\alpha\beta}$ can be related through the optical theorem to the $\gamma^* p$ cross section, as shown in the following diagram:

$$W_{\alpha\beta} \quad \propto \quad \sum_x \quad [\gamma^* p \to X \to \gamma^* p \text{ diagram}] \tag{3.43}$$

The cross section for a real photon is defined as

$$\sigma_\lambda^{\mathrm{tot}}(\gamma p \to \mathrm{X}) = \frac{4\pi^2\alpha}{K}\epsilon_\lambda^{\alpha^*}\epsilon_\lambda^\beta W_{\alpha\beta} \tag{3.44}$$

where K is the flux factor, ϵ_λ is the polarisation vector and λ is the helicity of the photon. The flux of real photons is $K = \nu$ and the allowed helicities are $\lambda = \pm 1$.

In order to extend the discussion to virtual photons, we need to know the polarization vectors and the flux of a virtual photon beam. To this end we will use the expressions of the polarization vectors derived for a massive vector meson:

$$\epsilon_{\pm 1}^\alpha = \pm\frac{1}{\sqrt{2}}(0; 1, \pm i, 0) \tag{3.45}$$

$$\epsilon_0^\alpha = \pm\frac{1}{\sqrt{Q^2}}(\sqrt{\nu^2 + Q^2}; 0, 0, \nu) \tag{3.46}$$

From parity conservation one can write (for the case where the incoming lepton is e or μ and the target proton is unpolarized) $\sigma_{+1} = \sigma_{-1}$. The two independent cross sections are defined as:

$$\sigma_T = \frac{1}{2}(\sigma_{+1} + \sigma_{-1}) \qquad \sigma_L = \sigma_0 \tag{3.47}$$

where σ_T and σ_L are the transverse and longitudinal $\gamma^* p$ cross section.

Flux of Virtual Photons The definition of the flux is a matter of convention, but the principle is that in the limit of approaching the real photon case ($Q^2 \to 0$), the flux K should be equal that of a real photon beam ($K \to \nu$). We shall mention two flux conventions, that of Gilman [46] and that of Hand [47]. Gilman adds the Q^2 of the virtual photon to the flux definition:

$$K^{\text{Gil}} = \sqrt{\nu^2 + Q^2} \tag{3.48}$$

Hand defines the flux K as that energy which a real photon would need in order to create the same final state. Thus K^{Hand} can be calculated using the following argument. The invariant mass squared of the hadronic final state for a $\gamma^* p$ interaction is given by:

$$W^2 = (p+q)^2 = m_p^2 + 2m_p\nu - Q^2 \tag{3.49}$$

The same quantity for a real photon of energy K^{Hand} is given by:

$$W^2 = m_p^2 + 2m_p K^{\text{Hand}} \tag{3.50}$$

Therefore

$$K^{\text{Hand}} = \nu - \frac{Q^2}{2m_p} \tag{3.51}$$

Clearly both definitions fulfil the requirement that when $Q^2 \to 0$ then $K \to \nu$.

The two structure functions $W_{1,2}(\nu, Q^2)$ are related to the total $\gamma^* p$ transverse and longitudinal absorption cross sections in the following way:

$$W_1(\nu, Q^2) = \frac{K}{4\pi^2\alpha}\sigma_T^{\gamma^* p} \tag{3.52}$$

$$W_2(\nu, Q^2) = \frac{K}{4\pi^2\alpha}\frac{Q^2}{Q^2+\nu^2}(\sigma_T^{\gamma^* p} + \sigma_L^{\gamma^* p}) \tag{3.53}$$

The Ratio $R = \sigma_L/\sigma_T$ The ratio of the longitudinal to transverse $\gamma^* p$ cross section carries information about the spin of the quarks in the quark–parton model. In order to see that we can use (3.53) to write:

$$R \equiv \frac{\sigma_L}{\sigma_T} = \frac{W_2}{W_1}(1 + \frac{\nu^2}{Q^2}) - 1 \tag{3.54}$$

One can use QED to calculate the exact expressions for the two structure functions W_1 and W_2 for the case of the scattering of two point–like spin–1/2 fermions $e\mu \to e\mu$:

$$W_2^{e\mu} = \frac{1}{\nu}\delta(1 - \frac{Q^2}{2m_\mu\nu}) \tag{3.55}$$

$$W_1^{e\mu} = \frac{Q^2}{4m_\mu^2\nu}\delta(1 - \frac{Q^2}{2m_\mu\nu}) \tag{3.56}$$

By substituting this result into (3.54) and using the δ–function condition, one gets:

$$R^{\mu} = \frac{4m_{\mu}^{2}}{Q^{2}} \overset{Q^{2}\to\infty}{\longrightarrow} 0 \tag{3.57}$$

Adopting this results for quarks, one expects R to approach 0 if quarks have spin 1/2. In other words:

$$\sigma_{L} \to 0 \qquad \text{for spin } \frac{1}{2} \tag{3.58}$$

How can one measure R experimentally? One can rewrite the cross section given by (3.42) in the form:

$$\frac{\mathrm{d}^{2}\sigma}{\mathrm{d}\Omega \mathrm{dE}'} = \Gamma(\sigma_{T} + \epsilon\sigma_{L}) \tag{3.59}$$

where the photon polarization ϵ is expressed as:

$$\epsilon = \left[1 + 2\frac{Q^{2}+\nu^{2}}{Q^{2}}\tan^{2}\frac{\theta}{2}\right]^{-1} \tag{3.60}$$

and the photon flux Γ is written in the form:

$$\Gamma = \frac{K\alpha}{2\pi^{2}Q^{2}}\frac{E'}{E}\frac{1}{1-\epsilon} \tag{3.61}$$

Thus the cross section can be written as:

$$\frac{\mathrm{d}^{2}\sigma}{\mathrm{d}\Omega \mathrm{dE}'} = \Gamma\sigma_{T}\left(1 + \epsilon\frac{\sigma_{L}}{\sigma_{T}}\right) = \Gamma\sigma_{T}(1 + \epsilon R) \tag{3.62}$$

The polarization ϵ is a function of ν, Q^{2} and θ. By keeping ν and Q^{2} fixed and by changing θ and E one expects a straight line when plotting the differential cross section as function of ϵ, as shown in Fig. 23. Thus one can fit directly the

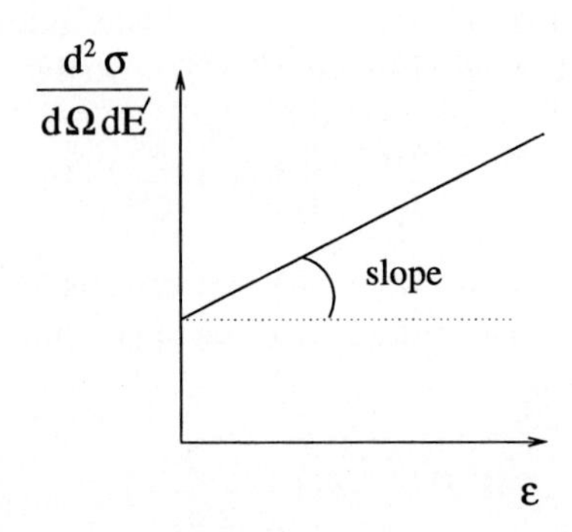

Fig. 23. *The differential cross section as function of the photon polarization ϵ for fixed ν and Q^{2}.*

intercept ($= \Gamma\sigma_T$) and the slope ($= R$) and once R is known, $\frac{d^2\sigma}{d\Omega dE'}$ depends only on one structure function.

We see therefore that measuring R not only gives information about the spin structure of the proton constituents but is also necessary if one wants to obtain the structure functions from the measured cross sections. Before describing how one determines the structure functions, one needs to take into account an additional important effect, namely radiative corrections, which are described in the next section.

3.3 Radiative Corrections

In order to determine the structure functions, one needs to know the Born cross section. However, the measured cross section includes in addition to the Born one contributions from a whole set of electroweak radiative processes. The radiation can come from the electron (either from the incoming or from the outgoing one), it can come from the hadron side, by having a quark radiate a photon and there are interference terms. On top of that there are loop and vertex corrections. In the following we present some of the diagrams included in the calculations of radiative corrections.

First, the radiation from the electron line. These contributions are the source of the largest corrections. In the diagrams (a) and (b) the simplest $\mathcal{O}(\alpha^3)$ graphs for real photon emission at the leptonic vertex and in (c) and (d) the $\mathcal{O}(\alpha^3)$ contributions from virtual photons associated with the leptonic vertex are shown.

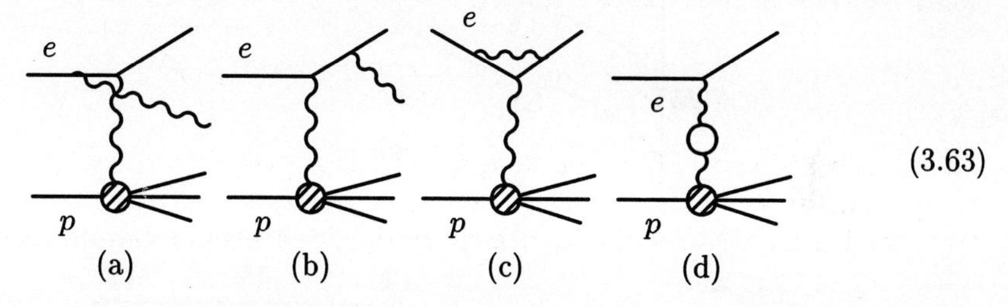

(3.63)

The radiative corrections coming from the hadron line are less important than the contributions of photons radiated off the electron. The following three diagrams have to be corrected for the radiation of a photon from quarks:

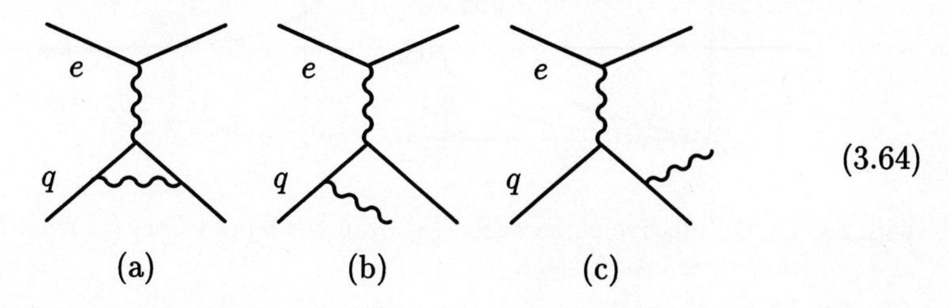

(3.64)

Finally there are the interference term corrections described diagramatically as:

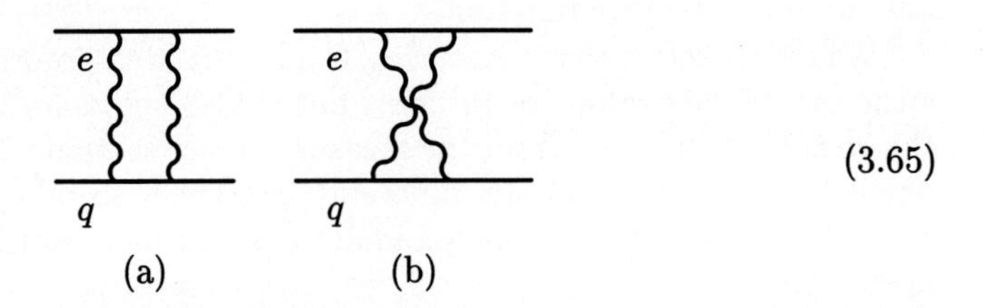

$$ (3.65) $$

One can get a feeling for the magnitude of these corrections [48] at HERA from the plots shown in Fig. 24. As one sees, the corrections depend strongly on the kinematical region and for small x and high y can be very large.

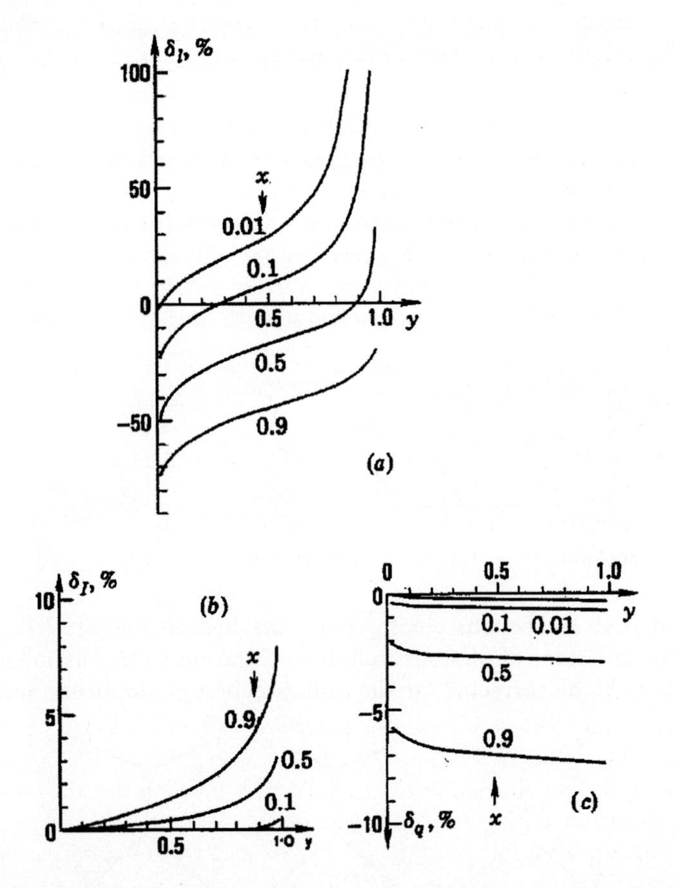

Fig. 24. *QED radiative corrections: (a) from the lepton line, (c) from the quark lines, and (b) from their interference.*

3.4 Experimental Determination of the Structure Functions

The expression of the Born cross section contains two structure functions, W_1 and W_2. One can use instead the structure function F_1 and F_2 which are related to the former ones as follows:

$$F_1 = W_1 \tag{3.66}$$

$$F_2 = \frac{\nu W_2}{m_p} \tag{3.67}$$

Thus one can write the Born cross section in the form:

$$\frac{\mathrm{d}^2\sigma^{\mathrm{Born}}}{\mathrm{dxdQ^2}} = \frac{4\pi\alpha^2}{xQ^4}\left[\frac{y^2}{2}2xF_1 + (1-y)F_2\right] \tag{3.68}$$

The first measurements of the structure functions showed that the data were compatible with the Callan–Gross [49] relation

$$2xF_1(x) = F_2(x). \tag{3.69}$$

With more precise DIS data covering a larger range of x and Q^2 a clear scale breaking was observed and thus one needs information about the difference between the two structure functions, defined as $F_L \equiv F_2 - 2xF_1$. By defining the ratio of the structure functions R_L as follows:

$$R_L \equiv \frac{F_L}{F_2} \tag{3.70}$$

one can express the Born cross section as:

$$\frac{\mathrm{d}^2\sigma^{\mathrm{Born}}}{\mathrm{dxdQ^2}} = \frac{4\pi\alpha^2}{xQ^4}F_2Y_+\left(1 - \frac{y^2}{Y_+}R_L\right); \qquad Y_+ = 1 + (1-y)^2 \tag{3.71}$$

However, in addition to the Born cross sections there are radiative processes. So the cross section which includes these radiative processes is expressed in the form:

$$\frac{\mathrm{d}^2\sigma^{\mathrm{rad}}}{\mathrm{dxdQ^2}} = \frac{4\pi\alpha^2}{xQ^4}F_2Y_+\left(1 - \frac{y^2}{Y_+}R_L\right)\left[1 + \delta_r(x,Q^2)\right] \tag{3.72}$$

where $\delta_r(x,Q^2)$ are the contributions coming from the radiative processes.

If one had an ideal detector which measures every outgoing particle from the reaction with a 100 % precision and with a background–free identification of the processes, this would be the cross section from which the structure function would be determined. In reality one measures something which is somewhat different than the above expression. The measured cross section is connected to the one in (3.72) in the following way:

$$\frac{\mathrm{d}^2\sigma^{\mathrm{meas}}}{\mathrm{dxdQ^2}} = \int \mathrm{dx'} \int \mathrm{dQ'^2}\, \frac{\mathrm{d}^2\sigma^{\mathrm{rad}}}{\mathrm{dx'dQ'^2}} \mathrm{A_{cc}(x',Q'^2)S(x,Q^2;x',Q'^2)} + \mathrm{background} \tag{3.73}$$

Here A_{cc} is the probability that an event which is produced at x', Q'^2 will be seen in the detector, which is thus a function of the geometry and efficiency of the detector, and S is a smearing function which gives the probability that an event which is produced at x', Q'^2 was measured at x, Q^2.

The experimental procedure to measure the structure functions includes thus the following steps:

- One selects a sample of events likely to be NC DIS events. This means that one needs to have a good electron finder [50] which has both high efficience and high purity for electron identification. This requirement usually results in the fact that only electrons having energies above a certain values can be identified.
- The values of x and Q^2 are determined by one of the methods described in the kinematic section.
- The background coming from non–DIS events has to be subtracted. An example of such a background could be photoproduction events, where the scattered electron remained undetected in the beam pipe, in which a π^0 was produced which decayed into two photons, one of which produced an electromagnetic shower and was mistakenly identified as an electron. This background can be measured in certain kinematic regions and has to be estimated in others.
- Using the luminosity measurements one gets the measured cross section, from which one has to unfold the one given by (3.72).
- One needs to apply the radiative corrections to get the Born cross section. In order to be able and do the calculations, one needs good parameterizations of structure functions at lower Q^2 down to the photoproduction region.
- Finally, in order to get the structure function F_2, one needs information on F_L, or equivalently on R_L. This ratio was measured [51, 52] in some of the fixed target experiments but is limited to relatively high x values. There is no measurements on R in the HERA x range and so far one needs to rely on its estimates calculated from QCD. This is one of the sources of the systematic errors of F_2.

We have already shown the latest results of F_2 as measured in the whole kinematic region, as function of x, for fixed values of Q^2 (Fig. 8), and which show the dramatic rise of the structure function as x decreases. Here we show the values of F_2 as function of Q^2, for fixed values of x (Fig. 25). In addition to the HERA measurement, the figure includes data from some fixed target experiments. A clear scale breaking with Q^2 is observed. The curves are the results of a QCD fit to the data.

3.5 Summary

In this chapter we discussed the following issues:

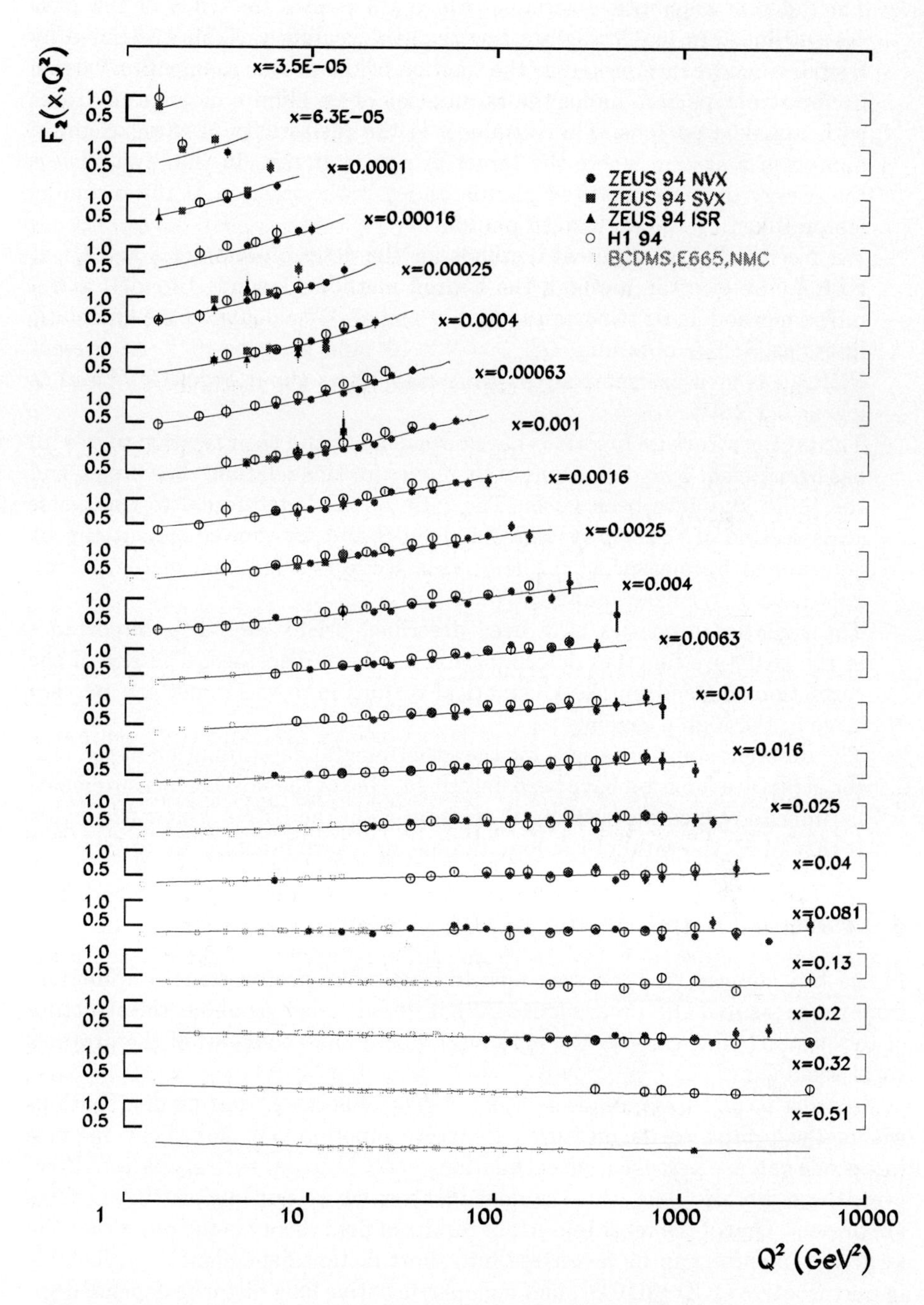

Fig. 25. *The structure function $F_2(x, Q^2)$ as function of Q^2, for fixed x values.*

- The different kinematical variables which are used in the study of DIS have been defined. In particular, we saw that the scaling variable x, defined by Bjorken, can be interpreted as the fraction of the proton momentum carried by the struck parton, under the assumption of an infinite momentum frame with massless partons. The variable y is the inelasticity of the exchanged photon in a system where the target proton is at rest. In that system ν is the energy of the exchanged photon and y is the fraction of the incoming lepton taken by the exchanged photon.
- We described three different methods for the determination of x and Q^2 at HERA: the electron method, the hadron method (Jacquet–Blondel) and a mixed method using two angles (double–angle). We calculated the kinematic limits at HERA obtaining $Q^2_{max} = 9 \times 10^4$ and $x_{min} \approx 10^{-6}$. At present HERA has measurements at Q^2 values as high as about 10000 GeV^2 and as low as 0.2 GeV^2.
- The proton structure functions have been defined and their relation to $\sigma(\gamma^* p)$ has been given. Two definitions of the fluxes for this relation, the Gilman and the Hand one have been given. The ratio of the longitudinal to transverse cross section of the $\gamma^* p$ system was defined and we showed that it can be determined by measuring the DIS cross section as function of the photon polarization, for fixed ν and Q^2.
- The radiative processes have been described briefly and their importance in the structure function determination has been discussed. The size of the corrections depend on the kinematical regions in x and y and can be very large in the high y region.
- The different steps necessary for the experimental determination of the proton structure function have been described. One of the missing measurements in order to reduce the systematic error on F_2 in the HERA kinematic region is that of R, the ratio of the longitudinal structure function F_L to F_2.

4 Parton Distributions in the Proton

In the last chapter we have seen how to obtain the proton structure function from the measured DIS cross section. What does it teach us about the structure of the proton? How can one use it to learn about the behaviour of the proton's constituents?

In order to do so, one needs a theory which connects the parton distributions within the proton to its measured structure function [53]. For short distance forces one can use perturbative calculations in QCD to get such a relation. However, there are also long distance dependencies where non–perturbative effects are present. One of the most important results of field theory is the proof that the structure function can be factorized into short distant dependencies, calculable in perturbative QCD (pQCD), and non–perturbative long distance dependences. This QCD factorization theorem will be the subject of the first section.

We will proceed with the DGLAP evolution equations of partons and discuss the structure function in the low–x region. This will lead to the BFKL evolution equation and to a short discussion about saturation. The last two sections will be devoted to parameterizations of parton distributions in the proton and to experimental determination of the gluon density in the proton.

4.1 The QCD Factorization Theorem

The QCD factorization theorem [54] discusses the situation where one measures an inclusive reaction, like the NC DIS one: $ep \to eX$. In this case one can prove that the structure functions can be factorized into short-distance dependences calculable in pQCD and into long-distance dependences which need to be taken from outside the theory. If we denote by F_a^{Vh} the structure function for a hadron h which is probed by a vector V, where a can be 1, 2 or 3 (in case of Z^0 exchange) the QCD factorization theorem allows to write the following equations:

$$F_{1,3}^{Vh}(x,Q^2) = \sum_{f,\bar{f},g} \int_x^1 \frac{\mathrm{d}z}{z} C_{1,3}^{Vi}\left(\frac{x}{z}, \frac{Q^2}{\mu^2}, \frac{\mu_\mathrm{F}^2}{\mu^2}, \alpha_S(\mu^2)\right) f_{\mathrm{i/h}}(z,\mu_F,\mu^2) \quad (4.1)$$

$$F_2^{Vh}(x,Q^2) = \sum_{f,\bar{f},g} \int_x^1 \mathrm{d}z\, \mathrm{C}_2^{\mathrm{Vi}}\left(\frac{\mathrm{x}}{\mathrm{z}}, \frac{\mathrm{Q}^2}{\mu^2}, \frac{\mu_\mathrm{F}^2}{\mu^2}, \alpha_\mathrm{S}(\mu^2)\right) \mathrm{f}_{\mathrm{i/h}}(\mathrm{z},\mu_F,\mu^2) \quad (4.2)$$

The coefficient functions C_a^{Vi} are independent of long distance effects and are computable in pQCD. The functions $f_{i/h}$ are the parton distribution functions which are specific to the hadron h but are universal as far as a and V are concerned. They are not calculable in pQCD but have to be measured experimentally.

There are two mass scales in the problem. One is the renormalization scale μ. The other one is the factorization scale μ_F. If we denote by k^2 the off–shellness, then for $k^2 > \mu_\mathrm{F}^2$ one has the coefficient functions while for $k^2 < \mu_\mathrm{F}^2$ one has the parton distribution functions. This can be pictured for the DIS case in the following diagram:

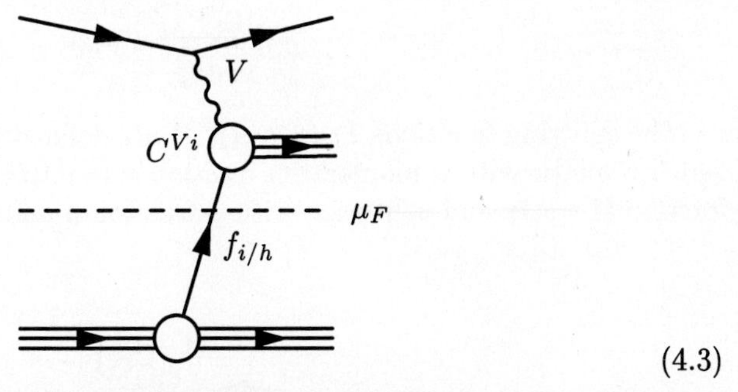

(4.3)

In the absence of any interaction (γ^* is absorbed by the quark i which continues) these functions are, in leading order, the following:

$$C_a^{\gamma i(0)}(x) = e_i^2\delta(1-x) \quad (4.4)$$

$$f_{i/i}^{(0)}(z) = \delta(1-z) \tag{4.5}$$

Beyond the leading order, there is considerable ambiguity and one has to specify in which scheme one works. There are usually two different schemes: the DIS and the $\overline{\text{MS}}$ schemes. In the DIS scheme, order by order in perturbative theory all corrections to F_2^{Vh} are absorbed into the distribution functions of q and $\bar{q}$ (for $\mu = \mu_F = Q$):

$$C_2^{Vq}(x) = e_q^2\delta(1-x) \tag{4.6}$$

$$C_2^{V\bar{q}}(x) = e_{\bar{q}}^2\delta(1-x) \tag{4.7}$$

$$C_2^{Vg}(x) = 0 \tag{4.8}$$

The $\overline{\text{MS}}$ scheme (modified minimal subtraction) follows from the idea of dimensional regularization by 'tHooft and Veltman [55].

4.2 The QCD Evolution Equation for Partons

Though, as we noted above, the parton distribution functions can not be calculated by pQCD, the theory provides a way to predict how these distributions should evolve with the scale Q^2, once they are given at an initial scale. Before we proceed to describe these evolution equations, we would like to note that the QCD factorization theorem was proven only in leading twist and thus the expression for the structure function includes also higher twist terms (note that we have somewhat simplified the notation):

$$F(x,Q^2) = \sum_i \int_x^1 \mathrm{dz}\, \mathrm{C_i}\left(\frac{\mathrm{x}}{\mathrm{z}}, \alpha_{\mathrm{S}}(\mu_{\mathrm{F}}^2), \frac{\mathrm{Q}^2}{\mu_{\mathrm{F}}^2}\right) \mathrm{f_i}(\mathrm{z}, \mu_{\mathrm{F}}^2) + \mathcal{O}\left(\frac{\Lambda^2}{\mathrm{Q}^2}\right) \tag{4.9}$$

where Λ is the QCD scale.

F is a measurable quantity and therefore can not depend on μ_F:

$$\mu_{\mathrm{F}}^2 \frac{\mathrm{dF(x, Q^2)}}{\mathrm{d}\mu_{\mathrm{F}}^2} = 0 \tag{4.10}$$

The splitting functions $P_{ij}(z, \alpha_S(\mu^2))$ are defined to represent the process in which a quark with a momentum fraction x radiates a parton of a momentum fraction $(1-z)x$ and continues with a fraction momentum zx:

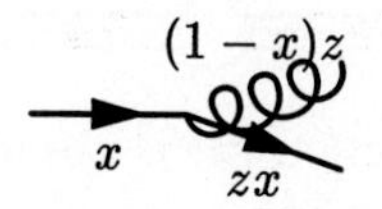

(4.11)

Then one can write the following equation for the parton distribution function f_i:

$$\frac{\mathrm{d}f_i(\mathrm{x},\mu^2)}{\mathrm{d}\ln\mu^2} = \sum_j \int_x^1 \frac{\mathrm{d}z}{z} P_{ij}(z, \alpha_S(\mu^2)) f_i(\frac{x}{z}, \mu^2) \tag{4.12}$$

This set of integro–differential equations is named DGLAP [56, 57, 58] after Dokshitzer, Gribov, Lipatov, Altarelli and Parisi.

The splitting functions P_{ij} give the probability of finding parton i in parton j. They can be expanded in orders of $(\alpha_S/2\pi)$:

$$P_{ij}(z,\alpha_S) = \sum_{n=1}^{\infty} \left(\frac{\alpha_S}{2\pi}\right)^n P_{ij}^{(n-1)}(z) = \frac{\alpha_S}{2\pi} P_{ij}^{(0)}(z) + \left(\frac{\alpha_S}{2\pi}\right)^2 P_{ij}^{(1)}(z) + \ldots \tag{4.13}$$

A similar expansion can be written for the coefficient functions C_i:

$$C_i(z,\alpha_S) = \alpha_S^p \left[C_i^{(0)} + \frac{\alpha_S}{2\pi} C_i^{(1)}(z) + \left(\frac{\alpha_S}{2\pi}\right)^2 C_i^{(2)}(z) + \ldots \right] \tag{4.14}$$

where the value of p depends on the initial process.

The partonic picture of the pQCD evolution for F_2 can be represented diagramatically by the following picture:

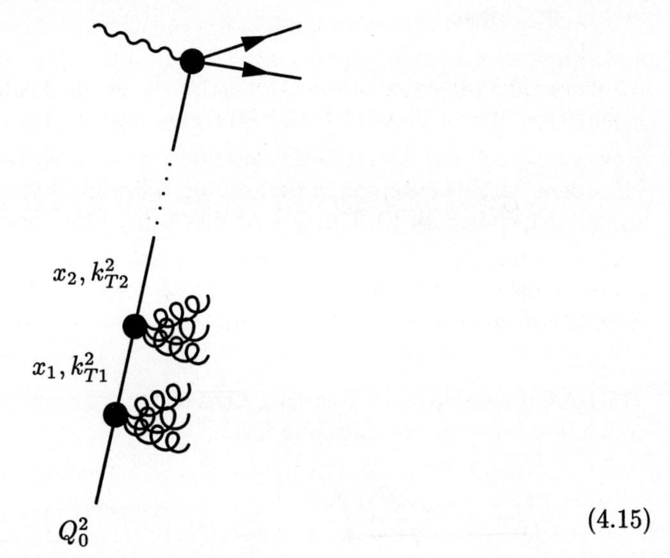

Each blob in the chain has a structure like this:

$$P(z,\alpha_S) = \frac{\alpha_S}{2\pi} P^{(0)} + \left(\frac{\alpha_S}{2\pi}\right)^2 P^{(1)} + \ldots \tag{4.16}$$

There is strong ordering in the transverse momenta k_T of each leg. If there are m steps in the chain, each having a transverse momentum square of k^2_{Ti}, then:

x_m, k^2_{Tm}

x_3, k^2_{T3}

x_2, k^2_{T2}

x_1, k^2_{T1}

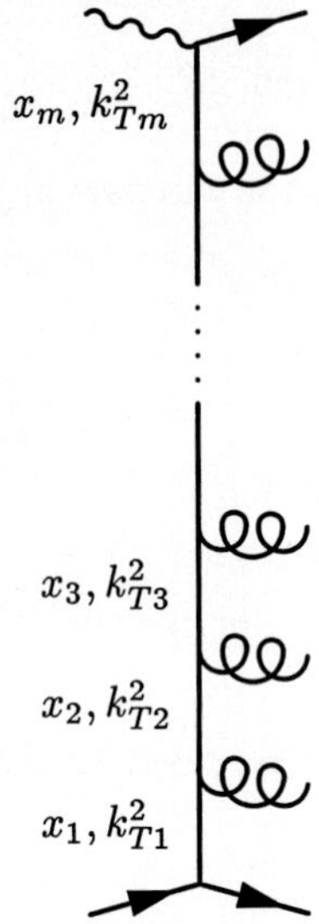

DGLAP

$$x_1 > x_2 > x_3 \ldots > x_m = x \tag{4.17}$$

$$k^2_{T1} \ll k^2_{T2} \ll \ldots \ll k^2_{Tm} = Q^2$$

If we take only the terms $C^{(0)}$ and $P^{(0)}$, we do leading order perturbation calculation. The terms $C^{(1)}$ and $P^{(1)}$ give next to leading order calculations. However by solving the DGLAP equations, we sum up the terms $(\alpha_S \ln \frac{Q^2}{\Lambda^2})^m$ to all orders. In this case one is performing a leading log approximation in $\ln Q^2$, usually shortened as LLA($\ln Q^2$) or just LLA. One should note that the splitting functions $P^{(n)}_{ij}(x) \sim (1/x) \ln x^{n-1}$, which gets logarithmically enhanced at low x. Thus, if one takes only the first two terms, $n = 1, 2$, one is restricted to moderate x only.

DGLAP Equations in Leading Order One can write the DGLAP equations in leading order in the following form:

$$\frac{\mathrm{dq_i(x, Q^2)}}{\mathrm{d\ln Q^2}} = \frac{\alpha_S(Q^2)}{2\pi} \int_x^1 \frac{\mathrm{dz}}{z} \left[\sum_j q_j(z, Q^2) P^{(0)}_{ij}\left(\frac{x}{z}\right) + g(z, Q^2) P^{(0)}_{ig}\left(\frac{x}{z}\right) \right] \tag{4.18}$$

$$\frac{\mathrm{dg(x, Q^2)}}{\mathrm{d\ln Q^2}} = \frac{\alpha_S(Q^2)}{2\pi} \int_x^1 \frac{\mathrm{dz}}{z} \left[\sum_j q_j(z, Q^2) P^{(0)}_{gj}\left(\frac{x}{z}\right) + g(z, Q^2) P^{(0)}_{gg}\left(\frac{x}{z}\right) \right] \tag{4.19}$$

The parton which is probed at the scale of Q^2 by the virtual vector meson, has a fractional momentum x which is the result of a chain of splitting which started from the parent parton. Since the longitudinal momentum of the daughter parton is always smaller or equal to that of the parent one, the integration is restricted to $x \leq z \leq 1$. Also if one neglects the masses of the quarks, the change in the distribution function depends only on the ratios of longitudinal momenta x/z. The sum j is over the quark flavours, and one assumes that the splitting functions are flavour independent. Thus:

$$P_{ij}^{(0)} = \delta_{ij} P_{ii} = \delta_{ij} P_{gg} \tag{4.20}$$

$$P_{gj} = P_{gq} \qquad P_{ig} = P_{qg} \tag{4.21}$$

Conservation of momentum for a parent quark and a gluon gives:

$$\int_0^1 \mathrm{dz}\,\mathrm{z}\,[\mathrm{P_{qq}(z)} + \mathrm{P_{gq}(z)}] = 0 \tag{4.22}$$

$$\int_0^1 \mathrm{dz}\,\mathrm{z}\,[2\mathrm{n_f}\mathrm{P_{qg}(z)} + \mathrm{P_{gq}(z)}] = 0 \tag{4.23}$$

where n_f is the number of flavours, and

$$P_{qq}(z) = P_{gq}(1-z) \tag{4.24}$$

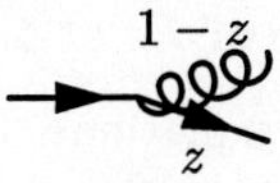

1 − z

z

$$P_{qg}(z) = P_{qg}(1-z) \tag{4.25}$$

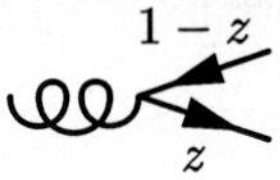

1 − z

z

$$P_{gg}(z) = P_{gg}(1-z) \tag{4.26}$$

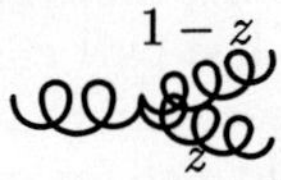

1 − z

z

In leading order the quark–parton model (QPM) relations between structure functions and parton distributions hold. So it is easy to derive the evolution equations for the structure functions. One usually defines the colour singlet combination, which evolves with gluons:

$$x\Sigma(x) = \sum_{i=1} n_f\,[xq_i(x) + x\bar{q}_i(x)] \tag{4.27}$$

The colour non–singlet combination, which evolves with quarks, is:

$$xV(x) = \sum_{i=1} n_f \left[xq_i(x) - x\bar{q}_i(x)\right] \tag{4.28}$$

Leading order splitting function: For completeness we give her the explicit form of the leading order splitting functions:

$$P^{(0)}_{qq}(x) = \frac{4}{3}\left[\frac{1+x^2}{(1-x)_+} + \frac{3}{2}\delta(1-x)\right] \tag{4.29}$$

$$P^{(0)}_{qg}(x) = (1-x)^2 + x^2 \tag{4.30}$$

$$P^{(0)}_{gq}(x) = \frac{4}{3}\frac{(1-x)^2+1}{x} \tag{4.31}$$

$$P^{(0)}_{gg}(x) = 6\left[\frac{x}{(1-x)_+} + \frac{1-x}{x} + x(1-x)\right] + \left[\frac{11}{2} - \frac{n_f}{3}\right]\delta(1-x) \tag{4.32}$$

where we used the following definition:

$$\int_0^1 \mathrm{dz}\,\frac{\mathrm{f(z)}}{(1-\mathrm{z})_+} \equiv \int_0^1 \mathrm{dz}\,\frac{\mathrm{f(z)}-\mathrm{f(1)}}{(1-\mathrm{z})} \tag{4.33}$$

4.3 The Behaviour of F_2 at Low x

What do the DGLAP equations tell us about the behaviour of the structure function at low x? We have already seen earlier that the structure function F_2^{ep} is connected to the $\gamma^* p$ cross section via the relation:

$$F_2^{ep}(x,Q^2) = \frac{K}{4\pi\alpha}\frac{Q^2\nu}{Q^2+\nu^2}\left(\sigma_T^{\gamma^* p} + \sigma_L^{\gamma^* p}\right) \tag{4.34}$$

The variable x can be expressed as:

$$x = \frac{Q^2}{Q^2+W^2-m_p^2} \tag{4.35}$$

and since we are discussing the region of low x, this means high W.

We have seen in the last part of the chapter 2 on Regge theory that the total photoproduction cross section behaves at high energies like $\sigma_{\mathrm{tot}}(\gamma p) \sim (W^2)^{0.08}$. Does this behaviour hold also for $\sigma_{\mathrm{tot}}(\gamma^* p)$? If it were so, this would mean that at low x we expect $F_2 \sim x^{-0.08}$. However, a look at the experimental data shown in Fig. 8 shows [59, 16] that $F_2 \sim x^{-0.3\div 0.4}$. Can such a behaviour be expected from the evolution equations which we presented in the last section?

Let us write again the DGLAP equations in a shorter notation:

$$\frac{\mathrm{dq_i}}{\mathrm{d}\ln \mathrm{Q}^2} = \frac{\alpha_S}{2\pi}\left[P_{qq}\otimes q_i + P_{qg}\otimes g\right] \tag{4.36}$$

$$\frac{\mathrm{dg}}{\mathrm{d}\ln \mathrm{Q}^2} = \frac{\alpha_S}{2\pi}\left[P_{gq}\otimes q_i + P_{gg}\otimes g\right] \tag{4.37}$$

but since

$$P_{gg}^{(0)}(x) = 6\left[\frac{x}{(1-x)_+} + \frac{1-x}{x} + x(1-x)\right] \tag{4.38}$$

$$P_{gq}^{(0)}(x) = \frac{4}{3}\frac{(1-x)^2+1}{x} \tag{4.39}$$

we see that gluons are produced most copiously at low x. Since the q_i are small at low x, the gluon evolution equation can be approximated by:

$$\frac{\mathrm{dg(x, Q^2)}}{\mathrm{d\ln Q^2}} \simeq \frac{\alpha_S(Q^2)}{2\pi}\int_x^1 \frac{\mathrm{dz}}{z} P_{gg}(\frac{x}{z}) g(z, Q^2). \tag{4.40}$$

Using $P_{gg}^{(0)} z \simeq 6/z$ one get the so–called double leading log approximation, where only terms proportional to $\ln\frac{1}{x}\ln Q^2$ are taken:

$$xg(x, Q^2) \sim \exp\left[\frac{48}{11-\frac{2}{3}n_f}\ln\frac{\ln\frac{Q^2}{\Lambda^2}}{\ln\frac{Q_0^2}{\Lambda^2}}\ln\frac{1}{x}\right]^{1/2} \tag{4.41}$$

where Q_0^2 is the starting scale for the Q^2 evolution. Numerically this expression has the same value as $\approx x^{-0.4}$.

This result however has a few problems: (1) it violates unitarity and (2) since in general the functions $P_{ab}^{(n)} \sim \frac{1}{x}\left(\ln^{n-1} x + \mathcal{O}(\ln^{n-2} x)\right)$ the series does not converge and thus higher orders are needed.

4.4 The BFKL Evolution Equation

The DGLAP equations give us a way to see how a parton distribution which is given at an initial scale Q_0^2 evolves to higher Q^2. Since Q^2 increases one has to resume the leading $\alpha_S \ln Q^2$ terms. When the *ep* center of mass energy is large, like at HERA, there is a second variable which becomes large, namely $1/x \sim s/Q^2$. In this case one has to resum the leading $\alpha_S \ln(1/x)$ contributions. The BFKL [60, 61, 62] equations do such a resummation. As one evolves to smaller x, there is no more strong ordering in k_T^2. The strong ordering is rather in x and therefore one considers here the unintegrated (over the gluon transverse momentum) gluon density distribution.

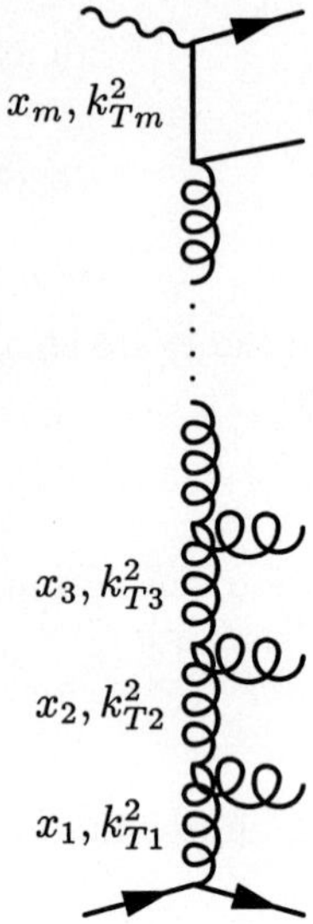

x_m, k_{Tm}^2

x_3, k_{T3}^2

x_2, k_{T2}^2

x_1, k_{T1}^2

BFKL (only gluon–gluon ladder)

$x_1 \gg x_2 \gg x_3 \ldots \gg x_m = x$

no ordering in k_{Ti}^2 (4.42)

(assume no evolution in Q^2)

$k_{T1}^2 \simeq Q^2$

At a given Q^2, gluons have a distribution in x and k_T, $f_g(x, k_T)$, which is related to $xg(x, Q^2)$ through:

$$xg(x, Q^2) = \int_0^{Q^2} \frac{\mathrm{d}k_T^2}{k_T^2} f_g(x, k_T^2) \tag{4.43}$$

Summing up all ladder diagrams in $\ln(1/x)$ gives the BFKL equation:

$$-x\frac{\partial f_g(x, k_T^2)}{\partial x} = 3\frac{\alpha_S}{\pi} k_T^2 \int_0^\infty \frac{\mathrm{d}k_T'^2}{k_T'^2} \left[\frac{f_g(x, k_T'^2) - f_g(x, k_T^2)}{|k_T'^2 - k_T^2|} + \frac{f_g(x, k_T^2)}{\sqrt{4k_T'^4 + k_T^4}} \right] \equiv$$
$$\equiv K \otimes f_g \tag{4.44}$$

where K is the BFKL kernel. Note that this equation relates only to gluon distributions, as also seen from the diagram. It does not discuss the Q^2 evolution. It is an equation which describes what happens to the gluon distribution when one starts from a distribution at x_0 (of the order of $\approx$ 0.01) and evolves to smaller x values.

The solution of the BFKL equation, after integrating over the gluon k_T, has an x dependence like:

$$xg(x, Q^2) \sim x^{-\lambda} \tag{4.45}$$

where, for fixed α_S, can be expressed as $\lambda = (3\alpha_S/\pi)4\ln 2$. Since $x^{-\lambda} \sim s^\lambda$, one obtains an energy dependence of the gluon density which is expected from the Regge theory at high energies. Thus BFKL succeeded to 'reggeize' the gluon and provide a connection between QCD and Regge theory. The exponent λ is usually said to be 0.5, though this requires $\alpha_S = 0.18$, which happens only at high Q^2.

If we assume that k_T is large, one gets a Q^2 dependence of the form:

$$xg(x, Q^2) \sim \sqrt{\frac{Q^2}{Q_0^2}} x^{-\lambda}, \tag{4.46}$$

which gives a stronger scaling violation than the one from the DGLAP equations ($\sim \ln Q^2$). Note that this solution is obtained in the leading log approximation (LLA) in $\ln(1/x)$.

These equations also have some problems: (1) the LLA solution violates unitarity, (2) higher order corrections are not yet available, (3) the integration on k_T starts from 0, thus one enters also the non-perturbative region, and (4) the equation doesn't have implicitly energy–momentum conservation.

Some Consequences from the BFKL Equation

BFKL Pomeron: We have seen in the Regge theory chapter 2 that at high energies the total cross section behaves like:

$$\sigma \sim s^{\alpha_{\mathbb{P}}(0)-1} \tag{4.47}$$

Since the cross section at high energies (low x) is driven by the gluons, and since the x dependence of the gluons is $\sim x^{-\lambda} \sim s^{\lambda}$, the Pomeron intercept comes out to be in this case:

$$\alpha_{\mathbb{P}}(0) = 1 + \lambda \approx 1.5 \tag{4.48}$$

in contrast to the result of $\alpha_{\mathbb{P}}(0) = 1.08$ as obtained by the analysis of Donnachie and Landshoff. Therefore one talks about the BFKL Pomeron which has an intercept of 1.5 and a DL Pomeron of intercept 1.08. Other names used in the literature are 'hard', 'perturbative', 'Lipatov' Pomeron (1.5) and the 'soft' Pomeron (1.08). We will discuss at a later stage the question of one or two Pomerons.

Hot spots: As we said above, the BFKL equations treat only the evolution in x. Since there is no evolution in Q^2, this means that the transverse area is fixed. Let us look at the schematic presentation of the proton with some partons inside, as shown in Fig. 26. When we evolve to lower x, the number of partons increases in a fixed area, leading to an increase in the local density. This phenomena is named 'hot spots' [63]. It still has to be seen at which value of x this should happen.

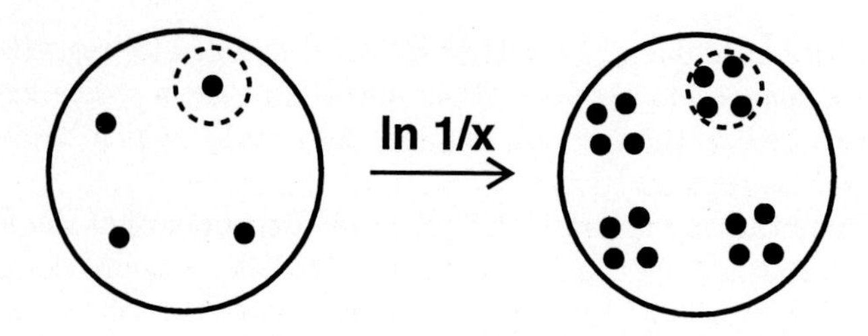

Fig. 26. *Diagram showing the increase in local density in an evolution in x.*

Jet in the proton region: In the DGLAP picture, with the strong k_T ordering, one expects that the large k_T jet would be near the γ^*, while near the proton direction there will be just the low k_T remnant jet. Since there is no strong k_T ordering in the BFKL dynamics, one can have a situation where the large k_T jet would be near the proton, in the proton direction, which will be balanced by a jet in the γ^* direction.

Signs for the BFKL Dynamics The behaviour of the structure function F_2 is not sensitive enough to tell the difference between the DGLAP dynamics and the BFKL one, at least not in the HERA kinematic region. Also both are compatible with a gluon density behaviour of $xg(x,Q^2) \sim x^{-\lambda}$ at low x. How then do we tell a BFKL type dynamics form a DGLAP one?

Let us look at the schematic presentation in Fig. 27. We start from a point

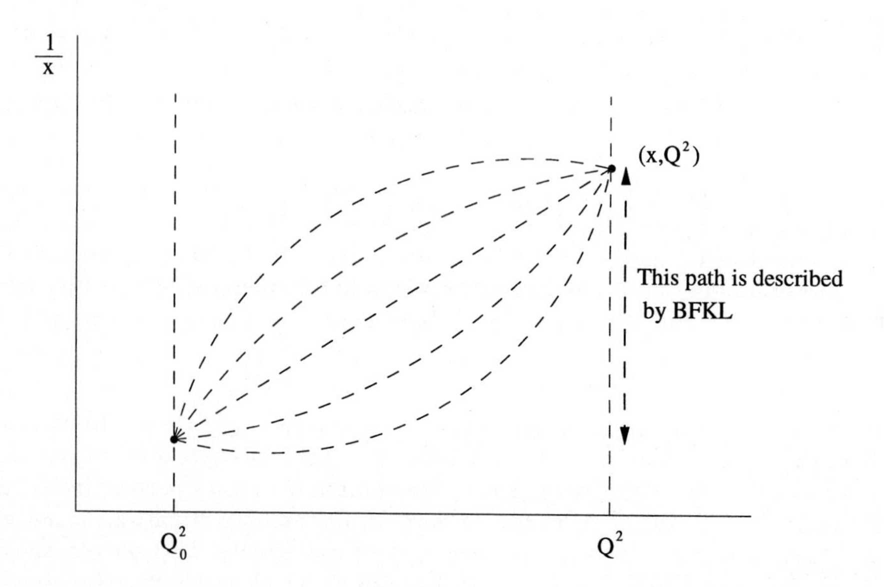

Fig. 27. *Diagram showing the evolution from point (x_0, Q_0^2) to (x, Q^2).*

with the coordinates (x_0, Q_0^2) and evolve to a point (x, Q^2). We can get from the one to the other in several ways. While the DGLAP equations describe the motion in the whole plain, if we give the initial conditions, the BFKL equation describes only the path along the $1/x$ axis, at a constant Q^2.

Clearly, the best way to see BFKL dynamics is to restrict the Q^2 evolution. Mueller and Navallet [64] pointed out that at low x one should look for a large transverse momentum jet near the proton direction. The large transverse momentum, compared to Q^2, guarantees a large k_T at the start of the evolution, thus forcing the rest to be an evolution only in x. Referring to Fig. 27 this would mean that we got in one step from (x_0, Q_0^2) to (x_0, Q^2). The rest would be the evolution from x_0 to x. Studying the energy behaviour of the cross section for such events and finding it rises steeply, like say $s^{0.5}$, should be a sign for the BFKL dynamics.

4.5 The CCFM Equation

For sake of completeness, we should mention the existence of a unified equation developed by Catani, Ciafaloni, Fiorani and Marchesini (CCFM) [65, 66, 67]. The CCFM equation, give the BFKL solution at low x and the DGLAP one at large x. It is based on the coherent radiation of gluons which leads to a strong angular ordering of gluon emissions.

4.6 Saturation

When the density of the partons becomes very large, the partons start overlapping and coherent effects are important. The partons interact with each other. When do these effects become important? In order to answer this, let us first explain what one means by the density of partons.

The quantity $xg(x, Q^2)$ is the gluon density per unit of rapidity. In order to see that lets start from the definition of rapidity. The rapidity y is :

$$y = \frac{1}{2}\ln\frac{E+p_z}{E-p_z} = \frac{1}{2}\ln\frac{(E+p_z)^2}{E^2-p_z^2} = \frac{1}{2}\ln\frac{(E+p_z)^2}{m_T^2} = \ln\frac{(E+p_z)}{m_T} \simeq \ln\frac{2p_z}{m_T} \tag{4.49}$$

The rapidity Δy of a parton with momentum $p_{zi} = xp$ relative to the proton is then

$$\Delta y = y_{\text{proton}} - y_{\text{parton}} = \ln\frac{2p}{m_T} - \ln\frac{2xp}{m_T} \simeq \ln\frac{2p}{2xp} = \ln\frac{1}{x} \tag{4.50}$$

and thus $\mathrm{dy} = \mathrm{dx/x}$. The number of gluons $\mathrm{dN_g}$ is given by

$$\mathrm{dN_g = g(x, Q^2)dx = xg(x, Q^2)\frac{dx}{x} = xg(x, Q^2)dy} \tag{4.51}$$

Therefore:

$$xg(x, Q^2) = \frac{\mathrm{dN_g}}{\mathrm{dy}} \tag{4.52}$$

meaning the number of gluons per unit of rapidity.

Let us estimate now the sizes of the gluon and of the proton. The proton size is usually taken to be $r_p \sim 1\,\mathrm{fm} \sim 5\,\mathrm{GeV}^{-1}$. The gluon radius is $r_g \sim 2/Q$, when the proton is probed at a scale of Q^2. The screening effects become important when the gluon density is of the order of the ratio of the square radiuses of the proton and the gluon:

$$xg(x,Q^2) \sim \frac{r_p^2}{r_g^2} = \frac{25\,\mathrm{GeV}}{\frac{4}{Q^2}} \simeq 6Q^2\ [\mathrm{GeV}^2] \tag{4.53}$$

In the section discussing the experimentally obtained gluon distributions we will show that for instance at $Q^2 = 20\,\mathrm{GeV}^2$, the gluon density reaches about 30 gluons per unit of rapidity at $x = 10^{-4}$. Since at this Q^2, according to (4.53) screening effects [68, 69] would be important at a density of about 120 gluons per unit of rapidity, one probably needs to go to much lower x values to observe screening.

4.7 Parton Parameterizations

In order to describe the hadronic processes at high energies it is necessary to know the individual parton distributions as function of x and Q^2. The basic formula for a generic high energy inclusive hadronic process $A + B \to C + X$ has the form:

$$\sigma(AB \to CX) = f_A^a \otimes \hat{\sigma}_{ab\to cX} \otimes f_B^b \tag{4.54}$$

where $\hat{\sigma}$ is the calculable hard cross section for the partonic subprocess, and $f_A^a(f_B^b)$ is the distribution function of parton $a(b)$ in hadron $A(B)$. In this notation, the gluon density distribution in the proton $xg(x,Q^2)$, would be f_p^g.

Since theory does not give absolute predictions for parton distributions, they have to be obtained from some experimental input and then the DGLAP equations allow to determine those parton distributions at any Q^2, even not accessible experimentally. However parton distributions are not directly measured in the experiment. It is the structure functions or hadronic cross sections that are measured.

One way of extracting the parton distributions from the measured data is based on the approach to introduce the parton distribution at the level of the global fit. It means that the structure functions are parameterized at some reference value Q_0^2 and then evolved numerically in Q^2 through the DGLAP equations in the kinematic regions where they are measured. A global fit is then performed to determine the best values for the starting parameters. A by–product of these fits performed on the singlet structure function F_2 is a parameterization of the gluon distribution at the reference scale Q_0^2. Because deep inelastic scattering does not constrain significantly the gluon distribution, a large variety of gluon behaviour is proposed in the literature. We will discuss the gluon density in a separate section.

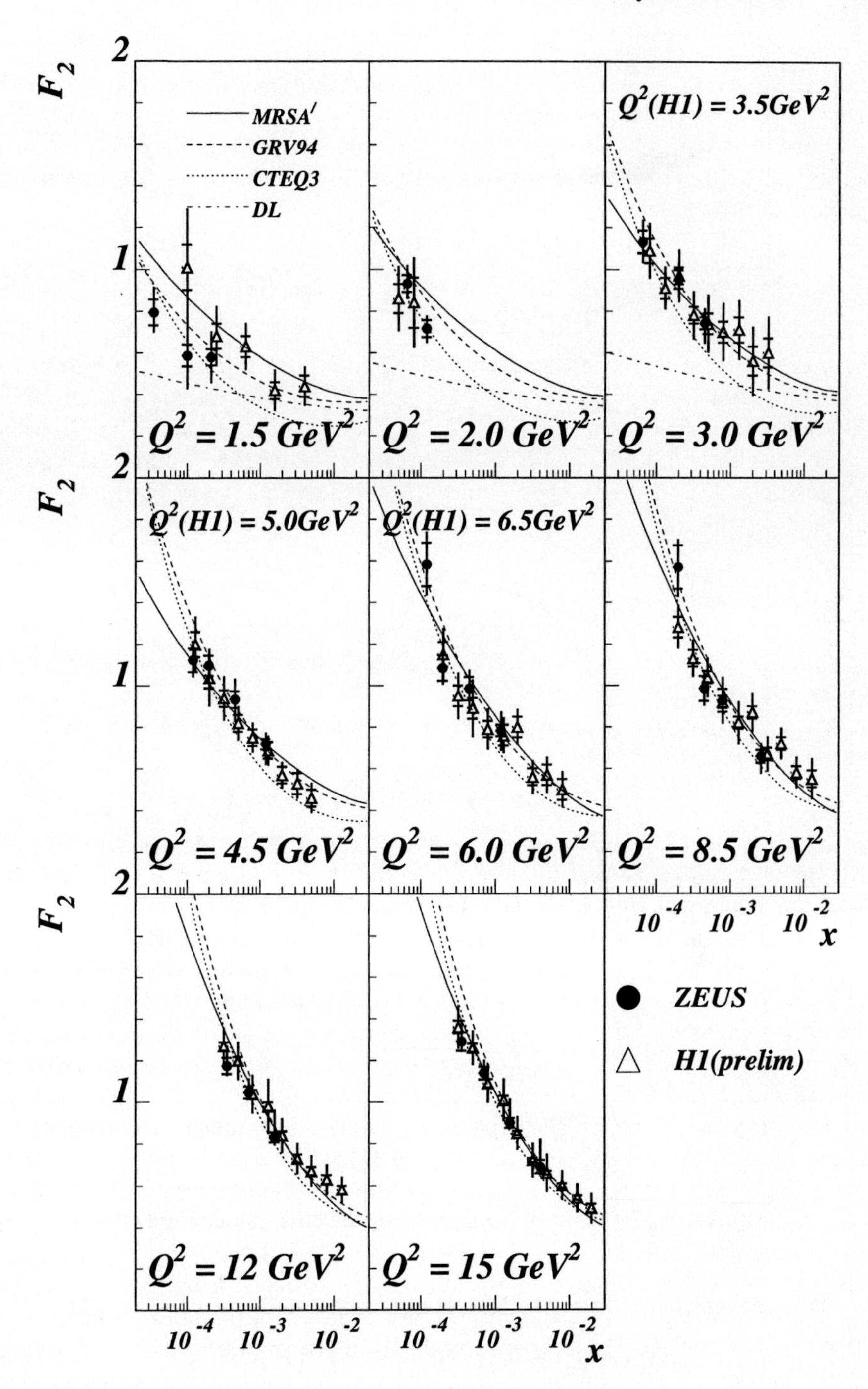

Fig. 28. *Parameterizations of parton distributions compared to ZEUS and H1 data.*

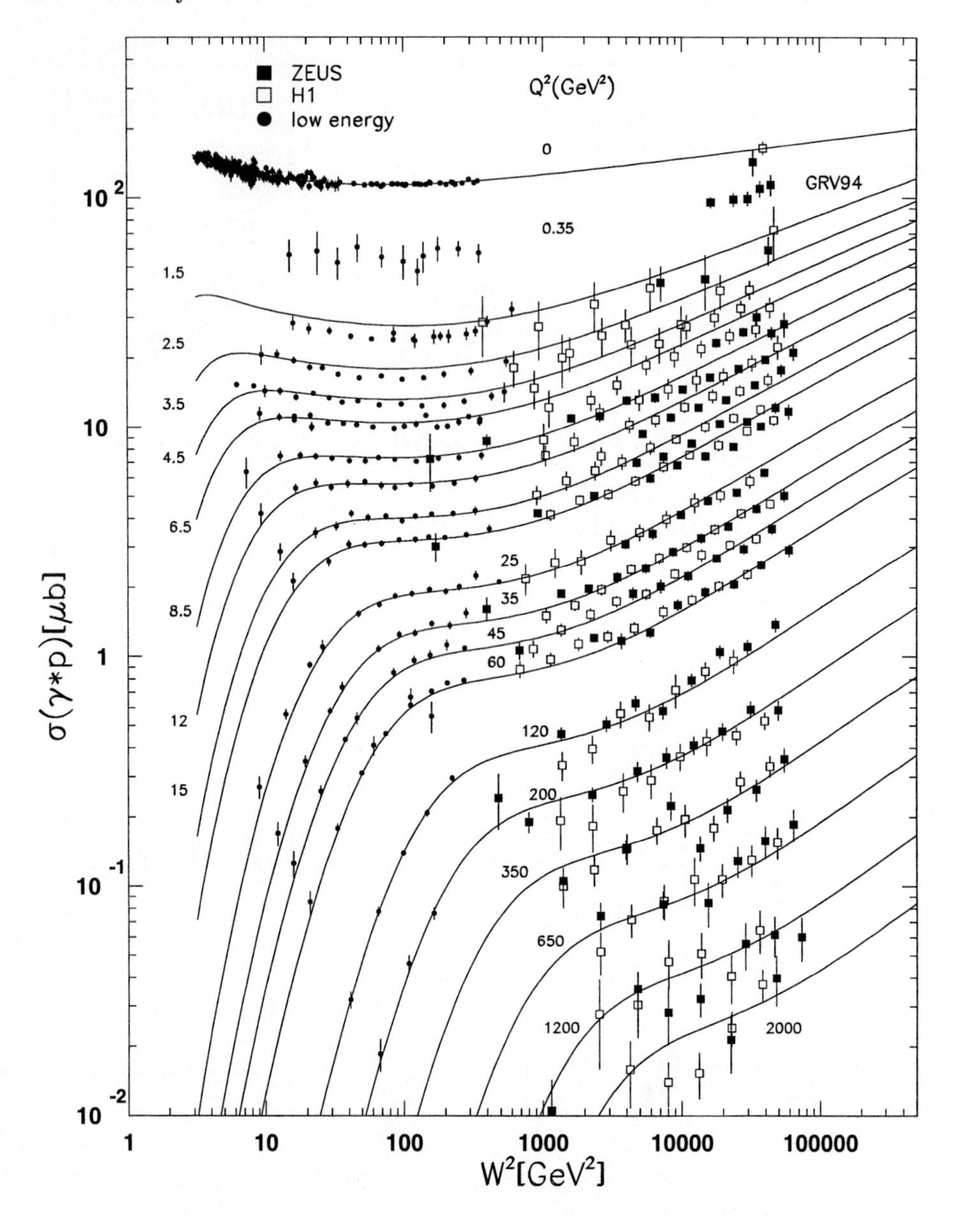

Fig. 29. *$\sigma_{tot}(\gamma^* p)$ vs W^2 compared to the GRV parameterization.*

It is conventional to use the following parameterization of $f_p^a(x, Q_0^2)$:

$$f_p^a(x, Q_0^2) = A_0^a x^{A_1^a} (1-x)^{A_2^a} P^a(x; A_3^a, \dots) \tag{4.55}$$

where $P^a(x)$ is a smooth function of x. Provided the functions are sufficiently flexible to accommodate the true distributions, the particular form of the pa-

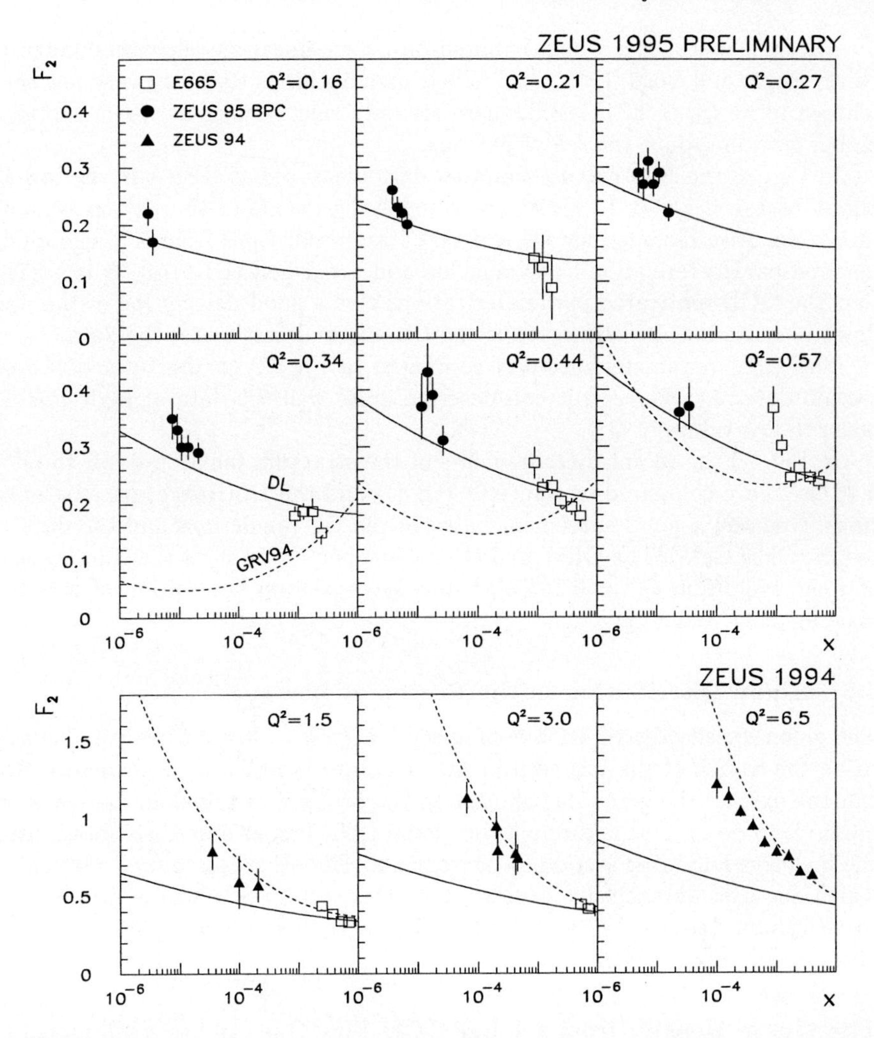

Fig. 30. *Low Q^2 measurements of the F_2 structure function at HERA, compared to the GRV and DL parameterizations.*

rameterization is, in principle, immaterial. The most frequent parameterizations used lately are those of Martin, Roberts and Stirling (MRS) [70, 71] and the CTEQ [72] collaboration, both of which use as a starting scale $Q_0^2 \simeq 4\,\mathrm{GeV}^2$. One example of a parameterization is the following:

$$xq_{\mathrm{NS}}(x, Q_0^2) = A_{\mathrm{NS}} x^{\delta_{\mathrm{NS}}} (1-x)^{\eta_{\mathrm{NS}}} \tag{4.56}$$

$$xq_{\mathrm{SI}}(x, Q_0^2) = A_{\mathrm{SI}} x^{\delta_{\mathrm{SI}}} (1-x)^{\eta_{\mathrm{SI}}} (1 + \epsilon_{\mathrm{SI}} \sqrt{x} + \gamma_{\mathrm{SI}} x) \tag{4.57}$$

$$xg(x, Q_0^2) = A_g x^{\delta_g} (1-x)^{\eta_g} \tag{4.58}$$

where NS and SI stand for the non–singlet and the singlet functions.

Another approaches which is based on a dynamical model is that taken by Gück, Reya, and Vogt (GRV) [73]. Their assumption is that at a very low scale (chosen to be $Q_0^2 \simeq 0.34\,\mathrm{GeV}^2$), there are only valence partons which evolve to higher Q^2 to produce the sea of partons.

In Fig. 28 the F_2 structure function data measured at HERA in the low Q^2 region of $1.5 < Q^2 < 15\ \mathrm{GeV}^2$ are compared to some of the parton parameterization. Also included for $Q^2 < 4\ \mathrm{GeV}^2$ are predictions from a Regge model inspired parameterization by Donnachie and Landshoff (DL) [74]. It is evident that the QCD motivated parameterizations give a good description of the data down to quite low Q^2 values, while the DL one underestimates the data.

The GRV parameterization is compared in Fig. 29 to the total $\gamma^* p$ cross section data. As one sees, it can describe quite well the data at high energies down to low values of Q^2.

In Fig. 30 the recent measurements of the structure function down to $Q^2 = 0.16\,\mathrm{GeV}^2$ are compared to the GRV (QCD) and the DL(Regge) parameterizations. One sees a good agreement between the DL predictions and the data up to $Q^2 = 0.57\,\mathrm{GeV}^2$. The GRV predictions are completely off at the starting scale of their evolution, $Q^2 = 0.34\,\mathrm{GeV}^2$, but gives a good description of the data starting at $Q^2 = 1.5\,\mathrm{GeV}^2$.

4.8 Gluon Distribution in the Proton at Low x

The gluon density distribution is of special interest at low x since it is believed to be the source of the rise seen in the structure function as x decreases. How can one extract the gluon distribution in the proton? In principle there are two methods to do so. One is through the global QCD fits, as described above, using the inclusive DIS cross section measurements. The other is a 'direct' method, in which one uses an exclusive process, the cross section of which is proportional to the gluon density. We will describe below both methods and show results obtained so far at HERA.

The Gluon Density from Global QCD Fits One can use a full global fit, using forms like in (4.58) to parameterize all the parton distributions, including the gluon one, and thus extract the gluon density distribution. At low x however, one can use the fact that the quark densities are much smaller than the gluon ones, to obtain the gluon density through approximate methods. One such method was provided both in leading order (LO) and in next to leading order (NLO) by Prytz [75, 76]:

$$\text{LO:}\quad xg(x,Q^2) \simeq \frac{\mathrm{d}F_2(\frac{x}{2},Q^2)}{\mathrm{d}\ln Q^2}\,\frac{1}{(40/27)\alpha_S/4\pi} \tag{4.59}$$

$$\text{NLO:}\quad xg(x,Q^2) \simeq \frac{\mathrm{d}F_2(\frac{x}{2},Q^2)}{\mathrm{d}\ln Q^2}\,\frac{1}{(40/27+7.96\alpha_S/4\pi)\alpha_S/4\pi} - \frac{(20/9)(\alpha_S/4\pi)N(\frac{x}{2},Q^2)}{(40/27+7.96\alpha_S/4\pi)} \tag{4.60}$$

where $N(\frac{x}{2}, Q^2)$ is a correction function which depends on the gluon density at large x ($x > 10^{-2}$), which is constrained by existing data. The resulting gluon

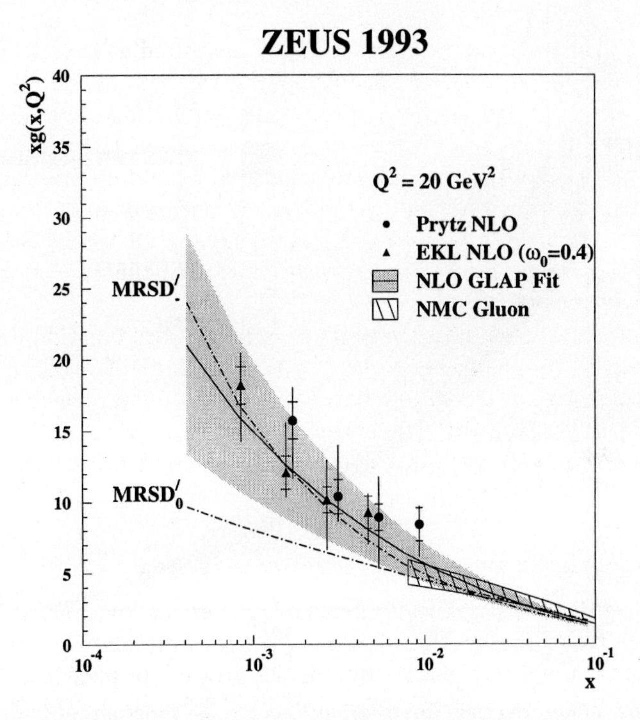

Fig. 31. *The gluon density distribution, $xg(x)$, as function of x at a fixed Q^2 of $20\,\mathrm{GeV}^2$, obtained from LO and NLO approximate methods (Prytz and EKL). The result of a global fit is shown for comparison.*

distribution extracted this way can be seen [77] in Fig. 31. In the same figure, results from another method (Ellis, Kunszt, Levin (EKL) [78]) and from a global QCD fit are shown for comparison. All methods give consistent results with each other.

A more recent extraction of both HERA experiments [16, 14], using a global QCD fit, is displayed in Fig. 32. The figure also shows how the higher statistics data yielded a narrower error band on the result.

Extracting the Gluon Density from Exclusive Processes This method is based on the fact that the cross section of some processes are proportional directly to the gluon density or to the square of the gluon density.

Two–jets production in DIS: In leading order, two jet events in DIS are produced either by photon gluon fusion or by QCD–Compton scattering, the diagrams of

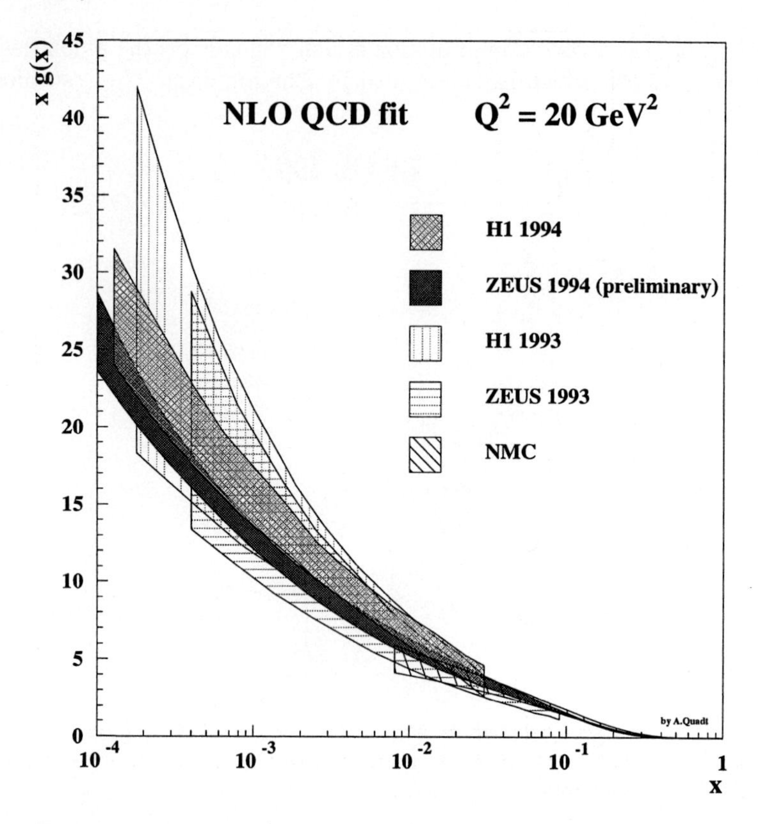

Fig. 32. *The gluon density distribution, $xg(x)$, as function of x at a fixed Q^2 of $20\,\mathrm{GeV}^2$, obtained from a NLO QCD analysis of F_2.*

which are:

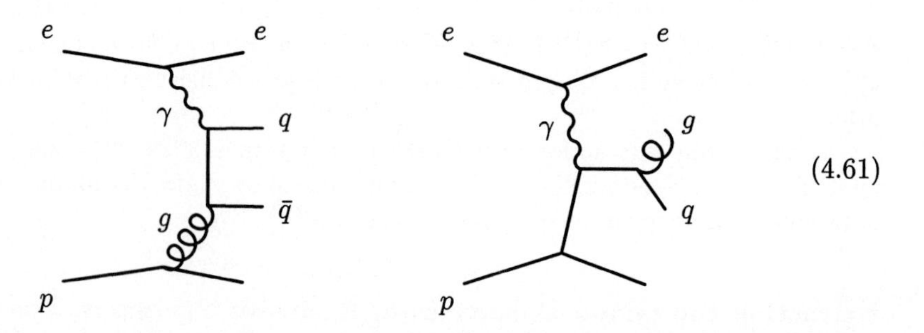

At small x the cross section is expected to be dominated by the photon–gluon fusion process. With this assumption one can extract [79] the gluon density, as shown in Fig. 33. This leading order extraction of the gluon density distribution is compared in the same figure to the results from the global QCD fits and the

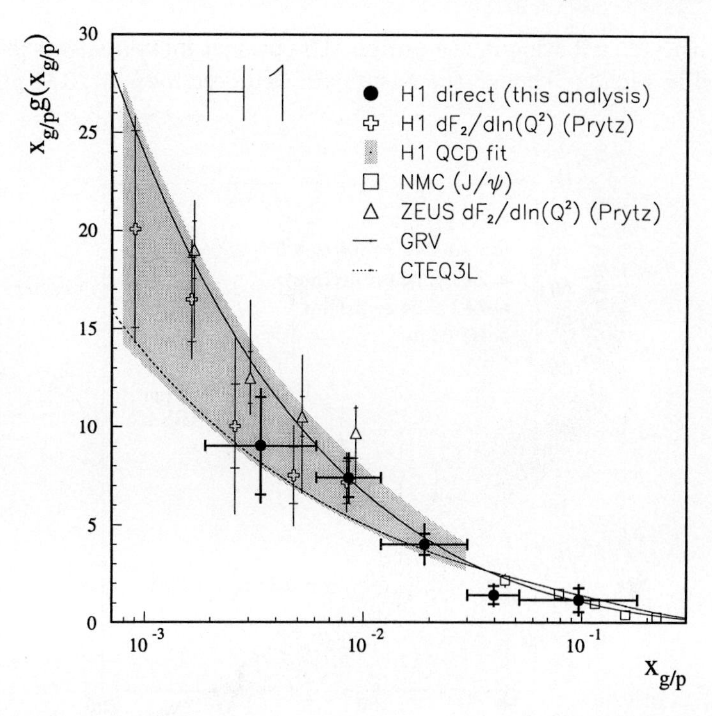

Fig. 33. *The gluon density distribution as function of x at $Q^2 = 30\,\mathrm{GeV}^2$ as determined from a leading order analysis of 2–jet events. The results are compared with those from a global QCD fit and from an approximate method.*

approximate methods, described above. Note the good agreement between the different gluon determinations, which provides a check on the universality of the gluon density.

Inelastic J/Ψ production: Inelastic J/Ψ production is the process $\gamma p \to J/\Psi X$ which is described diagramatically as follows:

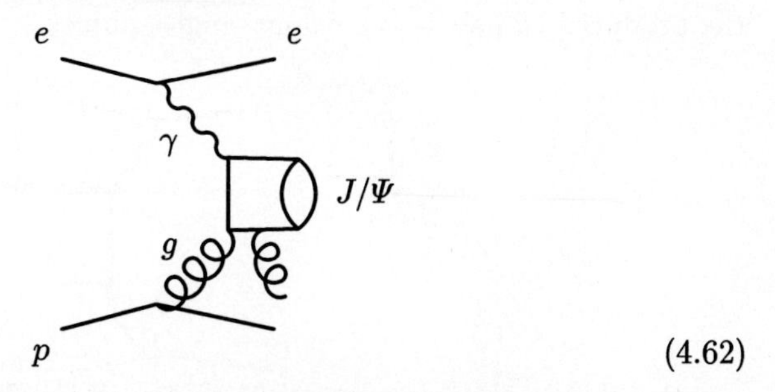

(4.62)

In this case a gluon from the proton interacts with the exchanged photon to produce a closed charm pair which radiate off a gluon and produce the colour singlet J/Ψ state. The cross section for this process is thus sensitive to the gluon density distribution in the proton. The present measurements [80, 81] are presented in Fig. 34. Though the results are well described by NLO pQCD cal-

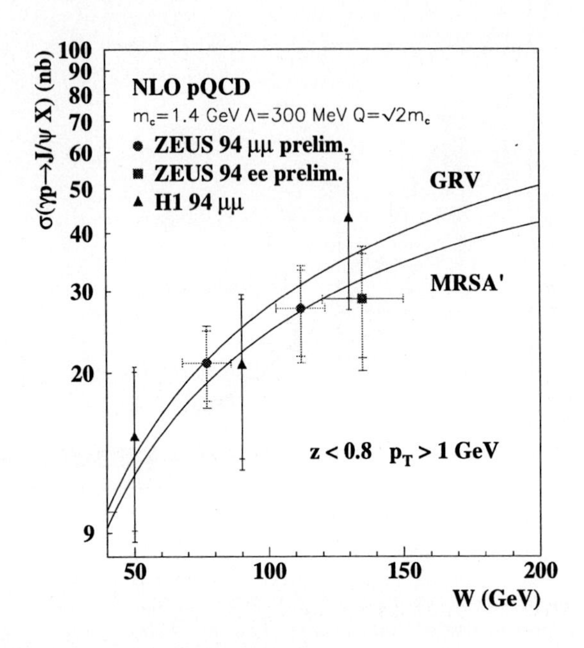

Fig. 34. *Inelastic J/Ψ cross section data as function of W compared with different choices of gluon distribution in the proton.*

culations, they are not yet precise enough to be able to distinguish between different gluon density shapes.

Open charm production: The open charm process, described in the diagram below, differs from that of the inelastic J/Ψ in that there is no combinations of the produced $c\bar{c}$ pair into a colour singlet object.

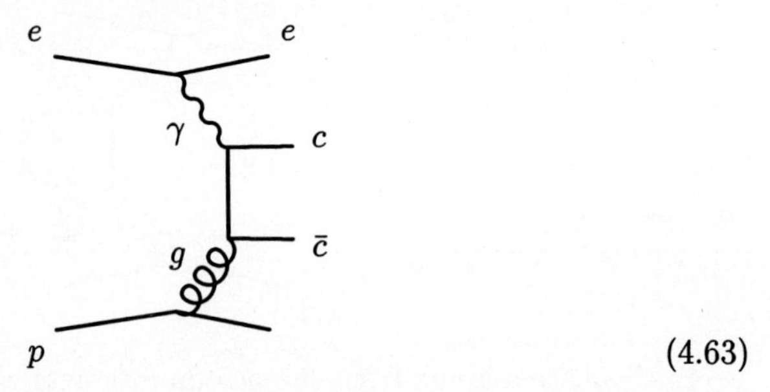

(4.63)

Instead, each charm quark hadronizes to produced a charm meson in the final state. By measuring for instance the inclusive D^* production one can obtain the

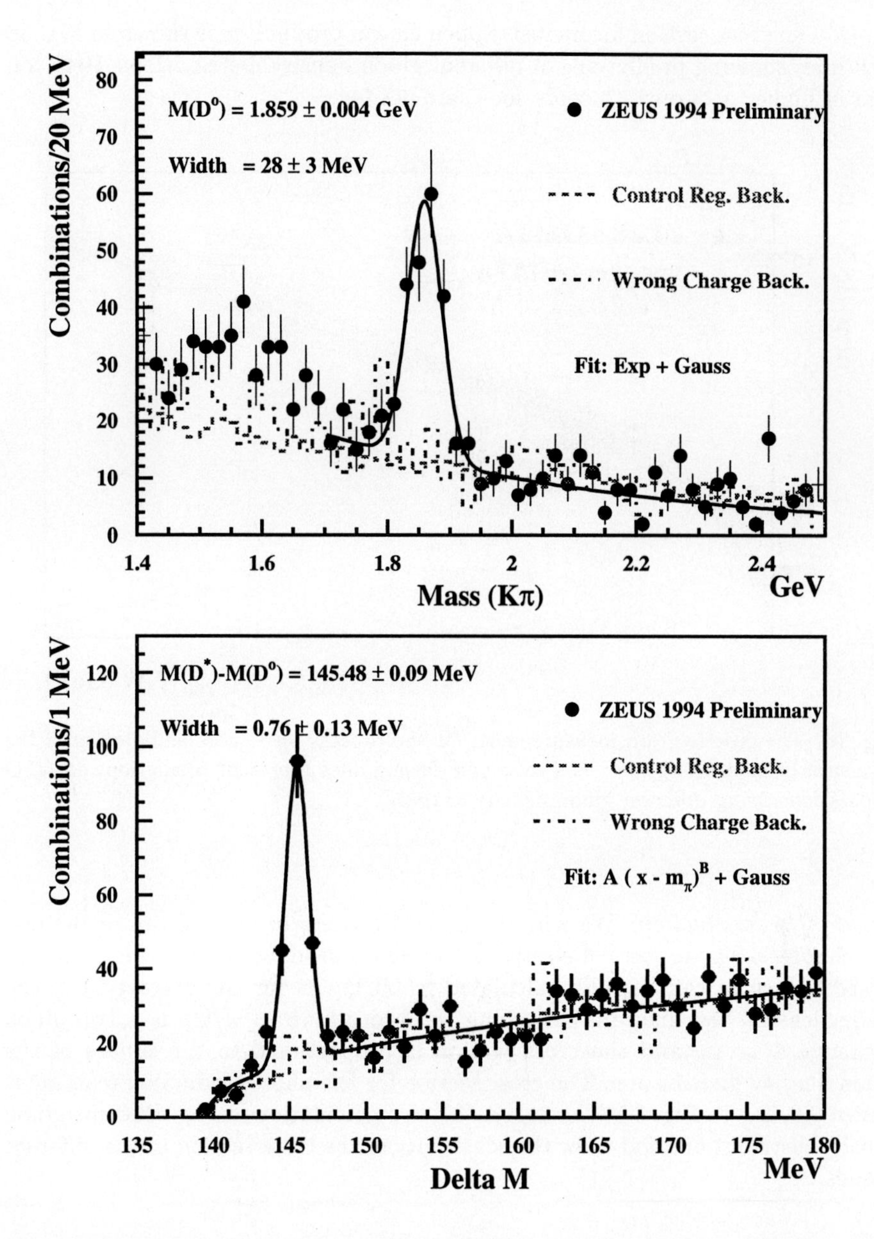

Fig. 35. *$M(K\pi)$ (top) and ΔM (bottom) distributions for events with $143 < \Delta M < 148$ MeV and $1.80 < M(K\pi) < 1.92$ GeV.*

cross section for the process $\gamma p \to c\bar{c}X$, which is sensitive to the gluon density. Figure 35 shows the D^0 signal seen [82, 83, 84, 85] directly from the $(K\pi)$ mass spectrum and the signal observed in the mass difference between the D^* and the D. The cross section for inclusive open charm production is shown in Fig. 36 with lines showing predictions of different gluon density distributions. Here too data of higher accuracy, to come soon, are needed.

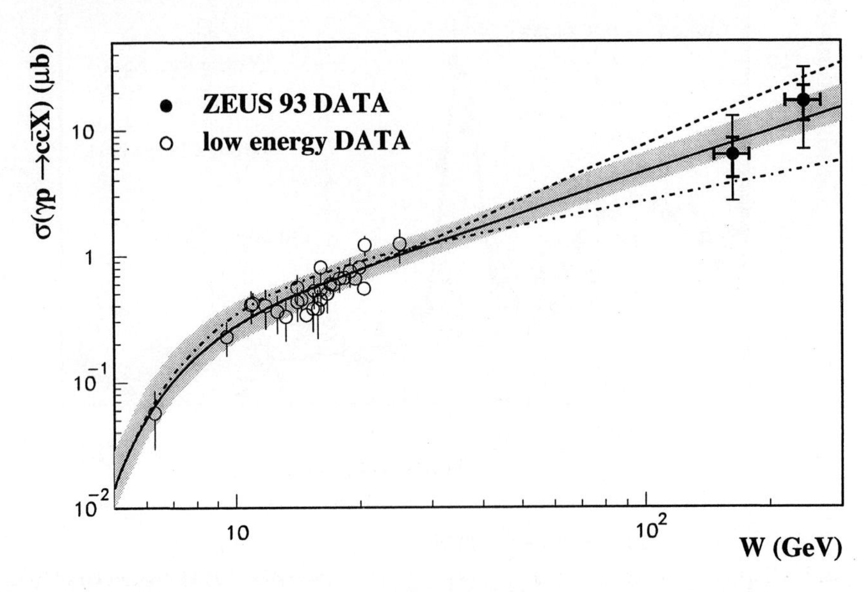

Fig. 36. *The cross section measurements for the process $\gamma p \to c\bar{c}X$ as function of the γp center of mass energy W. The solid and dashed lines represent predictions on NLO calculations using different gluon density shapes.*

Elastic J/Ψ production: We will discuss this process in more detail in the last two chapters. The process of elastic vector meson production in DIS is of special interest since in can be fully calculated in QCD. As for the extraction of the gluon density, the diagram describing the process $\gamma p \to J/\Psi p$ is a two gluon exchange diagram and the cross section is proportional to the square of the gluon density distribution. The cross section for the photoproduction reaction is shown [86, 80] in Fig. 37. The curves [87] are calculations using different gluon density distributions and show the sensitivity of the cross section to the different shapes.

4.9 Summary

In this chapter we have discussed the following issues:

– We have discussed the factorization theorem which holds for an inclusive process and which allows to factorize the structure function calculation into a part which is fully calculable in pQCD, the coefficient functions, and a

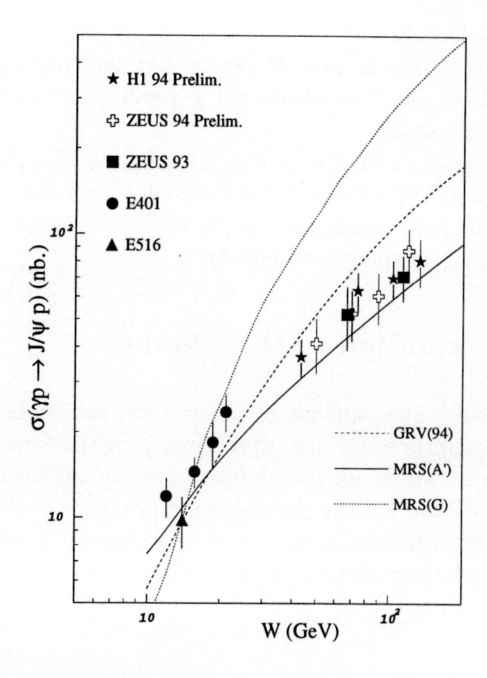

Fig. 37. *Elastic J/Ψ cross section data as function of W compared with different choices of gluon distribution in the proton.*

part which involves long distance effects and has to be obtained from the experiment, the parton distribution functions.

- Once we obtain the parton distribution functions at a given Q^2 scale, there are evolution equations which can predict the parton distribution functions at a higher scale. When the evolution is done in Q^2 one gets the DGLAP equations. These equations can describe the behaviour of the structure function F_2 at low x down to Q^2 values of $\approx 1.5\,\mathrm{GeV}^2$.
- The BFKL equation studies the evolution in x only. It predicts for the gluon density a behaviour of $x^{-\lambda}$, providing by this 'reggeization' a link between QCD and Regge theory. For a fixed α_S value, λ is of the order of 0.5. This introduces the 'BFKL Pomeron' as a trajectory with an intercept of 1.5.
- Presently it is not easy to find experimental signals for the BFKL mechanism. The behaviour of the structure function in the low x range provided by HERA is consistent with both the DGLAP and the BFKL approach. Signs of the BFKL mechanism can be obtained by studying the energy dependence of jets produced near the proton direction.
- We reviewed the extraction of parton distributions and discussed some of the parton parameterizations like the MRS, CTEQ and GRV. The GRV one is more than just a parameterization since it is based on a dynamical picture in which there are only valence partons at a very low scale and the sea is

then built by the evolution.
- The recent HERA data at low Q^2 shows that the Regge picture works well up to about $Q^2 \approx 1\,\mathrm{GeV}^2$, while the QCD parameterizations work well down to about the same scale.
- Finally we discussed methods for the extraction of the gluon density distribution from global QCD fits and from exclusive processes. The distributions from all methods give consistent results with each other, providing a check on the universality of the gluon density.

5 Parton Distribution in the Photon

This chapter describes the concept of the photon structure function. We will define the photon structure function after developing the formalism for that, and write the evolution equations for the photon, pointing out the differences to those of the proton. We will discuss the theoretical importance of the photon structure function and the experimental methods of extracting it from the data. Finally we will introduce some parameterizations of the photon structure function, both of real and of virtual photons.

5.1 Introduction

In the classification of elementary particles, the photon plays the role of a gauge and point–like particle, mediating electromagnetic interactions through its coupling to the charge of matter. Yet, it is well known from soft, low energy γp interactions that its behaviour can be similar to that of strongly interacting hadrons. The properties of those interactions are well described by the vector dominance model (VDM) [32], in which the photon turns first into a hadronic system with quantum numbers of a vector meson and then interacts with the target proton.

A nice justification for this model can be seen from the argument of Ioffe [88, 89]. We know from QED that a photon can fluctuate into a pair of virtual charged lepton like e^+e^-, which annihilate back to a photon. It can however also fluctuate in a quark–antiquark pair $q\bar{q}$. If the time of the fluctuation t_f is large compared to the time of the interaction t_{int} the interaction will occur between the $q\bar{q}$ pair and the proton, resulting in a hadronic interaction. We can estimate the fluctuation time by using the uncertainty principle. Assume a real photon ($Q^2 = 0$) with energy k interacts with a proton which is at rest. The energy difference ΔE at the vertex where the photon fluctuates into a $q\bar{q}$ pair having the same momentum as that of the photon and a mass of $m_{q\bar{q}}$ is:

$$\Delta E = (k^2 + m_{q\bar{q}}^2)^{\frac{1}{2}} - k \tag{5.1}$$

The Vector Dominance Model assumes that the fluctuation of the photon is into vector mesons, $m_{q\bar{q}} \simeq m_V$, where m_V is the vector mesom mass. For high energy photons, $k \gg m_V$ and one can approximate t_f by:

$$t_f \sim \frac{1}{\Delta E} \approx \frac{1}{k} \frac{1}{1 + \frac{m_V^2}{2k^2} - 1} = \frac{2k}{m_V^2} \tag{5.2}$$

The interaction time is of the order of the proton radius r_p, namely:

$$t_{\rm int} \approx r_p \tag{5.3}$$

For example, taking a $k = 10\,\mathrm{GeV}$ photon, and $m_V = m_\rho$, one gets $t_f \approx 7\,\mathrm{fm}$, while $t_{\rm int} \approx 0.8\,\mathrm{fm}$ and thus the condition $t_f \gg t_{\rm int}$ holds.

When instead of a real photon one has a virtual photon ($Q^2 \neq 0$), the fluctuation time is given by

$$t_f = \frac{2k}{Q^2 + m_V^2} \tag{5.4}$$

and thus as Q^2 increases, the fluctuation time becomes smaller and the photon behaves like a point–like object. However, as we shall see later, there are conditions for which even at high Q^2 the fluctuation time will be large (see next chapter).

We thus have [90] a picture of a two–component photon:

$$|\gamma> = |\gamma>_{\rm bare} + \text{coeff.} \times |\gamma>_h \tag{5.5}$$

in which the hadronic part is represented in VDM by

$$|\gamma>_h = a|\rho> + b|\omega> + c|\varphi> \tag{5.6}$$

This picture was verified in γp reactions by observing that for instance the reaction $\gamma p \to \pi^+\pi^- p$ is completely dominated by $\gamma p \to \rho^0 p$:

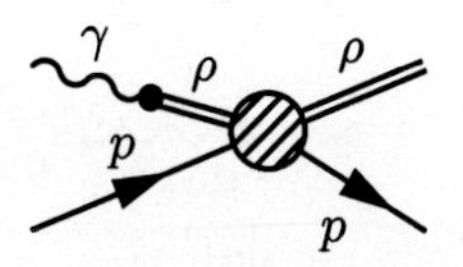

(5.7)

and also in e^+e^- experiments, by studying the reactions: $e^+e^- \to \pi^+\pi^-, \pi^+\pi^-\pi^0, K\overline{K}$, which showed the ρ^0, ω, φ:

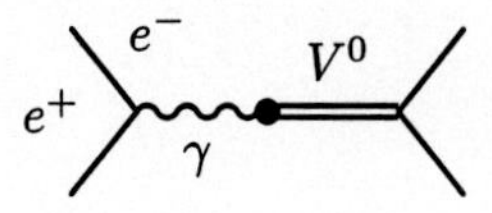

(5.8)

These reactions also provided a determination of the direct photon–vector meson coupling strength, $(4\pi/\gamma_V^2)$:

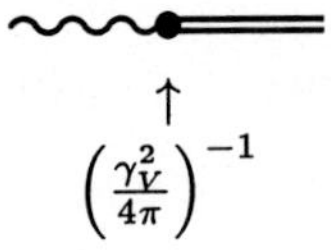

$$\uparrow \quad \left(\frac{\gamma_V^2}{4\pi}\right)^{-1} \tag{5.9}$$

and allowed to test the VDM prediction of their ratios:

$$\left(\frac{\gamma_\rho^2}{4\pi}\right)^{-1} : \left(\frac{\gamma_\omega^2}{4\pi}\right)^{-1} : \left(\frac{\gamma_\varphi^2}{4\pi}\right)^{-1} = 9:1:2 \tag{5.10}$$

The VDM predicted [90] many more relations, most of which were borne out by the data. However VDM is a model. It did not evolve from 'looking' at the photon with a probe. It related photoproduction reactions to hadron reactions by modelling the photon as a superposition of vector mesons, with direct photon–vector meson couplings which could be determined by experiment. This model worked very well at low p_T reactions.

How can one 'look' at the photon in a way similar to what has been done to the proton? The most natural way is to perform a deep inelastic scattering experiment on the photon [91] by studying $e\gamma$ reactions. First one produces a high energy photon beam by using a backscattered laser beam in a linear collider. A laser beam of about 1 eV colliding with a 0.25 TeV electron beam can produce a photon beam of about 0.2 TeV of energy:

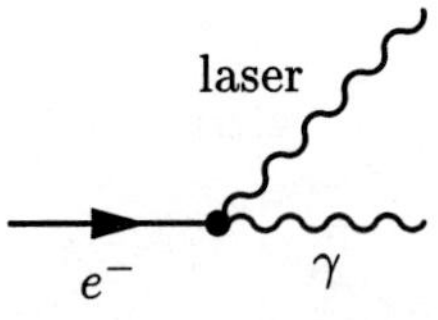

laser

e^- γ (5.11)

The resolution of such a beam can be of the order of $\Delta E_\gamma/E_\gamma \sim 10\,\%$. The high energy photon beam can then collide with another electron beam of energy 0.25 TeV giving a luminosity of the order of $\mathcal{L} \sim 10^{33}\,\mathrm{cm}^{-2}\,\mathrm{sec}^{-1}$:

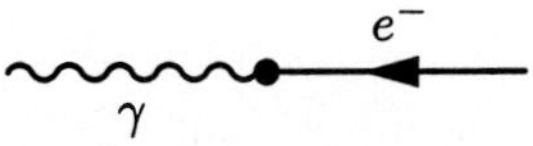

e^-

γ

(5.12)

It is also possible to collide two real photon beams this way when such a linear collider becomes available.

For the time being one has to use e^+e^- interactions leading to two–photon exchange.

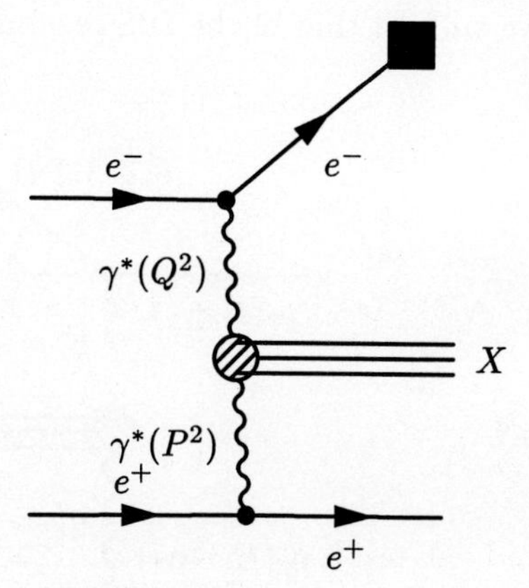

(5.13)

When one of the photons with very small virtuality ($P^2 \approx 0$) interacts with the other one with high virtuality (Q^2) ('single–tag' configuration), the interaction can be thought of as a DIS of one photon on the other, in which case the situation is similar to the probing of a proton by a highly virtual photon. It is therefore natural to introduce the notion of the photon structure function in analogy to the well–known proton one. We will introduce the formalism in the next section.

5.2 Formalism

The DIS for $e\gamma$ interaction [92] is depicted in the following diagram:

$$e\gamma \to eX \qquad d\sigma \sim \sum \left| \begin{array}{c} k \to k' \\ \gamma \\ \gamma \to X \end{array} \right|^2 \tag{5.14}$$

To stress the analogy with the proton case, we also show the DIS of the ep case:

$$ep \to eX \qquad d\sigma \sim \sum \left| \begin{array}{c} k \to k' \\ \gamma \\ p \to X \end{array} \right|^2 \tag{5.15}$$

We will start with the e^+e^- reaction, with the notations as defined in the diagram below, and develop the cross section formalism for the reaction $e^+e^- \to e^+e^-X$, from which we will get that of the DIS reaction $e\gamma \to eX$.

$ee \to eeX$ (5.16)

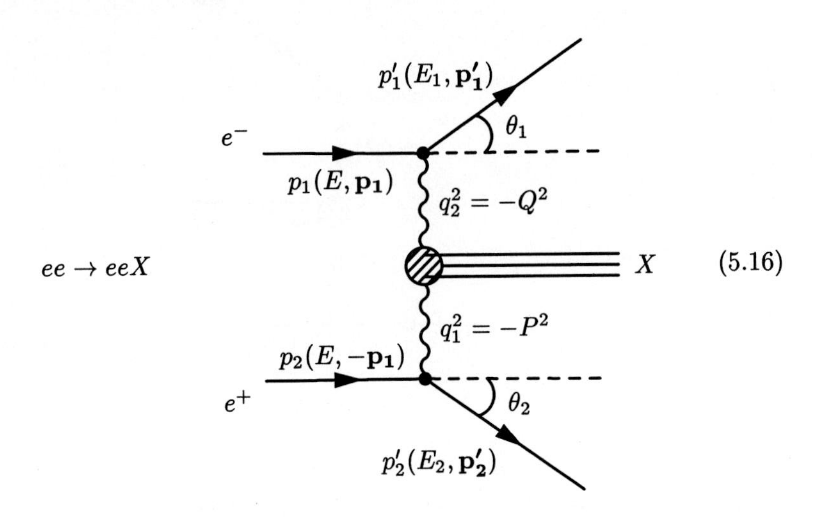

The matrix element for the reaction $e^+e^- \to e^+e^-X$ has 256 components. These can be reduced to 81 using gauge invariance and by further applying the optical theorem and P, T invariance one is left with the following 6 independent components: $\sigma_{tt}, \sigma_{tl}, \sigma_{lt}, \sigma_{ll}, \tau_{tt}, \tau_{tl}$. Here t stand for transversely and l for longitudinally polarized photon states, while τ denote their interference. Integrating over the scattering plane of the leptons and using the fact that the target photon is almost real and thus only transversally polarized, one has only the two independent components σ_{tt} and σ_{lt}.

The reaction $ee \to eeX$ can be viewed as a two–step process. In the first step the target photons are radiated by one of the electrons and are then probed in a DIS by highly virtual photons γ^* emitted by the second electron. In order to take into account the momentum spread of the target photons and their slight off–shellness, one uses the equivalent photon approximation (EPA). In this approach, described by the Weizsäcker–Williams formula [93, 94], the photons are assumed to be emitted real and their momentum spread is modified appropriately.

One can use the equivalent photon approximation (EPA) to write:

$$\mathrm{d}\sigma_{\mathrm{ee}\to\mathrm{eeX}} = \mathrm{d}\sigma_{\mathrm{e}\gamma\to\mathrm{eX}} \mathrm{f}_{\gamma/\mathrm{e}} \tag{5.17}$$

where

$$\mathrm{d}\sigma_{\mathrm{e}\gamma\to\mathrm{eX}} = \frac{\alpha \mathrm{E}'_1 \left(\mathrm{y}^2 - 2\mathrm{y} - 2\right) \mathrm{d}\mathrm{E}'_1 \mathrm{d}\Omega'_1}{2\pi^2 \mathrm{Q}^2 \mathrm{y}} \left(\sigma_{\mathrm{tt}} + \epsilon \sigma_{\mathrm{lt}}\right) \tag{5.18}$$

with

$$\epsilon = \frac{2(1-y)}{1+(1-y)^2} \tag{5.19}$$

where we have used the Hand definition of the flux. The factor $f_{\gamma/e}$ is the flux of the target photons, which is given by the Weizsäcker–Williams formula:

$$f_{\gamma/e} = \frac{\alpha}{\pi z}\left[(z^2 - 2z + 2)\ln\frac{E(1-z)\theta'_{1max}}{m_e z} - (1+z)\right] \tag{5.20}$$

where $z = E_\gamma/E$, m_e is the electron mass and θ'_{1max} is the limiting scattering angle of the tagged electron on the probing photon side.

5.3 Definition of Photon Structure Functions

We can now introduce [95] the notation:

$$F_1^\gamma \equiv \frac{Q^2}{4\pi^2\alpha}\frac{1}{2x}\sigma_{tt} \tag{5.21}$$

$$F_2^\gamma \equiv \frac{Q^2}{4\pi^2\alpha}(\sigma_{tt} + \sigma_{lt}) \tag{5.22}$$

where x is the Bjorken variable as defined earlier, and, because of the massless target, has the relation:

$$x = \frac{Q^2}{Q^2 + W^2} \tag{5.23}$$

and W is the $\gamma^*\gamma$ center of mass energy. With these definitions we can write the cross section for the $e\gamma$ DIS process as:

$$\frac{\mathrm{d}\sigma(\mathrm{e}\gamma \to \mathrm{eX})}{\mathrm{dxdy}} = \frac{4\pi\alpha^2 s}{Q^4}[(1-y)F_2^\gamma + xy^2 F_1^\gamma] \tag{5.24}$$

which is to be compared to the proton case, where

$$\frac{\mathrm{d}\sigma(\mathrm{ep} \to \mathrm{eX})}{\mathrm{dxdy}} = \frac{4\pi\alpha^2 s}{Q^4}[(1-y)F_2^p + xy^2 F_1^p] \tag{5.25}$$

As one can see, the similarity is complete. Therefore, F_i^γ can be treated as the structure functions of the target photon. The experimental conditions in the single–tag e^+e^- experiments are such that the accepted values of y are small. Typically the average value of the product xy^2 are of the order $< xy^2 >\simeq 0.01\ldots0.02$ and therefore the F_1^γ term is usually neglected in the expression (5.24).

Definition of Parton Distributions in the Photon We will again use the analogy to the proton case to define the parton distribution in the photon [96]. Let us look again at the proton case. The deep inelastic interaction of a probe with the proton is described by an incoherent sum of elastic scattering of the probe on free spin 1/2 quarks. This approach leads to the identification of the

F_2^p structure function as a sum of the contribution of all quarks and antiquarks that build up the proton:

$$F_2^p(x) = \sum_{i=1}^{2f} x e_i^2 q_i(x) \tag{5.26}$$

where $q_i(x)$ gives, by definition, the probability of finding a particular type of quarks (antiquarks) in the proton, and xe_i^2 is the elementary 'cross section' for the elastic scattering.

In contrast to the proton, for a photon target one can predict the structure function of the photon directly from the quark parton model (QPM). One can perform a full calculation of the cross section $\gamma^*\gamma \to X$ to the lowest order in α for the process

$$\gamma^* + \gamma \to q + \bar{q} \tag{5.27}$$

which is electromagnetic with known couplings. Note however that such an approach disregards possible contributions from the hadronic component of the photon, a point to be discussed later.

In QPM one can calculate F_2^γ through the 'box' diagram

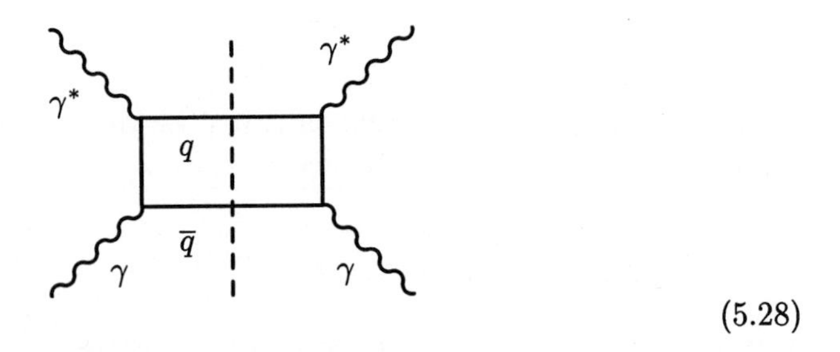

(5.28)

to get the expression [92] (for massive quarks):

$$F_2^\gamma\left(x, Q^2\right) = \frac{N_c \alpha}{\pi} \sum_{i=1}^{f} x e_{q_i}^4 \left\{ \left[x^2 + (1-x)^2\right] \ln \frac{Q^2 (1-x)}{m_{q_i}^2 x} + 8x(1-x) - 1 \right\} \tag{5.29}$$

where N_c is the number of quark colours.

By analogy to the proton case, one can think of F_2^γ as the sum of momentum–weighted densities of quarks 'inside' the photon:

$$F_2^\gamma\left(x, Q^2\right) = \sum_{i=1}^{2f} x e_i^2 q_i^\gamma\left(x, Q^2\right) \tag{5.30}$$

with:

$$q_i^\gamma\left(x, Q^2\right) = \frac{N_c \alpha}{2\pi} e_{q_i}^2 \left\{ \left[x^2 + (1-x)^2\right] \ln \frac{Q^2 (1-x)}{m_{q_i}^2 x} + 8x(1-x) - 1 \right\} \tag{5.31}$$

In spite of the complete analogy between the photon and the proton structure functions, there are important differences in their behaviour. In the QPM the proton structure function F_2^p is expected to fulfil Bjorken scaling, while F_2^γ manifests strong scaling violation even without the presence of gluon radiation. Thus, contrary to the character of scaling violation in the proton case, which yields a negative contribution at large x and a positive one at low x, the scaling violation for the photon is positive in the whole x region (already at the Born level). This can be seen in Fig. 38 where the photon structure function data are presented [14] as function of Q^2 for different x regions. The data show positive logarithmic scaling violation in all regions of x.

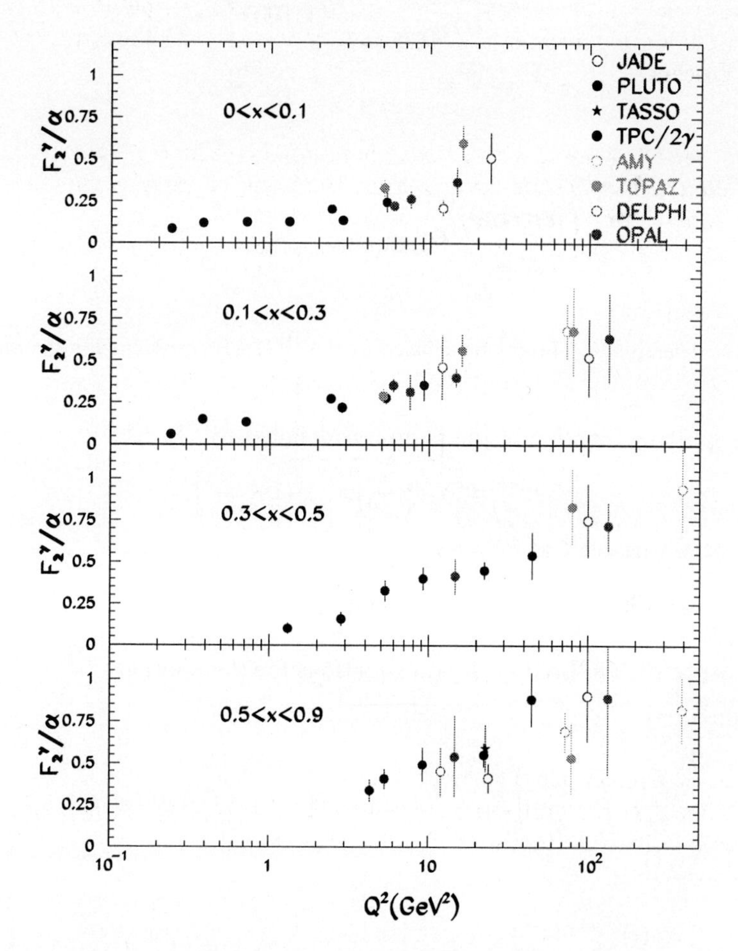

Fig. 38. *The photon structure function as function of Q^2 for different x intervals.*

Another difference between the proton and photon case is the x dependence. Simple counting rules predict that F_2^p should drop at large x, while F_2^γ is large in the high x region (see below).

5.4 Evolution Equations for the Photon

The DGLAP evolution equations for the photon [97, 98, 99] can be developed in a similar way to that of the proton. However, there is one basic difference: in the photon case there is an additional contribution coming from the splitting of the photon into a $q\bar{q}$ pair, as can be seen in Fig. 39.

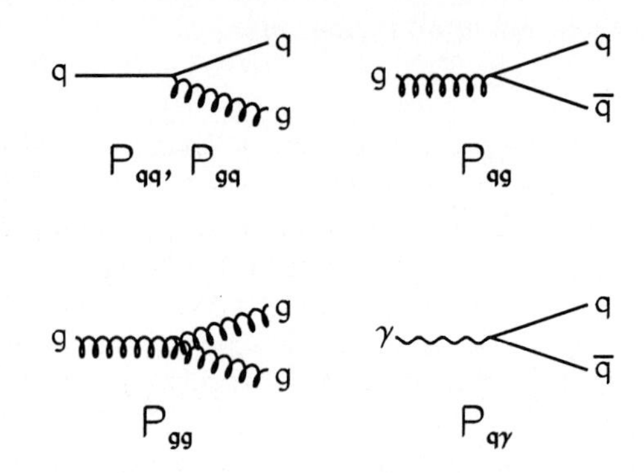

Fig. 39. *The splitting functions utilized in the DGLAP equations for the photon.*

The splitting function of the photon is denoted by

$$h^{\text{box}} = N_c e_q^2 \frac{\alpha}{2\pi}\left[x^2 + (1-x)^2\right] \tag{5.32}$$

Defining the variable t as follows:

$$t \equiv \ln \frac{Q^2}{\Lambda^2} \tag{5.33}$$

we can write the DGLAP evolution equations for the photon:

$$\frac{\mathrm{d}q_i^\gamma(\mathrm{x},\mathrm{t})}{\mathrm{dt}} = h^{\text{box}} + \frac{\alpha_S(t)}{2\pi}\int_x^1 \frac{\mathrm{d}x'}{x'}\left\{P_{qq}\left(\frac{x}{x'}\right) q^\gamma(x',t) + P_{gq}\left(\frac{x}{x'}\right) g^\gamma(x',t)\right\} \tag{5.34}$$

$$\frac{\mathrm{d}g^\gamma(\mathrm{x},\mathrm{t})}{\mathrm{dt}} = \frac{\alpha_S(t)}{2\pi}\int_x^1 \frac{\mathrm{d}x'}{x'}\left\{\sum_{q_i} P_{q_i g}\left(\frac{x}{x'}\right) q_i^\gamma(x',t) + P_{gg}\left(\frac{x}{x'}\right) g^\gamma(x',t)\right\} \tag{5.35}$$

In the case of the QCD evolution equation for the photon structure function, the h_{box} term introduces an inhomogeneity into all parton densities in the photon. This is different from the proton case, where all equations are homogeneous.

The solution to the set of equations (5.34) and 5.35 is given by a superposition of the general solution to the corresponding set of homogeneous equations and a particular solution of the inhomogeneous one.

The inhomogeneous solution is determined by h^{box} and thus depends only on the known point–like (pl) coupling of the photon to the quarks and quarks to gluons. This is why it is identified with the point–like contribution to the photon structure function. Since the homogeneous solution fulfils the hadron–like (had) evolution of the DGLAP equations, it is assigned to the hadron–like contribution to the photon structure function. One writes therefore:

$$F_2^\gamma = F_2^{\gamma,\text{pl}} + F_2^{\gamma,\text{had}} \tag{5.36}$$

The Resolved and Direct Photon Interactions This is perhaps the place to caution [100] the reader not to confuse between the point–like photon and the direct photon reactions. Whenever the interaction of a photon can be described as a two–step process in which the photon first resolves into partons and then one of the parton participates in the hard interaction, such a photon is called a resolved photon. The resolved photon includes both the point–like and the hadron–like part. In the other cases, when the photon interacts directly, all its energy participates in the hard interaction and we say that this is a direct photon interaction. Examples of diagrams describing the direct and resolved photon interactions are shown in Fig. 40.

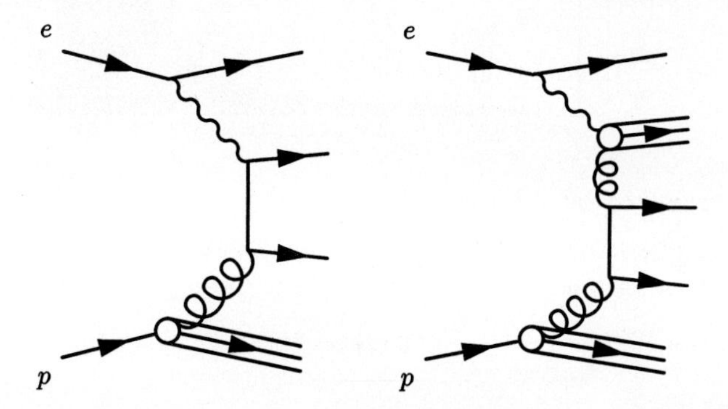

Fig. 40. *Diagrams representing direct and resolved photon interactions.*

This picture of resolved and direct photon was tested experimentally [101, 102] at HERA. One can define the quantity x_γ as the fraction of the photon momentum participating in the hard interaction. One expects then that for the direct photon processes $x_\gamma \sim 1$ while for the resolved ones $x_\gamma < 1$. A good way of estimating x_γ is to study two jet events. In this case one can calculate x_γ^{obs}, the experimentally observed quantity which is close to x_γ:

$$x_\gamma^{\text{obs}} = \frac{E_T^{j1} e^{-\eta^{j1}} + E_T^{j2} e^{-\eta^{j2}}}{2E_\gamma} \tag{5.37}$$

where E_T is the transverse energy of the jet, η its pseudorapidity and E_γ is the energy of the photon. The variable $x_\gamma^{\rm obs}$ is plotted in Fig. 41. The data [103] show an enhancement at high x and a distribution reaching down to low x values, as one expects from a sample of events produced by direct and resolved photon interactions. The data are compared to the distributions obtained from the sum of Monte Carlo generated events simulating direct and resolved photon processes. The agreement is quite good in most of the regions, except for the very low x region, which needs to be further studied.

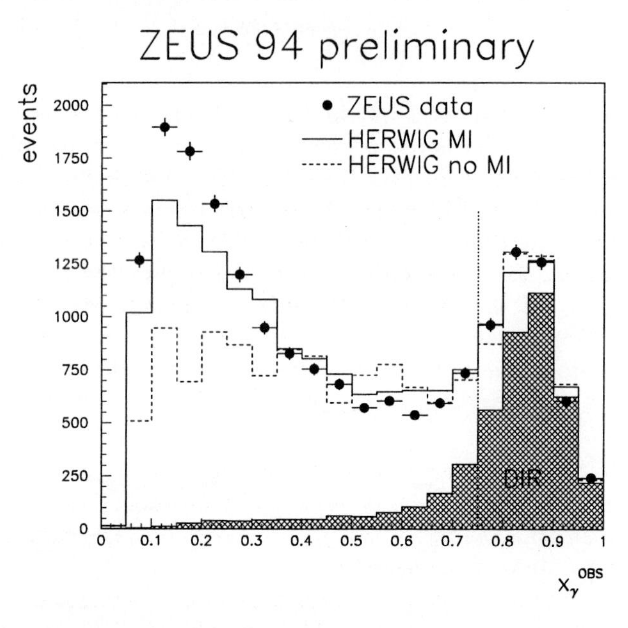

Fig. 41. *The x_γ distribution as obtained from two jet events, compared to direct and resolved photon interactions as simulated by Monte Carlo generators,*

5.5 The Theoretical Importance of F_2^γ

As is known from the proton case, the hadron–like contribution varies very slowly with Q^2. Thus in the high Q^2 limit:

$$F_2^\gamma \to F_2^{\gamma,\rm pl} = a(x) \ln \frac{Q^2}{\Lambda^2} \tag{5.38}$$

In this equation $a(x)$ is calculable in pQCD and therefore one can predict [104] both the shape and the magnitude of F_2^γ, resulting in the ability to determine the QCD scale Λ. Unfortunately, owing to many theoretical difficulties encountered in the calculation of F_2^γ, one of which is discussed below, the actual attempts to measure Λ through the study of F_2^γ have attained only a very limited success.

Higher Order Corrections We know from the proton case that in order to get reliable results, one needs to do at least a next to leading order calculation. What happens when one attempts to do it for the photon case? Can one continue this two component picture of a point–like and a hadron–like part of the photon structure function?

It would be too technical to develop here the next to leading log approximation. This is usually done by introducing the moments of the structure functions appearing in the evolution equation and by introducing the anomalous dimensions. We will just bring here the essence of such a calculation. The result is that when one continues to separate the structure function in a point–like and hadron–like part one gets:

$$F_2^{\gamma,\mathrm{pl}} < 0 \qquad \text{for } x \leq 0.1 \tag{5.39}$$

The situation gets worse with each order of calculation [105].

Since a structure function can not be negative, this means that the separation into a point–like and a hadron–like part gives unphysical results for the point–like part and thus the hadron–like part is needed in order to cancel the singularity. Since there is no way to calculate the hadron–like part in pQCD and it can not be neglected even at high Q^2, the absolute predictive power is lost.

5.6 The Experimental Extraction of F_2^γ

In principle, in order to measure F_2^γ one needs to measure the cross section as a function of Q^2 and x. Q^2 can be determined by measuring the tagged electron. This can be with an accuracy of $\Delta Q^2/Q^2 \sim 7 - 10\,\%$, depending on the energy and angular resolution of the detector.

In order to obtain x, one needs a good measurement of the total hadronic energy W, which together with Q^2 yield x according to (5.23). However, owing to finite resolution and limited acceptance of the detector, one measures W_{vis}, the visible hadronic energy, which is usual smaller than W, as can be seen in Fig. 42.

One needs thus to unfold the true result from the visible measurement. This procedure needs a Monte Carlo program having a good simulation of the detector and a good description of the structure of the event. One can see [106] for instance in Fig. 42(b) the improvement of the correlation between W_{vis} and the true W when using a better generator. However the simulation depends also on the fragmentation model used to generate the final state particles, which usually has problems with the description of the energy flow in the forward direction. Thus even the generator which gives a better correlation as far as the true W is concerned, fails to describe the energy flow in the forward direction, as seen in Fig. 43. Clearly more work is needed in this direction.

In spite of the problems described above, the photon structure function has been measured in a wide range of Q^2 values. However, at present, the statistics is limited and also the systematic errors are quite large. This can be see in Fig. 44

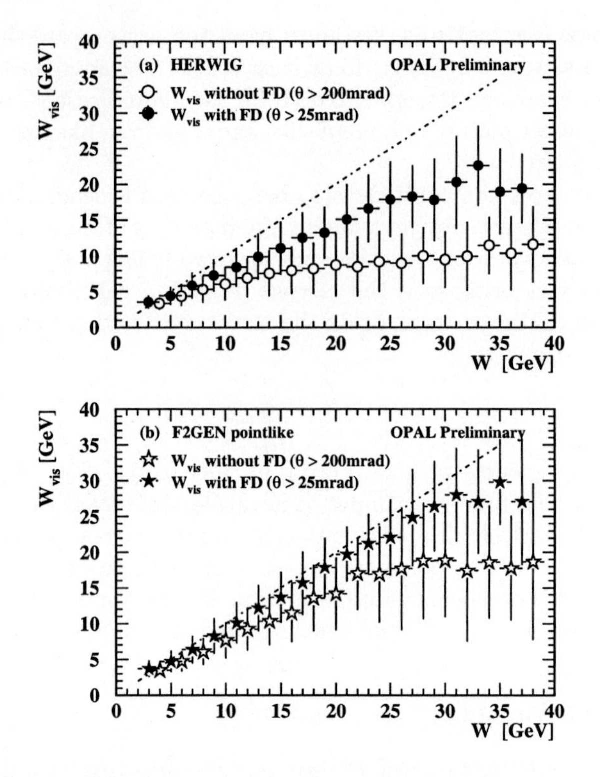

Fig. 42. *Comparison between the visible and the true W.*

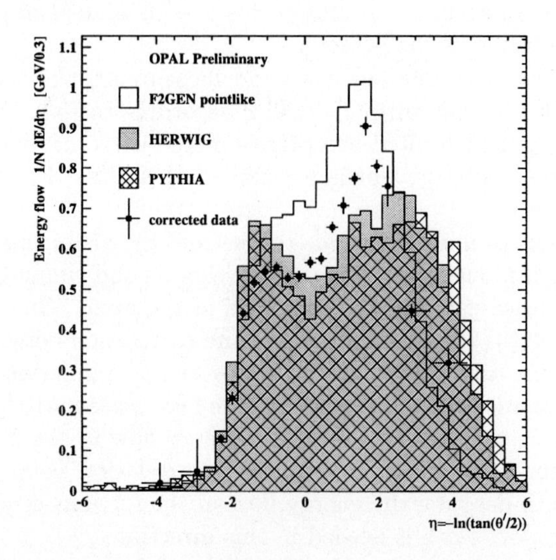

Fig. 43. *Comparison of the data on energy flow as function of the pseudorapidity with predictions of different Monte Carlo generators. The Q^2 values are in units of GeV^2.*

where a compilation [14] of all existing measurements of F_2^γ is presented as a function of x, for different Q^2 values.

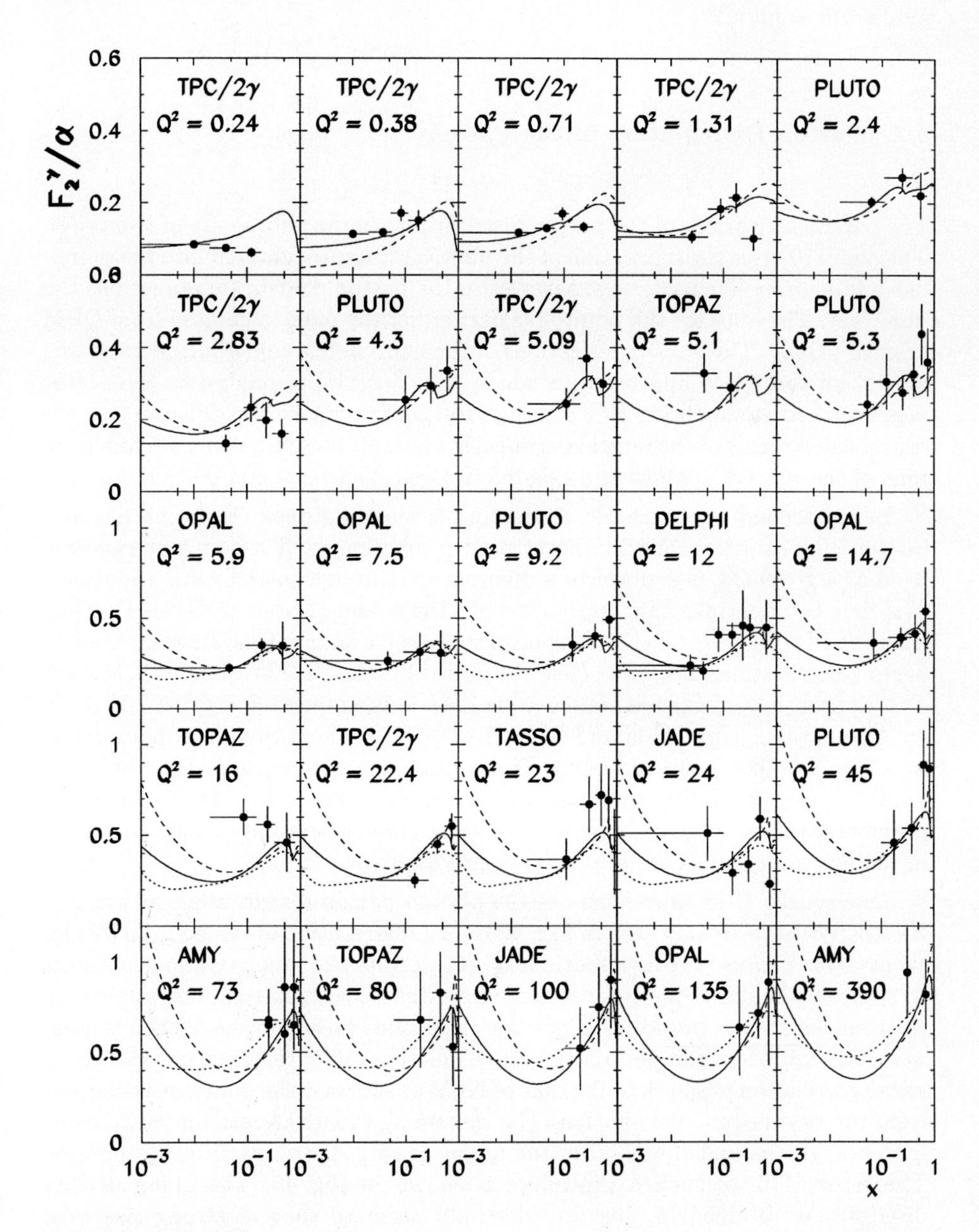

Fig. 44. *Compilation of all existing data on F_2^γ, compared to predictions of some parton parameterizations.*

The curves are the predictions of some of the parameterizations [107, 108, 109] of the parton distributions in the photon, to be discussed below. Note that there are very few measurements in the low x region, due to the experimental difficulties to isolate the photon–photon reactions from the e^+e^- annihilation final state at high W.

5.7 Parton Distribution in the Photon

The parameterizations of the parton distributions in the photon are of two types. The one [110] uses the separation of the photon structure function into the point–like and hadron–like parts to separate also the parton distribution functions the same way. They use for the point–like part either the ones calculated from QPM or from pQCD. The parameterizations of the hadron–like contribution are based mostly on the VDM approach, in which $F_2^{\gamma,\rm had}(x,Q^2)$ is related to the vector meson structure functions F_2^V. Through isospin invariance the different F_2^V are expressed in terms of the only experimental available mesonic structure function, that of the π^-, $F_2^{\pi^-}$, which is measured in Drell–Yan reactions.

In the second approach, no distinction is made between the point–like and hadron–like contributions to the structure function, and a parameterization, fixed at a given Q_0^2, is evolved to a different Q^2 through the DGLAP equations. The first to undertake this approach were Drees and Grassie (DG) [111]. They used the LLA modified DGLAP equations to evolve an input parameterization of parton distributions at $Q_0^2 = 1\,\mathrm{GeV}^2$ so that it fits the PLUTO data at $5.9\,\mathrm{GeV}^2$. This approach was extended later, using data in the range $1.3 < Q^2 < 100\,\mathrm{GeV}^2$ by Abramowicz, Charchula and Levy (LAC) [112], where the gluon parameters were also left free in the global fit. These LLA parameterizations have been extended to next to leading order, some of which are shown in Fig. 44. The different parameterizations obviously agree with each other where data exist, and differ in the low x region, where data are eagerly awaited.

One source of measurements of the photon parton distributions at low x is HERA. We have already seen in Fig. 41 the x_γ distribution obtained from two jet events. One can use the distributions obtained from the Monte Carlo generators to subtract the direct photon reactions from the x_γ distribution, being left with the resolved photon processes. Since the quark distribution in the photon is quite well constrained by the photon structure function measurements, one can use a parameterization prediction like that of GRV to subtract the quark distributions from the resolved x_γ distribution. The remaining events are attributed to come from the gluons and thus obtain the gluon density distribution in the photon. The result [113] of such a procedure is shown in Fig. 45. The gluon density distribution obtained in this way does not seem to show a strong rise as x decreases, but one needs more precise data and procedures to conclude something more definite.

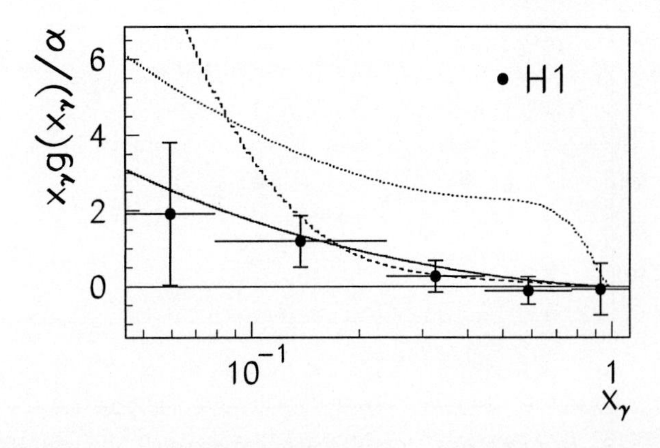

Fig. 45. *The estimated gluon density distribution in the photon, $x_\gamma g(x_\gamma)/\alpha$, compared to predictions of some parameterizations.*

5.8 Parton Distribution of a Virtual Photon

So far, we discussed the structure of real or quasi–real photons. The natural question that one is faced with is what happens to a virtual photon and whether it is legitimate to think that in DIS of charged leptons on protons it is indeed the structure of the proton that is probed and not some kind of convolution of both the structure of the target and of the probe [114]. The same question could be asked in the case of the deep inelastic $e\gamma$ scattering. We will return to this question in the last chapter.

There exists one result by PLUTO [115] from 1984 which measured the structure function of a target photon with virtuality of $P^2 \approx 0.4\,\text{GeV}^2$, at $Q^2 \approx 5\,\text{GeV}^2$. There is also an attempt [116] at HERA to measure the structure of a virtual photon. The electron calorimeter of the luminosity system taggs photons with a median virtuality of $P^2 \approx 10^{-5}\,\text{GeV}^2$. There is another calorimeter, the beam–pipe calorimeter, which taggs photons in the range $0.1 < P^2 < 0.6\,\text{GeV}^2$. By using these two taggers, one can isolate two jet events from quasi–real photons, and a similar sample from the virtual photon reactions. One can then reconstruct the x_γ^{obs} of the photon by using relation (5.37). The distribution of the two samples are shown [116] in Fig. 46. For real photons one sees the concentration of direct events at high x_γ^{obs}, and the low x enhancement coming from the resolved photon processes. For the higher P^2 region, one sees again the peak at high x from direct events, but also a contribution from the resolved photons at lower x. This shows that photons with virtualities in the range $0.1 < P^2 < 0.6\,\text{GeV}^2$ also have structure.

One can use an operational definition of direct photon reactions by the cut $x_\gamma^{\text{obs}} > 0.75$ and study the ratio of resolved photon to direct photon as function of the photon virtuality P^2. This is shown in Fig. 47, which seems to show a decrease of this ratio with increasing P^2, as expected. These data are preliminary and not yet corrected for acceptance effects, which are believed to cancel in the

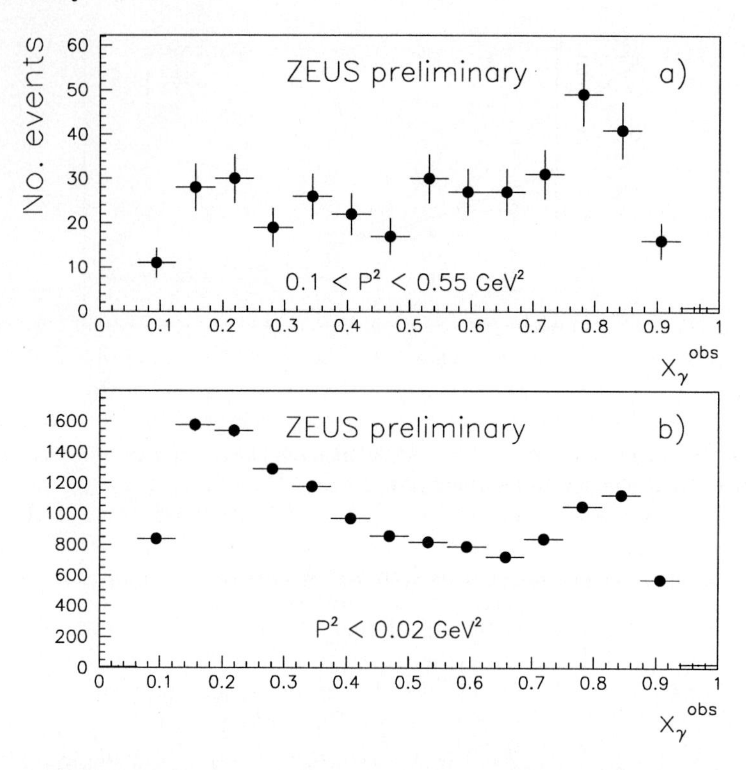

Fig. 46. *The $x_\gamma^{\rm obs}$ distribution (a) for photons of virtuality* $0.1 < P^2 < 0.6\,\mathrm{GeV}^2$ *and (b) for quasi–real photons.*

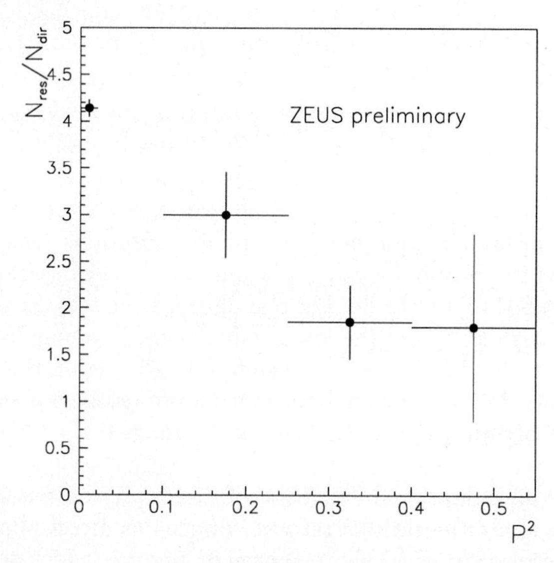

Fig. 47. *The ratio of resolved to direct photon reactions as a function of the photon virtuality* P^2.

ratio. Further study of this interesting question with more data and to higher P^2 values is in process.

5.9 Summary

In this chapter we studied the following subjects about the photon:

- We have introduced the concept of a photon with structure, where the structure is attained by the long fluctuation time of the photon into a $q\bar{q}$ pair before it interacts with the proton. Clearly the whole notion of the structure of the photon makes only sense when we view its interaction with another object.
- We developed the formalism of two photon reactions in e^+e^- collisions and related the process $e\gamma \to eX$ with that of $ee \to eeX$.
- We defined the photon structure function through its analogy to the proton structure function, defined the parton distribution functions in the photon and discussed the DGLAP evolution equation of the photon. These equations are inhomogeneous because of the splitting of a photon to a $q\bar{q}$ pair in addition to the splitting functions in the proton case.
- The point–like and the hadron–like parts of the photon structure functions have been described. Both parts are what is contained in the so–called resolved photon. In the direct photon processes the photon interacts directly with parton from the other projectile, while the resolved photon reactions are a two–step process in which the photon first resolves into its partons, one of which takes part in the hard interaction.
- The theoretical importance of F_2^γ was described as a potential source of determining the QCD scale parameter Λ. However due to problems in the low x region when one uses next to leading order corrections, this is not possible.
- We discussed the experimental procedure of obtaining F_2^γ from the data and the difficulties involved. At present both the statistical errors and the systematic ones are quite large. There is also very little data in the low x region. That is the reason why the different parameterizations of the parton distributions in the photon differ quite widely in the low x region.
- Finally, we discussed the question whether virtual photons also have structure. Preliminary data indicate that at least up to a virtuality of about $P^2 \approx 0.6\,\text{GeV}^2$, the photon seems to have structure.

6 Diffraction in DIS

The notion of diffraction in high energy physics is not easy to define. The dictionary defines it as 'the breaking up of a light beam into light and dark or coloured bands by passing it through a small opening'. We will describe what are the expected behaviour of a diffractive reaction in the introductory section after which we will make the connection to Regge theory and to the Pomeron which is the dominant trajectory exchanged in diffractive processes. Next we will discuss the different diffractive reactions in photoproduction ($Q^2 = 0$) and the discovery of the large rapidity gap events in DIS. The interpretation of these inclusive reactions as DIS on the Pomeron is presented and the partonic structure of the Pomeron is discussed.

6.1 General Introduction

The best example of a diffractive reaction is the process of elastic scattering $A + B \to A + B$ in which no quantum numbers are exchanged in the t channel. In the Regge language one the exchanged trajectory which has the quantum numbers of the vacuum the Pomeron trajectory. Thus one usually calls a process diffractive if its t channel amplitudes at high energies are determined by Pomeron exchanges. A diffractive process [28] has a total cross section practically independent of energy, a small real part of the forward scattering amplitude, and a forward peak in the differential cross section. Among other characteristics is the predominant conservation of s channel helicities of the scattered particles (to be discussed later). A non–diffractive process corresponds to exchanges with non–vacuum quantum numbers in the t channel and has a cross section decreasing with energy.

In addition to the elastic scattering, one can have inelastic diffractive processes. These include the single diffraction (SD) and the double diffraction (DD) reactions in which the beam and/or the target particles get excited into states with the same internal quantum numbers as those of the incoming particles.

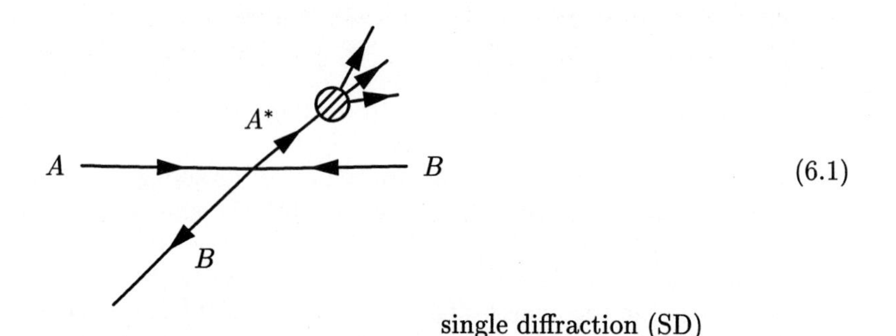

(6.1)

single diffraction (SD)

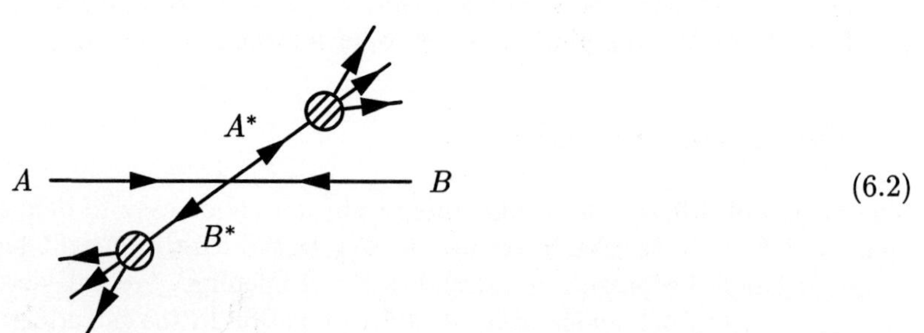

(6.2)

double diffraction (DD)

Some examples of inelastic diffractive processes are: $\pi^- p \to \pi^- N^*$ (SD), $\pi^- p \to a_1^- p$ (SD), $\pi^- p \to a_1^- N^*$ (DD).

In general, a diffractive reaction of the type $A + B \to X + B$ has in addition to all the above mentioned characteristic behaviours, a cross section which falls with the mass of X like:

$$\frac{\mathrm{d}\sigma}{\mathrm{dtdM_X^2}} \sim \frac{1}{M_X^2} \tag{6.3}$$

In such a diffraction reaction, B is a leading energetic particle such that $E_B \approx p_L$. In this case one has:

$$M_X^2 = (P - p_B)^2 = s + m_B^2 - 2E_B\sqrt{s} \approx s - 2p_L\sqrt{s} = (1 - \frac{s - 2p_L}{\sqrt{s}})s = (1 - x_F)s \tag{6.4}$$

where x_F is the Feynman x. Usually for diffractive reaction $x_F \geq 0.9$ which means $\frac{M_X^2}{s} \leq 0.1$. One can get this limit using a geometrical argument. The coherence is important for building up the forward peak and thus:

$$\frac{M_X^2}{s} \leq \frac{1}{2m_A R} \tag{6.5}$$

Since the radius R is of the order of $\sim 1\,\mathrm{fm} = 5\,\mathrm{GeV}^{-1}$, one gets the condition $\frac{M_X^2}{s} \leq 0.1$.

6.2 Diffraction and Regge Formalism

The Regge domain is defined as that where t is small and $\frac{s}{M_X^2} \to \infty$. In this case one can write the total, elastic and inelastic diffraction cross sections in the form:

$$\sigma_T^{ij} = \sum_k \beta_{ik}(0)\beta_{jk}(0)s^{[\alpha_k(0)-1]} \tag{6.6}$$

$$\frac{\mathrm{d}\sigma_{\mathrm{el}}^{\mathrm{ij}}}{\mathrm{dt}} = \sum_k \frac{\beta_{ik}^2(t)\beta_{jk}^2(t)}{16\pi} s^{2[\alpha_k(0)-1]} \tag{6.7}$$

$$\frac{\mathrm{d}^2\sigma^{\mathrm{ij}}}{\mathrm{dtdM_X^2}} = \sum_{k,l} \frac{\beta_{ik}(0)\beta_{jl}^2(t)g_{kll}(t)}{16\pi s} \left(\frac{s}{M_X^2}\right)^{2\alpha_l(t)} (M_X^2)^{\alpha_k(0)} \tag{6.8}$$

where the functions β and g are vertex functions and α is the Regge trajectory.

This can be illustrated with the following diagrams:

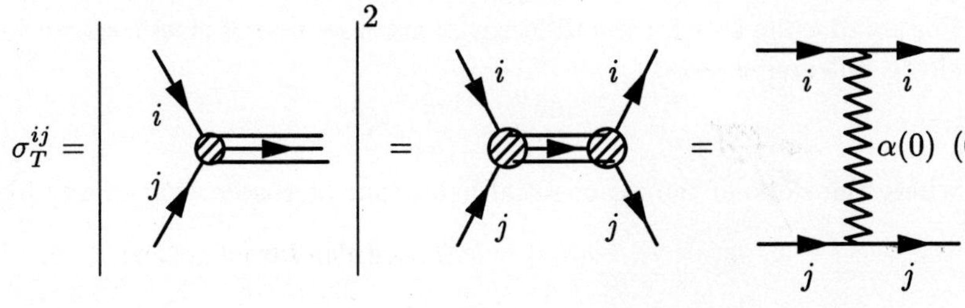

$\sigma_T^{ij} = \left| \; i, j \; \right|^2 = \; i \; i \; j \; j \; = \; i \; i \; \alpha(0) \; j \; j$ (6.9)

$$\frac{\mathrm{d}\sigma_{\mathrm{el}}^{\mathrm{ij}}}{\mathrm{dt}} = \tag{6.10}$$

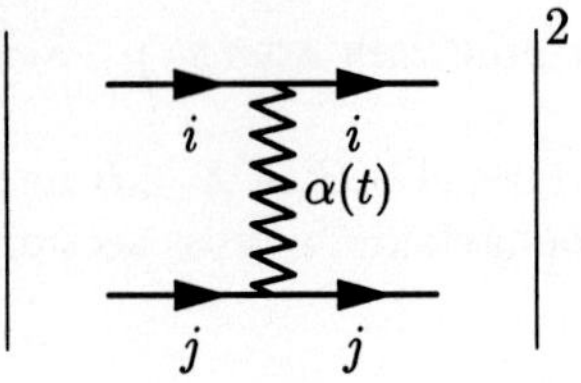

$$\frac{\mathrm{d}^2\sigma^{\mathrm{ij}}}{\mathrm{dtdM_X^2}} = \tag{6.11}$$

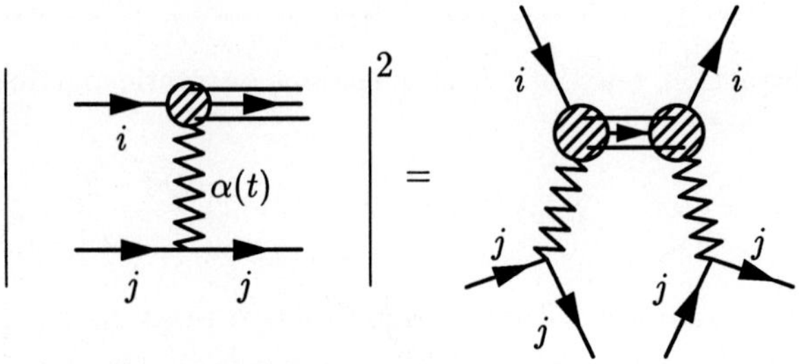

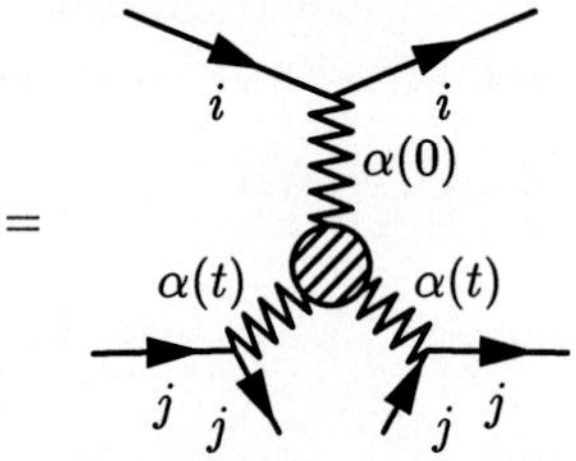

The Pomeron trajectory is the one dominating the diffraction processes. In order to describe the properties of diffraction, its trajectory has the form:

$$\alpha_{\mathbb{P}}(t) = \alpha_{\mathbb{P}}(0) + \alpha'_{\mathbb{P}}(t)t \approx 1 + \alpha' t \tag{6.12}$$

where for the present discussion we have assumed the Pomeron intercept to be 1 (taking $\epsilon = 0$).

- The total cross section can be expressed as:

$$\sigma_T^{ij} = \beta_{i\mathbb{P}}(0)\beta_{j\mathbb{P}}(0) = \text{const} \tag{6.13}$$

  If one uses an intercept of $1+\epsilon$, one gets a slowly rising cross section like s^ϵ.
- The elastic differential cross section is:

$$\frac{\mathrm{d}\sigma_{\mathrm{el}}^{\mathrm{ij}}}{\mathrm{dt}} = \frac{\beta_{i\mathbb{P}}^2(t)\beta_{j\mathbb{P}}^2(t)}{16\pi M_X^2} s^{\alpha' t} \tag{6.14}$$

  If one has a Pomeron intercept somewhat larger than 1, one has an additional factor of $s^{2\epsilon}$.
  For small t one gets for the differential cross section of elastic scattering the sharp diffractive peak:

$$\frac{\mathrm{d}\sigma_{\mathrm{el}}^{\mathrm{ij}}}{\mathrm{dt}} \approx \frac{\sigma_T^2}{16\pi} e^{b(s,t)t} \tag{6.15}$$

  where the slope of the exponential behaviour increases with energy like:

$$b(s,t) = b_0(t) + 2\alpha' \ln s \tag{6.16}$$

We have assumed that the scale in the logarithmic expression is $s_0 = 1\,\mathrm{GeV}^2$. The slope [30] of the Pomeron trajectory is $\alpha' \approx 0.25\,\mathrm{GeV}^{-2}$. The phenomena of the increase of the slope with energy is called shrinkage. The shrinkage can be seen in Fig. 48 for (a) pp elastic scattering and for (b) $\pi^- p$ elastic

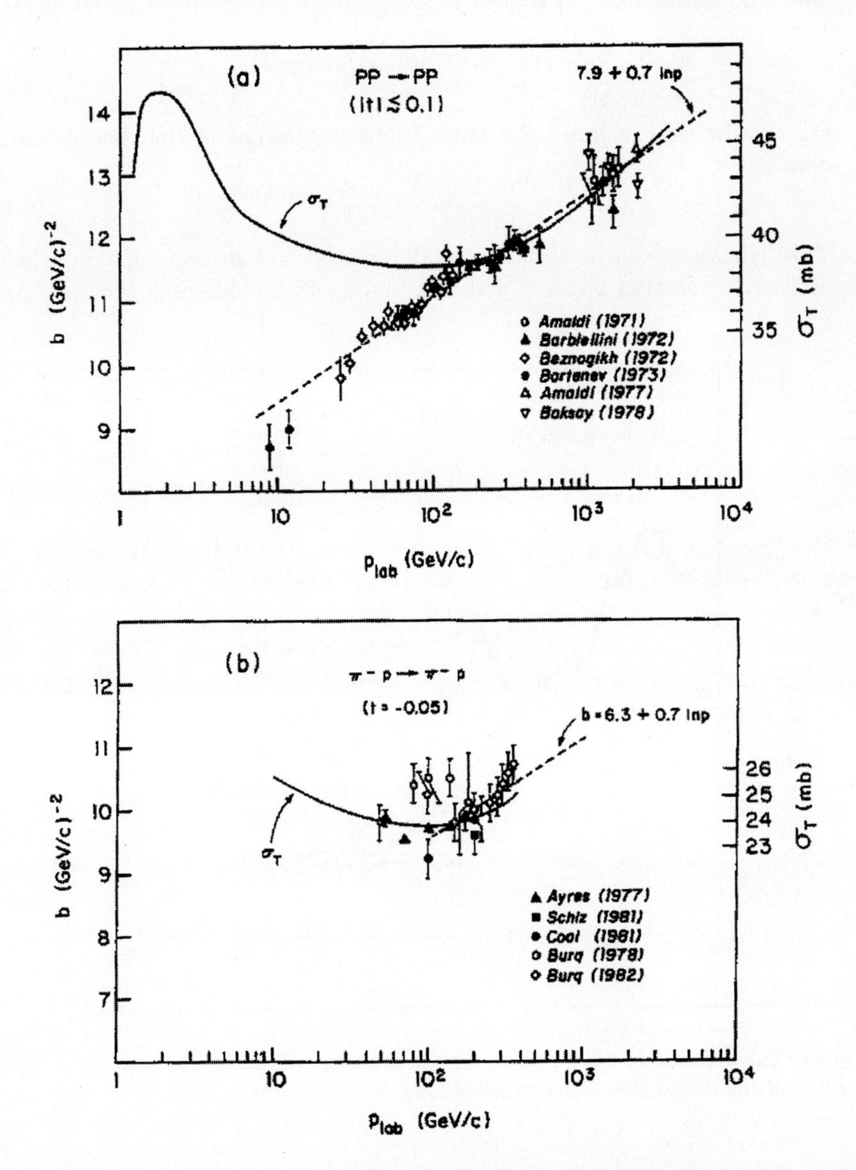

Fig. 48. *The dependence of the slope b on the laboratory momentum of the incoming projectile for (a) pp elastic scattering and for (b) $\pi^- p$ elastic scattering. On the right–hand side of the vertical axis the scale is the total cross section σ_T for (a) pp and (b) $\pi^- p$ reaction. The corresponding data are approximated by the solid line.*

scattering. One observes in both cases a rise of the slope with energy, as expected from expression (6.16). The same figure also shows the behaviour of the total pp and $\pi^- p$ cross sections, showing the slow increase with energy discussed in the former item.

– The differential cross section of inelastic single diffraction is given by [117, 118]:

$$\frac{\mathrm{d}^2\sigma^{\mathrm{ij}}(\mathrm{s}, \mathrm{M_X^2}, \mathrm{t})}{\mathrm{dtdM_X^2}} = \frac{\beta_{i\mathbb{P}}(0)\beta^2_{j\mathbb{P}}(t) g_{\mathbb{PPP}}(t)}{16\pi M_X^2} \left(\frac{s}{M_X^2}\right)^{2\alpha' t} \tag{6.17}$$

For small t we can make the same approximation as in the elastic case to obtain:

$$\frac{\mathrm{d}^2\sigma^{\mathrm{ij}}(\mathrm{s}, \mathrm{M_X^2}, \mathrm{t})}{\mathrm{dtdM_X^2}} \approx \frac{A}{M_X^2} e^{b_D(s,t)t} \tag{6.18}$$

This behaviour can be seen for the differential cross section of the inelastic diffractive reaction $pp \to Xp$ shown in Fig. 49 for different s values. As s

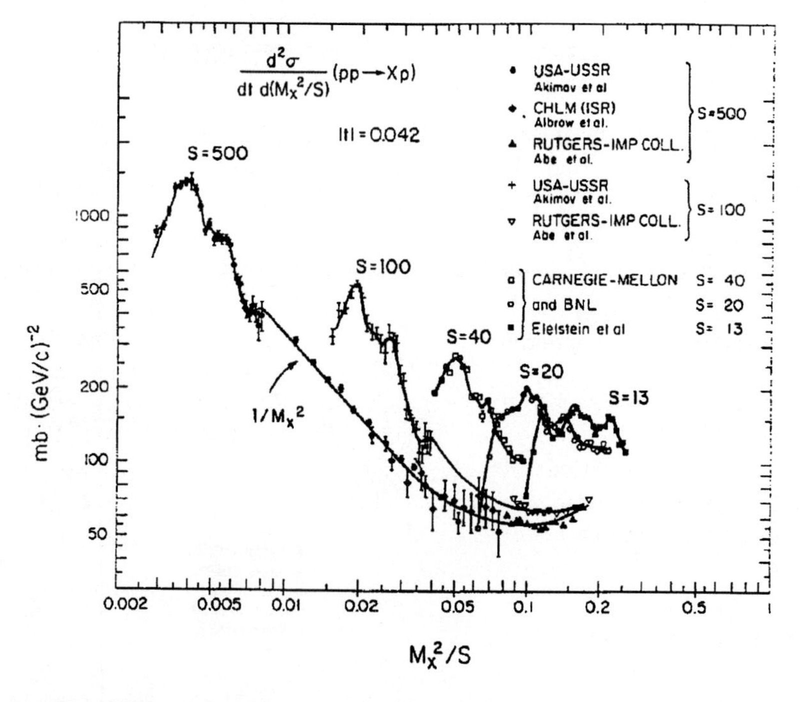

Fig. 49. *The differential cross section for the inelastic diffractive reaction $pp \to Xp$ as function of the scaled diffractive mass M_X^2/s.*

increases and one reaches the Regge domain, one sees a clear $1/M_X^2$ behaviour of the cross section. Here too the slope increases with energy like:

$$b_D(s,t) = b_{D,0}(t) + 2\alpha' \ln \frac{s}{M_X^2} \tag{6.19}$$

All hadronic reactions show an appreciable contribution of diffractive processes to the total cross section ($\sim$ 25–40 %), including double dissociation. Is this true also for photoproduction? Can one check it at HERA?

6.3 Diffraction in Photoproduction at HERA

What does one understand by diffractive processes in photoproduction? First one has the elastic scattering. In case of photoproduction the true elastic scattering process is $\gamma p \to \gamma p$. However this is an electromagnetic process. The hadronic elastic scattering in photoproduction is referred to the reaction $\gamma p \to V^0 p$, where V^0 are the neutral vector mesons. In the VDM picture this process is a two stage one. The photon first fluctuates into a virtual vector meson, which then scatters elastically from the target proton. With this in mind, the following processes contribute to diffractive photoproduction reactions:

$$\begin{aligned} \text{“elastic”} &: \gamma p \to Vp \;\; (V = \rho^0, \omega, \phi) \;\; (\text{not } \gamma p \to \gamma p) \\ \text{photon diffraction} &: \gamma p \to Xp \;\; (X \neq \rho^0, \omega, \phi) \\ \text{proton diffraction} &: \gamma p \to VY \;\; (Y \equiv \text{‘excited’ proton}) \\ \text{double dissociation} &: \gamma p \to XY \end{aligned}$$

This can also be illustrated in a VDM picture:

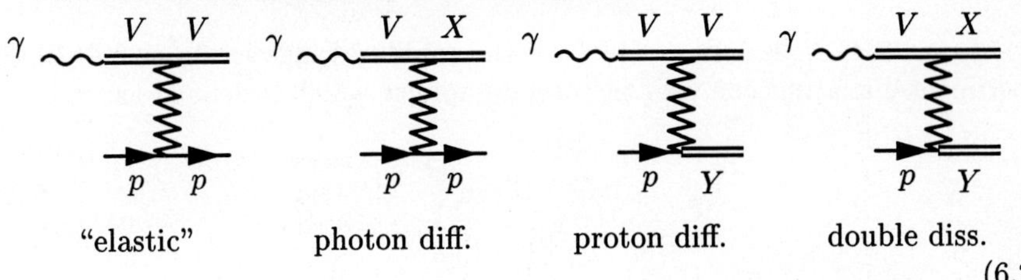

(6.20)

We have already seen how one measures photoproduction reactions at HERA in chapter 2. Is it possible to distinguish [119] diffractive from non–diffractive precesses in these reactions? In order to be able to do so, one needs a large rapidity phase space. How large is it at HERA?

Let us consider the reaction:

$$\gamma p \to Xp \tag{6.21}$$

In the γp center of mass system:

$p \longleftarrow \bullet \longrightarrow X$ (6.22)

In this system the maximum center of mass momentum is:

$$p_{\max}^{\text{cms}} \simeq \frac{\sqrt{s}}{2} = \frac{W}{2} = p^* \tag{6.23}$$

The maximum rapidity of X is $y_{\rm max}^{\rm cms}$ where the positive direction is taken as that of the proton:

$$y_{\rm max}^{\rm cms} = \frac{1}{2}\ln\frac{E+p^*}{E-p^*} = -\frac{1}{2}\ln\frac{(E+p^*)^2}{E^2-p^{*2}} \simeq -\frac{1}{2}\ln\frac{W^2}{M_X^2} \tag{6.24}$$

Therefore the rapidity of the diffractive system X and the diffractive proton can be given by:

$$y_X = -\frac{1}{2}\ln\frac{W^2}{M_X^2} + \text{boost to any system} \tag{6.25}$$

$$y_p = \frac{1}{2}\ln\frac{W^2}{M_p^2} + \text{boost to any system} \tag{6.26}$$

The rapidity span Δy is therefore:

$$\Delta y = \frac{1}{2}\ln\frac{W^2}{M_X^2} + \frac{1}{2}\ln\frac{W^2}{M_p^2} = \frac{1}{2}\left[\ln\frac{W^4}{M_X^2 m_p^2}\right] = \ln\frac{W^2}{M_X m_p} \tag{6.27}$$

For $M_X \simeq 10\,\text{GeV}$ and $W = 200\,\text{GeV}$ the rapidity range is $\Delta y = 8.4$. In the experimental analysis one uses the pseudorapidity which is defined as:

$$\eta = -\ln\tan\frac{\Theta}{2} \tag{6.28}$$

In Fig. 50 the rapidity distribution of the different photoproduction processes at HERA are shown [120, 121] together with the regions covered by the ZEUS detector. As one can see, much of the phase space is lost in the beam pipe. The detector covers only the rapidity range between -3.4 and $+3.8$.

Thus, in order to measure the total cross section one needs to correct for these losses. This requires the knowledge of the relative contribution of the different processes contributing to the total cross section. Since one can not measure them directly, one has to find variables whose distribution is sensitive to the different processes and fit the distributions to the combinations of cross sections which best describe the data. The H1 collaboration [122] used the variables $\eta_{\rm max}$ and $\eta_{\rm min}$ to determine the cross section of the different processes. The variable $\eta_{\rm max}$ ($\eta_{\rm min}$) is defined as the maximum (minimum) pseudo rapidity of all reconstructed charged tracks and all clusters in the calorimeter with energy larger than 400 MeV. The results of the H1 measurement, assuming that the DD cross section is in the range $0 < \sigma_{DD} < 40\,\mu$b, are shown in Table 2.

The total cross section is in good agreement with earlier measurements and with predictions of Regge motivated models [31, 18], as can be seen from Fig. 51.

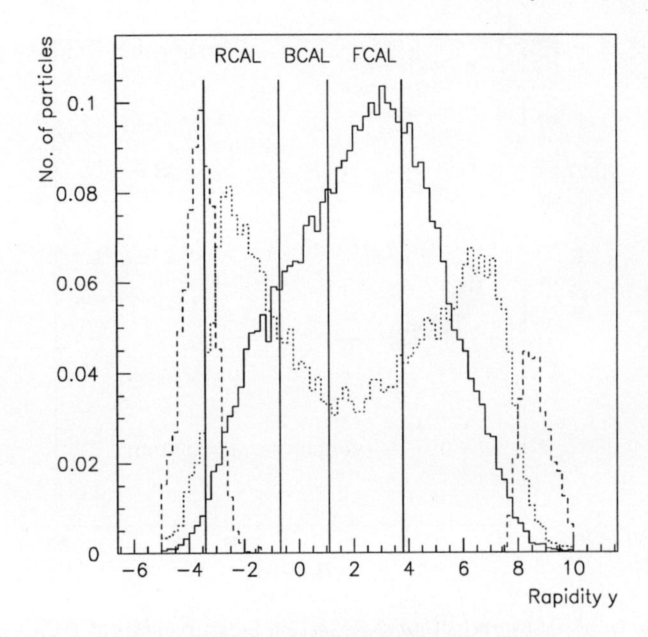

Fig. 50. *The rapidity distribution of the different photoproduction processes at HERA. The regions covered by the ZEUS detector are indicated.*

Table 2. *Cross–section for the different diffractive contributions at $W = 200$ GeV.*

process	cross section (μb)
$\sigma(\gamma p \rightarrow Vp)$	17.1 ± 4.3
$\sigma(\gamma p \rightarrow Xp)$	23.4 ± 11.3
$\sigma(\gamma p \rightarrow VY)$	8.7 ± 3.6
$\sigma(\gamma p \rightarrow XY)$	20 ± 20
diffractive (el + SD + DD)	69.2 ± 13.3
non–diffractive	96.1 ± 17.9
total	165.3 ± 11.2

Ratios of Cross Sections It is of interest to compare [59] the relative abundance of processes in γp with those in other hadronic reactions. To this end we present in Fig. 52(a) the ratio of the elastic to total cross section for $\bar{p}p$, πp, Kp and γp reactions. As expected, the ratio for $\bar{p}p$ is larger than for the πp and the Kp case. The γp ratio, though having large errors, is closer to the meson initiated reactions, in accordance with VDM expectations. It is however a bit on the low side, since at the HERA energies one would expect this ratio to be somewhat larger than the measured ones.

In Fig. 52(b) the ratio of single diffraction to total cross section is presented

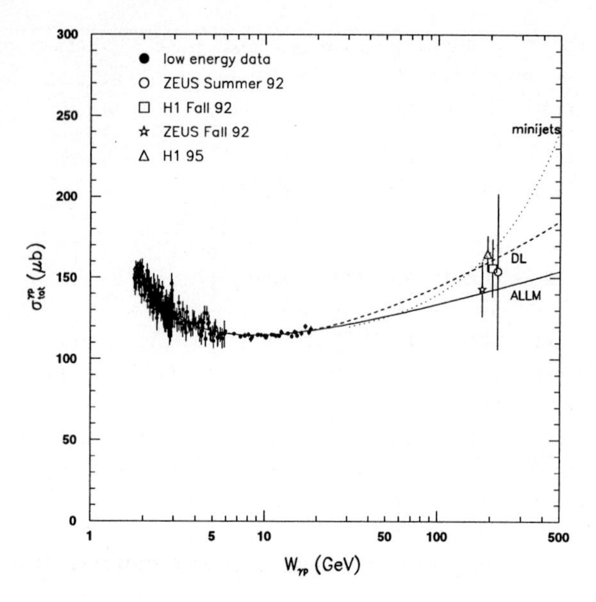

Fig. 51. *The total photoproduction cross section measurements at HERA and at lower energies, compared to different models.*

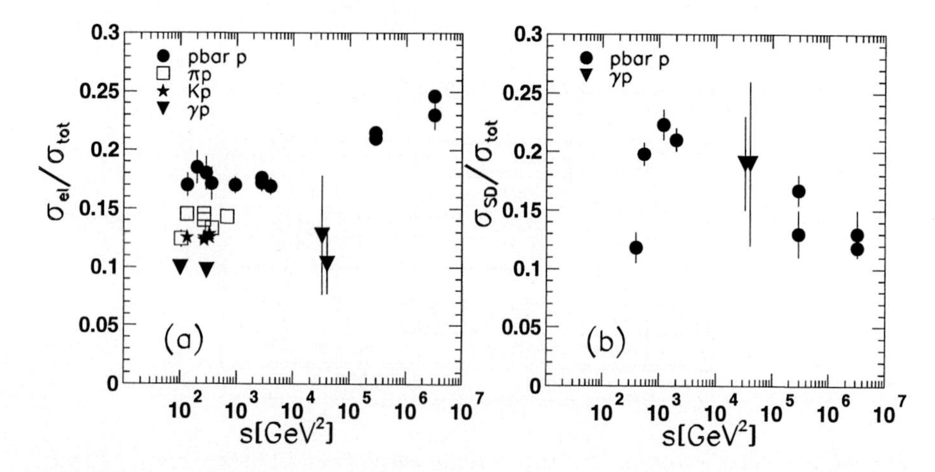

Fig. 52. *The contribution of the (a) elastic and (b) single diffraction processes relative to the total cross section for hadron–proton and γ–proton interactions.*

for $\bar{p}p$ and γp. The ratios for γp are somewhat larger than what one would naively expect from VDM. The slight deficiency of the elastic ratio and the access in the single diffraction case may be correlated and might be due to the way the processes are defined in photoproduction. The traditional definition of the elastic photoproduction reaction includes only the first three lightest vector mesons ρ^0, ω and ϕ. Higher vector mesons are included in the single diffraction channel. In view of the slight deviations of the photoproduction results from the expectations, one might have to redefine the exact meaning of elastic and single diffraction in photoproduction.

6.4 Large Rapidity Gap Events in DIS

The diffraction of the real photon can be clearly understood following the discussion presented in chapter 5, where a real photon can fluctuate into a $\bar{q}q$ pair and acquire a hadronic structure before interacting with the proton and thus producing in some of the time a diffractive process. Due to the shorter fluctuation time expected for the virtual photon case, it should behave like a point–like structureless object which does not diffract.

It thus came as a big surprise when events of the kind presented in Fig. 53 were discovered [20, 21] in DIS NC processes. These events had a large rapidity gap between the proton direction and the first observable particle produced in the collision. None of the Monte Carlo generators written for the HERA region could predict the frequency of these events, as was shown earlier in Fig. 10.

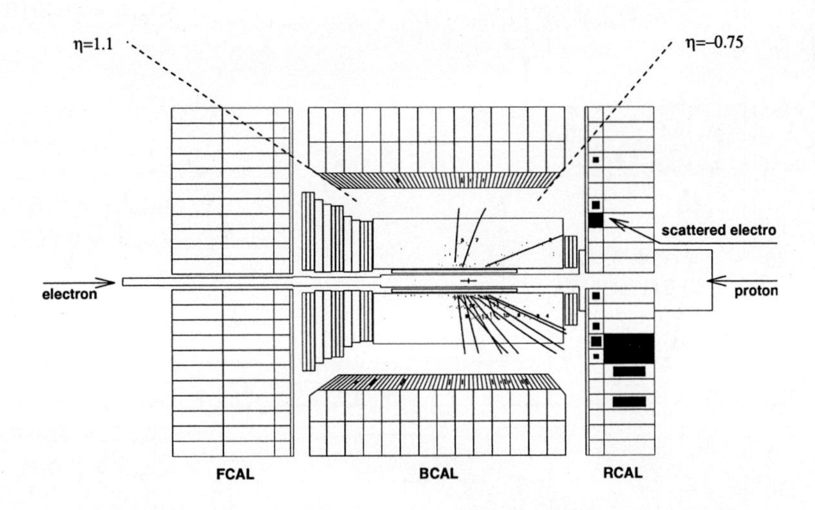

Fig. 53. *A DIS NC event in the ZEUS detector which has a large rapidity gap between the outgoing proton and the other produced particles in the ep collision.*

Are these large rapidity gap indeed diffractive? Why did one not expect earlier to see large rapidity gaps near the proton direction? In a DIS reaction the virtual photon hits one parton of the proton and produces what one calls the current jet. However due to the large colour forces the region between the current jet and the proton remnant is filled with radiated gluons and thus if one looks for instance at the energy flow [123, 124] as function of the pseudo rapidity, the forward region, which is the proton direction, is also filled with

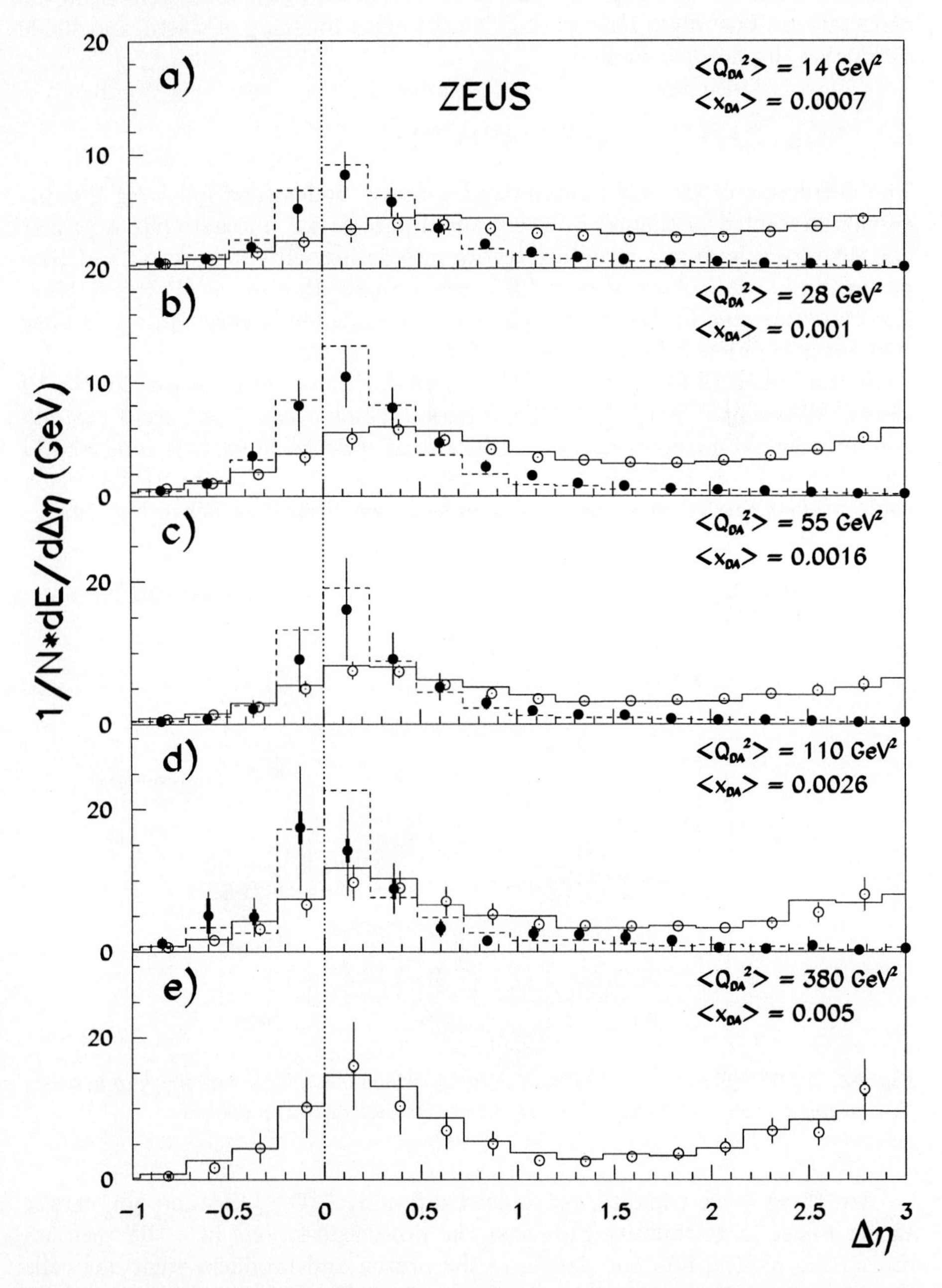

Fig. 54. *The energy flow of DIS events without a large rapidity gap (open dots) and those with a large rapidity gap ($\eta_{\max} < 1.8$, full dots), for different x, Q^2 bins.*

energy deposition in a regular DIS event. This can be seen in Fig. 54 where the energy flow is presented [123] for different x, Q^2 regions. The open data points are DIS events.

When however the virtual photon interacts with a colour singlet object, as would be the case in a diffractive process, the gluon radiation in the region between the current jet and the proton remnant is strongly suppressed and thus there should be no energy flow in the forward region. This is seen from the distribution of the full data points in Fig. 54, which have been selected as those events having $\eta_{\max} < 1.8$, meaning large rapidity gap events.

How can one be sure that these large rapidity gap events are due to a colour singlet object which is exchanged in diffractive reactions and not for instance due to the exchange of a pion, which is also a colour singlet object? One of the expected features of diffractive processes is a very slow energy dependence. Indeed the large rapidity gap events show this feature. In Fig. 55 one sees [125] on the right hand side the ratio of the large rapidity gap events to the inclusive DIS events as a function of Bjorken–x for constant Q^2 regions, which is equivalent of plotting the ratio as function of the γ^*p cms energy squared W^2. Indeed the

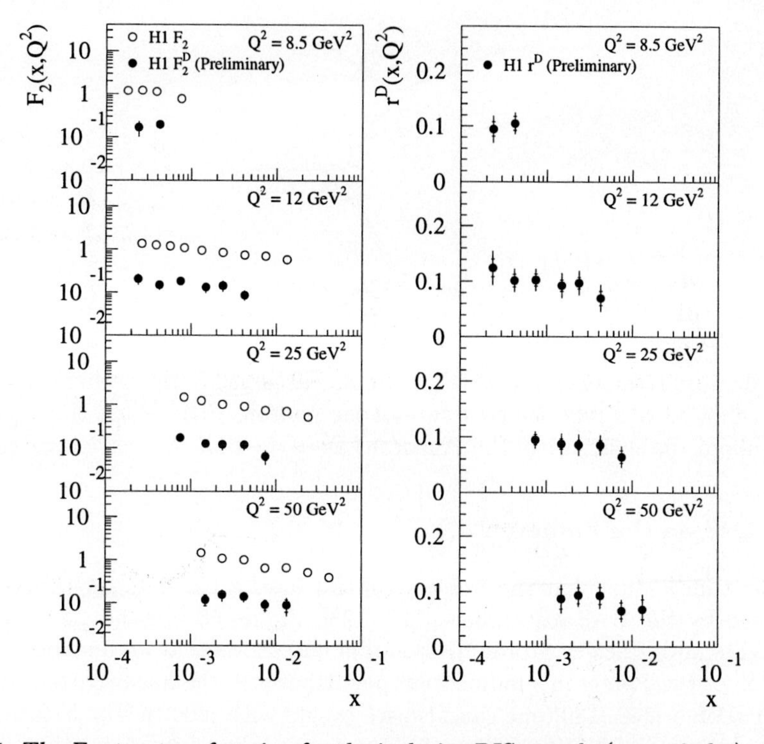

Fig. 55. *The F_2 structure function for the inclusive DIS sample (open circles) and for the large rapidity gap events (full dots) as function of Bjorken–x, for fixed Q^2 intervals. On the right hand side of the figure, their ratio is plotted as function of x for the same Q^2 intervals.*

ratio is very slowly changing with energy and seems to have roughly the same value in all the four Q^2 regions.

Another feature of the large rapidity gap events was that their M_X dependence was consistent with that expected from a diffractive process.

One could actually have anticipated the presence of diffraction in DIS, using the following argument. At high energies or equivalently in the low x region studied at HERA, the fluctuation time of the virtual photon into a state of mass $m_{\bar{q}q} \approx Q^2$ is [126]:

$$t_f \approx \frac{1}{2m_p x} \tag{6.29}$$

where m_p is the proton mass. Thus in the HERA regime, a photon of virtuality as high as $Q^2 \sim 2-3 \times 10^3$ GeV^2 can fluctuate into a $\bar{q}q$ pair, which will survive till arrival on the proton target. Therefore even highly virtual photons can produce diffractive processes which will look very similar to those in the real photon case.

Thus the large rapidity gap events have all the features expected from events produced in a diffractive process and one can interpret the interaction as that of a virtual photon interacting with a Pomeron, as described in the following diagram:

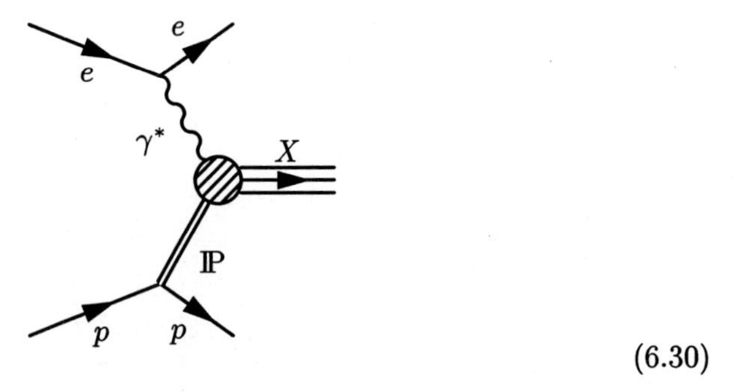

(6.30)

This diagram resembles that of the $\gamma^*\gamma$ case, discussed in the earlier chapter and which allowed to study the structure of the photon. Can we use this picture to learn about the structure of the Pomeron? Does the Pomeron have substructure?

6.5 DIS on the Pomeron

The first indication that the Pomeron might have a partonic substructure was reported by the UA8 experiment [127, 128]. Figure 56 presents the x of two–jet events in diffractive proton dissociation and shows that an unexpected large fraction of the Pomeron's momentum participates in the hard scattering.

At HERA [129, 130] one can also see events with jets. In Fig. 57(a) one can see an example of a DIS NC one–jet event which has a large rapidity gap. An example of a two–jet event with large rapidity gap is seen in Fig. 57(b).

In order to see whether these jets come from a hard scattering, one looks at the distribution of the transverse energy, which is shown in Fig. 58.

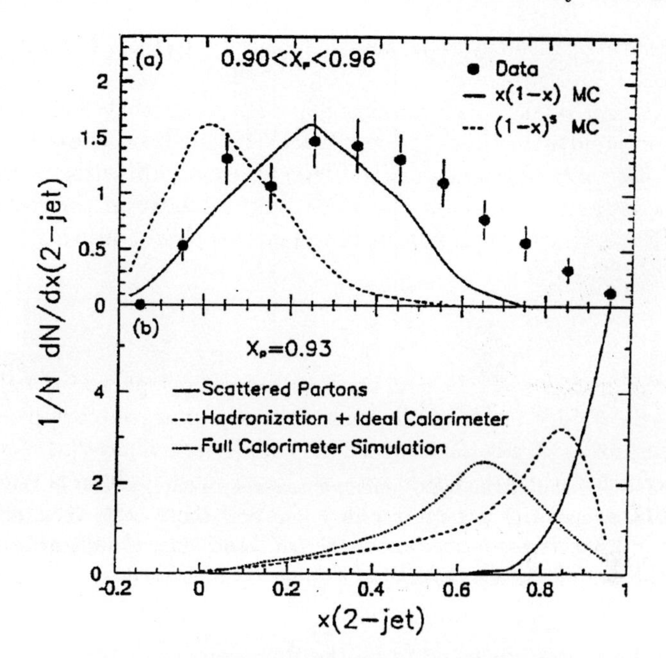

Fig. 56. *The x of two–jet events in diffractive proton dissociation in the UA8 experiment.*

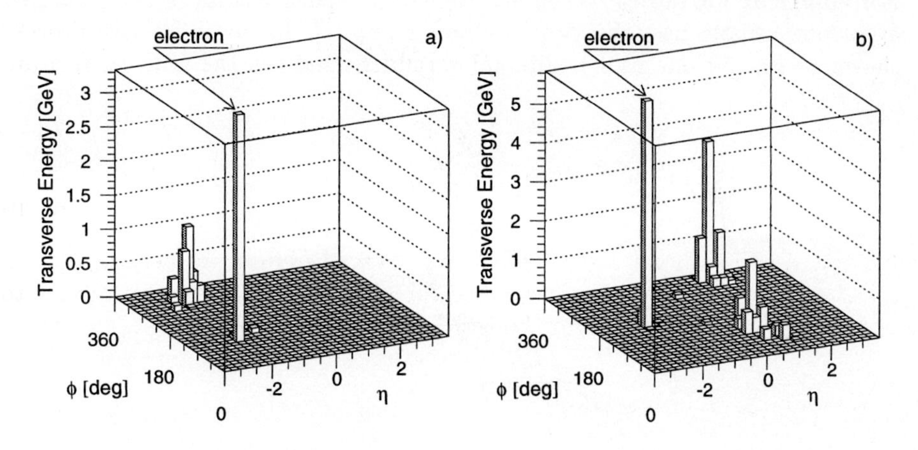

Fig. 57. *(a) Transverse energy deposition in η–ϕ space for a large rapidity gap event with one hadronic jet balancing the electron's transverse momentum. (b) A similar display for a large rapidity gap two–jet event.*

The observation of high E_T jets in the $\gamma^* p$ system for the large rapidity gap events where there is a noted absence of colour flow, indicate that a natural interpretation is the interaction of the virtual photon with partons in a colourless

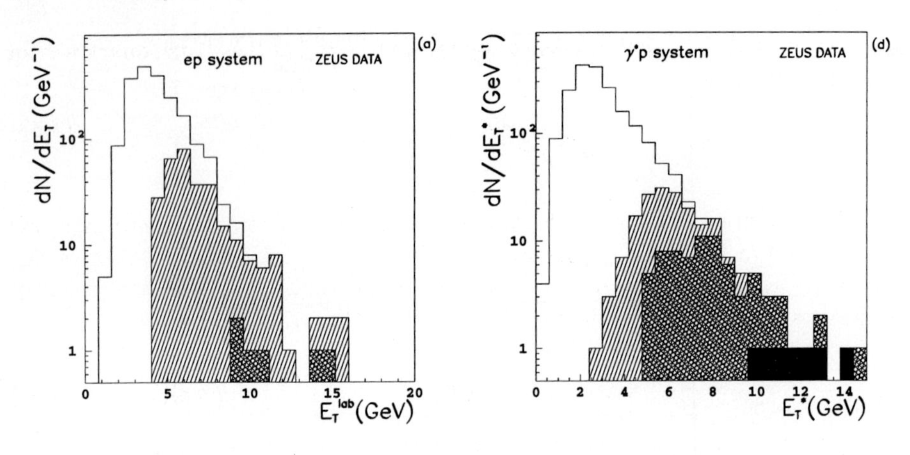

Fig. 58. *The distribution of the total hadronic transverse energy seen in the calorimeter E_T, for DIS events with a large rapidity gap and those with, in addition, ≥ 1 (hashed) and ≥ 2 jets (cross-hashed). On the left hand side, the quantity in the ep frame is presented while on the right hand side, in the $\gamma^* p$ frame.*

object inside the proton, believed to be the Pomeron.

Kinematical Variables When describing the inclusive cross section of a DIS event one usually uses the two variables x and Q^2. In the diffractive process shown in Fig. 59 one uses additional variables. One has the four momentum

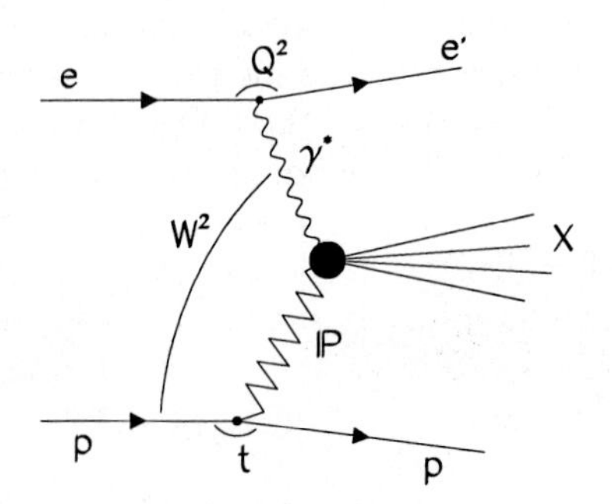

Fig. 59. *Diagram of a diffractive event.*

transfer squared at the proton vertex t defined as:

$$t = (P - P')^2 \tag{6.31}$$

The fraction of the proton momentum carried by the Pomeron is defined as:

$$x_{\mathbb{P}} = \frac{(P - P')q}{pq} \simeq \frac{M_X^2 + Q^2}{W^2 + Q^2} \tag{6.32}$$

Another variable is β, which is the momentum fraction of the struck quark within the Pomeron:

$$\beta = \frac{x}{x_{I\!P}} = \frac{Q^2}{M_X^2 + Q^2} \tag{6.33}$$

The Diffractive Structure Function With these kinematical variables one can define the diffractive structure function in a similar way to that of the inclusive DIS structure function, through the differential cross section. In order to do so we shall use the following four variables: β, Q^2, $x_{I\!P}$ and t.

$$\frac{\mathrm{d}^4\sigma_{\mathrm{diff}}}{\mathrm{d}\beta \mathrm{d}Q^2 \mathrm{d}x_{I\!P} \mathrm{d}t} = \frac{2\pi\alpha^2}{\beta Q^4}\left[\left(1+(1-y)^2\right)F_2^{D(4)} - y^2 F_L^{D(4)}\right](1+\delta_Z)(1+\delta_r) \tag{6.34}$$

where α is the electromagnetic coupling constant and the δ_i denote corrections due to Z^0 exchange and due to radiative corrections which are small in the measured range. The contribution of F_L to the diffractive cross section is not known but by restricting the measured y range to small values it can be neglected.

When t is not measured, an integration over t is performed and one determines $F_2^{D(3)}$ through the relation:

$$\frac{\mathrm{d}^3\sigma}{\mathrm{d}\beta \mathrm{d}Q^2 \mathrm{d}x_{I\!P}} \simeq \frac{2\pi\alpha^2}{\beta Q^4}[1+(1-y)^2]F_2^{D(3)}(\beta, Q^2, x_{I\!P}) \tag{6.35}$$

where one neglects the effect of F_L and the additional contributions noted above.

Factorization and Pomeron Structure Function The diffractive structure function defined above is describing the inclusive *ep* diffractive process. We can now go one step further and interpret the diffractive structure function as consisting of two parts. One in which a flux of Pomerons are emitted from the proton and another part due to the Pomeron structure function. This is reminiscent of the procedure taken when discussing the photon structure function. In order to do so, one has to assume that the Pomeron can be treated like a particle and abides to the factorization hypothesis. If so, we can define the Pomeron structure function in the following way:

$$F_2^{D(3)}(\beta, Q^2, x_{I\!P}) = f(x_{I\!P}) F_2^{I\!P}(\beta, Q^2) \tag{6.36}$$

where $f(x_{I\!P})$ is the function describing the flux of the Pomerons emitted from the proton. According to the Regge model the flux $f(x_{I\!P})$ should have an $x_{I\!P}$ dependence like:

$$f(x_{I\!P}) \sim \frac{1}{x_{I\!P}^n} \tag{6.37}$$

The exponent n is connected to the Pomeron trajectory through the relation:

$$n = 2\alpha_{I\!P}(t) - 1 \tag{6.38}$$

Since at present t is not measured, the exponent n gives a t–averaged slope of the Pomeron through relation (6.38). One can get the Pomeron intercept by assuming a diffractive slope and the slope of the Pomeron trajectory.

In order to check the factorization hypothesis, the diffractive structure function $F_2^{D(3)}$ is measured as function of $x_{\mathbb{P}}$ for fixed β and Q^2 intervals. If factorization holds, there should be one universal curve describing all data, up to a normalization factor. This is shown [125] in Fig. 60, where indeed one sees that in the range of variables presented in this figure the factorization hypothesis seems to be borne out by the data. The slope of the $x_{\mathbb{P}}$ dependence obtained by the H1 collaboration [125] is $n = 1.19 \pm 0.06(\text{stat}) \pm 0.07(\text{syst})$. The ZEUS collaboration [131] finds a slope of $n = 1.30 \pm 0.08(\text{stat})\ ^{+0.08}_{-0.14}(\text{syst})$. We will return to these results later.

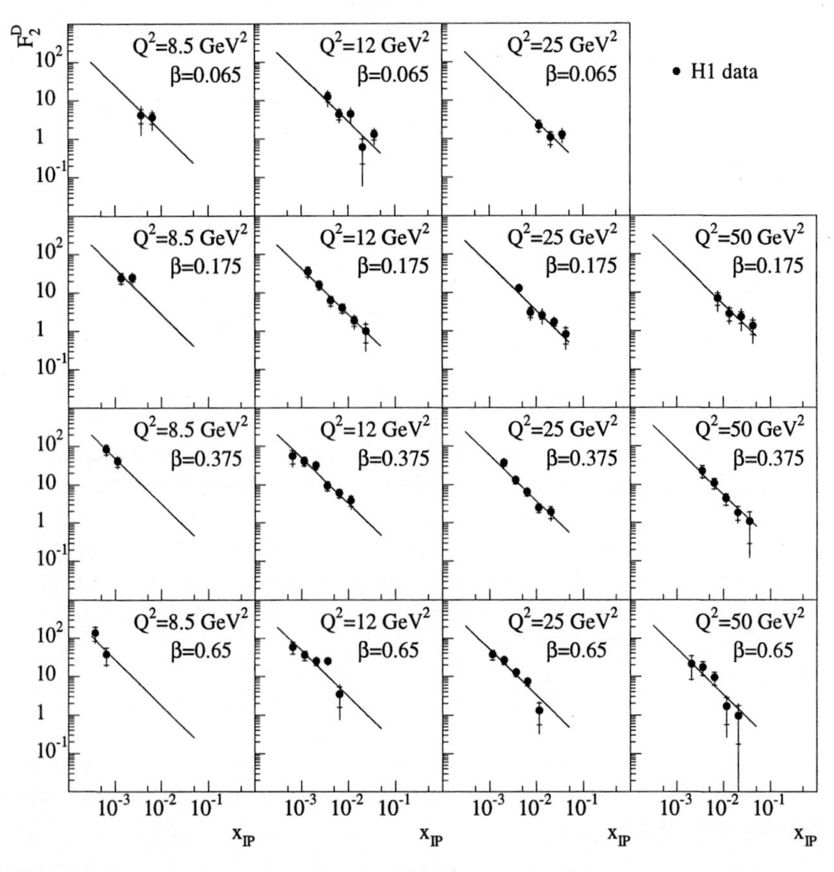

Fig. 60. *The diffractive structure function $F_2^{D(3)}$ as a function of $x_{\mathbb{P}}$ for fixed β and Q^2 intervals.*

The Partonic Structure of the Pomeron In an Ingelman–Schlein [132] type of model the Pomeron consists of partons like a regular particle. How can one get information about the quark and gluon contents of the Pomeron? We will discuss three methods to probe the partonic content of the Pomeron.

Assuming the momentum sum rule: If the Pomeron behaves like a regular particle which fulfills the momentum sum rule, one can use the flux normalization, either that of Donnachie and Landshoff [133, 134] or that of Ingelman and Schlein [132], and assume that the Pomeron consists only of quarks. In that case the quarks saturate all the momentum of the Pomeron. The predictions of this assumption can be seen [131] as the lines in Fig. 61, while the data are presented as dots.

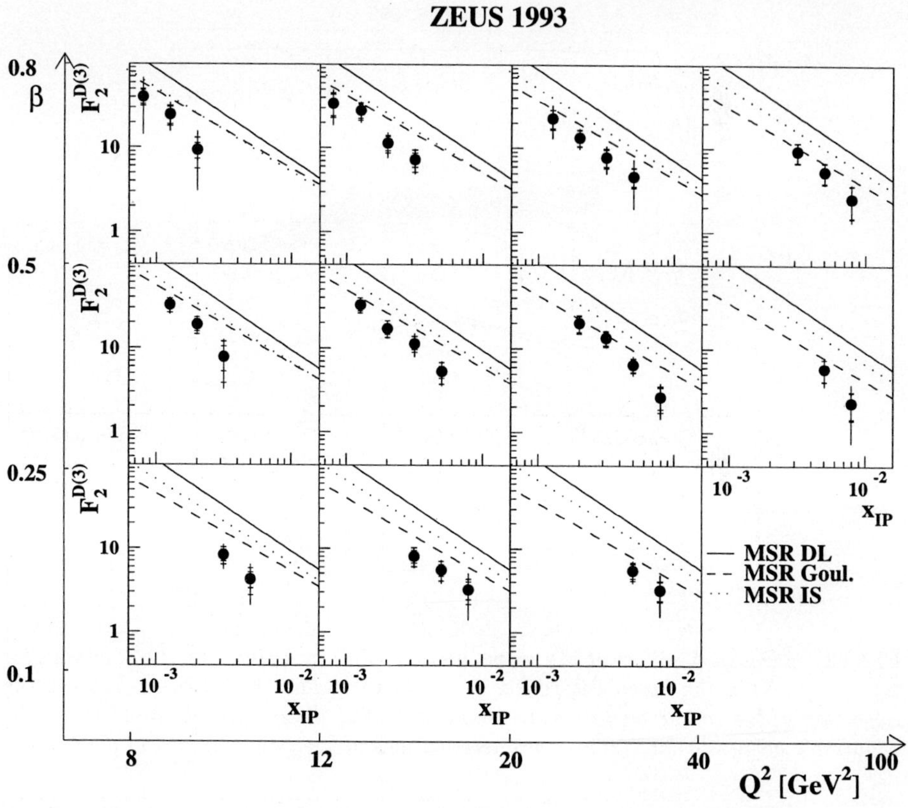

Fig. 61. *The results of* $F_2^{D(3)}$ *compared to an Ingelman–Schlein type model for which the momentum sum rule (MSR) for quarks within the Pomeron is assumed.*

Note that non–diffractive background, as well as a 15 % estimate of double dissociation has been subtracted from the data. As one sees, the data lies below the predictions indicating that the quarks carry only part of the momentum of the Pomeron. The amount depends on the expression used for the normalization of the flux.

Diffractive hard photoproduction: One can get information about the partonic content of the Pomeron by studying [135] inclusive jet cross sections for events with large rapidity gaps with respect to the proton direction from the reaction $ep \to \text{jet} + X$ with no detected electron in the final state, thus classified as photoproduction.

When one compares the measured cross sections with pQCD calculations of diffractive hard processes, as done in Fig. 62, one may conclude that the Pomeron consists of a large fraction of hard gluons. This conclusion is model dependent.

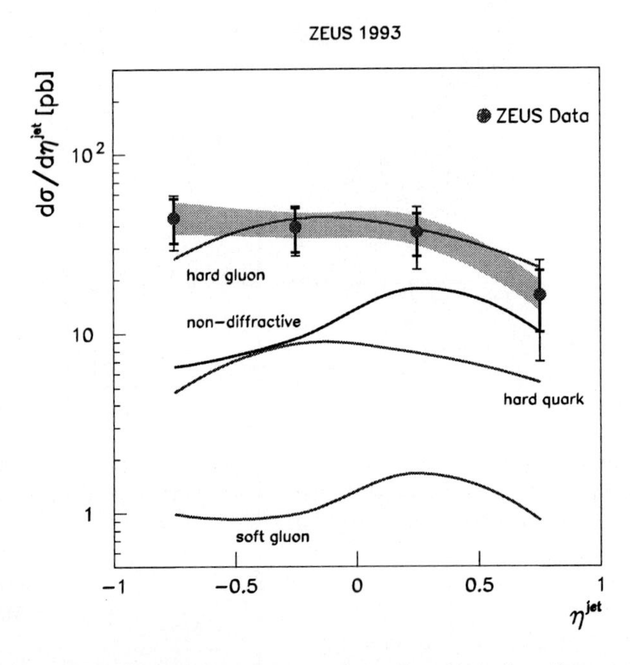

Fig. 62. *Measured differential cross section for inclusive jet production for* $E_T^{\text{jet}} > 8\,\text{GeV}$ *in the kinematic region* $Q^2 < 4\,\text{GeV}^2$. *The shaded band displays the uncertainty due to the energy scale of the jets. The lines are predictions using the POMPYT generator for various parameterizations of the Pomeron parton densities.*

However, if one combines the photoproduction measurement with the results on the diffractive structure function in deep inelastic scattering, discussed above, one finds experimental evidence for the gluon content of the Pomeron. One fits the photoproduction cross section to the expression:

$$\frac{\mathrm{d}\sigma}{\mathrm{d}\eta^{\text{jet}}} = \text{BG} + \Sigma_{\mathbb{P}} \left\{ c_g * (\text{hard gluons}) + (1 - c_g) * (\text{hard quarks}) \right\} \qquad (6.39)$$

where BG is the non–diffractive background, c_g is the fraction of hard gluons and $\sum_{\mathbb{P}}$ is the momentum sum of the Pomeron. The results of the fits are combined with those of the DIS diffractive structure function. This is shown [135] in Fig. 63 from which one may conclude that between 30 % and 80 % of the momentum of the Pomeron carried by partons is due to hard gluons.

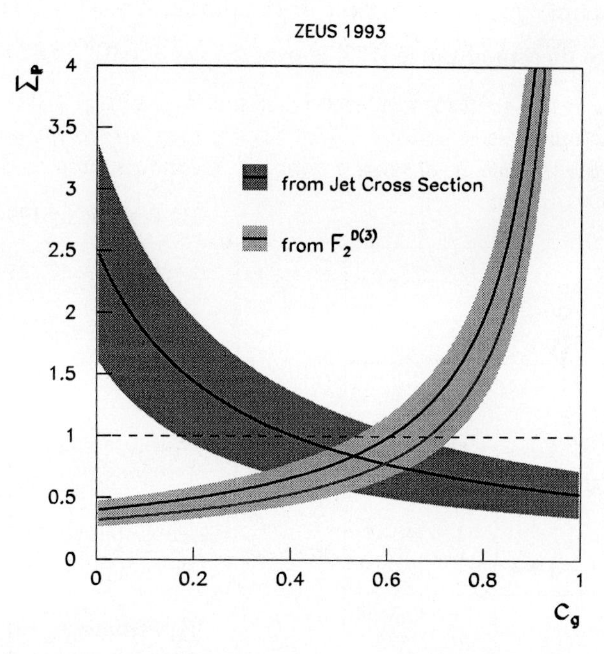

Fig. 63. *The plane of the variables $\Sigma_{\mathbb{P}}$ (momentum sum) and c_g (relative contribution of hard gluons in the Pomeron). The thick solid line displays the minimum for each value of c_g obtained from the χ^2 fit (the shaded area represents the $1\,\sigma$ band around these minima) to the measured $\mathrm{d}\sigma/\mathrm{d}\eta^{\mathrm{jet}}(\eta^{\mathrm{had}}_{\mathrm{max}} < 1.8)$ using the predictions of POMPYT. The constraint imposed in the $\Sigma_{\mathbb{P}} - c_g$ plane by the measurement of the diffractive structure function in DIS ($F_2^{D(3)}$) for two choices of the number of flavours (upper dot-dashed line for $\Sigma_{\mathbb{P}q} = 0.40$ and lower dot-dashed line for $\Sigma_{\mathbb{P}q} = 0.32$) is also shown. The horizontal dashed line displays the relation $\Sigma_{\mathbb{P}} = 1$.*

Note that this is independent of the normalization of the flux of Pomerons from the proton and does not rely on assumptions on the momentum sum of the Pomeron.

Evolution equation for the Pomeron structure function: The third method of getting information about the partonic composition of the Pomeron is to assume that one can apply the DGLAP equations also for the Pomeron structure function and perform a global QCD analysis like in the proton case, using the equations (4.36) and (4.37).

Assuming factorization it is possible to integrate $F_2^{D(3)}(\beta, Q^2, x_{\mathbb{P}})$ over the measured region of $x_{\mathbb{P}}$ to get a modified Pomeron structure function $\tilde{F}_2^{\mathbb{P}}(\beta, Q^2)$, where the tilde sign indicates that the Pomeron structure function is only for a limited $x_{\mathbb{P}}$ range:

$$\tilde{F}_2^{\mathbb{P}}(\beta, Q^2) \equiv \int_{x_{\mathbb{P}_1}}^{x_{\mathbb{P}_2}} F_2^{D(3)}(\beta, Q^2, x_{\mathbb{P}})\, \mathrm{d}x_{\mathbb{P}} \tag{6.40}$$

At present the measured range [125] is over $x_{\mathbb{P}_1} = 3 \times 10^{-4}$ and $x_{\mathbb{P}_2} = 0.05$.

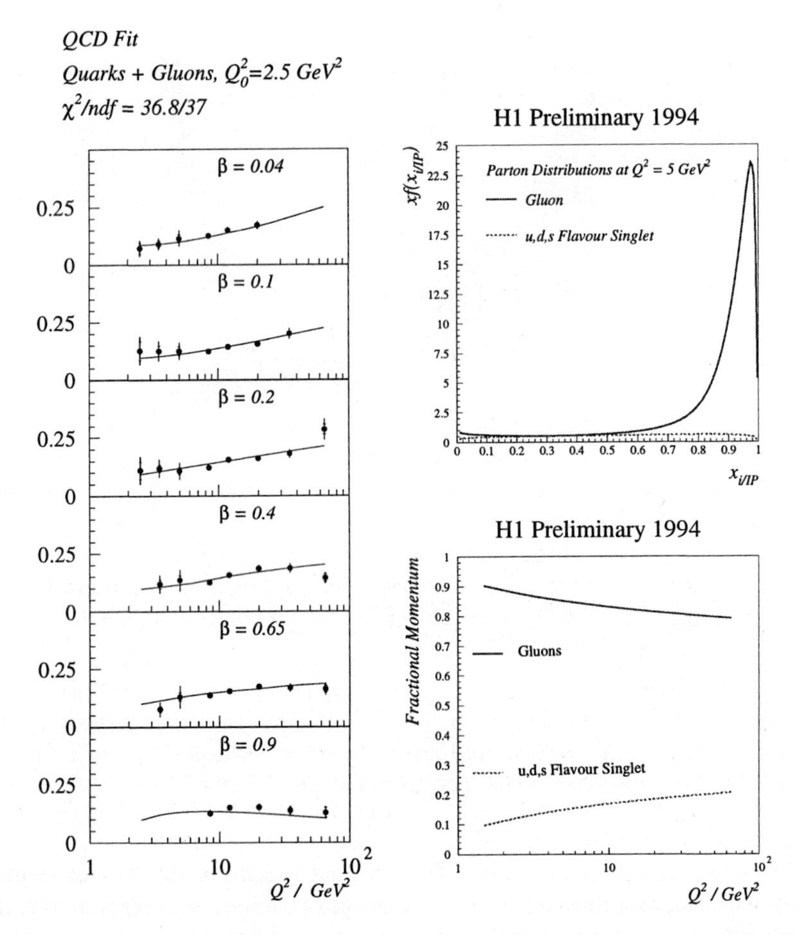

Fig. 64. *The Pomeron structure function $\tilde{F}_2^{\mathbb{P}}(\beta, Q^2)$ as function of Q^2 for fixed β regions (left hand side). The gluon momentum density distribution at a scale of $Q^2 = 5\,\mathrm{GeV}^2$ (upper right hand side). The evolution of the fractional momentum carried by the partons in the Pomeron as function of Q^2 (lower right hand side).*

The Pomeron structure function $\tilde{F}_2^{\mathbb{P}}(\beta, Q^2)$ is shown [136] in the left hand side of Fig. 64 as function of Q^2 for different β regions. At low β it shows the positive scaling violation, just like for the proton case. However at high β, it still shows a positive scaling violation, unlike the proton case and more like the photon case. The only way to get such a behaviour, assuming the homogeneous DGLAP equations to hold also for the Pomeron, is to assume a substantial gluon component in the structure of the diffractive exchange, as shown on the upper right hand side of Fig. 64. As Q^2 increases the fraction of the Pomeron momentum carried by the gluons decreases somewhat but still remains in excess of 80 %, as seen in the bottom right hand side of the figure.

The Pomeron Intercept from DIS Diffraction It is of interest to compare the Pomeron intercept as extracted from the DIS diffractive reactions to that obtained from photoproduction and from hadronic diffractive processes. We will present here two methods of obtaining the Pomeron intercept in DIS. One is using the relation (6.38) which connects the exponent n of the Pomeron flux with its trajectory. The other is to look at the W dependence of the differential cross section with respect to the diffractive mass M_X. One can of course also use the W dependence of other reactions, like the total $\gamma^* p$ cross section or that of exclusive vector meson production in DIS, to get information about the Pomeron intercept. This will be discussed in the next chapter.

$\alpha_{\mathbb{P}}$ *from* n*:* The exponent n of the Pomeron flux is related to the Pomeron trajectory through (6.38). In order to obtain the Pomeron intercept from the t–integrated value of n one usually assumes a diffractive slope of about 5–6 GeV^{-2} and a slope of the Pomeron trajectory $\alpha'_{\mathbb{P}} \sim 0.25$–$0.3\,\text{GeV}^{-2}$ [30].

The values presented [125, 131] in section 6.5 were based on low statistics data and thus had large errors for carrying out a meaningful comparison. The higher luminosity data allow a more detailed study of the behaviour of n. The value of n is shown [137] in Fig. 65(a) as function of β, integrated over the measured Q^2 range and in (b) as function of Q^2 integrated over the measured β range. The improved precision and enhanced kinematic range clearly reveals deviations from the universal factorization observed in Fig. 60. While there is no obvious dependence on Q^2, one sees that the value of n decreases significantly with β for $\beta \leq 0.3$, corresponding to large M_X values. One possible explanation [136] of this decrease, without abandoning the hypothesis of factorization, is to assume that in addition to the Pomeron there is a small contribution from the exchange of meson trajectories such as the $f_2^0(1270)$, for which one would expect $n \sim 0$. With this assumption one can explain the behaviour of n by a superposition of a Pomeron trajectory having $n_{\mathbb{P}} = 1.29 \pm 0.03(\text{stat}) \pm 0.07(\text{syst})$ and $n_M = 0.3 \pm 0.3(\text{stat}) \pm 0.6(\text{syst})$. From these values and from using the diffractive slope and Pomeron trajectory slope as mentioned above, one gets [136] a Pomeron intercept:

$$\alpha_{\mathbb{P}}(0) = 1.18 \pm 0.02(\text{stat}) \pm 0.04(\text{syst}) \tag{6.41}$$

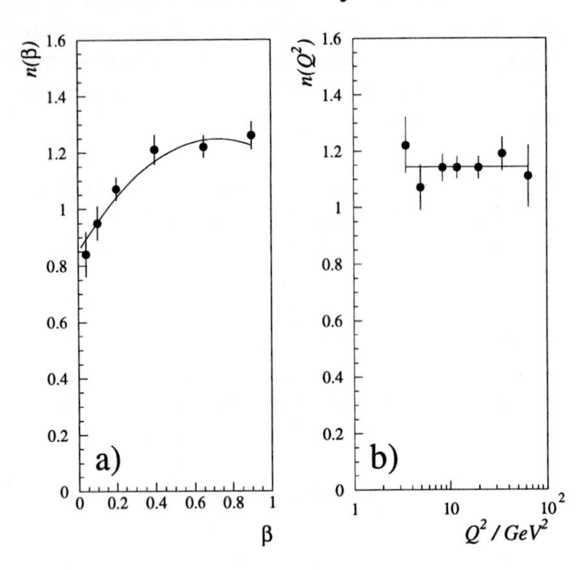

Fig. 65. *The dependence of the exponent of the Pomeron flux n on (a) β and on (b) Q^2.*

The intercept for the meson trajectory comes out from the fit to be $\alpha_M(0) = 0.6 \pm 0.1(\text{stat}) \pm 0.3(\text{syst})$, consistent with the value expected for the trajectory associated with the f_2 meson.

With the use of the leading proton spectrometer (LPS) [4] one can actually measure the t distribution of the diffractively produced DIS reactions in a limited β range. This was done [138] for the kinematic range $4 < Q^2 < 30\,\text{GeV}^2$, $70 < W < 210\,\text{GeV}$ and $0.02 < \beta < 0.4$ and is displayed in Fig. 66. The resulting value of the slope is $b = 5.9 \pm 1.3(\text{stat})^{+1.1}_{-0.7}(\text{syst})\,\text{GeV}^{-2}$. The LPS data have been used to extract the diffractive structure function and yielded [138] an exponent value of $n_{\mathbb{P}} = 1.28 \pm 0.07(\text{stat}) \pm 0.15(\text{syst})$ which can be converted to a Pomeron slope of

$$\alpha_{\mathbb{P}}(0) = 1.17 \pm 0.04(\text{stat}) \pm 0.08(\text{syst}) \tag{6.42}$$

in good agreement with the value in (6.41). Note that in case of the LPS measurement the outgoing proton is detected and thus one has no corrections for background or for double dissociation processes.

$\alpha_{\mathbb{P}}$ from the W dependence of the diffractive cross section: One can use the reaction $\gamma^* p \to XN$, where N is a nucleonic system with $M_N < 4\,\text{GeV}$, to measure the diffractive differential cross section $\text{d}\sigma^{\text{diff}}/\text{dM}_\text{X}$. This differential cross section has a W dependence which in the Regge model is given by $(W^2)^{(2\alpha_{\mathbb{P}}-2)}$. A novel method of extracting the diffractive cross section from the non–diffractive

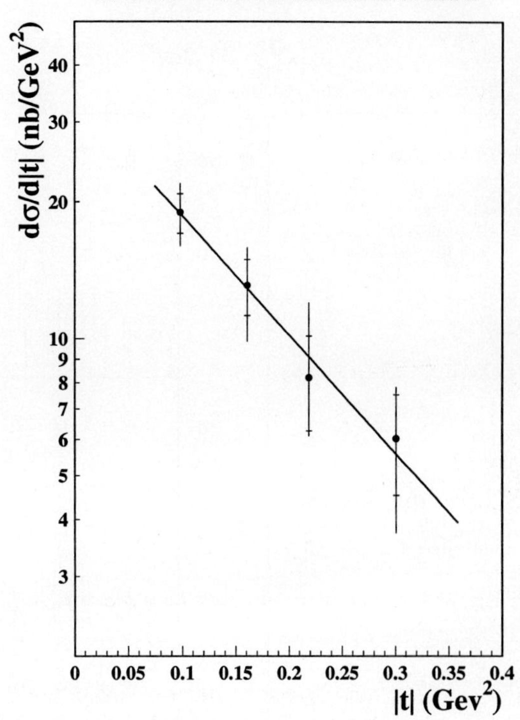

Fig. 66. *Differential cross section* $d\sigma/dt$ *for diffractive DIS events with a leading proton with a longitudinal momentum fraction* $x_L > 0.97$, *in the range* $4 < Q^2 < 30\,\text{GeV}^2$, $70 < W < 210\,\text{GeV}$ *and* $0.02 < \beta < 0.4$.

background is the realization that for the latter low $\ln M_X^2$ of the hadronic system observed in the detector are exponentially suppressed. This can be seen [139] in Fig. 67 for the W range of $W = 60 - 245\,\text{GeV}$ at $Q^2 = 31\,\text{GeV}^2$. One sees that the non–diffractive contribution moves to larger $\ln M_X^2$ values proportional to $\ln W^2$.

The differential cross section for the diffractive DIS reaction was determined as function of W for different M_X and Q^2 regions. In each (M_X, Q^2) bin, a fit of the form:

$$\frac{d\sigma^{\text{diff}}(\text{M}_\text{X}, \text{W}, \text{Q}^2)}{d\text{M}_\text{X}} \sim (W^2)^{(2\alpha_{\mathbb{P}}-2)} \tag{6.43}$$

was performed, yielding a series of values for the Pomeron intercept for each (M_X, Q^2) bin. One gets intercept values consistent with those mentioned in the earlier section.

How can one understand the Pomeron intercept values in DIS? They seem to be significantly higher than the value of 1.08 obtained from photoproduction or

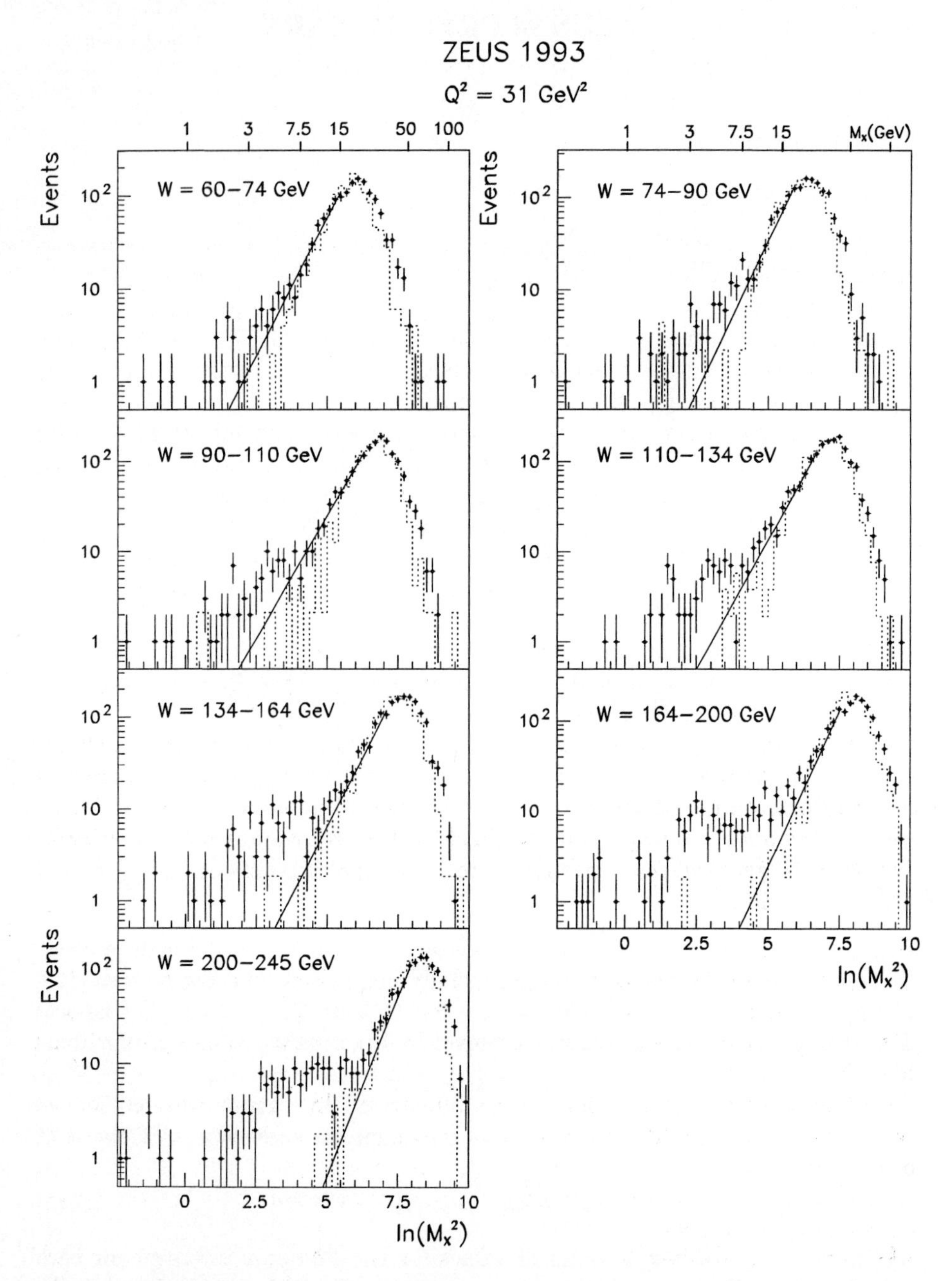

Fig. 67. *Distribution of* $\ln M_X^2$ *for the* W *intervals indicated at* $Q^2 = 31\,\mathrm{GeV}^2$. *The solid lines show the extrapolation of the non–diffractive background.*

hadronic reactions. Is it a different Pomeron? Are there two Pomerons? Does one approach the BFKL Pomeron of an intercept of 1.5 when measuring at higher Q^2 values? We will discuss these questions in the next chapter.

6.6 Summary

In this chapter we discussed the following issues:

- We defined some basic concepts connected with diffractive processes and showed their connection with the Regge formalism. In particular we discussed the trajectory intercept and its relation to the total and diffractive cross section. In addition we discussed the shrinkage of the elastic scattering slope.
- The diffractive phenomena is clearly observed in photoproduction reactions with a rate similar to that in hadronic reactions.
- Diffractive processes are present also in DIS reactions due to the fluctuations of highly virtual photons in the low x region. These processes are observed as a high rate of large rapidity gap events in DIS reactions. This opens up the possibility of studying DIS on the Pomeron.
- Observation of jets in DIS diffractive processes indicate that one could interpret them as the interaction of the virtual photon with partons within the Pomeron.
- We defined the diffractive structure function and checked that experimentally the factorization hypothesis holds in case of the Pomeron, over the measured kinematic region.
- The partonic content of the Pomeron was studied and evidence was shown for a large gluonic component in the Pomeron, carrying a large fraction of the Pomeron momentum.
- The Pomeron intercept as determined in DIS processes seems to be significantly larger than that determined from photoproduction and hadronic total cross section.

7 Interplay Between Soft and Hard Interactions

This chapter deals with the interrelations of soft and hard processes. We know how to calculate hard processes by using pQCD. However when we compare the calculation with data we would like to isolate only the hard part. Do we really know how to do it? Are the two processes completely separable? How can one define hard reactions? These questions will be discussed in this chapter. For completeness and easy reading, some of the arguments presented in the earlier chapters will be repeated.

7.1 Introduction

One of the aims of building HERA was to study the deep inelastic scattering (DIS) region with data at low x and high Q^2. Yet, recently efforts are being made to get to lower and lower Q^2 values in the low x region in order to study the transition from photoproduction to the DIS regime. The main motivation for looking at the transition region is the following: at $Q^2 = 0$ the dominant processes are of non–perturbative nature and are well described by the Regge picture. As Q^2 increases, the exchanged photon is expected to shrink, one expects perturbative QCD to take over and therefore to be able to make exact calculations to confront with data. What can one learn from the transition between soft processes (low virtuality) and hard processes (high virtuality)? Where does the change take place? Is it a sudden transition or a smooth one? The transition should shed light on the interplay between soft and hard interactions.

7.2 Operational Definition

It is not completely clear what one means by soft and hard interaction. One would have hoped that by going to the region of DIS one has a better way of probing the hard interactions. As a guideline to help distinguish the two, let us define some operational criteria for what we would consider as a soft and as a hard process. We can not do it in the most general terms, but let us concentrate on some selected measurements: total cross sections and elastic cross sections, the first being the most inclusive and the latter the most exclusive measurement we can make. At high energies, both these processes are dominated by a Pomeron exchange.

As discussed earlier, the total $\pi^{\pm}p, K^{\pm}p, pp, \bar{p}p$ and γp cross sections show a slow dependence on the center of mass energy W, consistent with the so–called soft pomeron [31], having a trajectory

$$\alpha_{\mathbb{P}(\text{soft})} = 1.08 + 0.25t \tag{7.1}$$

The hard or the perturbative Pomeron, also called the Lipatov Pomeron or the BFKL [60, 61, 62] Pomeron, is expected to have a trajectory

$$\alpha_{\mathbb{P}(\text{hard})} = 1 + \frac{12 \ln 2}{\pi} \alpha_S \tag{7.2}$$

The definition of the hard $\mathbb{P}$ is quite vague. First, the value of the intercept which is usually taken as 1.5 is a very rough estimate using the expression of the expected power of the reggeized gluon. Using a leading order calculation in $\ln 1/x$, the distribution of the momentum density of the gluon is expected to have the form $xg(x, Q^2) \sim x^{-\lambda}$ where $\lambda = \alpha_S/0.378$. Although usually the value of λ is taken to be 0.5 [140, 141], one should note that this requires a value of $\alpha_S = 0.18$, which happens only at large Q^2, whereas the BFKL calculation is expected to be valid for moderate Q^2 values. The second comment about the assumed hard $\mathbb{P}$ form is the fact that the slope of this trajectory is taken to be zero. The reason for this assumption can be understood intuitively by the fact that the slope is inversely proportional to the average transverse momentum square of hadrons, which is expected to be much larger in hard interactions compared to soft ones.

Table 3. *Expected behaviour of soft and hard processes.*

quantity	W dep	soft	hard
$\sigma_{tot}^{\gamma^* p}$	$(W^2)^{\alpha_{\mathbb{P}}(0)-1}$	$(W^2)^{0.08}$	$(W^2)^{0.5}$
slope b of $d\sigma/dt$	$\sim 2\alpha' \ln W^2$	shrinkage	no shrinkage
$\sigma(\gamma^* p \to Vp)$	σ_{tot}^2/b	$(W^2)^{0.16}/b$	$(W^2)^1$

Following the above definitions of the soft and the hard Pomeron, we have some expectations for the behaviour of the total $\gamma^* p$ cross section, $\sigma_{tot}^{\gamma^* p}$, and the elastic one, which in the HERA case is the reaction $\sigma(\gamma^* p \to Vp)$. These are presented in Table 3.

Before turning to the actual data, let us review some of the models relevant for the low x and low Q^2 region.

7.3 The Models for the Low x Low Q^2 Region

Donnachie and Landshoff (DL) Donnachie and Landshoff [31] found a simple Regge picture describing all hadron–hadron cross sections with a sum of two terms, that of a Pomeron exchange and that of a reggeon. They showed this picture to describe also real photoproduction cross sections. They extended the picture for virtual photons ($\gamma^*, Q^2 < 10\,\mathrm{GeV}^2$) to see what is the expected contribution of the non–perturbative mechanism to higher Q^2 [142, 74]. The main interest is in the low x region where the Pomeron dominates and thus the question of interest is what is the contribution of the 'soft' pomeron at intermediate Q^2.

Capella, Kaidalov, Merino, Tran–Than–Van (CKMT) In this picture [143, 144] there is no 'soft' or 'hard' Pomeron, there is just one 'bare' Pomeron. At low Q^2 absorptive corrections (rescattering) give a Pomeron with an effective intercept of $1+\Delta_0(\Delta_0 \sim 0.08)$. If one uses an eikonal approach, the bare intercept becomes $1+\Delta_1(\Delta_1 \sim 0.13)$. A more complete absorptive calculation results in $1+\Delta_2(\Delta_2 \sim 0.24)$. The absorptive corrections decrease rapidly with Q^2. They parametrize the data with this behavior of the Pomeron up to $Q^2 < 5\,\mathrm{GeV}^2$ and use it then as initial conditions to a pQCD evolution.

Badelek and Kwiecinski (BK) Badelek and Kwiecinski [145, 146] describe the low Q^2 region by using the generalized vector dominance model (GVDM): the proton structure function F_2 is represented by the contribution of a large number of vector mesons which couple to virtual photons. The low mass ones, ρ, ω, ϕ contribute mainly at low Q^2, while the higher mass are determined by the asymptotic structure function F_2^{AS} which is described by pQCD. The total structure function is given by a Q^2 weighted sum of the two components.

Abramowicz, Levin, Levy, Maor (ALLM) This parameterization [18, 19] is based on a Regge motivated approach extended into the large Q^2 regime in a way compatible with QCD expectations. This approach allows to parametrize the whole x, Q^2 phase space, fitting all the existing data.

Some General Comments The DL parametrization provides a good way to check to what value of Q^2 the simple 'soft' Pomeron picture can be extended. It is not meant to be a parameterization which describes the whole DIS regime. The CKMT and BK parametrizations are attempts to get the best possible presentation of the initial conditions to a pQCD evolution. The ALLM does not use the regular pQCD evolution equation but parametrizes the whole of the DIS phase space by a combination of Regge and QCD motivated parametrizations.

All parameterizations make sure that as $Q^2 \to 0$ also $F_2 \to 0$ linearly with Q^2.

Details of the Parametrizations

The DL parameterization: The proton structure function F_2 is given by

$$F_2(x, Q^2) \sim A\xi^{-0.0808}\phi(Q^2) + B\xi^{0.4525}\psi(Q^2), \tag{7.3}$$

where ξ is the rescaled variable

$$\xi = x\left(1 + \frac{\mu^2}{Q^2}\right), \tag{7.4}$$

with x being the Bjorken–x and the scale variable μ has different values for different flavors: for u and d quarks $\mu = 0.53\,\text{GeV}$, for the strange quark s, $\mu = 1.3\,\text{GeV}$ and for the charm quark c, $\mu = 2\,\text{GeV}$. The two functions $\phi(Q^2)$ and $\psi(Q^2)$ make sure that the structure function vanishes linearly with Q^2 as $Q^2 \to 0$,

$$\phi(Q^2) = \frac{Q^2}{Q^2 + a} \qquad \psi(Q^2) = \frac{Q^2}{Q^2 + b}. \tag{7.5}$$

The four parameters A, B, a and b are constrained so as to reproduce the total real photoproduction data,

$$\frac{A}{a}(\mu^2)^{-0.0808} = 0.604 \qquad \frac{B}{b}(\mu^2)^{0.4525} = 1.15. \tag{7.6}$$

In addition there is also a higher–twist term $ht(x, Q^2)$ contributing to the structure function,

$$ht(x, Q^2) = D\frac{x^2(1 - \xi)^2}{1 + \frac{Q^2}{Q_0^2}}, \tag{7.7}$$

with the parameters $D = 15.88$ and $Q_0 = 550\,\text{MeV}$.

The CKMT parameterization: Contrary to the DL parameterization, the CKMT assumes that the power behavior of x is Q^2 dependent,

$$F_2(x,Q^2) = Ax^{-\Delta(Q^2)}(1-x)^{n(Q^2)+4}\left(\frac{Q^2}{Q^2+a}\right)^{1+\Delta(Q^2)} + \\ +Bx^{1-\alpha_R}(1-x)^{n(Q^2)}\left(\frac{Q^2}{Q^2+b}\right)^{\alpha_R}, \tag{7.8}$$

where α_R is the Reggeon trajectory intercept, the power $n(Q^2)$ is given by

$$n(Q^2) = \frac{3}{2}\left(1+\frac{Q^2}{Q^2+c}\right) \tag{7.9}$$

and the power behavior of x is given by

$$\Delta(Q^2) = \Delta_0\left(1+\frac{Q^2}{Q^2+d}\right). \tag{7.10}$$

The constant parameters are determined by the requirement that F_2 and the derivative $\frac{\mathrm{d}F_2}{\mathrm{d}\ln Q^2}$ at $Q^2 = Q_0^2$ to coincide with that obtained from the pQCD evolution equations. They can do so at $Q_0^2 = 2\,\mathrm{GeV}^2$, provided a higher–twist term is added to that of pQCD,

$$F_2(x,Q^2) = F_2^{pQCD}(x,Q^2)\left(1+\frac{f(x)}{Q^2}\right) \tag{7.11}$$

for $Q^2 \geq Q_0^2$. The values of the parameters are: $A = 0.1502$, $a = 0.2631\,\mathrm{GeV}^2$, $\Delta_0 = 0.07684$, $d = 1.117\,\mathrm{GeV}^2$, $b = 0.6452\,\mathrm{GeV}^2$, $\alpha_R = 0.415$, $c = 3.5489\,\mathrm{GeV}^2$.

The BK parameterization: The proton structure function is written as the sum of two terms, a vector meson part (V) and a partonic part (par),

$$F_2(x,Q^2) = F_2^V(x,Q^2) + F_2^{\mathrm{par}}(x,Q^2). \tag{7.12}$$

The part representing the contribution from vector mesons which couple to the virtual photon is given by

$$F_2^V(x,Q^2) = \frac{Q^2}{4\pi}\Sigma_V\frac{M_V^4\sigma_V(W^2)}{\gamma_V^2(Q^2+M_V^2)^2}, \tag{7.13}$$

where $\gamma_V^2/(4\pi)$ is the direct photon vector meson coupling, W is the $\gamma^* p$ center of mass energy and σ_V is the total Vp cross section. The sum is over all vector meson satisfying $M_V^2 < Q_0^2$, where M_V is the mass of the vector meson and Q_0 is a parameter.

The partonic part of the structure function is given by the expression

$$F_2^{\mathrm{par}}(x,Q^2) = \frac{Q^2}{Q^2+Q_0^2}F_2^{AS}(\bar{x},Q^2+Q_0^2), \tag{7.14}$$

where the asymptotic structure function F_2^{AS} is given by pQCD at the scaled value of

$$\bar{x} = \frac{Q^2 + Q_0^2}{W^2 + Q^2 - M^2 + Q_0^2}, \tag{7.15}$$

where M is the proton mass. In practice the parameterization uses $Q_0^2 = 1.2\,\mathrm{GeV}^2$ and thus sums over the contribution of the 3 lightest vector mesons ρ, ω and ϕ.

The ALLM parameterization: This parameterization attempts to cover the whole x, Q^2 region above the resonances ($W^2 > 3\,\mathrm{GeV}^2$), at the expense of introducing more parameters than the other parameterizations. The proton structure function has the form

$$F_2(x,Q^2) = \frac{Q^2}{Q^2 + M_0^2}\left(F_2^{\mathcal{P}}(x,Q^2) + F_2^{\mathcal{R}}(x,Q^2)\right), \tag{7.16}$$

where M_0 is the effective photon mass. The functions $F_2^{\mathcal{P}}$ and $F_2^{\mathcal{R}}$ are the contribution of the Pomeron $\mathcal{P}$ or Reggeon $\mathcal{R}$ exchanges to the structure function. They take the form

$$\begin{aligned} F_2^{\mathcal{P}}(x,Q^2) &= c_{\mathcal{P}}(t) x_{\mathcal{P}}^{a_{\mathcal{P}}(t)} (1-x)^{b_{\mathcal{P}}(t)}, \\ F_2^{\mathcal{R}}(x,Q^2) &= c_{\mathcal{R}}(t) x_{\mathcal{R}}^{a_{\mathcal{R}}(t)} (1-x)^{b_{\mathcal{R}}(t)}. \end{aligned} \tag{7.17}$$

The slowly varying function t is defined as

$$t = \ln\left(\frac{\ln \frac{Q^2+Q_0^2}{\Lambda^2}}{\ln \frac{Q_0^2}{\Lambda^2}}\right). \tag{7.18}$$

The two scaled variables $x_{\mathcal{P}}$ and $x_{\mathcal{R}}$ are modified Bjorken–x variables which include mass parameters $M_{\mathcal{P}}$ and $M_{\mathcal{R}}$ which can be interpreted as effective Pomeron and reggeon masses:

$$\begin{aligned} \frac{1}{x_{\mathcal{P}}} &= 1 + \frac{W^2 - M^2}{Q^2 + M_{\mathcal{P}}^2}, \\ \frac{1}{x_{\mathcal{R}}} &= 1 + \frac{W^2 - M^2}{Q^2 + M_{\mathcal{R}}^2}. \end{aligned} \tag{7.19}$$

7.4 Comparison to Data

The Total $\gamma^* p$ Cross Section, $\sigma_{tot}^{\gamma^* p}$ The total $\gamma^* p$ cross section, $\sigma_{tot}^{\gamma^* p}$, can be related to the proton structure function F_2 through the relation

$$F_2(x,Q^2) = \frac{Q^2(1-x)}{4\pi^2\alpha}\frac{Q^2}{Q^2 + 4m_p^2 x^2}\sigma_{tot}^{\gamma^* p}(x,Q^2) \tag{7.20}$$

where the total $\gamma^* p$ includes both the cross section for the absorption of transverse and of longitudinal photons. In this expression the Hand [47] definition of the flux of virtual photons is used.

Figure 68 presents the dependence of $\sigma_{tot}^{\gamma^* p}$, obtained through ((7.20)) from the measured F_2 values [15, 16], on the square of the center of mass energy W^2, for fixed values of the photon virtuality Q^2. The new preliminary very low Q^2 measurements of the ZEUS collaboration [147], as well as those of the NMC collaboration [52] are included in the figure. Also shown are the measurements of the total real photoproduction cross sections. While the data below $Q^2 = 1\,\mathrm{GeV}^2$ show a very mild W dependence, the trend changes as Q^2 increases. Note that for higher values of Q^2 one sees the typical threshold behaviour for the case when $W^2 < Q^2$ [17]. The curves are the results of a new ALLM type parametrization which added to the earlier data used in the previous fit data from E665 [148] and the published HERA [149, 150] data.

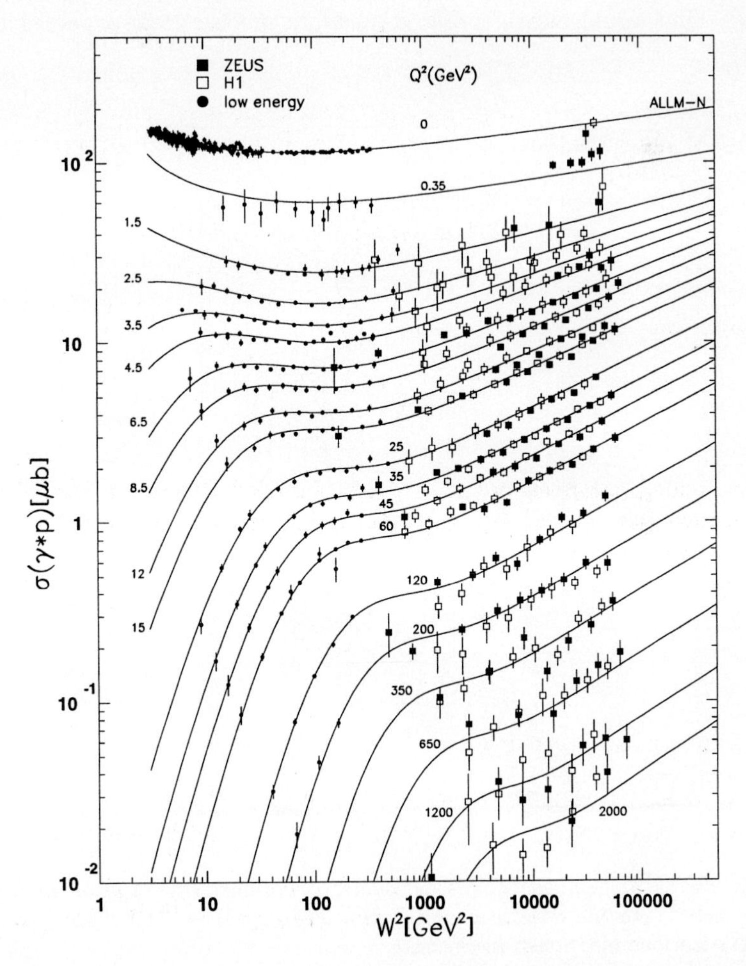

Fig. 68. *The total $\gamma^* p$ cross section as function of W^2 from the F_2 measurements for different Q^2 values. The lines are the expectations of a new ALLM type parametrization.*

Instead of comparing the data as presented in Fig. 68 with the different parameterizations, it is more economical as well as instructive to study the energy dependence of the $\gamma^* p$ cross section for fixed Q^2 values [59]. In order to see how the slope of the W dependence changes with Q^2, the cross section values in the region where $W^2 \gg Q^2$ were fitted to the form $\sigma_{tot}^{\gamma^* p} = \sigma_1 W^{2\Delta}$ for each fixed Q^2 interval. The resulting values of Δ from the fit are plotted against Q^2 in Fig. 69. Similar results have been obtain by the H1 collaboration [16] who use only their own data to fit the structure function measurements to the form $F_2 \sim x^{-\Delta}$. Also included in the figure are the recent preliminary results of the ZEUS collaboration [147] in the region $0.2 < Q^2 < 0.8\,\mathrm{GeV}^2$. One can see the slow increase of Δ with Q^2 from the value of 0.08 at $Q^2 = 0$, to around 0.2 for $Q^2 \sim 10 \ldots 20\,\mathrm{GeV}^2$ followed by a further increase to around $0.3 \ldots 0.4$ at high Q^2. One would clearly profit from more precise data, expected to come soon.

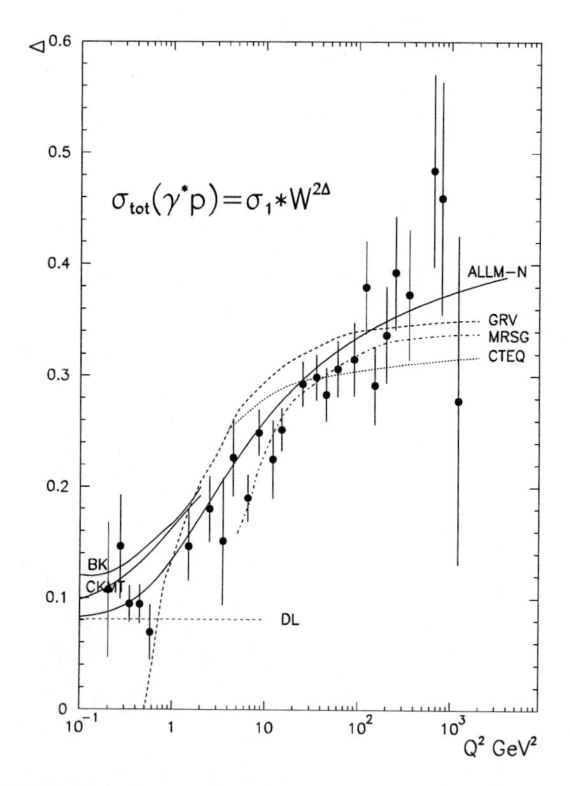

Fig. 69. *The Q^2 dependence of the parameter Δ obtained from a fit of the expression $\sigma_{tot}^{\gamma^* p} = \sigma_1 W^{2\Delta}$ to the data in each Q^2 bin. The curves are the expectations of the parameterizations mentioned in the text.*

The curves are the expectations of the DL, BK, CKMT, and the updated

ALLM parameterization, which includes also some of the recent HERA data in its fit. In addition, the expectations of the GRV [73], CTEQ [72] and MRSG [70] are also shown.

The DL parameterization can describe the data up to $Q^2 \sim 1\,\mathrm{GeV}^2$. All the others give in general the right features of the Q^2 behavior with a smooth transition from soft to hard interactions with an interplay between the two in the intermediate Q^2 range.

Vector Meson Production in γp and in $\gamma^* p$ Given the behaviour of the $\sigma_{tot}^{\gamma^* p}$ data, what kind of energy behaviour would one expect for the 'elastic' process $\gamma^* p \to Vp$ for real and virtual photons? In case of photoproduction we have seen that the total cross section follows the expectations of a soft DL type $\mathbb{P}$. Thus if one takes into account the shrinkage at the HERA energies, one expects $\sigma(\gamma p \to Vp) \sim W^{0.22}$. In case of DIS production of vector mesons in the range $Q^2 \sim 10\ldots 20\,\mathrm{GeV}^2$, the expectations are $\sigma(\gamma^* p \to Vp) \sim W^{0.8}$, since in this case one expects almost no shrinkage.

Figure 70 presents the measurements of the total and 'elastic' vector meson photoproduction cross sections as function of the γp center of mass energy W. As one can see, the high energy measurements of the total and the ρ [151, 152],

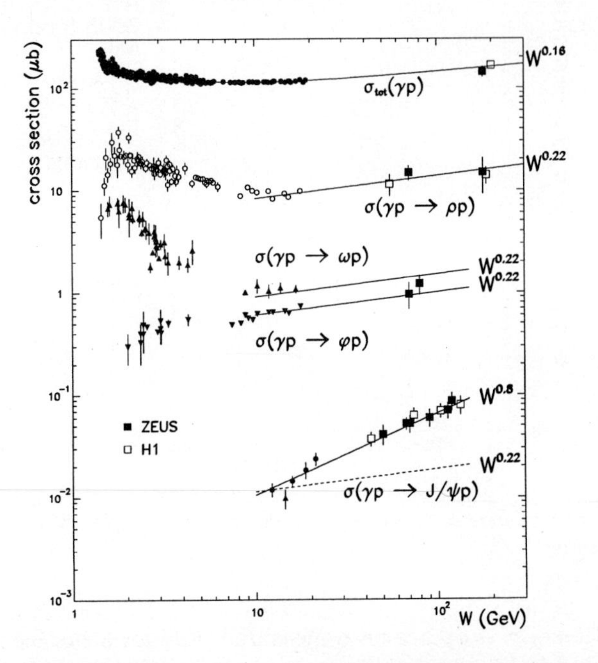

Fig. 70. *The total and 'elastic' vector meson photoproduction measurements as function of W, for the vector mesons ρ, ω, ϕ and J/Ψ. The curve to the total photoproduction cross section is the DL parametrization ($W^{0.16}$). The other lines are curves of the form $W^{0.22}$ and $W^{0.8}$.*

ω [153] and ϕ photoproduction [154] follow the expectations of a soft DL type Pomeron. However, the cross section for the reaction $\gamma p \to J/\Psi p$ [86, 155] rises much faster than the expected $W^{0.22}$ rise from a soft reaction. In fact, it can be well described by a power behaviour of $\sim W^{0.8}$. This surprising behaviour can be understood if one considers the scale which is involved in the interaction. In case of photoproduction reaction, the scale cannot be set by the photon since $Q^2 = 0$. The scale is set by the mass of the vector meson and by the transverse momentum involved in the reaction. Thus, for the lighter vector mesons the scale is still small enough to follow a soft behaviour. However, the mass of the J/Ψ is large enough to produce a scale which would be considered as a hard interaction.

The reaction $\gamma^* p \to \rho^0 p$ has been measured [156, 157] at six Q^2 values from 0.48 to 20 GeV^2 and is shown in Fig. 71. One observes that the W dependence gets steeper as Q^2 increases. In order not to be dependent on the normalizations of different experiments, the ZEUS data alone has been fitted to a W^a form. Though the data has quite large errors which is reflected in the large errors on the power a, one sees the trend of increasing a with Q^2.

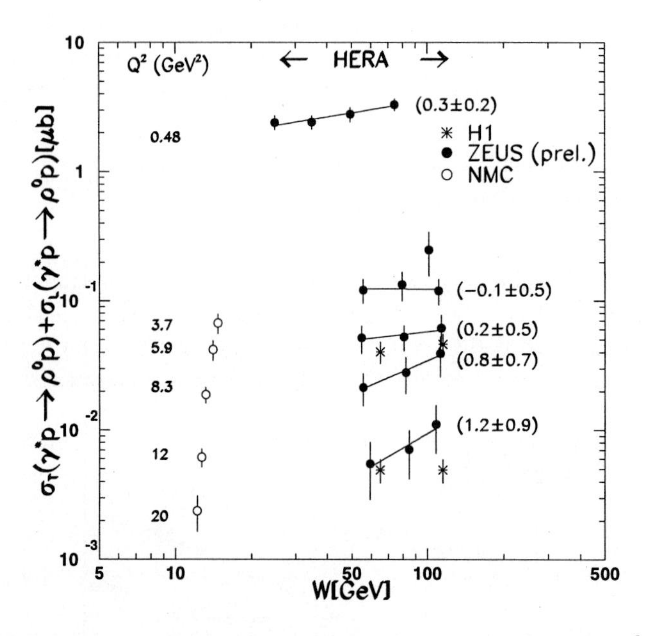

Fig. 71. *The dependence of the cross section for the reaction $\gamma^* p \to \rho^0 p$ on W, for different Q^2 values.*

The reaction $\gamma^* p \to \phi p$ has been measured [158] for Q^2 of 8.2 and 14.7 GeV^2 and is presented in Fig. 72. In this case, one has to use the lower W NMC data to get the slope of the energy dependence. It is steeper than that expected for a soft process and is compatible with the $W^{0.8}$ observed for the photoproduction J/Ψ case. For comparison, the $Q^2 = 0$ photoproduction with the shallow W

dependence is also shown.

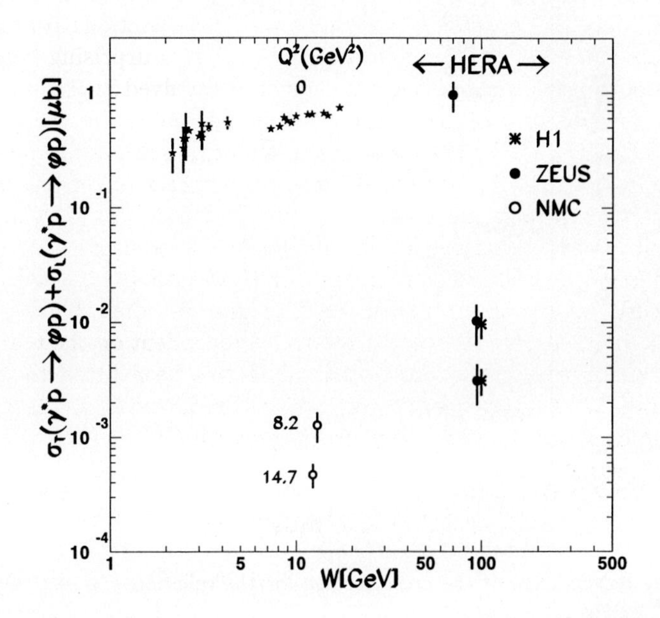

Fig. 72. *The dependence of the cross section for the reaction $\gamma^* p \to \phi p$ on W, for different Q^2 values.*

The J/Ψ vector meson already shows a steep W dependence for the photoproduction case. In Fig. 73 the photoproduction cross section is shown together with measurements of the reaction $\gamma^* p \to J/\Psi p$ at $Q^2 = 10$ and $20\,\mathrm{GeV}^2$.

These results are consistent with the Q^2 dependence of Δ as shown in figure 69.

The ratio of the higher mass vector mesons to ρ^0 cross sections is expected according to SU(4) to be:

$$\rho^0 : \omega : \phi : J/\Psi = 9 : 1 : 2 : 8 \tag{7.21}$$

This relation is quite badly broken in photoproduction for ϕ and for J/Ψ. For the case of the ϕ it is about 0.07 and for the J/Ψ it is somewhat W dependent and at the HERA W range it is about 0.005 for $Q^2 = 0$. As Q^2 increases one expects the SU(4) relations to be restored. For much higher Q^2 values one expects these relations to be broken again in the opposite direction.

In Fig. 74 the ratio $R(V^0/\rho^0)$ is presented as function of the vector meson mass squared M_V^2, for different Q^2 values as indicated next to the data points. One observes first that as the mass of the vector meson gets larger, the ratio becomes smaller and reaches a value of $< 10^{-3}$ for the Ψ'. As Q^2 increases the ratio get larger. It reaches close to the expected value of $2:9$ for the ϕ, close to 0.4 for the ρ', and ≈ 1 for the J/Ψ.

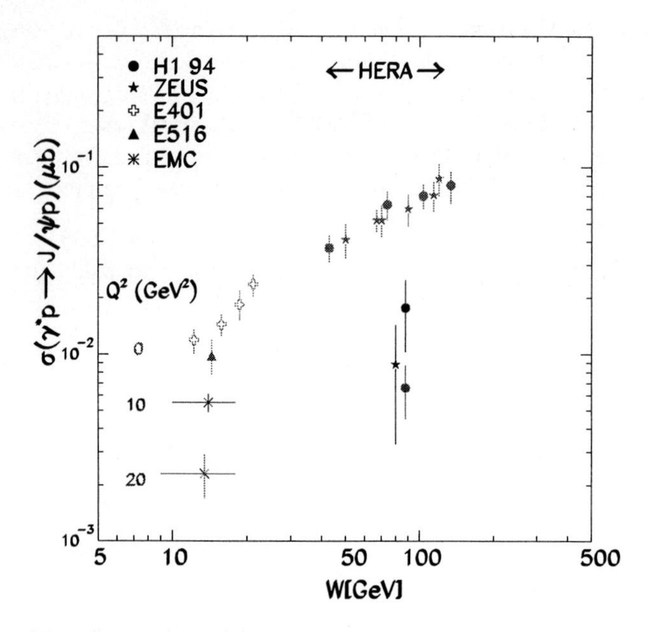

Fig. 73. *The dependence of the cross section for the reaction $\gamma^* p \to J/\Psi p$ on W, for different Q^2 values.*

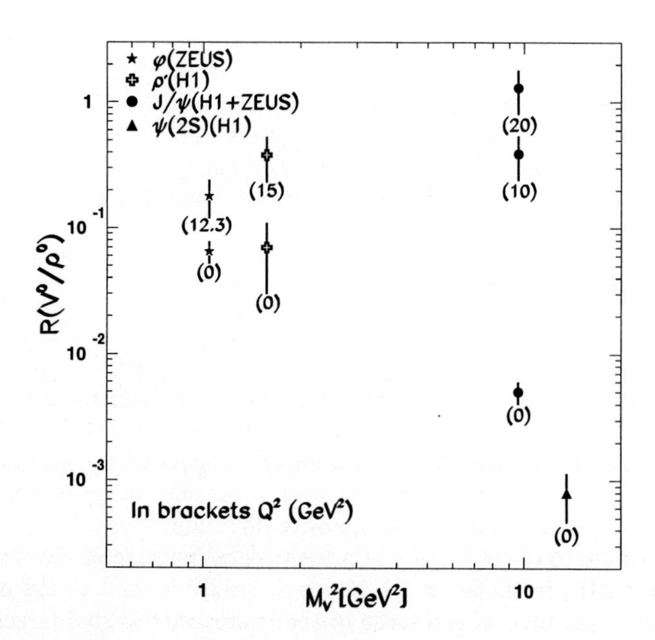

Fig. 74. *The ratio of the cross sections of vector mesons ϕ, ρ', J/Ψ and Ψ' to ρ^0 at different values of Q^2 as indicated in the figure.*

What can we learn from the behaviour of the slope? Does one see any shrinkage? It is not easy to conclude about that since there is no single experiment that has enough of a W range lever arm to measure shrinkage in one experiment. One thus is dependent on the systematics of different experiments. The photoproduction data of all three vector mesons ρ^0, ω and ϕ are consistent with shrinkage (see Fig. 75). What about the vector mesons produced in DIS?

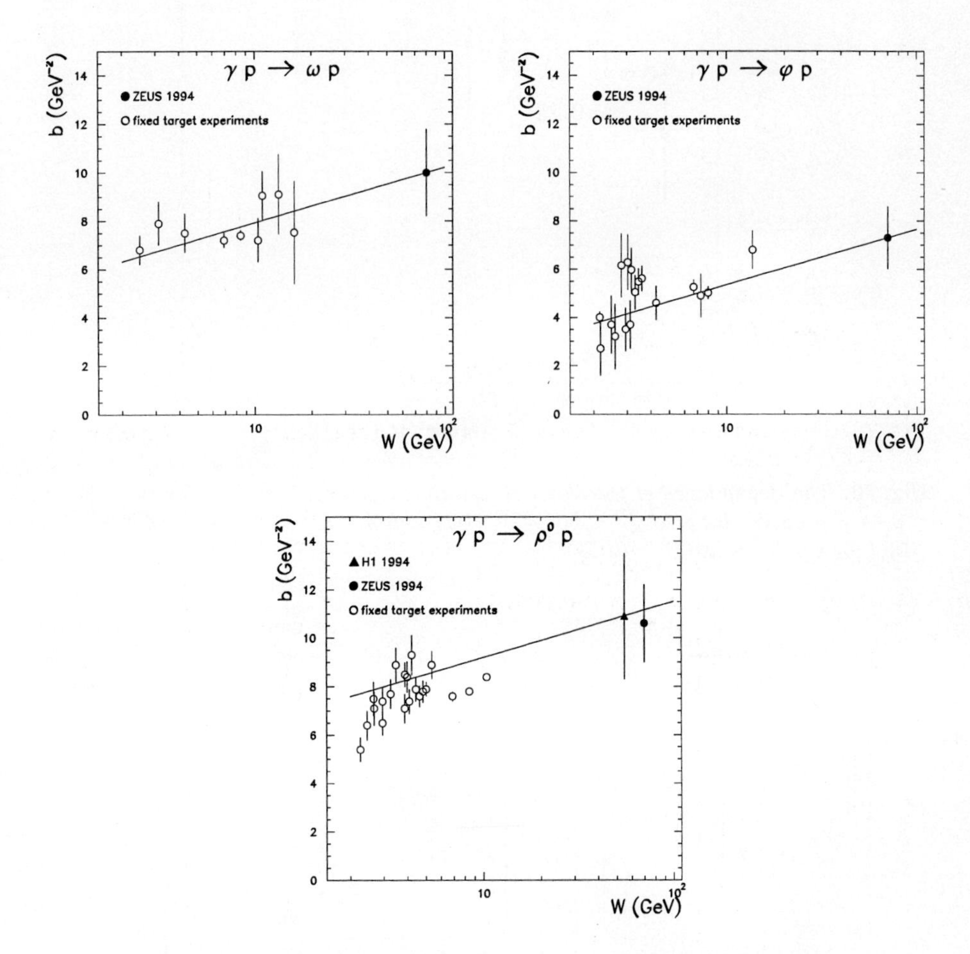

Fig. 75. *The dependence of the slope of the differential cross section*

The dependence of the slope of the differential cross section for the reaction $\gamma^* p \to \rho^0 p$ on W, is shown in Fig. 76 for $8 < Q^2 < 50\,\mathrm{GeV}^2$ (H1) and $5 < Q^2 < 30\,\mathrm{GeV}^2$ (ZEUS). The NMC data point is at $Q^2 \approx 10\,\mathrm{GeV}^2$. The HERA data alone can not, with the present measurement errors, distinguish between the shrinkage or non–shrinkage of the slope. Even with the addition of the NMC point the situation is not clear and one has to await more precise data.

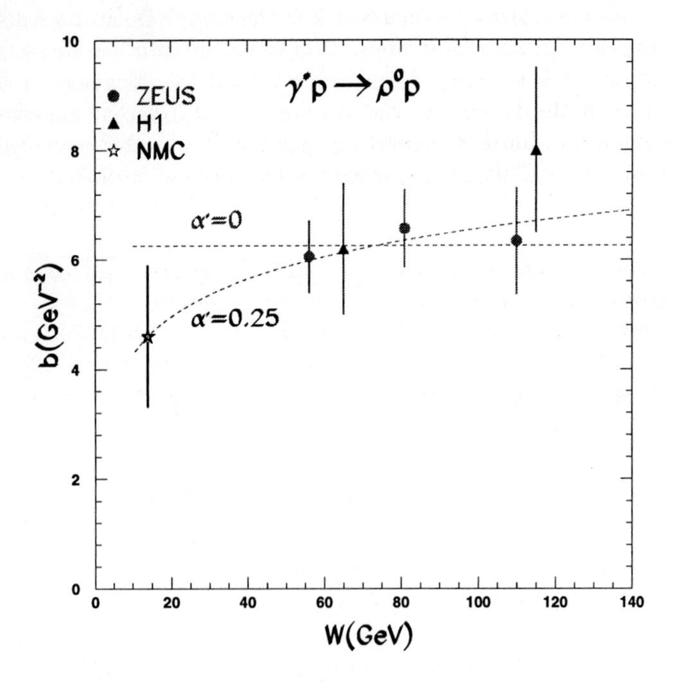

Fig. 76. *The dependence of the slope of the differential cross section for the reaction $\gamma^* p \to \rho^0 p$ on W, for $8 < Q^2 < 50\,\mathrm{GeV}^2$ (H1) and $5 < Q^2 < 30\,\mathrm{GeV}^2$ (ZEUS). The NMC data point is at $Q^2 \approx 10\,\mathrm{GeV}^2$.*

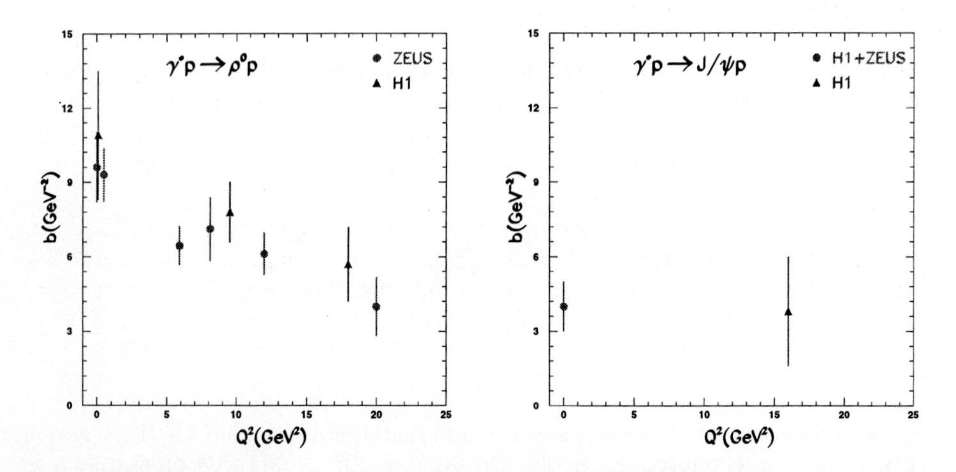

Fig. 77. *The dependence of the slope of the differential cross section for the reactions $\gamma^* p \to \rho^0 p$ (left) and $\gamma^* p \to J/\Psi p$ (right) on Q^2, for $< W >\approx 80\,\mathrm{GeV}$ (ρ^0) and $90\,\mathrm{GeV}$ (J/Ψ).*

One does however see a decrease of the slope with Q^2 in case of the exclusive ρ^0 production in DIS. The slope seems to decrease from a value of about $10\,\mathrm{GeV}^{-2}$ at $Q^2 = 0$ to about $5\,\mathrm{GeV}^{-2}$ at about $Q^2 = 20\,\mathrm{GeV}^2$. This result is consistent with the fact that as the scale gets larger, the reaction becomes harder and in case of hard processes all vector mesons are expected to have the same universal slope. This effect is nicely seen in the case of J/Ψ, where due to its large mass there is a hard scale already at $Q^2 = 0$, and thus the slope shows no change with Q^2.

It is worthwhile to note that the properties observed for vector mesons have a natural explanation in QCD, where vector meson production with a large scale can be described by an exchange mechanism of a Pomeron consisting of two gluon. For example, in the case of the model of Brodsky et al. [159] one expects that the differential ρ^0 cross section produced by longitudinal photons should be proportional to the gluon distribution in the proton:

$$\frac{\mathrm{d}\sigma}{\mathrm{dt}}(\gamma_L^* p \to \rho^0 p) \sim \frac{[\alpha_S(Q^2) x g(x,Q^2)]^2}{Q^6} C_\rho \tag{7.22}$$

Since at low x values $[\alpha_S(Q^2)xg(x,Q^2)]^2 \sim Q$ and since the k_T dependence of the ρ^0 wave function introduces [160] another $Q^{0.5}$ dependence, the expectations of the QCD calculation are that the data should have a Q^n dependence, where $n = 4.5 \ldots 5$. The ZEUS [156] experiment finds $n = 4.2 \pm 0.8^{+1.4}_{-0.5}$ and the H1 [157] experimental result is $n = 4.8 \pm 0.8$ (statistical error only). The x dependence of the ZEUS [156] measurement is consistent with their gluon determination from their F_2 measurement.

7.5 DIS Processes - Hard or Soft?

What have we learned from the behavior of the data with Q^2? What are we actually measuring? At low Q^2 the photon is known to have structure. Does F_2 still measure the structure of the proton? Bjorken [161] pointed out that physics is not frame dependent. The structure of the proton alone has no meaning. One has to study the $\gamma^* p$ interaction.

Let us look at the structure of a photon. It is a well known fact that real photon behave like hadrons when interacting with other hadrons. One way to understand this is by using the argument of Ioffe [88, 89]: the photon can fluctuate into a $q\bar{q}$ pair. The fluctuation time is given by

$$t_f = \frac{2E_\gamma}{m_{q\bar{q}}^2} \tag{7.23}$$

where E_γ is the photon energy in the rest system of the proton and $m_{q\bar{q}}$ is the mass of the $q\bar{q}$ system into which the photon fluctuates. The Vector Dominance Model assumes that the fluctuation of the photon is into vector mesons, $m_{q\bar{q}} \simeq m_V$, where m_V is the vector mesom mass. As long as $t_f \gg t_i$, where the interaction time $t_i \approx r_p$, with r_p being the proton radius, the photon interacts as if it were a hadron. When the photon becomes virtual with a negative square mass of Q^2, its fluctuation time becomes

$$t_f = \frac{2E_\gamma}{m_{q\bar{q}}^2 + Q^2} \tag{7.24}$$

and thus at low energies and moderate Bjorken x, the fluctuation time becomes small and the virtual photon behaves like a point–like structureless object, consistent with the DIS picture described above.

However, at high energies or equivalently in the low x region studied at HERA, the fluctuation time of a virtual photon can be expressed as

$$t_f \approx \frac{1}{2Mx} \tag{7.25}$$

where M is the proton mass. This can be derived easily from formula (5.1) assuming $m_{q\bar{q}}^2 \approx Q^2$ [126]. Thus in the HERA regime, a photon of virtuality as high as $Q^2 \sim 2 - 3 \times 10^3$ GeV^2 can fluctuate into a $q\bar{q}$ pair, which will survive till arrival on the proton target.

The photon can fluctuate into typically two configurations. A large size configuration will consist of an asymmetric $q\bar{q}$ pair with each quark carrying a small transverse momentum k_T (fig. 78(a)). For a small size configuration the pair is symmetric, each quark having a large k_T (fig. 78(b)). One expects the asymmetric large configuration to produce 'soft' physics, while the symmetric one would yield the 'hard' interactions.

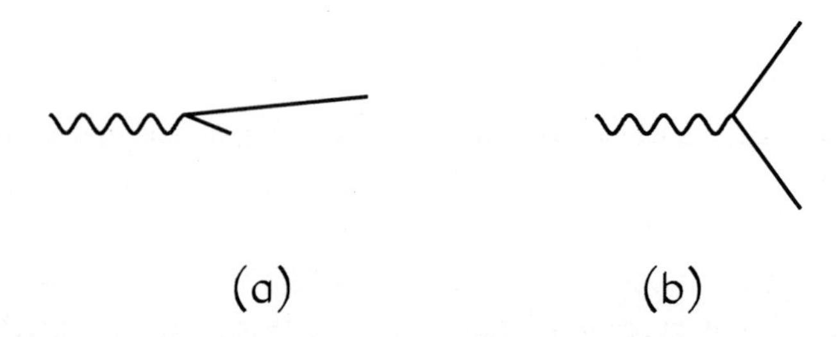

Fig. 78. *Fluctuation of the photon into a $q\bar{q}$ pair in (a) asymmetric small k_T configuration, (b) into a symmetric large k_T configuration*

In the aligned jet model (AJM) [162] the first configuration dominates while the second one is the 'sterile combination' because of color screening. In the photoproduction case ($Q^2 = 0$), the small k_T configuration dominates. Thus one has large color forces which produce the hadronic component, the vector mesons, which finally lead to hadronic non–perturbative final states of 'soft' nature. The symmetric configuration contributes very little. In those cases where the photon does fluctuate into a high k_T pair, color transparency suppresses their contribution.

In the DIS regime ($Q^2 \neq 0$), the symmetric contribution becomes bigger. Each such pair still contributes very little because of color transparency, but the phase space for the symmetric configuration increases. However the asymmetric pair still contribute also to the DIS processes. In fact, in the quark parton model (QPM) the fast quark becomes the current jet and the slow quark interacts with the proton remnant resulting in processes which look in the $\gamma^* p$ frame just like the 'soft' processes discussed in the $Q^2 = 0$ case. So there clearly is an interplay between soft and hard interactions also in the DIS region.

This now brings up another question. We are used by now to talk about the 'resolved' and the 'direct' photon interactions. However, if the photon always fluctuates into a $q\bar{q}$ pair even at quite large values of Q^2, what does one mean by a 'direct' photon interaction? To illustrate the problem, let us look at the diagram describing the photon–gluon fusion, which is usually considered in leading order a direct photon interaction and is shown in Fig. 79(a). An example of a resolved process is shown in Fig. 79(b) where a photon fluctuates into a $q\bar{q}$ pair with a given k_T, following by the interaction of one of the quarks with a gluon from the proton to produce a quark and a gluon with a given p_T.

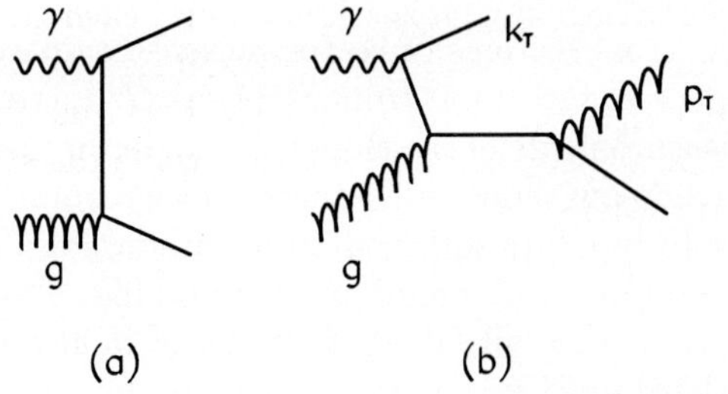

Fig. 79. *Diagrams describing examples of (a) 'direct' photon process, (b) 'resolved' photon process*

In the diagram shown in Fig. 79(b) there are two scales, k_T and p_T. The classification of the process as 'direct' or 'resolved' depends on the relations between the two scales. If $k_T \ll p_T$ we call it a resolved photon interaction, while in the case of $k_T \gg p_T$ one would consider this as a direct photon interaction. Practically in the latter case the p_T is too small to resolve the gluon and the quark jets as two separate jets, thus making it look like the diagram in Fig. 79(a). At low Q^2 the more likely case is that of $k_T \ll p_T$ and thus the resolved photon is the dominant component, while at high Q^2 the other case is more likely. A yet open question is how does one deal with the case where $k_T \sim p_T$.

7.6 Summary

We can summarize the results and discussions of this chapter in the following way:

- We have presented an operational definition of what we call soft and hard interactions by using the total cross section and the elastic process. The energy behaviour of the total and the elastic cross section is expected to be much steeper for hard interactions than for soft ones. In addition, the slope of the elastic differential cross section should shrink in the soft interactions while show no or very little shrinkage in case of hard interactions.
- The models describing the low–x low Q^2 region were discussed and compared to data. The energy behaviour of the total $\gamma^* p$ cross section shows a smooth transition from a shallow dependence at low Q^2 to a steeper one at higher Q^2.
- The vector meson exclusive production, which can be considered as the elastic processes for the photon case, follow the energy dependence behaviour of the total cross section. The W dependence get steeper for the ρ^0 and ϕ as Q^2 increases while it is already steep for the J/Ψ produced in the elastic photoproduction process.
- When a large scale is present, being the virtuality of the photon or the mass of the vector meson, the cross section is consistent with a rise driven by the rise of the gluon momentum density $xg(x, Q^2)$ with W. The Pomeron exchange mechanism described by two gluons gives results consistent with the data.
- The ratio of the cross sections of vector mesons compared to that of the ρ^0 is approaching the expectations from SU(4) as Q^2 increases.
- The present measurements of the slopes of the vector mesons are not precise enough to conclude anything about the shrinkage question.
- One would like to separate soft from hard interactions. However nothing is as soft as we would like nor as hard as we would like. There is an interplay of soft and hard processes at all values of Q^2. As Q^2 or any other scale increases, the amount of hard processes seems to increase. In order to resolve the hard processes one needs a good understanding of the soft fragmentation and hadronization. By combining various reactions one can try and extract the perturbative QCD part and to learn more about the interplay.
- The energy behavior of the $\gamma^* p$ cross section shows that there is a smooth transition between the Q^2 region where there is a mild energy dependence to that where the energy behavior is steeper. It happens somewhere in the region of about $1\,\mathrm{GeV}^2$. Does this tell us where soft interactions turn into hard ones? In order to understand the structure of the dynamics, one has to isolate in the transition region the specific configurations in k_T and p_T for a better insight of what is happening.

Acknowledgements

I would like to thank the organizers of the Banz lectures for setting up a special and very pleasant atmosphere during the lectures. I am thankful to Andreas Gute and to Manfred Ferstl from the University of Erlangen who took notes during the lectures and prepared the first draft with most of the Feynman diagrams.

Many thanks also to Stephan Boettcher from Tel Aviv University for editing the manuscript.

References

[1] R. G. Roberts, *The structure of the proton* (Cambridge University Press, 1990).
[2] B. H. Wiik, in *Proceedings of the Workshop Physics at HERA*, edited by W. Buchmüller, G. Ingelman (Hamburg, Germany, 1991), p. 1.
[3] H1 Collab., I. Abt et al., *DESY 93-103* (1993).
[4] ZEUS Collab., M. Derrick et al., *DESY 1993* (1993).
[5] W. Bartel et al., in *Proceedings of the Workshop Future Physics at HERA*, edited by G. Ingelman, A. D. Roeck, R. Klanner (Hamburg, Germany, 1996), p. 1095.
[6] HERMES Collab., K. Coulter et al., *DESY-PRC 90/01* (1990).
[7] HERA-B Collab., T. Lohse et al., *DESY-PRC 94/02* (1994).
[8] G. Ingelman, R. Rückl, *Phys. Lett.* **B 201** (1988) 369.
[9] H1 Collab., I. Abt et al., *Phys. Lett.* **B 324** (1994) 241.
[10] ZEUS Collab., M. Derrick et al., *Phys. Rev. Lett.* **75** (1995) 1006.
[11] ZEUS Collab., M. Derrick et al., *Z. Phys.* **C 67** (1995) 81.
[12] H1 Collab., T. Ahmed et al., *Phys. Lett.* **B 346** (1995) 415.
[13] ZEUS Collab., M. Derrick et al., *Phys. Lett.* **B 363** (1995) 201.
[14] H. Abramowicz, in *Proceedings of the International Conference on High Energy Physics* (Warsaw, Poland, 1996), (and references therein).
[15] ZEUS Collab., M. Derrick et al., *Z. Phys.* **C 72** (1996) 399.
[16] H1 Collab., S. Aid et al., *Nucl. Phys.* **B 470** (1996) 3.
[17] A. Levy, U. Maor, *Phys. Lett.* **B 182** (1986) 108.
[18] H. Abramowicz et al., *Phys. Lett.* **B 269** (1991) 465.
[19] A. Marcus, Energy dependence of the $\gamma^* p$ cross section, Master's thesis, Tel Aviv University, 1996, TAUP 2350-96.
[20] ZEUS Collab., M. Derrick et al., *Phys. Lett.* **B 315** (1993) 481.
[21] H1 Collab., T. Ahmed et al., *Nucl. Phys.* **B 249** (1994) 477.
[22] M. Gell-Mann, Y. Ne'eman, *The Eightfold Way* (Benjamin, New York, 1964).
[23] E. Leader, Lecture Notes, 1987.
[24] P. D. B. Collins, A. D. Martin, *Hadron Interactions* (Adam Hilger, 1984).
[25] G. F. Chew, S. C. Frautschi, *Phys. Rev. Lett.* **8** (1962) 41.
[26] M. Froissart, *Phys. Rev.* **123** (1961) 1053.
[27] A. Martin, *Nuovo Cimento* **42** (1966) 930.
[28] K. Goulianos, *Phys. Rep.* **C 101** (1983) 169, (and references therein).
[29] I. Y. Pomeranchuk, *Sov. Phys. JETP* **7** (1958) 499.
[30] G. Jaroszkiewicz, P. V. Landshoff, *Phys. Rev.* **D 10** (1974) 170.
[31] A. Donnachie, P. V. Landshoff, *Phys. Lett.* **B 296** (1992) 227.
[32] J. J. Sakurai, *Ann. Phys.* **11** (1960) 1.

[33] G. A. Schuler, J. Terron, in *Proceedings of the Workshop Physics at HERA*, edited by W. Buchmüller, G. Ingelman (Hamburg, Germany, 1991), p. 599.
[34] H1 Collab., T. Ahmed et al., *Z. Phys.* **C 66** (1995) 525.
[35] D. Kisielewska et al., *DESY-HERA 85-25* (1985).
[36] J. Andruszkow et al., *DESY 92-066* (1992).
[37] K. Piotrzkowski, *DESY-F35D-93-06* (1993).
[38] ZEUS Collab., M. Derrick et al., *Phys. Lett.* **B 293** (1992) 465.
[39] H1 Collab., T. Ahmed et al., *Phys. Lett.* **B 299** (1992) 374.
[40] F. Eisels, in *Proceedings of the International Europhysics Conference on High Energy Physics*, edited by J. Lemonne, C. V. Velde, F. Verbeure (Brussels, Belgium, 1995).
[41] A. Caldwell, in *Proceedings of the 17th International Symposium on Lepton-Photon Interactions*, edited by Z. Zhi-Peng, C. He-Sheng (Beijing, China, 1995), p. 505.
[42] S. Bentvelsen, J. Engelen, P. Koijman, in *Proceedings of the Workshop Physics at HERA*, edited by W. Buchmüller, G. Ingelman (Hamburg, Germany, 1991), p. 23.
[43] F. Jacquet, A. Blondel, in *Proceedings of the study of an ep facility for Europe*, edited by U. Amaldi (1979), p. 391.
[44] M. N. Rosenbluth, *Phys. Rev.* **79** (1950) 615.
[45] E. Leader, E. Predazzi, *An Introduction to Gauge Theories and the New Physics* (Cambridge University Press, 1982).
[46] F. J. Gilman, *Phys. Rev.* **167** (1968) 1365.
[47] L. N. Hand, *Phys. Rev.* **129** (1963) 1834.
[48] H. Spiesberger et al., in *Proceedings of the Workshop Physics at HERA*, edited by W. Buchmüller, G. Ingelman (Hamburg, Germany, 1991), p. 798.
[49] C. G. Callan, D. Gross, *Phys. Rev. Lett.* **22** (1969) 156.
[50] H. Abramowicz, A. Caldwell, R. Sinkus, *Nucl. Instrum. Meth.* **A 365** (1995) 569.
[51] L. W. Whitlow et al., *Phys. Lett.* **B 250** (1990) 193.
[52] NMC Collab., M. Arneodo et al., *hep-ph/9610231*, accepted for publication in Nuc. Phys. B.
[53] H. Abramowicz et al., *DESY 90-019*
[54] J. C. Collins, D. E. Soper, G. Sterman, in *Perturbative Quantum Chromodynamics*, edited by A. H. Mueller (World Scientific, Singapore, 1989).
[55] G. 't Hooft, M. Veltman, *Nucl. Phys.* **B 50**.
[56] V. N. Gribov, L. N. Lipatov, *Sov. J. Nucl. Phys.* **15** (1972) 438,675.
[57] Y. L. Dokshitzer, *Sov. Phys. JETP* **46** (1977) 641.
[58] G. Altarelli, G. Parisi, *Nucl. Phys.* **B 126** (1977) 298.
[59] A. Levy, in *Proceedings of the International Europhysics Conference on High Energy Physics*, edited by J. Lemonne, C. V. Velde, F. Verbeure (Brussels, Belgium, 1995), p. 700.
[60] E. A. Kuraev, L. N. Lipatov, V. S. Fadin, *Sov. Phys. JETP* **44** (1976) 443.
[61] E. A. Kuraev, L. N. Lipatov, V. S. Fadin, *Sov. Phys. JETP* **45** (1977) 199.
[62] Y. Y. Balitski, L. N. Lipatov, *Sov. J. Nucl. Phys.* **28** (1978) 822.

[63] A. H. Mueller, *Nucl. Phys.* **B 307** (1988) 34.
[64] A. H. Mueller, N. Navelet, *Nucl. Phys.* **B 282** (1987) 727.
[65] M. Ciafaloni, *Nucl. Phys.* **B 296** (1988) 49.
[66] G. Marchesini, *Nucl. Phys.* **B 445** (1995) 49.
[67] S. Catani, F. Fiorani, G. Marchesini, *Nucl. Phys.* **B 336** (1990) 18.
[68] L. V. Gribov, E. M. Levin, M. G. Ryskin, *Phys. Rep.* **C 100** (1983) 1.
[69] E. Gotsman, E. Levin, U. Maor, *Phys. Lett.* **B 379** (1996) 186.
[70] A. D. Martin, R. G. Roberts, W. J. Stirling, *Phys. Lett.* **B 354** (1995) 155.
[71] A. D. Martin, R. G. Roberts, W. J. Stirling, *RAL-TR-96-037* (1996).
[72] R. Brock et al., *Rev. Mod. Phys.* **67** (1995) 157.
[73] M. Gluck, E. Reya, A. Vogt, *Z. Phys.* **C 67** (1995) 433.
[74] A. Donnachie, P. V. Landshoff, *Z. Phys.* **C 61** (1994) 139.
[75] K. Prytz, *Phys. Lett.* **B 311** (1993) 286.
[76] K. Prytz, *Phys. Lett.* **B 332** (1994) 393.
[77] ZEUS Collab., M. Derrick et al., *Phys. Lett.* **B 345** (1995) 576.
[78] R. K. Ellis, Z. Kunszt, E. M. Levin, *Nucl. Phys.* **B 420** (1994) 517.
[79] H1 Collab., S. Aid et al., *Nucl. Phys.* **B 449** (1995) 3.
[80] H1 Collab., S. Aid et al., *Nucl. Phys.* **B 472** (1996) 3.
[81] ZEUS Collab., M. Derrick et al., ICHEP96, pa02-047.
[82] ZEUS Collab., M. Derrick et al., *Phys. Lett.* **B 349** (1995) 225.
[83] ZEUS Collab., presented by J. Bulmahn, ICHEP96, pa02-028.
[84] H1 Collab., S. Aid et al., *Nucl. Phys.* **B 472** (1996) 32.
[85] H1 Collab., S. Aid et al., *Z. Phys.* (1996), to be published.
[86] ZEUS Collab., M. Derrick et al., *Phys. Lett.* **B 350** (1995) 120.
[87] M. G. Ryskin et al., *hep-ph/9511228* (1995).
[88] B. L. Ioffe, *Phys. Lett.* **B 30** (1969) 123.
[89] B. L. Ioffe, V. A. Khoze, L. N. Lipatov, *Hard Processes* (North–Holland, 1984), p. 185.
[90] T. H. Bauer et al., *Rev. Mod. Phys.* **50** (1978) 261.
[91] J. C. Sens, in *Proceedings of the VIIIth International Workshop on Photon-Photon Collisions*, edited by U. Karshon (Shoresh, Israel, 1988), p. 143.
[92] V. M. Budnev et al., *Phys. Rep.* **C 15** (1975) 181.
[93] C. F. von Weizsäcker, *Z. Phys.* **88** (1934) 612.
[94] E. J. Williams, *Phys. Rev.* **45** (1934) 729.
[95] C. Berger, W. Wagner, *Phys. Rep.* **C 146** (1987) 1.
[96] H. Abramowicz et al., *Int. J. Mod. Phys.* **A 8** (1993) 1005.
[97] R. J. D. Witt et al., *Phys. Rev.* **D 19** (1979) 2046.
[98] C. Peterson, T. F. Walsh, P. M. Zerwas, *Nucl. Phys.* **B 229** (1983) 301.
[99] M. Glück, E. Reya, *Phys. Rev.* **D 28** (1983) 2749.
[100] A. Levy, *J. Phys.* **G 19** (1993) 1489.
[101] ZEUS Collab., M. Derrick et al., *Phys. Lett.* **B 297** (1992) 404.
[102] H1 Collab., T. Ahmed et al., *Phys. Lett.* **B 297** (1992) 205.
[103] ZEUS Collab., M. Derrick et al., *Phys. Lett.* **B 354** (1995) 163.
[104] E. Witten, *Nucl. Phys.* **B 120** (1977) 189.
[105] G. Rossi, *Phys. Rev.* **D 29** (1984) 852.

[106] OPAL Collab., presented by J. Lauber, ICHEP96, pa03-007.
[107] M. Gluck, E. Reya, A. Vogt, *Phys. Rev.* **D 45** (1992) 3986.
[108] L. E. Gordon, J. K. Storrow, *Z. Phys.* **C 56** (1992) 307.
[109] G. A. Schuler, T. Sjostrand, *Z. Phys.* **C 68** (1995) 607.
[110] J. H. Field, F. Kapusta, L. Poggioli, *Phys. Lett.* **B 181** (1986) 362.
[111] M. Drees, K. Grassie, *Z. Phys.* **C 28** (1985) 451.
[112] H. Abramowicz, K. Charchula, A. Levy, *Phys. Lett.* **B 269** (1990) 450.
[113] H1 Collab., T. Ahmed et al., *Nucl. Phys.* **B 445** (1995) 195.
[114] A. Levy, in *Proceedings of the Workshop on Deep Inelastic Scattering*, edited by G. D'Agostini, A. Nigro (Rome, Italy, 1996).
[115] PLUTO Collab.,Ch. Berger et al., *Phys. Lett.* **B 142** (1984) 119.
[116] ZEUS Collab., M. Derrick et al., EPS-ICHEP95, EPS-0384.
[117] A. H. Mueller, *Phys. Rev.* **D 2** (1970) 2963.
[118] R. D. Field, G. C. Fox, *Nucl. Phys.* **B 80** (1974) 367.
[119] M. Kasprzak, Ph.D. thesis, Warsaw University, 1996, DESY F35D-96-16.
[120] B. Burow, Ph.D. thesis, University of Toronto, 1994, DESY F35D-94-01.
[121] M. Krzyzanowski, Ph.D. thesis, Warsaw University, 1997.
[122] H1 Collab., S. Aid et al., *Z. Phys.* **C 69** (1995) 27.
[123] ZEUS Collab., M. Derrick et al., *Phys. Lett.* **B 338** (1994) 483.
[124] H1 Collab., S. Aid et al., *Z. Phys.* **C 70** (1996) 609.
[125] H1 Collab., T. Ahmed et al., *Phys. Lett.* **B 348** (1995) 681.
[126] H. Abramowicz, L. Frankfurt, M. Strikman, *DESY 95-047*
[127] UA8 Collab., R. Bonino et al., *Phys. Lett.* **B 211** (1988) 239.
[128] UA8 Collab., A. Brandt et al., *Phys. Lett.* **B 297** (1992) 417.
[129] ZEUS Collab., M. Derrick et al., *Phys. Lett.* **B 332** (1994) 228.
[130] H1 Collab., T. Ahmed et al., *Nucl. Phys.* **B 435** (1995) 3.
[131] ZEUS Collab., M. Derrick et al., *Z. Phys.* **C 68** (1995) 569.
[132] G. Ingelman, P. E. Schlein, *Phys. Lett.* **B 152** (1985) 256.
[133] A. Donnachie, P. V. Landshoff, *Phys. Lett.* **B 191** (1987) 309.
[134] A. Donnachie, P. V. Landshoff, *Nucl. Phys.* **B 303** (1988) 634.
[135] ZEUS Collab., M. Derrick et al., *Phys. Lett.* **B 356** (1995) 129.
[136] H1 Collab., presented by J. P. Phillips, ICHEP96, pa02-061.
[137] A. Metha for the H1 Colaboration, in *Proceedings of the Conference on Hard Diffractive Scattering* (Eilat, Israel, 1996), p. 710.
[138] ZEUS Collab., presented by G. Barbagli, ICHEP96, pa02-026.
[139] ZEUS Collab., M. Derrick et al., *Z. Phys.* **C 70** (1996) 391.
[140] J. C. Collins, J. Kwiecinski, *Nucl. Phys.* **B 316** (1989) 307.
[141] J. Kwiecinski, A. D. Martin, P. J. Sutton, *Phys. Rev.* **D 44** (1991) 2640.
[142] A. Donnachie, P. V. Landshoff, *Nucl. Phys.* **B 244** (1984) 322.
[143] A. Kaidalov, L. Ponomarev, K. A. Ter-Martirosyan, *Sov. J. Nucl. Phys.* **44** (1986) 468.
[144] A. Capella et al., *Phys. Lett.* **B 337** (1994) 358.
[145] J. Kwiecinski, B. Badelek, *Z. Phys.* **C 43** (1989) 251.
[146] B. Badelek, J. Kwiecinski, *Phys. Lett.* **B 295** (1992) 263.

[147] Q. Zhu for the ZEUS Colaboration, in *Proceedings of the Workshop on Deep Inelastic Scattering*, edited by G. D'Agostini, A. Nigro (Rome, Italy, 1996).

[148] A. V. Kotwal for the E665 Colaboration, in *Proceedings of the XXXth Recontres de Moriond, QCD and High Energy Interactions* (Moriond, 1995).

[149] ZEUS Collab., M. Derrick et al., *Z. Phys.* **C 65** (1995) 379.

[150] H1 Collab., T. Ahmed et al., *Nucl. Phys.* **B 439** (1995) 471.

[151] ZEUS Collab., M. Derrick et al., *Z. Phys.* **C 69** (1995) 39.

[152] H1 Collab., S. Aid et al., *Nucl. Phys.* **B 463** (1996) 3.

[153] ZEUS Collab., M. Derrick et al., *DESY 96-159*, accepted for publ. in Z. Phys. C.

[154] ZEUS Collab., M. Derrick et al., *Phys. Lett.* **B 377** (1996) 259.

[155] H1 Collab., S. Aid et al., *Nucl. Phys.* **B 472** (1996) 3.

[156] ZEUS Collab., presented by J. Bulmahn, ICHEP96, pa02-028.

[157] H1 Collab., S. Aid et al., *Nucl. Phys.* **B 468** (1996) 3.

[158] ZEUS Collab., M. Derrick et al., *Phys. Lett.* **B 380** (1996) 220.

[159] S. J. Brodsky et al., *Phys. Rev.* **D 50** (1994) 3134.

[160] L. Frankfurt, W. Koepf, M. Strikman, *Phys. Rev.* **D 54** (1996) 3194.

[161] J. D. Bjorken, in *Proceedings of the Workshop on Deep Inelastic Scattering*, edited by A. Levy (Eilat, Israel, 1994), p. 151.

[162] J. D. Bjorken, in *Proceedings of the International Symposium on Electron and Photon Interactions at High Energies* (Cornell, 1971), p. 281.

Subject Index

Bold face refers to entire sections or subsections starting at the page indicated.

Druck: Strauss Offsetdruck, Mörlenbach
Verarbeitung: Schäffer, Grünstadt

New Series m: Monographs

Vol. m 1: H. Hora, Plasmas at High Temperature and Density. VIII, 442 pages. 1991.

Vol. m 2: P. Busch, P. J. Lahti, P. Mittelstaedt, The Quantum Theory of Measurement. XIII, 165 pages. 1991. Second Revised Edition: XIII, 181 pages. 1996.

Vol. m 3: A. Heck, J. M. Perdang (Eds.), Applying Fractals in Astronomy. IX, 210 pages. 1991.

Vol. m 4: R. K. Zeytounian, Mécanique des fluides fondamentale. XV, 615 pages, 1991.

Vol. m 5: R. K. Zeytounian, Meteorological Fluid Dynamics. XI, 346 pages. 1991.

Vol. m 6: N. M. J. Woodhouse, Special Relativity. VIII, 86 pages. 1992.

Vol. m 7: G. Morandi, The Role of Topology in Classical and Quantum Physics. XIII, 239 pages. 1992.

Vol. m 8: D. Funaro, Polynomial Approximation of Differential Equations. X, 305 pages. 1992.

Vol. m 9: M. Namiki, Stochastic Quantization. X, 217 pages. 1992.

Vol. m 10: J. Hoppe, Lectures on Integrable Systems. VII, 111 pages. 1992.

Vol. m 11: A. D. Yaghjian, Relativistic Dynamics of a Charged Sphere. XII, 115 pages. 1992.

Vol. m 12: G. Esposito, Quantum Gravity, Quantum Cosmology and Lorentzian Geometries. Second Corrected and Enlarged Edition. XVIII, 349 pages. 1994.

Vol. m 13: M. Klein, A. Knauf, Classical Planar Scattering by Coulombic Potentials. V, 142 pages. 1992.

Vol. m 14: A. Lerda, Anyons. XI, 138 pages. 1992.

Vol. m 15: N. Peters, B. Rogg (Eds.), Reduced Kinetic Mechanisms for Applications in Combustion Systems. X, 360 pages. 1993.

Vol. m 16: P. Christe, M. Henkel, Introduction to Conformal Invariance and Its Applications to Critical Phenomena. XV, 260 pages. 1993.

Vol. m 17: M. Schoen, Computer Simulation of Condensed Phases in Complex Geometries. X, 136 pages. 1993.

Vol. m 18: H. Carmichael, An Open Systems Approach to Quantum Optics. X, 179 pages. 1993.

Vol. m 19: S. D. Bogan, M. K. Hinders, Interface Effects in Elastic Wave Scattering. XII, 182 pages. 1994.

Vol. m 20: E. Abdalla, M. C. B. Abdalla, D. Dalmazi, A. Zadra, 2D-Gravity in Non-Critical Strings. IX, 319 pages. 1994.

Vol. m 21: G. P. Berman, E. N. Bulgakov, D. D. Holm, Crossover-Time in Quantum Boson and Spin Systems. XI, 268 pages. 1994.

Vol. m 22: M.-O. Hongler, Chaotic and Stochastic Behaviour in Automatic Production Lines. V, 85 pages. 1994.

Vol. m 23: V. S. Viswanath, G. Müller, The Recursion Method. X, 259 pages. 1994.

Vol. m 24: A. Ern, V. Giovangigli, Multicomponent Transport Algorithms. XIV, 427 pages. 1994.

Vol. m 25: A. V. Bogdanov, G. V. Dubrovskiy, M. P. Krutikov, D. V. Kulginov, V. M. Strelchenya, Interaction of Gases with Surfaces. XIV, 132 pages. 1995.

Vol. m 26: M. Dineykhan, G. V. Efimov, G. Ganbold, S. N. Nedelko, Oscillator Representation in Quantum Physics. IX, 279 pages. 1995.

Vol. m 27: J. T. Ottesen, Infinite Dimensional Groups and Algebras in Quantum Physics. IX, 218 pages. 1995.

Vol. m 28: O. Piguet, S. P. Sorella, Algebraic Renormalization. IX, 134 pages. 1995.

Vol. m 29: C. Bendjaballah, Introduction to Photon Communication. VII, 193 pages. 1995.

Vol. m 30: A. J. Greer, W. J. Kossler, Low Magnetic Fields in Anisotropic Superconductors. VII, 161 pages. 1995.

Vol. m 31: P. Busch, M. Grabowski, P. J. Lahti, Operational Quantum Physics. XI, 230 pages. 1995.

Vol. m 32: L. de Broglie, Diverses questions de mécanique et de thermodynamique classiques et relativistes. XII, 198 pages. 1995.

Vol. m 33: R. Alkofer, H. Reinhardt, Chiral Quark Dynamics. VIII, 115 pages. 1995.

Vol. m 34: R. Jost, Das Märchen vom Elfenbeinernen Turm. VIII, 286 pages. 1995.

Vol. m 35: E. Elizalde, Ten Physical Applications of Spectral Zeta Functions. XIV, 228 pages. 1995.

Vol. m 36: G. Dunne, Self-Dual Chern-Simons Theories. X, 217 pages. 1995.

Vol. m 37: S. Childress, A.D. Gilbert, Stretch, Twist, Fold: The Fast Dynamo. XI, 410 pages. 1995.

Vol. m 38: J. González, M. A. Martín-Delgado, G. Sierra, A. H. Vozmediano, Quantum Electron Liquids and High-T_c Superconductivity. X, 299 pages. 1995.

Vol. m 39: L. Pittner, Algebraic Foundations of Non-Commutative Differential Geometry and Quantum Groups. XII, 469 pages. 1996.

Vol. m 40: H.-J. Borchers, Translation Group and Particle Representations in Quantum Field Theory. VII, 131 pages. 1996.

Vol. m 41: B. K. Chakrabarti, A. Dutta, P. Sen, Quantum Ising Phases and Transitions in Transverse Ising Models. X, 204 pages. 1996.

Vol. m 42: P. Bouwknegt, J. McCarthy, K. Pilch, The W_3 Algebra. Modules, Semi-infinite Cohomology and BV Algebras. XI, 204 pages. 1996.

Vol. m 43: M. Schottenloher, A Mathematical Introduction to Conformal Field Theory. VIII, 142 pages. 1997.

Vol. m 44: A. Bach, Indistinguishable Classical Particles. VIII, 157 pages. 1997.

Vol. m 45: M. Ferrari, V. T. Granik, A. Imam, J. C. Nadeau (Eds.), Advances in Doublet Mechanics. XVI, 214 pages. 1997.

Vol. m 46: M. Camenzind, Les noyaux actifs de galaxies. XVIII, 218 pages. 1997.

Vol. m 47: L. M. Zubov, Nonlinear Theory of Dislocations and Disclinations in Elastic Body. VI, 205 pages. 1997.

Vol. m 48: P. Kopietz, Bosonization of Interacting Fermions in Arbitrary Dimensions. XII, 259 pages. 1997.

Vol. m 49: M. Zak, J. B. Zbilut, R. E. Meyers, From Instability to Intelligence. Complexity and Predictability in Nonlinear Dynamics. XIV, 552 pages. 1997.

Vol. m 50: J. Ambjørn, M. Carfora, A. Marzuoli, The Geometry of Dynamical Triangulations. VI, 197 pages. 1997.

Lecture Notes in Physics

For information about Vols. 1–461
please contact your bookseller or Springer-Verlag